Book Part I_Phys
Nutshell_ Preface etc

PHYSICS IN A NUTSHELL -
COMPANION FOR SUCCESS
IN COMPETITIVE TESTS

(A Guide on Formulae, Plots and self-explanatory diagrams)

(Revised & Enlarged Edition)

By

Prof. S. Devanarayanan, Ph.D.; D.Sc., Dip (Uppsala)
& Ajith Shankar Devan, B.Tech.,MBA (UD, USA)

and

3 Sets of Practice Tests, each of 100 MCQs,

At Three different levels

with Answer keys

[for Entry Level College (Sophomore), Undergraduate

and Graduate Level Students]

Original Edition, ISBN – 10.1517784417

January 2016

Second Edition, ISBN -13: 978-1981791453

ISBN – 10: 1981791450

December 2017

CreateSpace Independent Publishing, USA

(Amazon.com)

DEDICATION

This work is dedicated to *my*

Syananthapureesa, the Deity of Thiruvananthapuram

|| Saadhu veera tvayaa Prushtamavataara-kathaam Hare:
Yattvam pruchhasi Martyaanaam Mrutyu-paasa-visaataneem ||
(*Maitreya said to Vidura*; *Srimad Bhaagavatam, Sk* : 3; *Chp* 14, *Sl* 4)

|| Yada tu sarva-bhuteshu Darushvagnimiva stitam
pratichaksheeta Maam Loko jahyaathrhyeva kasmalam ||
(SrimadBhagavatam: Sk III. Ch 9. Sl 32)

	OM Namo Bhagavate Vaasudevaya	
	Ahamevaasamevaagre naanyadyat sadasatparam
Paschaadaham yadetaccha yo f vasishyeta so f smyaham	
	Riter f tham yatpratiyeta na pratiyeta	chaatmani
Tatvidyaatmano maayaam yatha ff bhaaso yatha tama:	
	Yatha mahaanti bhutaani bhuteshoocchaavacheshvanu
Pravishtaanya-pravishtaani tatha teshu na teshvaham	
	Etaavadeva jigjnaasyam Tatva jigjnaasunaa ff tmana:
Anvaya vyatirekaabhyaam yatsyaat sarvatra sarvadaa	
{|| Srimad *Bhaagavatam* (*Sk II*; *Chapter* 9: *Verses* 32 - 35)
Satya - Sanaa tan *a - Dharmam vijayetetaram* ||}
- Bhagavan, the Supreme GodHead, spoke to Brahma the Creator,
the Bhaagavatam in a Nutshell.

PREFACE

This book "PHYSICS IN A NUTSHELL: Companion for Success in Competitive Tests" contains a succinct and cogent coverage of the material dealt with for any Competitive Test like the Entrance Level test for admission to Professional Courses in a University, such as AIEEE, JEE (IIT- Screening & Main Tests), State Level SET by the Kerala, TNPCC and other similar level tests, as well as for All UPSC Tests involving General Studies. The book is also intended for both the teachers, students and Question Paper setter of undergraduates and for Civil Service Examinations of UPSC, GATE, GRE (for students seeking MS courses in USA), NAT tests conducted by UGC..

Covers in 42 Chapters all the compulsory material required for any advanced course in Physics at the B.Tech, M.B BS., UPSC conducted appropriate Civil Service Examinations and JEEs. Important definitions, formulae and principles / laws are aimed at revision purposes. It will serve as a revision book at the end of Higher Secondary School courses as well as at the start of a graduate course. Plots and illustrative schematic diagrams of relevant material have been provided; so that the contents will be self explanatory.

In order to give an idea of where a candidate stands a **Practice Test 1** (for Juniors) consisting of 100 multiple choice questions is included. Worked out solutions are separately provided for verification and evaluation. Additional tests **Practice Test 2 & 3** (for Seniors) are included with 100 multiple choice questions each. Answers to these tests are also included separately.

This compulsory text covers all the material required for the revised Higher Grade Physics courses, including a revision of Grade material which needs to be taken to the Higher Standard. To assist with problem-solving there are a large number of problems with fully worked-out solutions. Important definitions, formulae, and laws are highlighted for revision purposes. Further descriptions of essential experiments have been added.

This book may be used throughout the course as substitute for a set of notes, as a running summary, or for help with problem-solving. It will also serve as a revision book at the end of the course.

Students using this book should feel confidently prepared for the Higher Secondary Grade Examination, and be in possession of a sound base for further study.

I thank Mrs. Chitra Devanarayanan for her deep interest in the completion of the book.

Any suggestions and criticism on the contents is welcome. Error or blemish, if any, may kindly brought the author's attention through the e-mail: ajitsdevan@gmail.com or chsd1976@gmail.com

15th April 2014.   S. Devanarayanan

                                                            Ajith Shankar Devan

PREFACE of this Edition

This Revised Edition of the book: "PHYSICS IN A NUTSHELL: Companion for Success in Competitive Tests", has been brought up after getting suggestions from the readers. Changes and additional matter in the text of most of 42 Chapters in the book have been made. The author's (SD) research and teaching experience In courses in Quantum mechanics, Nuclear Physics, Solid State Physics, Spectroscopy and Electronics in the University at Post-graduate level have brought this book to the present unique level. The font size is changed from 10 to 11, with reduced sizes of several illustrative diagrams to reduce the size of the book. Further, the contents are expanded wherever necessary, for example, with Physics in sports, for higher utility of the book.

July 2017   S. Devanarayanan

CONTENTS

Title of the Book
Dedication
Prefaces 1 & 2

Bernoullis flow theorem - Applications, Airlift, Spinning ball, Siphon, Venturi meter, Airborne balloon, velocity of liquid surface wave, 9. Floatation - Archimede's principle, CG, CM, CP, Metacentre, Equilibrium of ship – Hydrometer- hydrostatic pressure, Various Pumps, 10. Surface Tension –Quincke method, Liquid drops method, Jaeger method, LaPlace's law, Liquid bubble, special cases, ST examples, 11. Capillarity – seven Applications.

atmosphere, Critical angle, Total reflection, Reversibility principle, Optical path, Fermat's principle, 5. Wave optics – Huygen's principle, 6. Newton's visible spectrum, Spectra, Deviation & Dispersion, Prism, Stoke's formula, Angle of minimum deviation, Newton's spectral colours, Variation of refractive index, Dispersive power, Achromatic prism, Direct vision spectroscope, Cauchy's Relation, Single prism spectrometer, Schuster's method, Various types of prisms, Rainbow, Primary colours, Colour Mixing, Constant deviation spectrometer, Emission spectrum, Hartmann equation, Zeeman effect, Absorption spectra of CCl_4 , Determination of λ and $\delta\lambda$ by Michelson's Interferometer, 7. Ray Optics. Optical images, Plane, Spherical mirrors, Mirror formula, Newton's formula, Image formation, Concave & Convex Mirrors, Uses, Sign convention, Defects of mirrors.

Applications, uses, 5. Voltage sources – Pb-acid battery, Leclanche, Daniel, Cd cell, Conventional, Primary & Secondary batteries, Button cell.

1. Electric current I , Transport equation, Drift velocity, Resistance (R), Ohm's law, – Resistivity, Temp dependence of R , in series & parallel, Potential divider, Uses of resistors, EMF & R , 2. Electric energy & Power, Kirchhoff's Point & Loop rules, 3. RC circuits, Discharge circuit, 4. Six Methods of measuring R , DC potentiometer, its resistance, calibration, Wheatstone bridge, Maxwell-Wein, Carey Foster, Metre Bridge, Kelvin double bridge.

1. Sources of $\vec{H}$ - Laplace's Law, Biot-Savart Law, RH rule, 2. $\vec{H}$ of current loop, at a point & Centre, Circular coil, 3. Ampere's law, $\vec{H}$ on the axis of Solenoid, Forces of interaction between two conductors carrying current I in $\vec{H}$, 4. Magnetostatics, Force on a dipole, Force on a current carrying conductor, between two conductors, Gradient coils, 5. Motion of charges in EM field, 6. Gauss's Law in a $\vec{H}$, 7. Faraday's Law of EM Induction, Lenz' Law, Motional EMF, Generators, Dynamos, AC generator, Torque,   8. Galvanometers, Moving-coil, T.G., Ballistic type, Dead-beat, Galvanometer as Ammeter & Voltmeter, Sensitivity of G, Measurement of resistance of G, Electrodynamometer, 9. Faraday's Law of Electrolysis, Electro-plating.

1. Inductance, Faraday's law, Lenz' law, Self induced EMF, Helical coil, Toroid, 2. DC RL Circuits, Current in LR circuit, Energy transfer, Mutual inductance, Transformers, 3. Magnetic field in Matter, Magnetic substances, Dia-, Para-Magnetism, Quincke's method, Curie's Law, Ferro-magetism, Magnetization, Anti-ferro-, Ferri-magnetism, Hysteresis by expt, 4. Heydweiller's bridge, Kirchoff's method, Kelvin's double bridge, Mathissen's method, Rayleigh's bridge, Remington's bridge, Stroude and Oate's, 5. AC circuits, Advantages of 120 V, 60 Hz AC Power,LC oscillations, The vectors of AC, Purely R, purely L, purely C circuits, Phase relationships, Differences between DC and AC, Damping in series RLC, Impedance, Resonant circuits – series and parallel, Impedance bridge, 6. Heat sources using electricity – Microwave Oven, TWT, Induction Heating, 7. Network Theory, Network Theorems, Superposition, Compensation, Thevenin, Norton, Maximum Power Transfer, Reciprocity, Four-teminal networks, T-Network (Star), π -network (Delta), Network terminations, Ladder networks, Wave Filters, LP, HP, BP & Band Stop, Constant Ladder type,

1. Cathode Rays – Electric discharge of Gas, Cathode rays, Gas discharge tube at various pressures, 2. Thomson's tube $\frac{q}{m_e}$, charge determination & quantization, Millikan experiment, 3.

Quantization of energy, Wien's law, R-J law, UV Catastrophe, Planck's radiation law, Thomson tube, Quantization of EM radiation, 3.   Quantization of energy, Planck hypothesis, UV catastrophe, Planck distribution law, 4. Discovery of photon, Threshold equation, 5. Photons as particles, Balmer spectrum, 6. Thomson discovery 7. Special -theory of relativity, Michelson-Moreley expt, Time dilation, Length contraction, Mass enery equivalence, Rest mass.

Television, Camera, colour mixing, one chip CCD camera, modulating, Transmission, 8.
Antenna, Dipole, Radiating power distribution pattern, Different types of antenna,

Chapter 1

INTRODUCTION –
UNITS OF MEASUREMENTS, ESTIMATION OF
ERRORS, INDIAN SCIENTISTS

"The latest authors, like the most ancient, strove to subordinate the phenomena of nature to the laws of mathematics" -   Isaac Newton
"There are not many but only One. Who sees variety and not the unity wanders from death to eath"
                                                                    - The Upanishads

1.1.   INTRODUCTION TO STUDENTS

1.1.1   CHOOSING THE CORRECT ANSWER KEY TO MCQ s. Solving Problems:
A basic part of a physics examination is solving problems effectively.
Solving of Problems in physics serves two purposes in questions in any examination paper:
 (a) Solving problems is useful and practical in it.
 (b) Solving problems enable the examiner to judge if you think about the ideas and concepts, and applying of the concepts helps one to evaluate the student's understanding them. But knowing how to do a problem- even to begin it- may not always seem easy. After reading the problem through carefully more than once, spend a moment to try to understand what physics principle might be involved.
I will spend a little time now summarizing how to approach problems.

1.1.2 PROCEDURE FOR SOLVING PROBLEMS
Diagram the situation and list the Data
 1. Select the appropriate formula
 2. Substitute therein
 3. Solve on calculator
 4. Do the unit! Dimensional analysis to check the result.

        The solving of problems often involves creativity. – Each problem is different. Nonetheless an outline of the general approach to solving problems is as follows: At first it is very important to know, by heart, the definitions, terminology and the basic principles and the laws that apply
 (1) Keep in mind that you have only a limited time of 1 minute to identify the correct answer key.
 (2) Read the written problems carefully even leaving out a word.
 (3) Draw an accurate picture or diagram of the situation which is most crucial step.
 (4) Write down what quantities are "known" or "given" and then what you want to know.
 (5) Think about what principles, definitions or equations relate the quantities involved If you find an applicable equation that involves only known quantities and the desired

unknown, solve the equation algebraically for the unknown.. In many instances, several sequential calculations, and / or a combination of equations, may be needed..

(6) Think carefully about the result that you obtain: Is it reasonable? Does it make sense? According to your own intuition?

(7) Be sure to keep track of units; an equal sign implies the units on either side must be the same. If the units do not balance, a mistake has no doubt been made. This will serve as check on your solution. That it tells only if you are wrong.

(8) The use of dimensional analysis can also serve as a check for many problems.

(9) Remember that the slope of a curve at a point (slope of the tangent to the curve at that point) is required in graphical analysis.

(10)   Now choose the correct answer key.

1.2.   UNITS OF MEASUREMENTS
The value of a physical quantity consists of two things – a **number** combined with a **unit**. In order that scientists and engineers communicate / exchange ideas a common System of units is adopted.

1.2.1   **SI Systems of Units**
        System International d'Unites, Paris in 1960 proposed the **SI System of Units**. It has 7 symbols and base units corresponding to 7 independent physical quantities: L (**metre**), M (**kg**), T (**s, second**), A (**Amp**), K (**temp.** Kelvin), amount of substance (**mole**) luminous intensity (**Candela**, Cd), charge (**Coulomb**, C) $= I A s = 6.2418$ e , in vacuum.

1.2.2   *LENGTH* (L):
Metre (m) is the unit of length in S.I. System.
"The distance between the two marks on a Platinum-Iridium bar kept at $0\,°C$ in the International Bureau of Weight and Measures in Paris "based in terms of the length of wavelength λ in vacuum of a particular spectral line of Krypton-86 from $2p^{10}$ to $5d^5$.

$$\boxed{1\,m = 1650\,763.73\,\text{Å}\ \ \text{Kr}}$$

1.2.3   *MASS* (M):
A piece of Platinum-Iridium kept under standard conditions at Sevres, near Paris, France. ($\sim 10^{-3}\ m^3$ of distilled water at 4^0 C).
Kilogram (kg) is the unit of mass in **S.I. System**.

"Kilogram is defined as the mass of a platinum cylinder placed in the International Bureau of   Weight and Measures in Paris."

1.2.4   *TIME* (T)
The frequency of any periodic event, such as the mechanical oscillation of a pendulum or quantum oscillation of an atomic dipole, can be adopted to define the unit of time, the SECOND (s).
For centuries: unit of time was

$$1\ \sec ond\ =\ \frac{1}{86400}\ \text{Mean Solar Day}$$

In 1949: Time of N atoms in molecular ammonia NH_3 to make $2.387\ x\ 10^{10}$ oscillations (based on the inversion transition at ~24 *GHz*)

In 1956, the International Union and the International Committee of Weights and Measures recommended Ephemeris Time, based on Earth's rotation around the Sun..

In 1967, in terms of atomic time: quartz crystals resonant frequencies calibrated relative to the Ephemeris Time

$$1\ s = 1/315569259747 \text{ of the year } 1900$$

Second is the unit of time in S.I .System, *viz.*,
The **time period of Cs-133 atoms, *i.e.*" one second, is equal to 9,192,631,770 periods of vibrations of Cs-133 atoms."**

$1\ s = 9192631770$ periods of the radiation
corresponding to the transition between the two-hyperfine levels
of the ground state of the Cs-133 atom

(Now being considered: Narrow optical transitions in Hg-199 and its single ion optical clock measuring time with an anticipated precision of one part in 10^{18} (Physics Today, March 2001).

1.3. **SI Units** (The International System of Units)
       The SI uses a single basic unit for each physical quantity, and multiples and fractions of this unit are formed by adding a prefix. For example, the SI unit for length is the metre (*m*). The *milli – metre* (*mm*), *centi – metre* (*cm*), and *kilo – metre* (*km*) are formed by adding the prefixes *milli –* (*m*), *centi –* (*c*), and *kilo* (*k*) to *metre* . The same prefixes are used for all physical quantities. The prefixes and their symbols are listed in Table A.1.

1.3.1 The following rules should be observed when using SI symbols of units:
       *Do not put a period after a symbol. Therefore,

$$"7\ km." \text{ is incorrect, but } "7\ km" \text{ is correct}$$

*The symbol for a unit is a lowercase letter except when the unit is derived from a proper name (*i.e.*, Newton, Pascal, Joule, *etc.*). Thus, the symbol for the meter is *m*, whereas the symbol for the Newton is *N*. (An exception is the symbol for *L* for litre.)
A symbol with the prefix is treated as a new unit which can be raised to a power without using brackets. Thus, the notation

$$" cm^3 " \text{ means } "10^{-2}m^3 "$$
$$(10^{-2}m)^3 = 10^{-6}m^3, \text{ and not } 10^{-2}m^3$$

1.4 **Accuracy** means that a measurement is close to the accepted value.
     Precision means that consistent results are obtained. A measurement can be precisely inaccurate.

1.5 USE OF DIMENSIONS TO DERIVE EQUATIONS
      An example, the period of a simple pendulum depends on the three quantities, *viz.*, mass *m* of bob, length ℓ of string and gravity *b*. *The equation of the time period, T, can be written as* $T \propto m^x \ell^y g^z$

      Dimensions for the period is $M^0 L^0 T^1$.
      Equating, $L : 0 = y + z$, *and* $T : 1 = -2z$.

This means $x = 0$, $y = \frac{1}{2}$, $z = -\frac{1}{2}$

Hence $\frac{1}{2}$ become $T \propto \sqrt{\frac{\ell}{g}}$

1.6. THE THIRTEEN DISTINGUISHED INDIAN SCIENTISTS

1. Srinivasa Ramanujan, FRS, (1887 -1920), world renowned mathematical genius, who continues to amaze the world, comparable only with L. Euler and Carl Jacobi. He had neither a pass in the School Examination nor a basic University Degree. He is famous for contributions in number theory, mathematical analysis, infinite series, Ramanujan conjecture, compiled mostly identities and equations, investigated the $\sum \frac{1}{n}$ series, theory of numbers, discovered functions looking like "mock modular forms", partition function. Ramanujan did not differentiate between formal proof and apparent truth based on intuition

2. Prof. Sir Chandrasekhara.VenkataRAMAN (NOBEL Laureate in Physics, 1930) Physicist (Vibrations, Sound & Scattering of Light); FRS, Bharat Ratna Awardee for 1954.

3. Prof. Meghnad N. SAHA, FRS, an eminent Astrophysicist who contributed the theory of thermal ionization (1920) helpful to explain stellar spectra. He propounded the SAHA Equation.

4. Satyendra Nath BOSE, working with Albert Einstein, developed Bose-Einstein Statistics, and "boson", subatomic particle, is named after him. later on now Bose-Einstein Condensate.

5. M. Visveswarayya, Civil Engineer, Bharat Ratna Awardee for 1955.

6. Dr. Homi Jahangir BHABHA Physicist (Father of Indian Nuclear Programme).

7. Dr. Har Gobind Khorana, Biochemist (Nobel Prize in Medicine, 1968).(Indian origin American)

8. Dr. Vikram Ambalal SARABHAI, Physicist (Father of Indian Space Programme)

9. Prof. Subramanyan CHANDRASEKHAR, Astro-physicist (NOBEL Laureate in Physics, 1983) (India born American)

10. Prof.Gnanasundaram.Narayana.RAMACHANDRAN Physicist (Crystallography, Molecular Biophysics). (Father of Molecular Biology in India).

11. Avul Pakir Jainulabdeen ABDUL KALAM Aeronautic Engineer,(Missile man of India) Bharat Ratna Awardee for 1997. (President of India, 2002-07)

12. Dr. Venkataraman RAMAKRISHNAN, Physicist (Nobel Prize in Chemistry, 2009) (India born American).

13. Prof. CNR Rao (Chintamani Nagesa Ramachadra RAO), Chemist (Solid-State and Structure), Bharat Ratna Awardee for 2013.

1.7.1 The PYRAMID OF SCIENCE

A discernible hierarchy (not of social value or of intellectual power) exists in Science. It is known (Leon Lederman, 1993) that there exists a pyramid of science. The base of the pyramid is mathematics, as it does not depend on any other disciplines. Physics (and astronomy) lies on the next layer of the pyramid, because it relies on mathematics. Next on the upper layer falls chemistry, which depends on physics, and not the vice-versa. Akin to this is physical chemistry, mathematical physics. Next comes biological sciences, (include biochemistry and biophysics), which invariably are dependent on both chemistry and physics. Further upper layers of the pyramid become

5

increasingly blurred and less definable, as one reaches physiology, medicine, psychology, biotechnology, *etc.* In a nutshell one may accept the old saying that physicists defer only to mathematicians, whereas mathematicians defer only to GOD.

1.7.2. <u>Can humans colonize any Planet other than the Earth?</u> Physical conditions are important.
The first condition for comfort living in the Universe is the planet should lie in the **Goldilocks Zone.** It is an important question that is yet to receive adequate scientific attention even over the past 50 *yrs* or more of space exploration. The fundamental question is how to survive in conditions different than Earth? It is a fact one has to accept that a human body is composed of 11 common Elements from the Periodic Table with 15 trace elements and remaining elements. That means the human body is formed of Earth. Now human, an organic life, requires for his survival essentially Air, Fire, Space and water, in addition to Earth. Further, acceleration due to gravity of Earth is essential, atmospheric pressure, protection from subatomic particles (Interstellar radiation) arriving at the Earth's upper atmosphere and their prevention by Earth's magnetic field, Experiments with rats in space orbits did not result in production of babies. Radiation damage occurs in the gonads, the ovaries and the testes. Lack of required gravity causes loss in the mass of human muscle resulting in inability bear one's own body weight. Female mice are reported to have stopped ovulating. Microgravity has affected hormone levels in both males and females. How to live in micro-gravity or very weak gravity conditions? Now how to survive at least a month without space-suits, liquid water and air? (fivethirtyeight.com). Scientists should not forget that without atmospheric oxygen one will die breathless, eyes of every human will get dried up resulting in loss of eyesight. How can one cultivate in low gravity without liquid water and have food?
In Taithireeya Upanishad II, Sloka 3, (*Srimad Bhagavad Gita : Ch. VII, Sl* 4) it is seen the formation of the Primordial FIVE Subtle Elements (Pancha Bhutas) as follows:

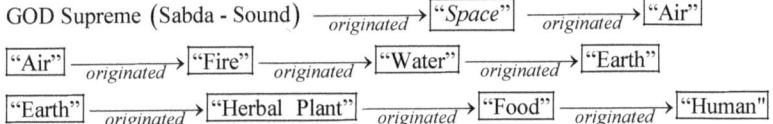

GOD Supreme (Sabda - Sound) $\xrightarrow[originated]{}$ "*Space*" $\xrightarrow[originated]{}$ "Air"

"Air" $\xrightarrow[originated]{}$ "Fire" $\xrightarrow[originated]{}$ "Water" $\xrightarrow[originated]{}$ "Earth"

"Earth" $\xrightarrow[originated]{}$ "Herbal Plant" $\xrightarrow[originated]{}$ "Food" $\xrightarrow[originated]{}$ "Human"

Panchikarana (Ref. Adi Shankara) involves one $\frac{1}{2}$ of the original subtle element to be mixed up with $\frac{1}{8}$th part each of the other original subtle elements to produce the gross element of the subtle element contributing its own one $\frac{1}{2}$. When gross elements are produced then consciousness enters into these elements as their presiding deities, then comes the feeling of egoism (I-ness) identifying with the body. Gross elements solidify and assume forms as per their fundamental qualities.

Chapter 2

MEASUREMENTS

"Vyasochhishtam Jagat - Sarvam"–
means There is nothing remains that is not taught by Saint Veda Vyaasa
- Hindu Dharma
|| *Anur dvau Paramaanu syat trasarenustraua* : *smruta* :
Jaalaarkkrashmyvagata : *Khamevaanupa* tan *nagat* ||
(Sreemad Bhaagavata, Sk III, Chap 11, Sl 5)
(Time and size defined)

2.1   MEASURE OF LENGTH

2.1.1.   The Vernier and Vernier Calipers:

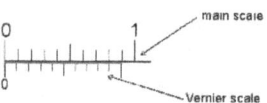

Fig 2.1 Main and Vernier scales

2.1.1.1. What is Zero-error? (to be found from readings on scale)

2.1.1.2 What is Least Count?

> Minimum measurement that can be made by a measuring
> device is known as "LEAST COUNT

Smaller is the magnitude of least count of a measuring instrument, more precise the measuring instrument is.
A measuring instrument cannot measure anything whose dimensions are less than the magnitude of least count.
Least Count of Vernier Calipers = 0.01 *cm*
Least Count of Micrometer Screw gauge = 0.001 *cm*

$$\text{Least count (Vernier Calipers)} = \frac{\text{Minimum measurement on main scale}}{\text{Total number of divisions on Vernier scale}}$$

$$\text{Least count (Screw Gauge)} = \frac{\text{Minimum measurement on main scale}}{\text{Total number of divisions on Circular scale}}$$

2.1.2. The Slide Calipers

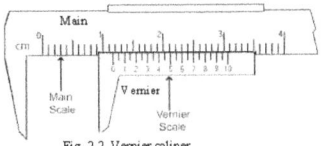

Fig 2.2 Vernier caliper

Determine zero-error and average of repeated measurements.

2.1.2. The Screw Gauge

Determine i) zero-error, ii) pitch, and iii) Least count, to get the dimension of the object

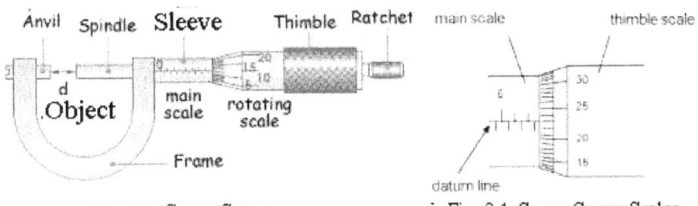

Fig 2.3 Screw Gauge   Fig 2.4 Screw Gauge Scales

2.1.2.1 What is PITCH?

This is the perpendicular distance between two consecutive threads of the screw gauge or spherometer.

$$\text{Pitch} = \frac{\text{Distance traveled on main scale}}{\text{Total number of rotations}}$$

2.1.3. The Spherometer

Its three legs lie on the vertices of an equilateral triangle of side a. Δr is the elevation of the screw point from plane surface as a result of curvature of the surface having radius of curvature r.

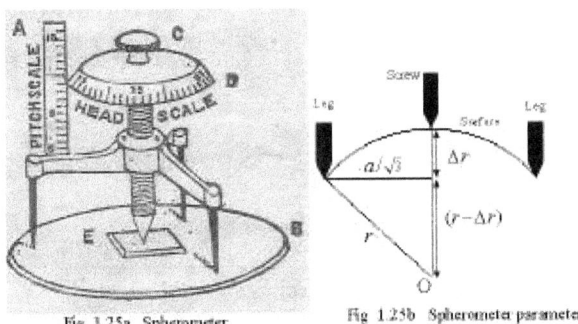

Fig 1.25a Spherometer   Fig 1.25b Spherometer parameters

Radius of curvature r of a spherical surface

Put $a = \ell$; $\Delta r = h$ and $r = R$.

$$R = \frac{h^2 + (\ell^2/3)}{2h}$$ or $$R = \frac{\ell^2}{6h} + \frac{h}{2}$$

Measure for both sides of the big convex lens the curvature radii R_1 and R_2. Focal length f of this lens

$$\frac{1}{f} = (n-1)\left\{\frac{1}{R_1} + \frac{1}{R_2}\right\}$$

2.1.4. The Optical Lever

It is a convenient device to magnify a small displacement and thus to make possible an accurate measurement of the displacement.

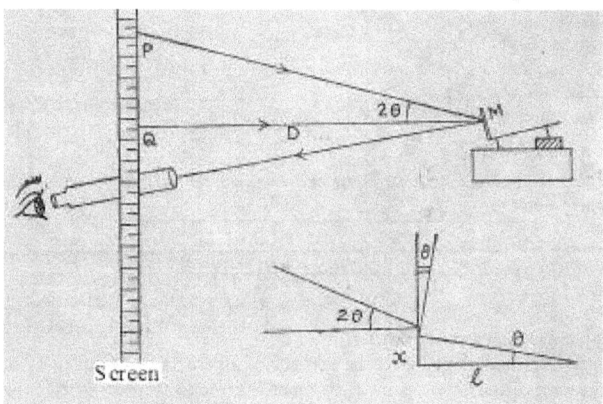

θ = Angle through which the mirror M tilted when the object of thickness $\Delta\ell$ is introduced below the rear leg of the lever, d = distance between the rear leg and ine between front two legs. s = shift in the reading of the telescope, D= distance between the screen and mirror M.

$$\theta = \frac{x}{\ell}; \quad 2\theta = \frac{s}{D}$$

$$x = \frac{\ell \cdot s}{2D},$$

2.2. SIGNIFICANT FIGURES, Standard Form

Significant figures (s.f.) are the number of meaningful digits in a numerical quantity. The accuracy of the method of measurement employed will determine the number of significant figures used. Examples are:

0.0067324	five significant figures
6.7324×10^{-9}	five significant figures
6.4×10^{-6}	two significant figures
6.73	Three significant figures

Zeros at the beginning and end of a number are not counted. But zeros at the middle of a number are counted. For example, 5.04, 50400, 0.0504 all have three significant figures.

Standard form: Useful to write very large and very small physical quantities $a \times 10^{\pm n}$, where a is a number between 1 to 10, and the index n is an integer 0 or non-zero.
Example

$$(11.37\ m)\ (6.9\ m)\ \neq\ 76.84\ m^2\ =\ 77\ m^2$$.

A length 2 *cm* observed with an instrument of precision 10 μm is written correctly as

$$(20.00\ mm \pm 0.01\ mm) \text{ or as } 2.000 \times 10^1 mm$$

2.3. CALCULATION OF ERRORS:
Q = A physical quantity whose determination involves two quantities to a and b be measured.

Fig 2.6 Physical quantity ab

2.3.1 Sum $Q = a + b$, or difference
If $a = 16.5\ cm \pm 0.1\ cm \equiv (16.5\ \pm 0.1)\ cm$
$b = (25.4 \pm 0.1)\ cm$
$\quad Q_{average} = b - a = 8.9\ cm$
$\quad Q_{maximun} = b - a = 9.1\ cm$
$\quad Q_{minimun} = b - a = 8.7\ cm$.
Error, $\Delta Q = \Delta a + \Delta b = (0.1\ cm + 0.1)\ cm = 0.2\ cm$
$\quad Q = Q_{av} \pm \Delta Q = 8.9\ cm \pm 0.2\ cm$
$\quad \%$ error in $Q = \dfrac{\Delta Q}{Q_{av}} 100\% = 2.2\%$

2.3.2 Product or quotient of two physical quantities. $Q = ab$ or
ΔQ, Δa, Δb, then
$Q_{max} = Q_{av} + \Delta Q = ab + (a\Delta b + b\Delta a)$
i.e., $\Delta Q = (a\ \Delta b + b\ \Delta a)$
Fractional error, $\dfrac{\Delta Q}{Q} = \dfrac{\Delta a}{a} + \dfrac{\Delta b}{b}$

2.3.3   If $Q = ab^n$

$$\boxed{\frac{\Delta Q}{Q} = \frac{\Delta a}{a} + \frac{n\Delta b}{b}}$$

2.3.4   Random error - Standard deviation s

Variance $s^2 = \dfrac{\Sigma(x - <x>)^2}{(n-1)}$

Standard deviation of a mean of n observations, s_m

$s_m = s/\sqrt{n}$

In t-distribution, the static t is given by

$t = $ (measured value - true value)$/s_m$.

2.3.5.   Statistical best fitting a straight line, by regression of y on x

The best straight line is one such that the squares of deviation of every point measured parallel to the y-axis and summed for all points is a minimum.

2.3.6.   **Curve fitting** (the Polynomial Curve)

a) If the independent variable is not equally spaced, a regression relatiojn may be calculated with x and $z = x^2$ as two variables to give $Y = ax + bz$.

b) If the independent variable is taken at regular intervals, without issing values, a polynomial curve may be fitted to the data.

2.3.7.   The Chi-squared test.

For Poisson distribution this is used.

2.3.8.   The Method of Least Squares for the best representative curve

The method provides the overall rationale for the placement of the line of best fit among the data points being studied. The method is a regression analysis.

Step 1: Calculate the mean of the x -values and the mean of the y -values.

Step 2: Calculate the slope of the line of best fit:

Step 3: Compute the y-intercept of the line

Step 4: Use the slope m and the y -intercept bb to form the equation of the line.

Example: The data points (chosen for illustration) when plotted on a coordinate plane appears as in Fig.

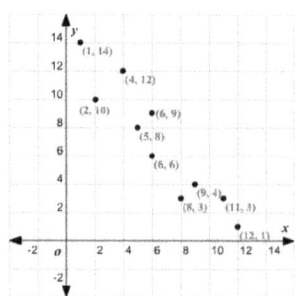

The slope of the line is -1.1 and the y -intercept is 14.0 .

Use the slope and y -intercept to form the equation of the line of best fit.

Therefore, the equation $y = - (1.1) x + 14.0$.

Draw the line on the scatter plot.

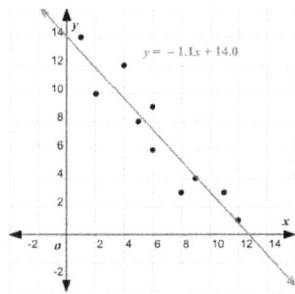

2.4. **Use of Dimensions to derive equations**:

Example: To find the period t of a Simple Pendulum.

Let m, ℓ and g are mass & length of bob, and acceleration due to gravity.

$t \propto m^{\alpha} \cdot \ell^{\beta} \cdot g^{\delta}$.; α, β, δ are unknown powers to be found.

$t = k\ m^{\alpha} \cdot \ell^{\beta} \cdot g^{\delta}$;

which has the <u>Dimensional form</u>

$$\boxed{[M]^0\ [L]^0\ [T]^{1} = [M]^{\alpha}\ [L]^{\beta}\ [L]^{\delta}\ [T]^{-2\delta}}$$;

$[M] : 0 = \alpha,$; $[L] : 0 = \beta + \delta$; $[T] : 1 = -2\delta$;

This means $\alpha = 0,$; $\beta = \frac{1}{2}$, $\delta = -\frac{1}{2}$

$$\boxed{t = k\sqrt{\left(\frac{\ell}{g}\right)}}$$

2.5. STATISTICS

Mean (Average) value $\bar{x}$ and Variance s^2

The mean of n values x_i of a random variable x is

$$\bar{x} = \frac{1}{n}\sum_{i=1}^{n} x_i$$

$$s^2 = \frac{1}{n}\sum_{i=1}^{n}(x_i - \bar{x})^2 = \frac{1}{n}\sum_{i=1}^{n}x_i^2 - \bar{x}^2 .$$

If a random variable has p4robability density function f ,

the expectation (or mean) $\mu = \int_{-\infty}^{\infty} t f(t)\ dt$

Variance $\sigma^2 = \int_{-\infty}^{\infty}(t - \mu)^2 f(t)\ dt$

Standard deviation $= \sqrt{\text{Variance}}$.

For discrete distributions, with $p_k =$ probability that the x takes the value x_k ,

$$\mu = \sum_k P_k x_k$$

$$\sigma^2 = \sum_k P_k (x_k - \mu)^2$$

Statistical (Probability) Distributions

Depending on the number of events statistical distribution functions appropriate may be used to describe physical events. A large number of problems in physics xcan be approximately by the binomial, Poisson and Gaussian distributions.

a) Discrete (Binomial) Distribution

This function specifies the number of times (x) that an event occurs in n independent trials, p being the probability of the event occurring in a single trial.

Mean, $\mu = np$

Variance, $\sigma^2 = np(1 - p)$

$\sqrt{n p (1 - p)}$ = standard deviation

$$f(x) = \frac{n! \, p^x (1 - p)^{n-x}}{x! \, (n - x)}$$

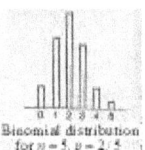

Binomial distribution
for $n = 5$, $p = 2/5$

b) Poisson distribution

Distribution with parameter λ (> 0)

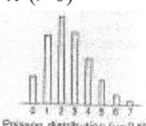

Poisson distribution ($\mu \approx 2.5$)

$$f(k) = \frac{e^{-\lambda} \lambda^k}{k!}$$

λ = mean (average) number of events per interval.

$\sqrt{\lambda}$ = standard deviation

Mean μ = variance σ^2.

The Poisson is the limit of the binomial, as $n \to \infty$. with $p = \frac{\mu}{n}$.

c) Continuous distribution

Rectangular (Uniform) distribution

$$f(t) = \begin{cases} 1/(b - a), & \text{if } a \le t \le b \\ 0 & \text{, otherwise} \end{cases}$$

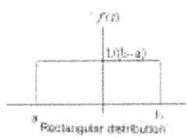

$$\mu = \tfrac{1}{2}(a+b), \quad \sigma^2 = \tfrac{1}{r2}(b-a)^2$$

d) Exponential distribution
The standard exponential distribution $f(x)$

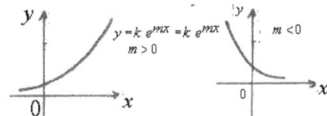

$$f(x) = e^{-x}, \text{ for } x \geq 0.$$

e) Gaussian (Normal) distribution
The probability density function $f(x)$ is a bell-shaped curve.

σ =standard deviation, mean μ and variance σ^2.

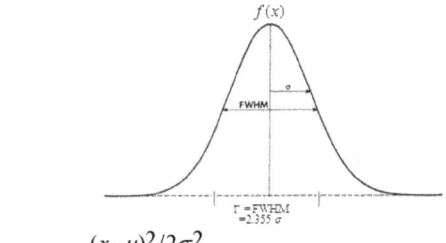

$$f(x) = \frac{e^{-(x-\mu)^2/2\sigma^2}}{\sqrt{2\pi}\,\sigma}, \quad (-\infty < x < \infty)$$

f) Standard Cauchy's distribution

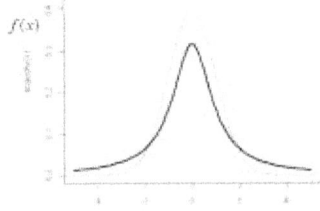

$$f(x) = \frac{1}{\pi(1+x^2)}, \quad -\infty < x < \infty.$$

g) The Student (or 't') distribution

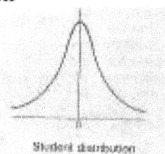

Student distribution

$$f(t) = \frac{\frac{\Gamma(\frac{1}{2}v+1)}{(\pi v)^{1/2}\Gamma(\frac{1}{2}v)}}{(1+\frac{t^2}{v})^{\frac{1}{2}v+\frac{1}{2}}}, \qquad (-\infty < t < \infty).$$

It is used o compute confidence intervals for an estimate of the population mean calculated from sample data, of unknown mean and variance.

-0-0-0-0-0-0-

Chapter 3

MATHEMATICAL PRELIMINARIES

I am compelled to fear that science will be used to promote the power of dominant
groups rather than to make men happy. ~Bertrand Russell, Icarus, or the Future of Science, 1925

|| *Chhaayaayaa* : *Kardhamo Jagne Devahootya* : *Pati* : Pr *abhu* :
Manaso Dehataschedam Jagne Viswakruto Jagat ||
(Sreemad Bhaagavata, Sk III, Chap 12, Sl 27)
(Origin of Trigonometry from Shadow)

3.1.   **GRAPHS**

Graphs are extremely useful in physics for finding and confirming relationships between different variables. Various large number of graphs present experimental results or theories in physics. It is customary to choose the <u>independent variable in the X-axis</u> and the dependent variable in the Y- axis to plot a graph relating these two quantities.

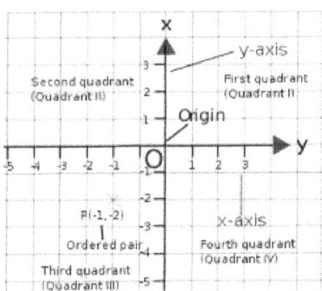

Fig. 3.1 Cartesian Coordinates
(Rectangular Coordinate System)

3.2.1   Co-ordinate Geometry
Slope (or gradient) of a curve
Often a point in a plot between two variable quantities, say x and t, is not a straight line.

The distance between two points (x_1, y_1) and (x_2, y_2)

$$d = \sqrt{[(x_1 - x_2)^2 + (y_1 - y_2)^2]}$$

The slope, also called gradient, of a curve at a point is the slope of the tangent at that point.

$$\boxed{\text{Slope of a curve} \; = \; \lim_{\delta t \to 0} \frac{\Delta x}{\Delta t} \; = \; \frac{dx}{dt} \; = \frac{(y_2 - y_1)}{(x_2 - x_1)}}$$

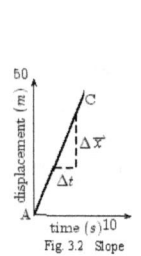

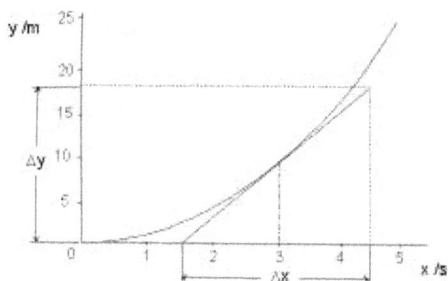

Fig. 3.2 Slope

3.2.2. Basic Trigonometry

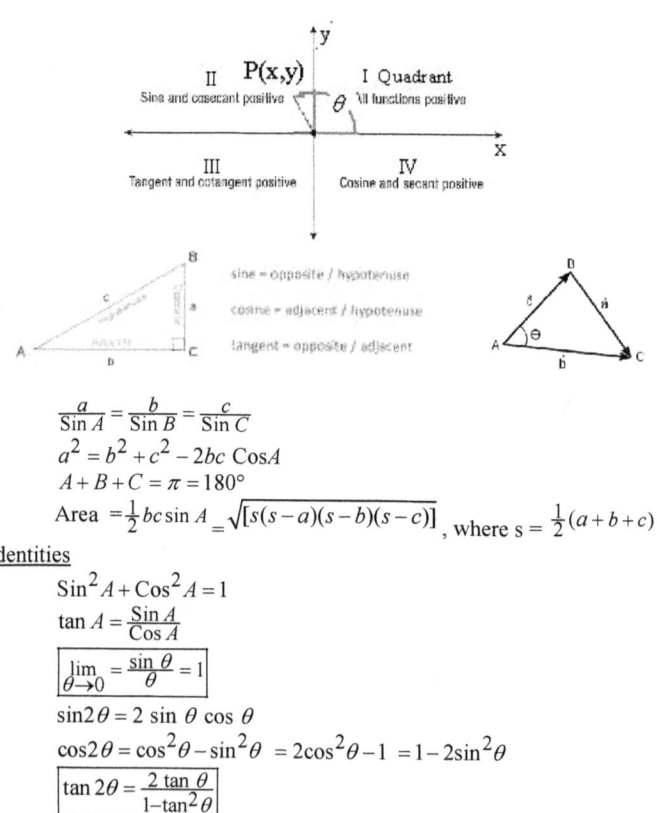

$$\frac{a}{\sin A} = \frac{b}{\sin B} = \frac{c}{\sin C}$$

$$a^2 = b^2 + c^2 - 2bc \cos A$$

$$A + B + C = \pi = 180°$$

Area $= \frac{1}{2} bc \sin A = \sqrt{[s(s-a)(s-b)(s-c)]}$, where $s = \frac{1}{2}(a+b+c)$

Identities

$$\sin^2 A + \cos^2 A = 1$$

$$\tan A = \frac{\sin A}{\cos A}$$

$$\boxed{\lim_{\theta \to 0} = \frac{\sin \theta}{\theta} = 1}$$

$$\sin 2\theta = 2 \sin \theta \cos \theta$$

$$\cos 2\theta = \cos^2 \theta - \sin^2 \theta = 2\cos^2 \theta - 1 = 1 - 2\sin^2 \theta$$

$$\boxed{\tan 2\theta = \frac{2 \tan \theta}{1 - \tan^2 \theta}}$$

3.2.3. Hyperbolic Functions

$$\cosh x = \tfrac{1}{2}(e^x + e^{-x}); \quad \sinh x = \tfrac{1}{2}(e^x - e^{-x})$$

$$\tanh x = \frac{(e^x - e^{-x})}{(e^x + e^{-x})} = \frac{(1 - e^{-2x})}{(1 + e^{-2x})};$$

$$\operatorname{sech} x = \frac{2}{(e^x + e^{-x})}; \quad \operatorname{cosech} x = \frac{2}{(e^x - e^{-x})}$$

3.2.4. Inverse Hyperbolic Functions

$$\cosh^{-1} x = \ln [x + \sqrt{x^2 - 1}] \ (|x| \geq 1); \sinh^{-1} x = \ln [x + \sqrt{x^2 + 1}]$$

3.2.5 Significance of area under a Non-linear graph:

Distance traveled by a body in motion having non-linear graph, say velocity-time graph, can be determined from the area of the graph.

3.3.1 Straight Line (Linear):

Straight line is the most useful form. Its features are:

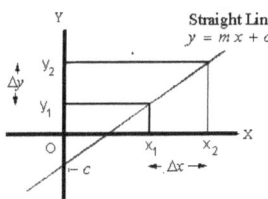

Fig. 3.3 Straight line

When $x = 0$, the intercept on the y-axis is '$-c$',
When $y = 0$, the intercept on the x-axis is '$-c/m$',

The **slope** of the line is $\dfrac{\Delta y}{\Delta x} = m$

Example, $y = mx + c$.

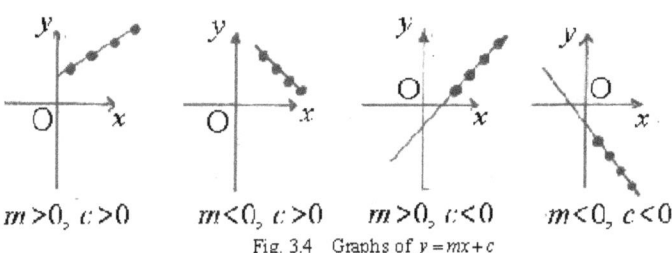

$m > 0, c > 0 \qquad m < 0, c > 0 \qquad m > 0, c < 0 \qquad m < 0, c < 0$

Fig. 3.4 Graphs of $y = mx + c$

In physics of classical motion, $s = u + at$ is a *straight line* motion of a body moving at a uniform speed u getting accelerated at constant acceleration a.

3.3.2 Basic Quadratic:

$$y = m x^2 + c$$

Example: Variation of the kinetic energy of a body in motion with a velocity.

$$T = \frac{1}{2} m v^2 + V$$

$$S = u t + \frac{1}{2} g t^2 .$$

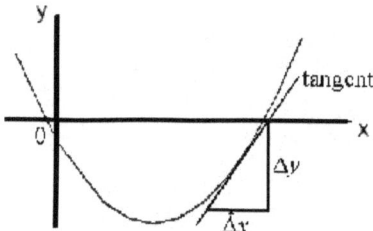

Fig.3.5 Graph of displacement

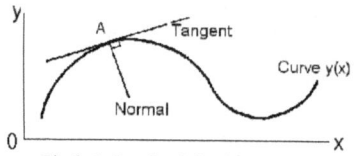

Fig.3.6 Graph of displacement

3.3.3 Graphical comparison of Linear, Square, Cubic, Quadratic, Square root and Absolute Functions.

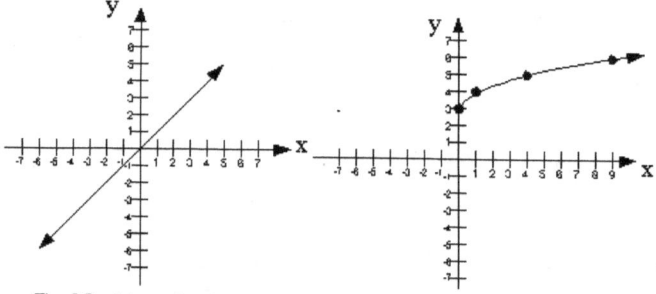

Fig. 3.7a Linear function   Fig. 3.7b Square root function

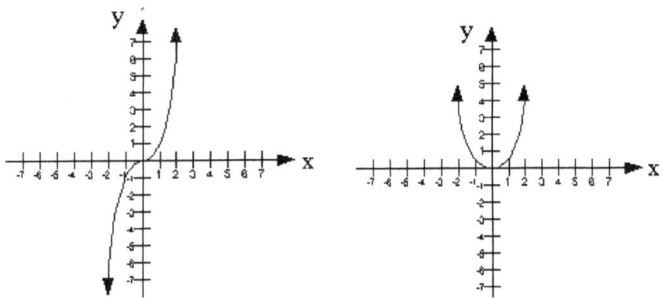

Fig. 3.7c: Cubic function   Fig. 3.7d Quadratic function

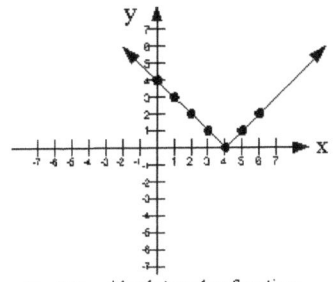

Fig. 3.7e Absolute value function

3.3.4 Exponential increase:

$$y = k\, e^{mx}$$

Increase of Pressure of air with depth $y(= P_h) = P_0\, e^{-(mg/kT)h}$

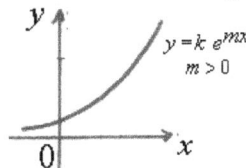

$y = k\, e^{mx}$
$m > 0$

Fig. 3.8a Exponential inrease curve

3.3.5 Exponential Decrease:

Example: Radioactive decay Activity, $N = N_0\, e^{-\lambda t}$;

Voltage across a Capacitor in an RC circuit, $V_t = V_0\, e^{-t/RC}$

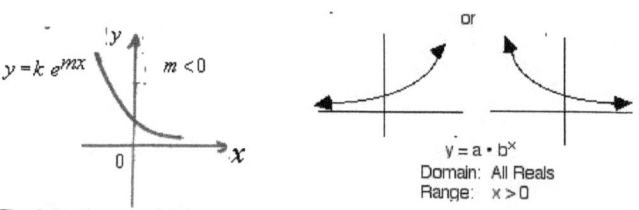

Fig. 3.8b Exponential decrease curve   Fig. 3.8c Exponential curves

3.3.6.1. Length of a Plane curve, s
($y = f(x)$ between $x = a$ and $x = b$)

$$s = \int_a^b \left[1 + (\tfrac{dy}{dx}) \right]^{1/2} dx \qquad \text{(Coordinate form)}$$

$$s = \int_\alpha^\beta \left[1 + (\tfrac{dr}{d\theta}) \right]^{1/2} d\theta \qquad \text{(Polar form)}$$

3.3.6.2. Plane area

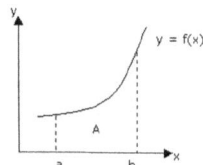

$$= \tfrac{1}{2} \int_a^b y \, dx$$

$$= \tfrac{1}{2} \int_\alpha^\beta r^2 d\theta$$

3.3.6.3.   Radius of curvature of a Plane curve, at point (x , y)

$$p = \left[1 + (\tfrac{dy}{dx}) \right]^{1/2} / (\tfrac{d^2 y}{dx^2})$$

3.3.6.4.   Mean and RMS values of y, of $y = f(x)$, between $x = a$ and $x = b$

$$\text{Mean value} = \frac{1}{b-a} \int_a^b y \, dx \; ; \; \text{RMS} = \sqrt{\frac{1}{b-a} \int_a^b y^2 \, dx}$$

3.4.1   Properties of logarithms –

$\boxed{p = \log_a x}$ if and only if $x = a^p$ ($a > 0$).

$\boxed{\log_a a^x \;=\; \ln x / \ln a}$

* $\boxed{\log_a 1 = 0}$, because $a^0 = 1$.

* No matter what the *base* is, as long as it is legal, the log of 1 is always 0. That's because logarithmic curves always pass through (1, 0)

* $\boxed{\log_a a = 1}$, because $a^1 = a$

* Any value raised to the first power is that same value.

$\boxed{\log_a a^x \;=\; x}$

* The log base a of x and a to the x power are inverse functions. Whenever inverse functions are applied to each other, they inverse out, and you're left with the argument, in this case, x.

* $\log_a x = \log_a y$ implies that $x = y$

* If two logs with the same base are equal, then the arguments must be equal.

$\log_a x = \log_b x$ implies that $a = b$.

$\log_b x = \log_a x \cdot \log_a b$

- Exponential function: $y = a^x = x$, $a > 0, a \neq 1$.

  $$\boxed{a^x = e^{x \log a}}$$

- $\log x$ means it represents the <u>common logarithm</u> $\log_{10} x$ ..

- $\ln x$ represents the <u>natural logarithm</u> $\log_e x \equiv \ln x$, (Napierian base).

- $\boxed{\lg x = \log_{10} x}$

- $\boxed{\text{lb}x = \log_2 x}$

- $\boxed{\log_a xy = \log_a x + \log_a y}$

- $\log_a x^n = n \log_a x$

3.4.2   Logarithmic function:
$$y = \log_a x, \; a > 0, \; a \neq 1.$$

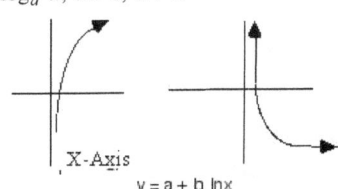

X-Axis

$y = a + b \ln x$
Domain: $x > 0$
Range: All Reals

Fig. 3.9 Logarithmic curve

3.4.3.1 <u>Log - log curve</u>
$$y = k \, e^{c \, x}$$
Taking natural logarithms
$$\text{Ln } y = \ln k + c x$$
\Example: Current I through a silicon *pn* diode varies as
$$I_V = I_0 \, e^{eV/k_B T} \; ;$$
Is expressed as $\text{Ln } I = \ln I_0 + \left(\dfrac{e k_B}{T} \right) V$

3.4.3.2 $y = k \, x^2$

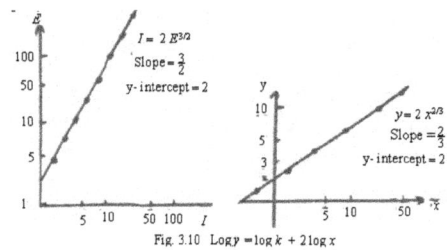

Fig. 3.10 $\log y = \log k + 2 \log x$

Taking logarithms,

$$\log y = \log k + 2 \log x$$

Example: Simple Harmonic Motion

$$\omega = \sqrt{k/m} = (k/m)^{1/2}; \ m = k \ \omega^2$$

$$T = = 2\pi\sqrt{\frac{m}{k}}; \ m = (\frac{k}{4\pi^2}) T^2;$$

$$T = = 2\pi\sqrt{\frac{\ell}{g}}; \ \ell = (\frac{g}{4\pi^2}) T^2.$$

Consider $y = \log x$, if $x = 10^{2y}$

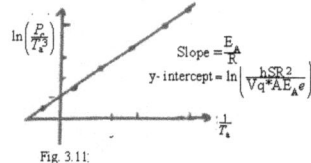

Fig. 3.11

3.4.3.3 Log – log graph for a power relation

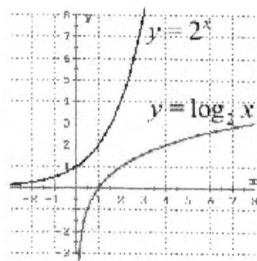

Fig. 3.12 Log-log graph

3.4.3.4 Comparison

Comparison of Exponential and Logarithmic Functions		
	Exponential	Logarithmic
Function	$y = a^x$, $a > 0$, $a \neq 1$	$y = \log_a x$, $a > 0$, $a \neq 1$
Domain	all reals	$x > 0$
Range	$y > 0$	all reals
intercept	$y = 1$	$x = 1$
increasing	when $a > 1$	when $a > 1$
decreasing	when $0 < a < 1$	when $0 < a < 1$
asymptote	y-axis	x-axis
continuous	yes	yes
smooth	yes	yes

3.5.1 QUADRATIC EQUATION

$$a x^2 + bx + c = 0$$

Solution (Roots) $x = \dfrac{-b \pm \sqrt{b^2 - 4ac}}{2a}$; Determinant $D = \sqrt{b^2 - 4ac}$.

Zeroes are x. If p and q are roots, Sum of roots $= p + q = \dfrac{-b}{a}$, Product $pq = \dfrac{c}{a}$;

Equations is $(x - p)(x - q) = 0$, i.e., $x^2 - (\text{Sum of roots})x + \text{Product of roots} = 0$.

Angle between two straight lines

$$y = m_1 x + c_1 \text{ and } y = m_2 x + c_2 \text{ is } \alpha = \tan^{-1} \frac{m_2 - m_1}{1 + m_1 m_2}$$

$$A_1 x + B_1 y + C_1 = 0 \text{ and } A_2 x + B_2 y + C_2 = 0 \text{ is } \alpha = \tan^{-1} \frac{A_1 B_2 - A_2 B_1}{B_1 B_2 + A_2 A_1}.$$

Condition that the two are parallel: $A_1 B_2 = A_2 B_1$,

Condition that the two are perpendicular: $A_1 A_2 = -B_1 B_2$

3.4.2. NUMERICAL ANALYSIS

1) Solution of Algebraic equation $f(x) = 0$, with x_n = nth estimate,

$$x_{n+1} = x_n - \frac{f(x_n)}{f'(x_n)} \quad \text{(Newton's method)}$$

$$x_{n+1} = \frac{x_{n-1} f(x_n) - x_n f(x_{n-1})}{f(x_n) - f(x_{n-1})} \quad \text{(Secant method)}$$

2) Numerical Integration (with n=even positive integer),

$$\int_a^{a+nh} f(x)dx \approx \frac{h}{3}\left(f(a) + f(a + nh) + 2\sum_{i=1}^{\frac{n}{2}-1} f(a + 2ih) + 4\sum_{i=1}^{\frac{n}{2}} f(a + (2i - 1)h) \right)$$

3.5 CONICS

Equations of 2-D surfaces in the horizontal and vertical axes are, respectively,

a) Circle: $x^2 + y^2 = a^2$ {general equation, (i) $(x-h)^2 + (y-k)^2 = r^2$; centre at (h,k)} and radius r (ii) $x^2 + y^2 + 2gx + 2fy + c = 0$, with centre

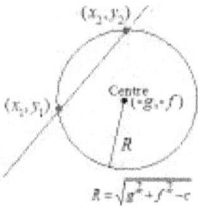

at $(-g, -f)$, radius $\sqrt{(g^2 + f^2 - c)}$}.

b) Ellipse: $\dfrac{x^2}{a^2} + \dfrac{y^2}{b^2} = 1$, eccentricity $\varepsilon = \sqrt{1 - \dfrac{b^2}{a^2}}$

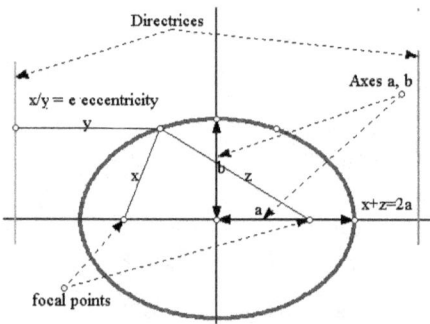

c) Parabola: $y^2 = 4ax$,

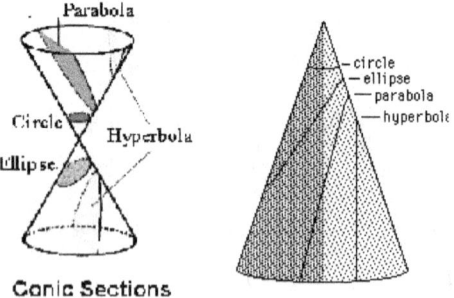

Fig. 3.13 Conical sections

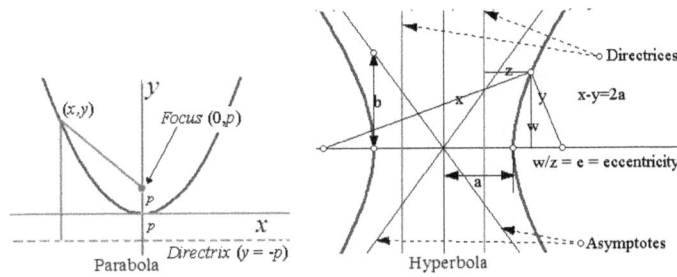

Parabola

Directrix $(y = -p)$

Hyperbola

Its general equation with axis parallel to y-axis is $y = ax^2 + bx + c$ d) d0

d) **Hyperbola**: $\frac{x^2}{a^2} - \frac{y^2}{b^2} = 1$, eccentricity $\varepsilon = \sqrt{1 + \frac{b^2}{a^2}}$

3.6.1. **SCALARS and VECTORS,** Definitions:
A fairly simple way handling motion in 2- or more Dimensions is possible by using vectors.
A quantity is a <u>scalar</u> if it has *magnitude* only.
A quantity is a <u>vector</u> if it has *both* magnitude and direction.

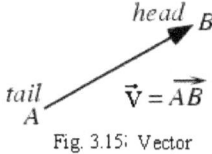

Fig. 3.15: Vector

3.6.2 **Multiplication** *of vectors*:

3.6.3 Principle of Superposition

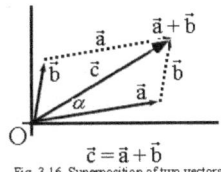

$\vec{c} = \vec{a} + \vec{b}$

Fig. 3.16 Superposition of two vectors

$$c = \sqrt{a^2 + b^2 + 2\,ab\,\cos\alpha}$$
$$\tan\alpha = \frac{b\,Sin\alpha}{a\,b\,Cos\alpha}$$

3.6.3.1 **Unit vectors**
A <u>unit vector</u> is a *dimensionless* vector of magnitude 1. The three unit vectors $\hat{i}$, $\hat{j}$, $\hat{k}$ form the basis vectors of the axes of a XYZ (Cartesian) coordinate system.

3.6.3.2 Resolution of Vectors

Any vector in 3-Dimensions is expressed in terms of its components and unit vecotors,

$$\vec{a} = a_x\hat{i} + a_y\hat{j} + a_z\hat{k}$$

3.6.4 Scalar product *(Dot product)*: of two vectors $\vec{a}$ & $\vec{b}$
∴

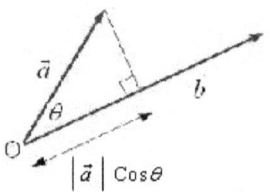

Fig. 3.17 Scalar product $\vec{a}\vec{b}$

$$\vec{c} = |\vec{a}||\vec{b}|\cos\theta = |\vec{b}|\, x, \text{ Projection of } \vec{a} \text{ on } \vec{b}$$

where $|\vec{a}|$ and $|\vec{b}|$ denote *magnitudes* of $\vec{a}$ & $\vec{b}$:

3.7.5 VECTOR product of $\vec{a}$ and $\vec{b}$

$$\vec{a} \wedge \vec{b} = \vec{c}, \text{ a third vector normal to both } \vec{a} \text{ and } \vec{b}$$

$$\text{Area} = |\vec{a}||\vec{b}|\sin\vartheta$$

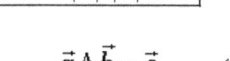

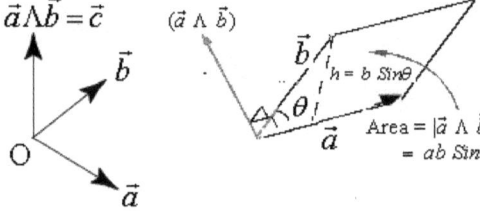

Fig. 3.18 Cross product $\vec{a}\wedge\vec{b}$

3.7. DETERMINANTS $|A|$ and MATRICES

In many of the physical problems can be conveniently dealt by using array of mathematical quantities called *determinants* and *matrices*.

3.7.1. A determinant is an arrangement of N^2 quantities into a square array with N rows and N columns, where N is called the *order* of the determinant. Thus the arrays

$$\begin{vmatrix} b_{11} & b_{12} \\ b_{21} & b_{22} \end{vmatrix} \quad \text{and} \quad \begin{vmatrix} a_{11} & a_{12} & a_{13} \\ a_{21} & a_{22} & a_{23} \\ a_{31} & a_{32} & a_{33} \end{vmatrix}$$

are determinants, the first of order 2 and the second of order 3. A determinant is designated by a symbol $|A|$ enclosed between two vertical lines (or by two braces), each element a_{ij} (a_{ij}) will have two subscripts. The first subscript (i) defines the row and the second (j) specifies the column in which the element a_{ij} appears. This means a_{ij} belongs to i^{th} row and j^{th} column of $|A|$. Every determinant has a numerical value.

3.7.2.   The minor A_{ij} of an element a_{ij} is the determinant of $(N-1)^{th}$ order obtained when the ith row and j^{th} column of the original determinant $|A|$ are struck out.. The *co-factor* of an element a_{ij} in a determinant $|A|$ is the signed *minor* of the element a_{ij}; the sign being $(-1)^{i+j}$.

Co-factor of $a_{ij} = (-1)^{i+j}$ [*minor* of the element a_{ij}].

3.7.3.   The value of determinant

system	array	matrix	determinant
Ax +By = C	A B C	$\begin{bmatrix} A & B & C \\ D & E & F \end{bmatrix}$	$\begin{vmatrix} de & ant \\ min & ter \end{vmatrix} =$
Dx +Ey = F ;	D E F ;	,	(de)(ter) - (min)(ant)

i.e., $\quad |A| = \sum_{i=1 \; or \; j=1}^{N} (-1)^{i+j} (\text{Minor of } a_{ij})$

Example 1.1

$$|A| = \begin{vmatrix} 5 & 10 & 8 \\ 10 & 2 & -2 \\ 8 & -2 & 11 \end{vmatrix}$$

$$= 5 \begin{vmatrix} 2 & -2 \\ -2 & 11 \end{vmatrix} - 10 \begin{vmatrix} 10 & -2 \\ 8 & 11 \end{vmatrix} + 8 \begin{vmatrix} 10 & 2 \\ 8 & -2 \end{vmatrix}$$

$$= 5\,(22 - 4) - 10\,(110 + 16) + 8\,(-20 - 16) = -1538.$$

3.7.4.   Properties of Determinants:
   (i)   The value of a determinant changes sign when two rows or two columns are interchanged.
   (ii)   If two rows are identical or two columns are identical, the value of the determinant is zero.
   (iii)   Solving for x and y

constant colum replaces
desired variable column

If $\quad \begin{matrix} Ax + By = C \\ Dx + Ey = F \end{matrix}$ then $x = \dfrac{\begin{vmatrix} C & B \\ F & E \end{vmatrix}}{\begin{vmatrix} A & B \\ D & E \end{vmatrix}}$ and $y = \dfrac{\begin{vmatrix} A & C \\ D & F \end{vmatrix}}{\begin{vmatrix} A & B \\ D & E \end{vmatrix}}$

same denominator,
coefficients columns

$$x = \frac{CE - BF}{AE - BD} \quad \text{and} \quad y = \frac{AF - CD}{AE - BD}$$

3.7.5. Matrix Methods – Eigen value problem

Consider the operation of rotation $\hat{R}$ on a vector $\vec{r}$ through an angle θ ihe anti-clockwise sense to yield the new vector $\vec{r}^{\,'}$, such that

$$|\vec{r}| = |\vec{r}^{\,'}|.$$

In the matrix notation this kind of mathematical transformation is given as

$$\hat{R}\,\vec{r} = \vec{r}^{\,'}.$$

Example.

Given the matrix $\hat{R} = \begin{pmatrix} 2 & -1 \\ -1 & 2 \end{pmatrix}$ find the eigen values λ_i and eigen vectors of $\hat{R}$.

$\boxed{Step\ \#1}$: Form the secular equation using the unit matrix $\hat{E} = \begin{pmatrix} 1 & 0 \\ 0 & 1 \end{pmatrix}$

$$\det(\hat{R} - \lambda\hat{E})\ \vec{r} = 0$$

$\boxed{Step\ \#\ 2}$.   To find the eigen values

i.e. $\begin{vmatrix} 2 - \lambda & -1 \\ -1 & 2 - \lambda \end{vmatrix} = 0$

This gives $(2 - \lambda)^2 - 1 = 0$. i.e., $\lambda^2 - 4\lambda + 3 = 0$, or

$$\lambda_1 = +1 \text{ and } \lambda_2 = +3$$

$\boxed{Step\ \#\ 3}$. To find the eigen vectors: $\vec{r}_1 = \hat{r}_1$ and $\vec{r}_2 = \hat{r}_2$

$\det(\hat{R} - \lambda\hat{E})\ \hat{r}_1 = 0$ gives $\begin{pmatrix} 2 - \lambda & -1 \\ -1 & 2 - \lambda \end{pmatrix}\begin{pmatrix} x \\ y \end{pmatrix} = \begin{pmatrix} 0 \\ 0 \end{pmatrix}$,

$$\vec{r}_1 = x\begin{pmatrix} 1 \\ 1 \end{pmatrix}$$

Similarly using $\lambda_2 = +3$ the secular equation becomes

$$\det(\hat{R} - \lambda\hat{E})\ \hat{r}_2 = 0$$

$$\vec{r}_2 = x\begin{pmatrix} 1 \\ -1 \end{pmatrix}$$

$\boxed{Step\ \#\ 4}$.   Normalization

$$\hat{r_1} = \frac{1}{\sqrt{2}} \begin{pmatrix} 1 \\ 1 \end{pmatrix} \text{ and } \hat{r_2} = \frac{1}{\sqrt{2}} \begin{pmatrix} +1 \\ -1 \end{pmatrix}$$

Step # 5 . The orthogonal matrix $\hat{X}$ that diagonalizes $\hat{R}$ is constructed with each eigen vector representative of one column of it.

$$\hat{X} = \begin{pmatrix} 1/\sqrt{2} & 1/\sqrt{2} \\ 1/\sqrt{2} & -1/\sqrt{2} \end{pmatrix}$$

Step # 6 : Find $\hat{X}'$ the transpose of $\hat{X}$

$$\hat{X}' = \begin{pmatrix} 1/\sqrt{2} & 1/\sqrt{2} \\ 1/\sqrt{2} & -1/\sqrt{2} \end{pmatrix}$$

Step # 7 :   Find the diagonalized matrix of $\hat{R}$

$$\hat{X}' \hat{R} \hat{X} = \begin{pmatrix} 1/\sqrt{2} & 1/\sqrt{2} \\ 1/\sqrt{2} & -1/\sqrt{2} \end{pmatrix} \begin{pmatrix} 2 & -1 \\ -1 & 2 \end{pmatrix} \begin{pmatrix} 1/\sqrt{2} & 1/\sqrt{2} \\ 1/\sqrt{2} & -1/\sqrt{2} \end{pmatrix}$$

$$= \begin{pmatrix} 1 & 0 \\ 0 & 3 \end{pmatrix} = \hat{R}_d$$

Thus the eigen values are the diagonal elements of the diagonalized matrix $\hat{R}_d$
In practice, there are several different ways to diagonalizing a given matrix.

3.7.6.   SUMMATION AND PRODUCT NOTATION
The equation of the form

$$y = a_1 + a_2 + a_3 + \ldots\ldots\ldots + a_i + \ldots\ldots + a_n$$

is written as

$$y = \sum_{i=1}^{n} a_i$$

The product equation of the form

$$z = a_1 . a_2 . a_3 \ldots\ldots a_i \ldots a_n$$

$$z = \prod_{j=1}^{n} a_j$$

Sums of powers of the Natural numbers

$$\sum_{r=1}^{n} r = 1 + 2 + 3 = \ldots + n = \tfrac{1}{2} n(n+1)$$

$$\sum_{r=1}^{n} r^2 = 1^2 + 2^2 + 3^2 \ldots + n^2 = \tfrac{1}{6} n(n+1)(2n+1)$$

$$\sum_{r=1}^{n} r^3 = 1^3 + 2^3 + 3^3 \ldots + n^3 = \tfrac{1}{4} n^2 (n+1)^2$$

3.8.1.   PERMUTATION

$\boxed{P(n,r) =^{n} P_r =_n P_r}$ is read as n is the number of things to choose from, and one choose r of them. (no repetition, only order matters). *i.e.,* Number of permutations of n objects taken r at a time. $\boxed{^{n}P_r = \frac{n!}{(n-r)!}}$, $\boxed{0! = 1}$

$$n! = (n)(n-1)(n-2)(n-3)......1$$

Example: How many ways can the first and second rank be awarded among 10 students in an examination in the best performance?

Answer: $^{10}P_2 = \frac{10!}{(10-2)!} = (10)(9) = 90$.

3.8.2 COMBINATIONS

$C(n,r) =^{n} C_r =_n C_r$ is read as the number of combinations of n objects chosen r at a time (*i.e.,* Binomial coefficient).

$$\boxed{^{n}C_r = \frac{n!}{r!\,(n-r)!} =^{n} C_{n-r}} \qquad \boxed{^{n}C_r = \frac{^{n}P_r}{r!}}$$

Example: If 16 balls in a pool are there, find the number combinations of the balls are taken 3 at a time.

Answer: $^{16}C_3 = \frac{16!}{3!\,(16-3)!} = \frac{16!}{3!\,13!} = 560 =^{16} C_{13}$.

3.9.   PROGRESSIONS

3.9.1. Arithmetic Series

$$a,\ (a+d),\ (a+2d),....,[a+(n-1)d],\ , etc$$

Sum of n terms in the series is S_n.

$$S_n = \frac{n}{2}[1^{st}\text{ term} + n^{th}\text{ term}],$$

Example: Sum of first 100 natural numbers $= 1+2+3+..... +100 = \frac{100}{2}(1+100) = 5050$.

3.9.2. Geometric Series

$$a,\ ar,\ ar^2,....,ar^{n-1},\ , etc$$

$$\text{Sum } S_n = \frac{a\,(1-r^n)}{(1-r)}.$$

Example: A bacterium undergoes binary fission every 30 *mts*. The number of bacteria at the end of I day $= (1)\dfrac{(1-2^{48})}{(1-2)}$.

Example: N cells of yeast fill up 1 l of volume in 60 units of time; G.P. is $(1+\frac{1}{2}+\frac{1}{2^2}+.....+\frac{1}{2^{60}})$.   The time taken for yeast cells to fill $\frac{1}{8}l = 57$ units of time.

3.9.3. Harmonic Series (Inverse AP)

$$\frac{1}{a},\ \frac{1}{a+d},\ \frac{1}{a+2d},\ \cdot\ \cdot,\ \frac{1}{a+(n-1)d},\ \cdot$$

3.9.4. Arithmetic –Geometric Series

$$a,\ (a+d)\,r,\ (a+2d)\,r^2,.....,[(a+(n-1)d]\,r^{n-1},\ , etc$$

$$S_n = \frac{a}{(1-r)} + \frac{d\,r}{(1-r)^2}$$

3.9.5. Binomial series

$$(1+x)^n = 1 + nx + \ldots + {}^n C_r\, x^r + \ldots \quad {}^n C_r = \frac{n!}{r!\,(n-r)!} = {}^n C_{n-r}.$$

3.9.6. Taylor's Series

$$f(a+x) = f(a) + xf'(a) + \frac{x^2}{2!}f''(a) + \ldots + \frac{x^{n-1}}{(n-1)!}f^{(n-1)}(a) +$$

3.10. Angle Degree *versus* radian

$$Angle = \frac{Arc}{Radius}\,Radian$$

Arc length $= r\ \theta^c$

$$\boxed{\pi^c = 180^0}$$

3.10. Basic properties and formulae

Property	Formula
(1) Circumference of circle	$= 2\pi r$
(2) Area of circle	$= \pi r^2$
(3) Curved surface of Cylinder	$= 2\pi r h$
(4) Area of surface of Sphere	$= 4\pi r^2$
(5) Area of rectangle	$= $ (Length)(breadth)
(6) Area of Triangle	$= \frac{1}{2}$ (Base)(Altitude)
(7) Area of Parallelogram	$=$ (Base)(Altitude)
(8) Area of surface of Cone	$= \pi r \ell$
(9) Area of Trapezium	$= \frac{1}{2}(a+b)h$
(10) Area of Triangle $= \sqrt{s(s-a)(s-b)(s-c)}$ $s = \frac{1}{2}$ (Perimeter)	
(11) Area of a Sector	$= \frac{1}{2}r^2\theta$
(12) Volume of a Cylinder	$= \pi r^2 h$
(13) Volume of a sphere	$= \frac{4}{3}\pi r^3$
(14) Volume of a Cone	$= \frac{1}{3}\pi r^2 h$
(15) Volume of Rectangular Block	$=$ (L)(B)(Height)

3.11.1. $(1+x)^{-1} = 1 - x + x^2 - x^3, \quad + (-1)^r x^r + \ldots .$

3.11.2. $(1-x)^{-1} = 1 + x + x^2 - x^3, \quad + x^r + \ldots$

3.11.3. $e^{\pm x} = 1 \pm \frac{x}{1!} + \frac{x^2}{(2)!} \pm \frac{x^3}{(3)!} + . \pm . +$

3.11.4. $\log(1+x) = x - \frac{x^2}{2} + \frac{x^3}{3} - \frac{x^4}{4} + \ldots$

3.11.5. Gamma function $\Gamma(x)$

$$\Gamma(x) = \int\limits_{x=0}^{\infty} u^{x-1} e^{-u}\, du$$

$$\Gamma(n+1) = n! \quad (n = \text{integer}); \quad \Gamma(\tfrac{1}{2}) = \sqrt{\pi}\,.$$

3.12.   CONCEPTS on Matter

3.12.1. Matter: Something that occupies space and has mass.
Energy: The ability to do work.

Matter (m) and energy (E) are interchangeable. $\boxed{E = m\,c^2}$.

Hypothesis: Scientific Guess.
Theory: An idea with much supporting evidence.
Law or Principle: It is proved; no exceptions.

3.12.2. Phases of Matter

There are FOUR states of matter

i)   **Solids** : have shape and volume.
ii)   **Liquids**: have volume only.
iii)   **Gases**: have neither a definite shape nor volume.
iv)   **Plasmas**: have neither definite shape nor volume.

But the using the Laws of Physics, in March 2017, (reported by the physicist Wolfgang Ketterle of MIT) Both US and Swiss scientists have created an incredible (weird) substance, named **Super-solid**, which behave nothing like the ones known so far from chemistry. Supersolid has both the qualities of a normal crystalline solid and frictionless flow of a liquid at the same time. Supersolid resulted when atoms are turned into a 'Bose-Einstein Condensate', a hyper cold gas made from atoms with even numbers of electrons.

-0-o-0-o-0-o-0-o-0-o-0-

Chapter 4

DYNAMICS - I
KINEMATICS (ONE-DIMENSIONAL MOTION)

"Mechanics is the paradise of the mathematical sciences, because by means of it one comes to the fruits of mathematica" Leonardo da Vinci

4.1. INTRODUCTION

4.1.1 THREE BASIC STEPS FOR SOLVING PROBLEMS IN MECHANICS

Before continuation of further treatments of topics, students may follow the three important preliminary steps for solving problems in mechanics:

1. **Choose a coordinate system:** The x-axis runs parallel to the plane, where the x direction is positive or negative. and the y-axis runs perpendicular to the plane, where up is the positive y direction.
2. **Draw free-body Ask yourself how the system will move:** .
3. **diagrams:** The two forces acting on the body are the force of gravity, acting straight downward, and the normal force, acting perpendicular to the plane, along the y-axis. The result is a free-body diagram

4.2. ONE-DIMENSIONAL (RECTILINEAR) MOTION

Matter : has mass and occupies space.
Mass : Quantity of matter measured by inertia.
Inertia: Resistance to change in motion.
Density : Mass/Volume

An object (physical body) is in motion, relative to another, when its *position*, measured relative to the second body, is changing with *time*.

4.2.1. Describing Motion

4.2.1.1 PHYSICAL QUANTITIES:

TIME, SPEED, AND VELOCITY have magnitude and dimensions.

Time, t	Unit second (s)	(Scalar)	$(M^0 L^0 T^1)$
Distance, s	Unit metre (m)	(Scalar)	$(M^0 L^1 T^0)$
Displacement, $\vec{r}$	Unit metre (m)	(Vector)	$(M^0 L^1 T^0)$

Vector $\vec{r}$ in a plane, $\boxed{\vec{r} = (\hat{i}\, x + \hat{j}\, y)}$

34

| Speed, $v = \dfrac{\text{Distance travelled}}{\text{Time taken}} = \dfrac{s}{t}$ | Unit $m\ s^{-1}$ | (Scalar) | $(M^0\ L^1\ T^{-1})$ |

| Average Speed, $<v> = \dfrac{\text{Total Distance travelled}}{\text{Total Time taken}}$ | Unit $m\ s^{-1}$ (Scalar) $(M^0\ L^1\ T^{-1})$ |

| velocity, $\vec{v} = \dfrac{\text{Displacement}}{\text{Time taken}} = \dfrac{d\vec{r}}{dt}$ | Unit $m\ s^{-1}$ (Vector) $(M^0\ L^1\ T^{-1})$ |

| Instantaneous velocity, $\vec{v} = \dfrac{\text{Displacement}}{\text{Time taken}} = \dfrac{d\vec{r}}{dt} = \lim_{\delta t \to 0} \dfrac{\delta x}{\delta t}$ |

Consider the case of a *rectilinear motion* (motion along a straight line) taking place from A to B. The magnitude of change in the position vector, $\Delta \vec{x}$, is as shown.

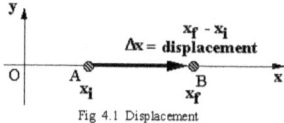

Fig 4.1 Displacement

4.2.1.2 **Graphs of Motion**

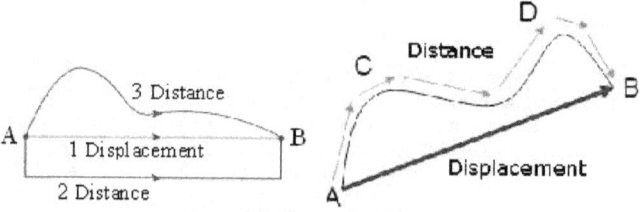

Fig 4.2 Displacement & distance

4.2.1.3 **Displacement – time graphs**

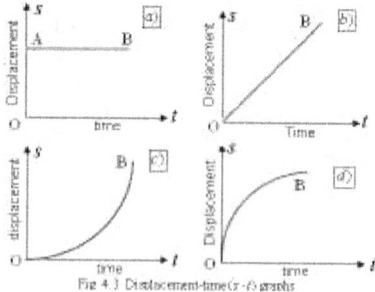

Fig 4.3 Displacement-time (s-t) graphs

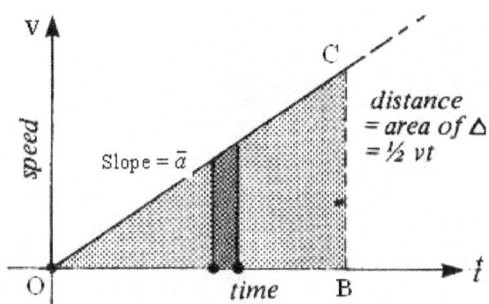

Fig 4.4 Speed-time (v-t) graph

Magnitude of $\vec{v}$	$= \boxed{\lvert \vec{v} \rvert = v = \sqrt{(v_x^2 + v_y^2)}}$
Direction angle, θ	$\boxed{\theta = \tan^{-1}\left(\dfrac{v_y}{v_x}\right)}$
Components,	$\boxed{v_x = \lvert \vec{v} \rvert \cos\theta, \text{ and } v_y = \lvert \vec{v} \rvert \sin\theta}$

Mass, *m*

| Mass, m | Unit | *kg* | (scalar) | $(M^1 \, L^0 \, T^0)$ |

(**Tipler**, in his book, defines mass as an intrinsic property of an object (body) that measures its resistance to acceleration.

4.2.1.4. <u>Inertia</u>

A **free particle** is not subject to any interaction. It should be either completely <u>isolated,</u> or else the only particle in the planet Earth.

Law of Inertia states that a free particle always moves with a velocity or or without acceleration **Earth is not an inertial frame**, as it is always rotating, and it interacts with the Sun.

4.3. **Newton's First Law of Motion**: (Law of Inertia)

<u>Definition</u>: Everybody persists in its state of rest or of uniform motion in a right (straight) line unless it is compelled to change that state by an external force impressed <u>thereon</u> (Isaac Newton's *Principia*, 1686).

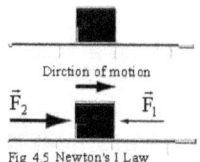

Fig 4.5 Newton's I Law

This law reinforces the idea of inertia.
The process of observation involves an interaction between the observer and the particle; *i.e.* no free particle can be observed.

Forces are Balanced

$$a = 0 \ m \ s^{-2}$$

Objects at Rest   Objects in Motion
(v = 0 ms^{-1})   (v ≠ 0 ms^{-1})

Stay at Rest   Stay in Motion
                    (Same speed and Direction)

Fig 4.5

4.4. Newton's Second Law of Motion:

Definition: The rate of change of momentum ($\vec{p} = m \ \vec{v}$) of a body varies directly as the force $\vec{F}$ causing the change and takes place in the same direction as the force.

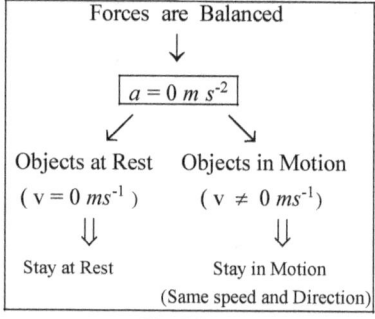

Fig 4.6 Newton's II Law

$$\vec{F} \propto \frac{d(mv)}{dt}$$

$$\vec{F} = k \frac{d(mv)}{dt}$$, where k=1, in SI units.

Force $\left(\vec{F}\right)$	*Unit Newton* $(N = m\ s^{-2})$	(Vector)	$(M^1\ L^1\ T^{-2})$

4.5.1 <u>**Momentum (Linear)**</u> $\vec{p}$

$\vec{p} = m\ \vec{v}$	Unit $kg\ ms^{-1}$	(vector)	$(M^1\ L^1\ T^{-1})$

Uniform motion, $\vec{v}$ = v = constant
Initial velocity = $\vec{u}$,
Final velocity = $\vec{v}$,

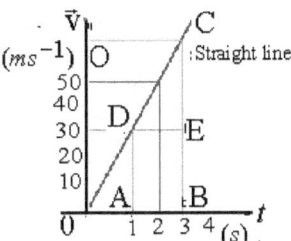

Fig 4.7 Velocity-time $(\vec{v} - t)$ graph

Average velocity,

Average velocity, $<\vec{v}>$ = $\dfrac{\text{Initial velocity + Final velocity}}{2}$

Average velocity, $<\vec{v}>$ = $\dfrac{\text{Displacement}}{\text{Time taken}}$ = $\dfrac{\vec{u} + \vec{v}}{2}$	Unit $m\ s^{-1}$ (Vector) $(M^0\ L^1\ T^{-1})$

4.5.2 <u>**Acceleration**</u> a = Time rate of change of velocity,

$\vec{a} = \dfrac{\Delta \vec{v}}{\Delta t}$	Unit ms^{-2}	(vector)	$(M^0 L^1 T^{-2})$

$\vec{a} = \hat{i}\ \dfrac{d\vec{v}_x}{dt} + \hat{j}\ \dfrac{d\vec{v}_y}{dt} + \hat{k}\ \dfrac{d\vec{v}_z}{dt}$

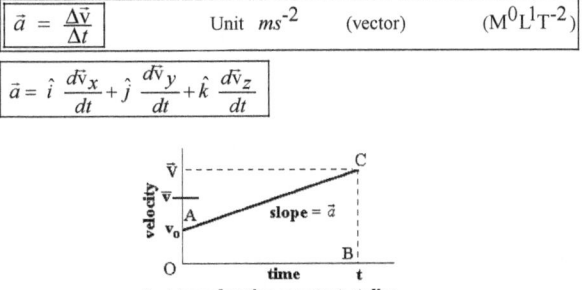

Fig 4.8 **acceleration constant st. line**

4.5.3 <u>**Uniform acceleration**</u> **in straight line,**

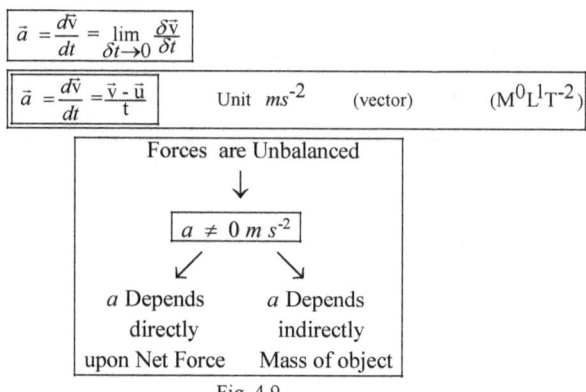

$$\vec{a} = \frac{d\vec{v}}{dt} = \lim_{\delta t \to 0} \frac{\delta \vec{v}}{\delta t}$$

$$\vec{a} = \frac{d\vec{v}}{dt} = \frac{\vec{v} - \vec{u}}{t} \qquad \text{Unit } ms^{-2} \qquad \text{(vector)} \qquad (M^0 L^1 T^{-2})$$

Forces are Unbalanced

↓

$$a \neq 0 \, m \, s^{-2}$$

↙ ↘

a Depends directly upon Net Force

a Depends indirectly Mass of object

Fig 4.9

4.5.4 Relation between Position and Velocity of a Particle

$$\int_{v_0}^{v} v \, dv = \int_{x_0}^{x} a \, dx \quad \Rightarrow \quad \frac{1}{2}[v^2 - v_0^2] = \int_{x_0}^{x} a \, dx$$

4.5.5 A Graphical Relationship between Displacement, Velocity and Acceleration

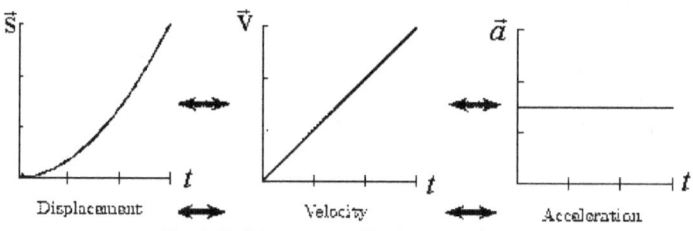

Fig 4.10 Displacement ($\vec{s}$)-, Velocity ($\vec{v}$)-, Acceleration ($\vec{a}$)-time (t) graphs

Given a s – t graph like the one on the left, one can plot the corresponding v – t graph by

remembering that the slope of a s –t graph gives the velocity. Similarly, one can plot an acceleration-time graph from the gradient of the v – t graph

Speed may be constant, say an object in orbit; but since it is changing direction in orbit, it is accelerating.

4.6. The THREE *Equations of Motion*: VUSAT Equations

(For uniformly accelerated rectilinear motion)

$$\vec{v} = \vec{u} + \vec{a} \, t$$

VUSAT # 1

a) Uniform velocity, $\vec{v} = $ constant.

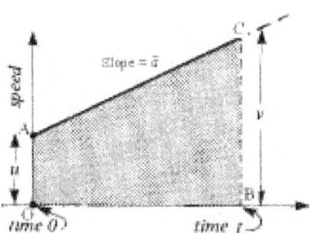

4.6.1 *Distance traversed in time t*

$$s = \vec{u}\,t + \frac{1}{2}\vec{a}\,t^2 \; ;$$

$$\boxed{\text{VUSAT \# 2}}$$

= area under the v *versus* t graph;

$$s = \int_{t_1}^{t_2} v\,dt$$

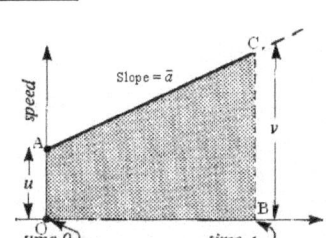

Fig 4.11 Speed-time graph

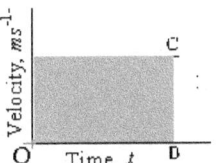

Fig 4.12 Velocity-time graph

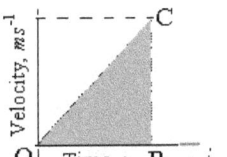

Fig 4.13 Velocity-time graph

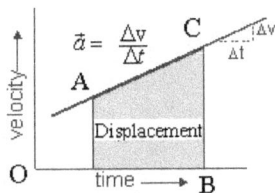

Fig 4.14 Velocity-time graph

$$\boxed{v^2 = u^2 + 2\,a\,s}\;;$$

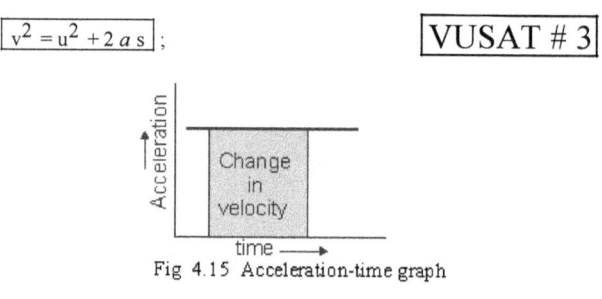

Fig 4.15 Acceleration-time graph

$$\boxed{\vec{F} = m\,\vec{a}} \qquad \text{Unit } kg \text{ m s}^{-2} = 1\,N \quad \text{(Vector)} \qquad (M^1 \, L^1 \, T^{-2})$$

4.6.2. Displacement in n^{th} second:

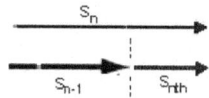

Fig 4.16 Displacement

S_n be the total displacement during n seconds.

S_{n-1} be the total displacement during the first (n -1) of those n seconds.

The total displacement in t seconds is given by:

$$S_{n-1} = \vec{u}\,(n-1) + \frac{1}{2}\vec{a}\,(n-1)^2$$

$$S_n - S_{n-1} = \vec{u} + \frac{1}{2}\vec{a}\,(2n-1)$$

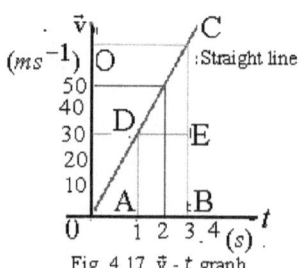

Fig 4.17 $\vec{v}$ - t graph

4.7. Newton's Third Law of Motion:

Definition: To every action there is an equal and opposite reaction.

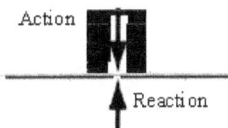

Fig 4.18 Newton's III Law

This law can be used to explain
i)   the generation of lift by a wing of an aero plane
ii)   the production of thrust by a jet engine,
iii)   spinning ball being deflected up,
iv)   air foil deflecting up, *etc*.
Mass is defined using this law:

Mass of a body $\quad m = (1\,kg)\dfrac{a_{1kg}}{a_b}$.

4.7.1 Jet Engine.

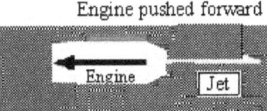

Fig 4.19 **Jet Engine**

Fuel	Exhaust gas velocity (ms^{-1})
Hydrogen & Oxygen	5800
Acetylene & Oxygen	5500
Petro & Oxygen	5000
Kerosene & Oxygen	5000
Alcohol & Oxygen	4850
Smokelesw Gunpowder	3500
Black Gunpowder	2600

4.7.2. Air Foil

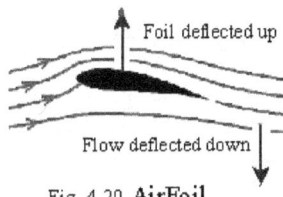

Foil deflected up

Flow deflected down

Fig 4.20 **AirFoil**

4.8. Applications of Newton's Laws of Motion

1st Law and 3rd Law Problems: Statics
2nd and 3rd Law problems: Dynamics, Universal Gravitation Problems

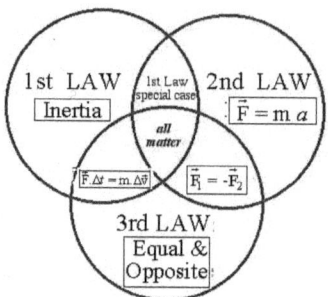

Fig 4.21 Applications of Newton's Laws

4.9. Linear Momentum, $\vec{p}$

For a system composed of several particles,

Total momentum, $\vec{P} = \sum \vec{p}_i$

4.9.1. Principle of Conservation of (Linear) Momentum

Definition: For two interacting particles, the interaction produces an exchange of momentum;
Momentum 'lost' by one particle = Momentum 'gained' by the other particle,

$$\vec{P} = \sum \vec{p}_i = \vec{p}_1 + \vec{p}_2 = \text{constant, for an isolated system}$$

The **Conservation Principle of Momentum** is *one of the **most fundamental and universal principles of physics**.*

4.9.2 Weight, W – the Force of Gravity on a body

$$\boxed{\vec{W} = m\vec{g}} \qquad \text{Unit } kg\ m\ s^{-2} = 1\ N \qquad (\text{vector}) \qquad (M^1\ L^1\ T^{-2})$$

4.9.3. Impulse (I), Work (W), Power (P) and Energy (E):

Impulse, $I = (\text{infinitely large } \vec{F})(\text{infinitesimally small time } \delta t \text{ for which } \vec{F} \text{ acts})$

$$\boxed{I = \vec{F} \bullet \delta t} \qquad \text{Unit } N\text{-}s \qquad (\text{vector}) \qquad (M^1\ L^1\ T^{-1})$$

$$\boxed{I = (\vec{p} - \vec{p}_o) = \vec{F} \bullet \delta t = m\ (\vec{v} - \vec{u}) = \text{Finite}}$$

4.9.3.1 Impulse momentum Equation $\boxed{I = \vec{F} \bullet \delta t = m \bullet \delta \vec{v}}$

$\vec{F}$ is not known as a function of time, $\vec{F} = \vec{F}(\vec{r})$, where $\vec{r} = \vec{r}(t)$

The area under the $\vec{F}$ *versus* t curve is Impulse or change of momentum.

4.9.4 **Work done**, $W = (Force, \vec{F}) \cdot (Displacement, \vec{s}$ in the direction of $\vec{F})$
Unit of work is Joule (J)

$$\boxed{W = \vec{F} \cdot \vec{r}\, Cos\theta} \qquad \text{Unit } N\ m\ =\ J \quad \text{(scalar)} \qquad (M^1L^2T^{-2})$$

4.9.4.1. Work done by the Earth on a satellite going round it in a circular orbit, if the Earth exerts a force of 10,0000 N on satellite $W = \vec{F} \cdot \vec{s} = 0$.

4.10. **ENERGY (E)** is the capacity to do work

4.10.1 Energy, E   Unit Joule (J) $\boxed{1\ eV = 1.6021 \times 10^{-19}\ J}$. (Scalar) ($M^1L^2T^{-2}$)

Potential energy (V) of a body is energy possessed by virtue of its gravitational position or state of strain. Here $\vec{F}$ is the force due to gravity

4.10.2 **Kinetic energy** (T) of a body is its energy by virtue of motion.
             = Work done on the particle,

K.E. $\boxed{T = \vec{F} \cdot \vec{s} = \frac{1}{2} m\, v^2}$.

4.10.3 P.E. $\boxed{V = \vec{F} \cdot \vec{s} = \int_{o}^{h} (mg)\, dr = m\, g\, h}$

Table Typical Energy values (in J)	
1) Moon light on face for 1-second	10^{-3}
2) Pressing down a typewriter key	1
3) House brick lifting up to shoulder level	30
4) Burning a match stick	1000
5) P.E. of a person at the top of 1st Stair	1500
6) K.E. of a car traveling at 110 *kmph*	5×10^5
7) Electrical energy of a fully charged car battery	2×10^6
8) Chemical energy in a day's food intake	11×10^6
9) Chemical energy in a litre of petrol	35×10^6
10) 1st Atomic bomb	10^{13}
11) Very severe Earth Quake	10^{20}
12) Earth's annual share of the Sun's heat	10^{25}
13) Rotational K.E. of the Earth	10^{29}

4.10.4. **POWER (P)**

$$\boxed{P = \frac{W}{T}} \qquad \text{Unit } W\ =\ J\ s^{-1} \qquad \text{(scalar)} \qquad (M^1\ L^2\ T^{-3})$$

Table   Power used by an adult male of mass 75 *kg*
Activity   Power used (in *W*)
----
1) Sleeping   83
2) Sitting   120
3) Walking (4.8 *kmph*)   265
4) Cycling (15 *kmph*)   410
5) Tennis play   440
6) Swimming (1.5 *kmph*)   475
7) Skating   535
8) Climbing up stairs (116 steps / min)   685
9) Cycling (21.3 *km / h*)   700
10) Basket ball   800

4.10.5. **PRINCIPLE OF CONSERVATION OF ENERGY**:
For a closed system (*i.e.*, when the forces are conservative), Energy can neither be created nor be destroyed; but can be transformed from one kind to another without loss.

Total energy of a conserved system, $\boxed{E = T + V = \text{constant}}$

4.11.   **Types of Collision**:
(i) Elastic Collision: Kinetic energy and momentum both are conserved.
   Eg. A moving nucleus deflected by another nucleus.
(ii) Inelastic Collision: Total Momentum is conserved, but the kinetic energy usually decreases, being converted into Potential or other forms of energy.
   Eg. In a complete inelastic collision the two bodies join together.
(iii) Explosion: Total momentum is conserved, but the kinetic energy increases. .

4.12.   EXAMPLES
4.12.1. **ATWOOD'S MACHINE**

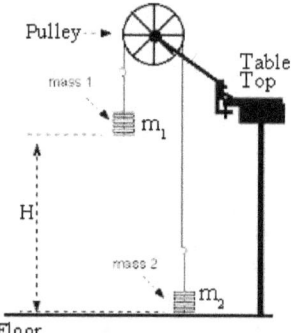

Fig 4.22 **Atwood's machine**

$\vec{T}$ = Tension in the string, $\vec{T} = M\,g$

$$\vec{a} = \left(\frac{\text{Mass difference}}{\text{Total mass}}\right)g = \left(\frac{m_2 - m_1}{m_1 + m_2}\right)g$$

$$\frac{2}{M} = \frac{1}{m_1} + \frac{1}{m_2}$$

Also, using a stop watch, obtain average value of time t for m_2 to reach the floor.

Then $\boxed{a\ (cm.s^{-2}) = \frac{2H(cm)}{t(s)^2}}$.

4.12.2 Trolley

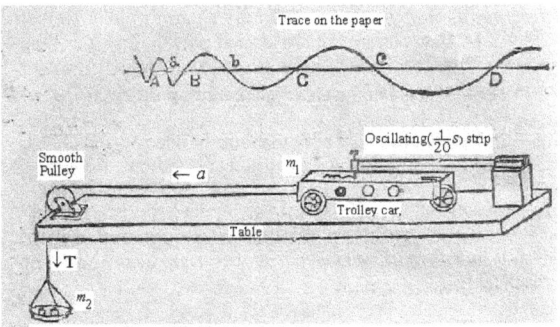

Fig 4.23 Fletcher's Trolley

$$\boxed{m_2 < \frac{1}{20}m_1}$$

$$m_1 g - T = m_1 a\ ;\ \ T = m_2 a$$

$$\boxed{\vec{a} = \left(\frac{m_1}{m_1 + m_2}\right) g}$$

Measure the lengths, AB, BC, CD, ab, bc, cd, *etc.*, correct to a tenth of a *mm* to calculate the accelerations, and take the average value of a. Newton's II Law can be verified.

$$+*\%+\%\&+*\%+*+*\%+$$

Chapter 5

DYNAMICS 2 –
TWO-DIMENSIONAL MOTION
(CURVINEAR MOTION)

"Imagination is more important than knowledge"   Albert Einstein

5.   **PROJECTILES (Object projected at angle of motion)**

In rectilinear motion both v and a have the same or opposite directions.
In curvilinear motion both v and a have different directions, since the body is projected at an angle.

5.1.   Velocity and Acceleration
Contrary to 1-D motion, v and a need not be in the same direction. Many of the   motions that occur in nature are confined to a plane.

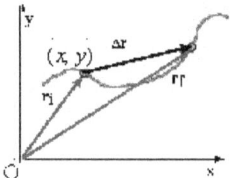

$$\vec{r} = x\,\hat{i} + y\,\hat{j}$$

$$\vec{v} = \lim_{\Delta t \to 0} \frac{\Delta r}{\Delta t} = \frac{d\vec{r}}{dt}$$

$$\vec{a} = \lim_{\Delta t \to 0} \frac{\Delta v}{\Delta t} = \frac{d\vec{v}}{dt}$$

5.1.1   **A body in 2D motion**: **Constant Acceleration: Projectile Motion**

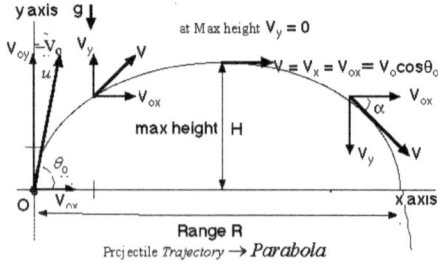

Projectile *Trajectory* → *Parabola*

u = velocity of projection
θ_O = angle of projection

$$\vec{v} = \vec{u} + \vec{a}\, t$$

For a projectile, the components of acceleration are, $\vec{a}_x = 0$,$\vec{a}_y = -g$

5.1.2   Using VUSAT # 3
Maximum height attained, H:

$$H = \frac{u^2\, Sin^2\theta_O}{2\,g}\ ;$$

5.1.3   Using VUSAT # 2
T = Time of flight required for the projectile to return to the ground.

$$T = \frac{2\,u\, Sin\theta_O}{g}$$

5.1.4   Range (horizontal distance traversed) R:

$$R = (u\,cos\theta_O)\left(T = \frac{2\,u\, Sin\theta_O}{g}\right)$$

$$R = \frac{u^2\, Sin2\theta_O}{g}\ ;\quad Unit\quad m\ (scalar)\quad (M^0 L^1 T^0)$$

For the same value of the set u and R, there are two values of angles, θ_O and $\theta_O\,'$ such that

$$\theta_O = (\tfrac{1}{2}\pi - \theta_O\,')$$

If $u^2 > R\,g$,

$$\theta_O = (\tfrac{1}{2}\pi - \theta_O\,')$$

5.1.5   Trajectory, the path of the projectile, in vacuum, is a parabola, expressed mathematically by the equation.

$$y^2 = 4\left(\frac{u^2\, Cos^2\theta_O}{2\,g}\right) x$$

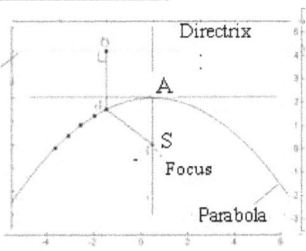

with latus rectum $\left(\dfrac{2\,u^2\, Cos^2\theta_O}{g}\right)$, with focus at a distance $\left(\dfrac{u^2\, Cos^2\theta_O}{2\,g}\right)$ from its

vertex.; whereas the **directrix** is horizontal and $\left(\dfrac{u^2\, Cos^2\theta_O}{2\,g}\right)$ above the vertex.

5.2.1. Relation between angles in projectile motion

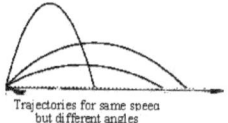

Trajectories for same speed
but different angles

5.2.2 If the projectile first clears the top of a wall of height h at a distance a from the point of projection,

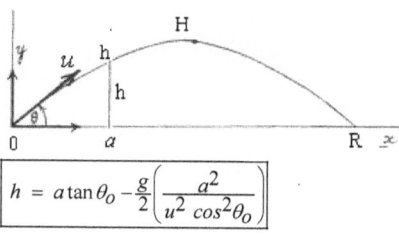

$$h = a\tan\theta_0 - \frac{g}{2}\left(\frac{a^2}{u^2\cos^2\theta_0}\right)$$

5.2.3 Relation between the elevation angles θ_0 and β of the highest point in the trajectory

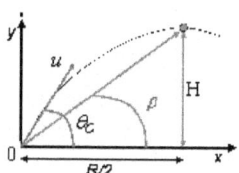

$$T = \frac{2u\,Sin(\theta_0-\beta)}{g\,Cos\beta}$$

$$R = \frac{2u^2\,Sin(\theta_0-\beta)\,Cos\theta_0}{g\,Cos^2\beta}$$

$$Tan\,\beta - \frac{2H}{R}$$

5.2.4 Monkey problem

A monkey is sitting on a tree along the line of sight of a riffle which is aimed at the monkey.

. α = angle of elevation a gun should fire the target plane flying at an altitude h and speed u is given by

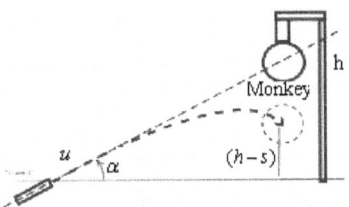

Trajectory of a shot fired aimed at a monkey

$$\tan\alpha = \sqrt{\frac{2gh}{u^2}}.$$

After $t = \tau$ the monkey is at the height of

$$(h - s) = h - \left(\frac{u^2 Sin^2\theta_0}{2g}\right)$$ from the ground.

5.3. IMPACT (COLLISION, IMPINGE)

During an impact the fundamental principles to hold good are:
1) Conservation of linear momentum,
2) Tangential velocity remains unchanged,
3) Newton's Law of Impact.

Whenever two bodies impinge, their relative velocity along the common normal, after impact, bears a constant ratio to the relative velocity before impact, along the common normal, and is opposite in sign; the constant is called the coefficient of restitution or the coefficient of velocity ε.

$$\varepsilon = -[(v_1 - v_2)/(u_1 - u_2)] \ll 1$$, always.

5.3.1 For perfectly elastic collisions, $\varepsilon = 1$

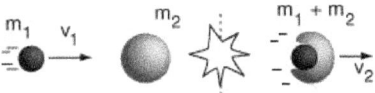

For perfectly inelastic collision, $\varepsilon = 0$.

5.3.2 Elastic Collisions

The final velocities of the two colliding bodies of masses, m_1 and m_2, in 1-D, with m_2 at rest

$$v_1' = v_1 \frac{m_1 - m_2}{m_1 + m_2}$$

$$v_2' = v_1 \frac{2m_1}{m_1 + m_2}$$

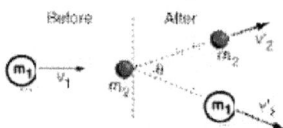

Elastic Collisions – Target Initially at Rest'

Type of collision	Diagram				What happens	Conserv quantity
perfectly inelastic	m_1 $v_{1,i}$ $P_{1,i}$	m_2 $v_{2,i}$ $P_{2,i}$	$m_1 + m_2$ v_f P_f		The two objects stick together after the collision so that their final velocities are the same.	momentu
elastic	m_1 $v_{1,i}$ $P_{1,i}$	m_2 $v_{2,i}$ $P_{2,i}$	m_1 $v_{1,f}$ $P_{1,f}$	m_2 $v_{2,f}$ $P_{2,f}$	The two objects bounce after the collision so that they move separately.	momentu kinetic en
inelastic	m_1 $v_{1,i}$ $P_{1,i}$	m_2 $v_{2,i}$ $P_{2,i}$	m_1 $v_{1,f}$ $P_{1,f}$	m_2 $v_{2,f}$ $P_{2,f}$	The two objects deform during the collision so that the total kinetic energy decreases, but the objects move separately after the collision.	momentu

5.3.3 **APPLICATIONS**: of the Law of Conservation of Momentum
Conservation of linear momentum and of kE in <u>two-body collisions</u>
Relative velocities during Elastic collisions

$$m_1 u_1 + m_2 u_2 = m_1 v_{u1}' + m_2 v_2 ';$$
$$m_1 u_1^2 + m_2 u_2^2 = m_1 v_1'^2 + m_2 v_2'^2$$

$$\boxed{\text{Relative velocities,} \quad u_1 - u_2 = -(v_1 - v_2)}\,,$$

changes sign, but keeps the same magnitude.

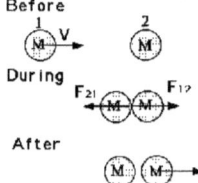

5.3.4.1 Generally,

$$\boxed{v_1' = v_1 \frac{m_1 - m_2}{m_1 + m_2} + v_2 \frac{2\,m_1}{m_1 + m_2}}$$

$$\boxed{v_2' = v_2 \frac{m_1 - m_2}{m_1 + m_2} + v_1 \frac{2\,m_1}{m_1 + m_2}}$$

5.3.5.1 In 2-Dimensions,

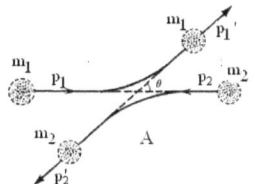

5.3.5.2 Newton's cradle. Collision in a straight line,

a series of suspended ball bearings can be made to collide with each other.

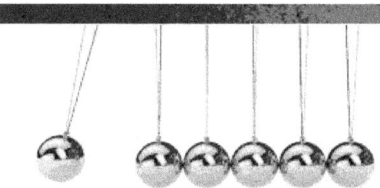

5.4   Inelastic Collisions

5.4.1   In 1-Dimensions

$$m_1 u_u + m_2 u_2 = [m_1 + m_2] \, v$$

$$m_1 u_u^2 + m_2 u_2^2 = [m_1 + m_2] \, v^2$$

$$v_2' = v_2 \, \frac{m_1 v_1 - m_2 v_2}{m_1 + m_2}$$, in 1-D

A Perfectly Inelastic collision

The loss in kE after the impact here will appear as heating the bodies, energy of sound, or the like.

5.4.2   In 2-Dimensions

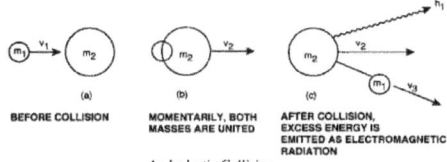

An Inelastic Collision

5.5 OBLIQUE COLLISIONS (Glancing Blow) is Snooker game.

$$m_1 u_1 = m_1 v_1' \, Cos\,\theta_1 + m_2 v_2' \, Cos\,\theta_2 \text{ , for the x-direction.}$$

$$0 = m_1 v_1' \, Sin\theta_1 + m_2 v_2' \, Sin\theta_2 \text{ , for the y-direction.}$$

Conservation of kE must be taken in to consideration.

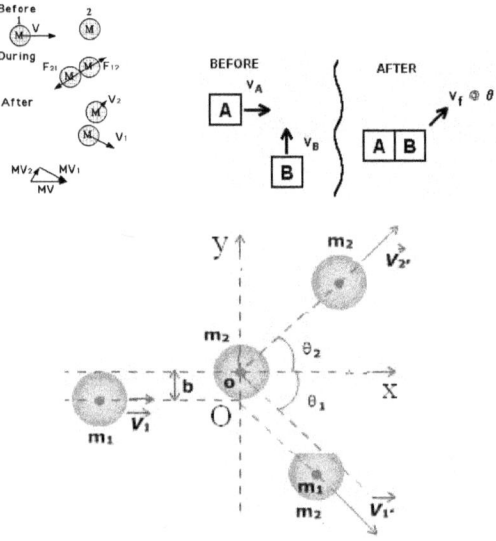

v_1 and v_2 are at right angles since the two colliding bodies have the same mass.

5.6.1 ROCKET MOTION

Rocket engines can generate thrusts in excess of 1 MN by rapidly ejecting a stream of gas through a nozzle. Liquid Oxygen and liquid fuels are burnt at $500 \ kgs^{-1}$ in the combustion chamber, with exhaust gas at $v_{gr} = -2 \ kms^{-1}$.

Momentum of the exhaust gas / s = $(500 \ kgs^{-1})(-2 \ kms^{-1})$

Force acting on the rocket $= + 1.0x \ 10^6 \ N$

Thrust $+1.0x \ 10^6 N$ $M(a+g)$

Exhaust gas

$2 \ kms^{-1}$ Mg

:

v_{rp} - velocity of the observer in the launch pad

v_{gr} -velocity of the exhaust gases w.r.t. the rocket

v_{gp} -velocity of the exhaust gases w.r.t. the launch pad

$$v_{gp} = v_{gr} + v_{rp}$$

Δm - Masses of the gases exhausted

M - Mass of the rocket

For rocket equilibrium

$$dv_{rp} = v_{gr} \frac{dM}{M}$$

For a solid or liquid fuel propulsion systems (rocket) the greater the velocity ω of the exhaust gases the greater is their momentum and hence of the rocket. If u and M_0 are the initial velocity and mass of the rocket, velocity ω of the rocket of mass M is

$$V = u - \omega \ln \frac{M}{M_0}$$

This gives the change in velocity of the rocket, and shows that larger the value of v_{gr} the better the rocket propulsion.

Fuel	Exhaust gas velocity, ms^{-1}
Hydrogen & Oxygen	5800
Acetylene & Oxygen	5500
Petrol & Oxygen	5000
Keroene & Oxygen	5000
Alcohol & Oxygen	4850
Smokeless gun powder	3500
Black gun powder	2600

Example: For Kerosene in oxygen $\omega = 5000\ ms^{-1}$.

5.6.1.1. Given a rocket, its payload and fuel having total weight M kg.

For such an orbiting rocket to the required thrust to lift from the rocket platform is

$$F = M\ a = M\ (-g) = -\ M(kg)\ g(ms^{-2})\ N.$$

6.2   Hose pipe:

A = cross sectional area of hose pipe,

v = velocity of water jet from hose

ρ = density of water

$$\boxed{\text{Force on wall} = \text{Rate of change of momentum}, \rho\ v^2 A}$$

5.6.3   Sand falling on Conveyer Belt:

v = velocity of belt

m = mass of sand falling on belt / sec

$$\boxed{\text{Force on wall} = \text{Rate of change of momentum}, m\ v}$$

5.6.4   Helicopter:

M = mass of helicopter

R = radius of a rotator blade

v = velocity of the column of air moving vertically due to rotation of blades

ρ = density of air

For the helicopter hovering above ground is to be stationary,

$$Mg = \pi R^2 v^2 \rho$$

5.6.5 Painful to be hit by a hailstone than by a raindrop!

Force in hailstone hitting bounces; whereas that by raindrop does not bounce.

5.6.6 A ball of mass m falling from rest from an altitude (h) and hits a ball

of mass M bounces from ground to a height (h')

Coefficient of restitution

$$\varepsilon = \sqrt{\frac{h'}{h}} \; .$$

5.6.7 Shooting guns / Rifles

+*+*+*+*+*+*+

Chapter 6

DYNAMICS 3 –
UNIFORM CIRCULAR MOTION

"Success s a lousy teacher. It seduces smart people into thinking they can't lose" - Bill Gates

6. **UNIFORM CIRCULAR MOTION**:

6.1 Defining uniform Circular Motion
$\vec{v}$ = linear velocity (tangential to a point in the circle
$\vec{\omega}$ = angular velocity
r = radius of the circle.

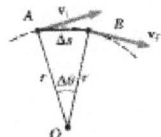 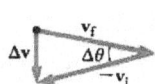

6.2 **Equations of Circular Motion**

6.2.1 Relation between linear speed $\vec{v}$ and angular speed ω :

$$\omega = \frac{v}{r}$$

If the body goes from c to d in time *t*,

$$\text{Angle, } \theta \text{ (in radians)} = \frac{\text{Arc of circle, } s}{\text{Radius of circle, } r}$$

$$s = r\,\theta$$

$$v = \frac{s}{t} = r\frac{\theta}{t}$$

$$\vec{v} = -\,\vec{r} \wedge \vec{\omega}$$

6.2.2 **Formulae for centripetal acceleration (Acceleration normal to the tangent)**
(Normal Acceleration)

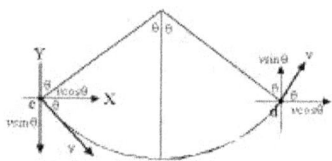

The instantaneous acceleration is

$$\vec{a}_{centre} = \vec{v} \wedge \vec{\omega} = [-(\vec{r} \wedge \vec{\omega})] \wedge \vec{\omega} = -\frac{v^2}{r}$$

and the **negative sign** indicates that it acts in towards the centre of the circle. Putting

$$v = r\,\omega \,,$$

$$\vec{a} = -r\,\omega^2$$

Acceleration along the tangent, at any point $\boxed{\vec{a}_{tangent} = 0}$

6.2.3   Centripetal force, Formula

By Newton's second law of motion

$$\vec{F} = m\,\vec{a}$$

$$\Rightarrow \vec{F} = -\frac{m\,v^2}{r} = -m\,r\,\omega^2$$

Period of revolution $\boxed{T\,(s) = -\sqrt{\dfrac{4\,m(kg)\,r(m)\,\pi^2}{\vec{F}(N)}}\ s}$

6.2.4   Formulae for periodic time and frequency

Periodic time P

$$P = \frac{\text{Length of one complete Orbit}}{\text{Speed}} = \frac{2\pi r}{v} = \frac{2\pi}{\omega}$$

Frequency f

$$f = \frac{\text{Number of orbits}}{\text{One second}} = \frac{1}{P} = \frac{v}{2\pi r} = \frac{\omega}{2\pi}$$

6.2.5   Centrifugal Force, $\vec{F}_{CF}$

There is nothing like Centrifugal force.

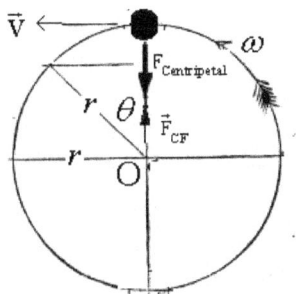

The Moon revolves around the Earth in (assumed) circular orbit. Draw neat labeled diagram and indicate the following in the diagram.

(i) Centripetal force.
(ii) Centri-fugal force.
(iii) Centri-fugal reaction.
(iv) Inertial frame of reference.
(v) Non-inertial frame of reference

(i) $\vec{F}_{CP}$ is the Centri-petal force on Moon,

(ii) $\vec{F}_{CF}$ is the Centri-fugal force on Moon,

(iii) $\vec{R}_{CF}$ is the centrifugal reaction on Earth,

(iv) I is the Inertial frame,

(v) N is the Non-inertial frame.

6.3 **APPLICATIONS** of Circular Motion:

System	Force that makes it move in a path other than a straight line
1) Centrifuge -	reaction at the walls.
2) Gramaphone needle -	friction with grooves
3) Aircraft Banking -	lift on the wings (Bernoulli effect)
4) Planetary Orbits -	gravitation
5) Electron orbit, say in an atom -	Electro-static Force
6) Car (% Bicycle) cornering -	Friction at wheels
7) Car (and Bicycle) cornering on Banked track -	component of gravity
8) Whirling a Body (Object on String) -	Tension on string
9) Rotating liquid surface -	gravity
10) Governors of Steam engines -	Tension in bars
11) Variation of g wih latitude -	gravity
12) Conical Pendulum -	Tension in the string
13) Motion of a railway carriage along - a circular track	Component of gravity
14) Well of Death (Motion of a particle on - a smooth vertical circle)	gravity

6.3.1 Whirling a Body:

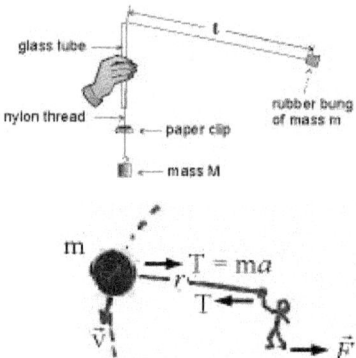

There is **no** centrifugal force appearing in this problem.

6.3.3 Conical Pendulum:

T= tension on the string ; ℓ is length of the string
t = period of rotation

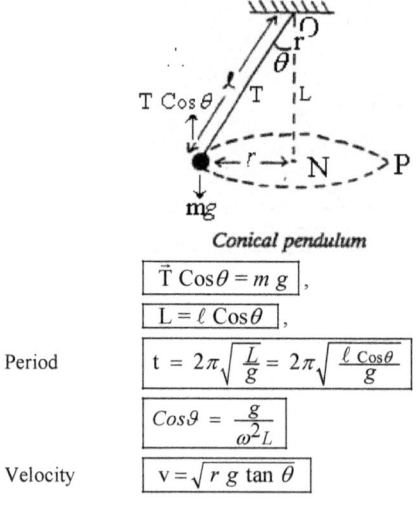

Conical pendulum

$$\vec{T} \cos\theta = m\,g ,$$

$$L = \ell \cos\theta ,$$

Period
$$t = 2\pi\sqrt{\frac{L}{g}} = 2\pi\sqrt{\frac{\ell \cos\theta}{g}}$$

$$\cos\vartheta = \frac{g}{\omega^2 L}$$

Velocity
$$v = \sqrt{r\,g\,\tan\theta}$$

6.3.4 Turntable (The Centrifuge and Washing machine)

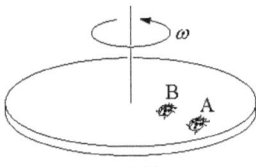

Turn Table

A centrifuge is a device that spins liquid samples at high speeds and thus creates a strong centripetal force causing the denser materials to travel towards the bottom of the centrifuge tube more rapidly than they would under the force of normal gravity.

$$\vec{F} = -\frac{m\,v^2}{r} = -\,m\,r\,\omega^2$$

So the bigger the mass of the object and the faster it goes, the more force is needed to keep it turning

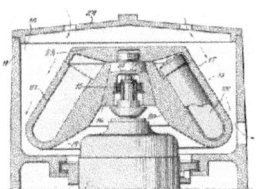

One of the most common uses for centrifuges is in separating mixtures of things. A washing machine is a mixture of clothes and water and the spinning drum separates those very efficiently. Laboratory centrifuges are used to separate things like blood, which consists of red blood cells suspended in plasma (a yellowish fluid). Put some blood in a test-tube and spin it at high speed and these two components are separated very quickly, with the plasma at the top of the tube and the red blood cells at the bottom. (They travel to the bottom because they are heavier, so need more centripetal force to push them round in a circle. The force comes from the bottom of the tube pushing inward against the blood cells clumped there).

6.3.5 Car turning (cornering) on a Level (Flat) road:

$2b$ = separation of the two front wheels,
h = diameter of the wheels
Using moments about CG

Velocity Maximum, $\boxed{\bar{v}_{max} = \sqrt{b\,r\,g\,/\,h} = \sqrt{\mu\,r\,g}}$

As $\dfrac{M\,\bar{v}_{max}{}^2}{r} = \mu\,M\,g$

μ = Coefficient of static friction between the wheel tire and the road.

6.3.6 Car on a banked track

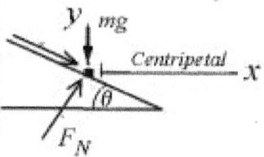

$$\vec{F}_N \; Cos\theta = mg \; ; \; \text{and} \; \vec{F}_N \; Sin\theta = \frac{m \, v^2}{r}$$

$$\boxed{Tan \; \theta = \frac{v^2}{r \, g}} \; ,$$

where v = maximum speed.

6.3.7 Motion of a Cyclist cornering

$$\boxed{Tan \, \vartheta = \frac{v^2}{r \, g}}$$

6.3.8 Making a coin of mass *m* stick to the palm of one's hand as he accelerates it (to *a*) quickly downwards

$$\boxed{\vec{F}_{Downward} , \; m \, \vec{a} \; > \; \vec{F}_{Gravity} , m \, \vec{g}} \quad \text{for the coin not to fall}$$

6.3.9 Motion of a Railway carriage along a Circular Track

$$\boxed{Tan \, \vartheta = \frac{v^2}{r \, g}}$$

$$\boxed{x = a \; \vartheta}$$

a is the width of the Gauge of railway
x upward tilt of one side of the carriage.

6.3.10 Motion of a Particle on a Smooth Vertical Circle {Loop the loop}

Conservation of energy is the principle to be used.

$$\boxed{v^2 = u^2 \pm 2 \, g \, h} \; ,$$

"+ sign" for downward / "-sign" for upward motion.

A toy car goes in a loop-the-loop around a circular track with radius R. The minimum speed the car must have at the top of the loop

Total mechanical Energy of the train = KE + PE
Conservation of total ME at initial and final time; consider friction; centripetal acceleration,

$$\boxed{a_{cenpet} = \frac{v^2}{r}} \; ;$$

Roller Coaster Motion

a) Top: $\vec{F} = \vec{F}_N$; $\vec{F} = m\ [g - \dfrac{v^2}{R}]$

b) Bottom $\vec{F}_N = mg$

c) Otherwise: $\vec{F} = m\ [g + \dfrac{v^2}{R}]$

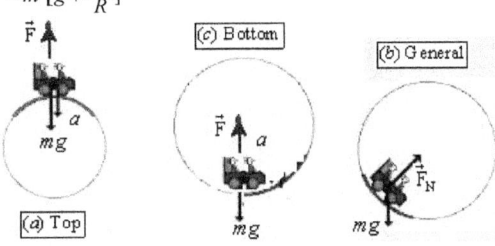

d) Loop the Loop

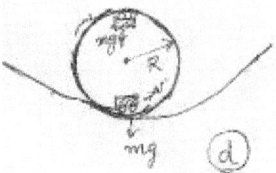

(i) At the Top: As $\vec{F} \to 0$, $\boxed{\vec{v}_{min} = \sqrt{g\ R}}$

(ii) At the Bottom: $\boxed{\vec{F} = m\ [g + \dfrac{v^2}{R}]}$

$v = \sqrt{\dfrac{F \pm mg}{m} R}$

Roller coasters today employ clothoid loops rather than the circular loops of earlier roller coasters. If the radius is reduced at the top of the loop, the centripetal acceleration is increased sufficiently to keep the passengers and the train from slowing too much as they move through the loop. A large radius is kept through the bottom half of the loop, thereby reducing the centripetal acceleration and the gravity acting on the passengers.

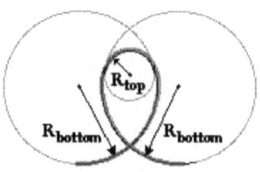

The Clothoid Loop

$$\boxed{R_{bottom} \gg R_{top}}$$

1) At the top: $\boxed{\vec{F} = m\,[g - \dfrac{v^2}{R}]}$

2) At the Bottom: $\boxed{\vec{F} = m\,[g + \dfrac{v^2}{R}]}$

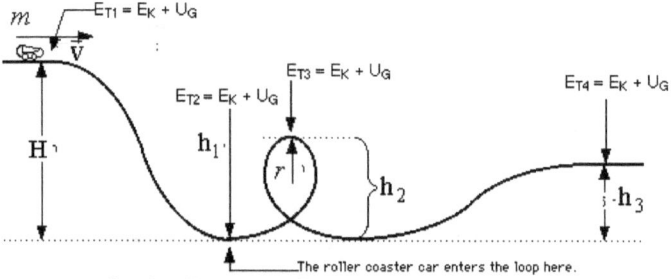

The roller coaster car enters the loop here.

Free body diagram of Irregular Roller coaster

6.3.11 Motion of a Ring threaded on a Smooth Vertical Circle

$$u_{min} = 2\sqrt{a\,g}$$

6.3.12 Motion of a Particle tied to a String in a Vertical Circle

$$\boxed{u_{min} = \sqrt{5\,a\,g}}$$

is the critical velocity required for the particle to reach the topmost poit in the circular path and not leave the circular path.

6.3.13 Tension in a Rotating Ring

T_o Maximum tension / cm^2

m area of cross section; ρ density of ring

$$T = m\,r^2\,\omega^2 ; \quad \boxed{v_{max} = r\,\omega_{max} = \sqrt{\dfrac{T_o}{\rho}}}$$

6.3.14.1 Motorcycle and **Wall of Death**

Wall of death is a carnival show featuring a woodent cylinder of 10 m diameter. The most obvious appeal of this stunt is the fact that the motor cycle-rider is speeding around on a vertical path without falling.

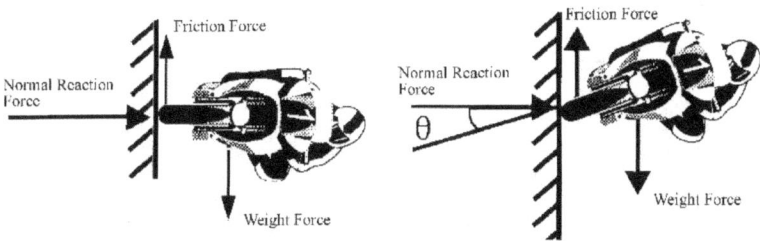

Rider has mass m in circle of radius R, The slower the speed of the rider to keep upright position (For the rider minimum coefficient of friction of the wall $\mu \approx 0.5$) L the separation between the front wheels of car, H height of CM of car from surface.
Both the following two conditions must apply simultaneously.

$$v^2 \geq \frac{2 R g H}{L}$$

$$v^2 \geq \frac{R g}{\mu}$$

6.3.14.2. Why do customarily runners have to run counter-clockwise direction since 1913?

Structure of the human body has the most precious heart on the left side. For humans and animals, running counter-clockwise makes the centrifugal force in the body to act from left to right. When the body loses equilibrium, it has a strong tendency to fall toward the heart side. This also explains why most riders find it easier to corner to the left than to the right. Superior venecava, the principal vein, carries the impure blood from the upper half of the body to the heart's right atrium assisted by heart suction. The centrifugal force, due to running in clockwise direction, will make the centrifugal force to impede suction and tire the athlete. Apart from this, it is also argued that when an athlete runs in counter-clockwise direction, he encounters only left turns and as a matter of fact left turns are easier than right turns, as explained above.

6.3.15 **Aircraft Banking**:

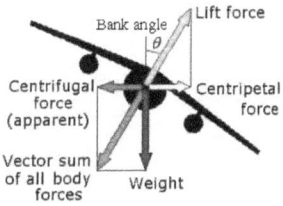

$$Tan\ \theta = \frac{v^2}{r\ g}$$

6.3.16 **Cyclist cornering** & railway carriage along a circular track

$$\text{Tan } \theta = \frac{v^2}{r\,g}$$

6.3.17 **Rotating Liquid Surface**:

If x is the distance from the Axis of rotation to the point in this liquid,

$$\text{Tan } \theta = \frac{\omega^2 x}{g}$$

6.3.18 **Governors of Steam engines**:

H = vertical distance of the collar sliding in the shaft from the top.
# of rotations (n) of the light rod with weight at its tip

$$n = \frac{1}{2\pi} \sqrt{\frac{g}{H}}$$

6.4.1 **RELATIVE MOTION**

Two observers moving with constant velocity with respect to each other measures different values for the velocity of an object. The difference between the two measurements is equal to the velocity their velocity relative to each other.

6.4.2 **VECTOR RELATION** *between* **v**, **r** *and* **ω** *in Circular Motion*:

$$\vec{v} = \vec{r} \wedge \omega$$

Period of Earth = 365 days
γ = angle corresponding to one day between points P' and P" = 1^0,
Time taken to move through angle γ with ω

$$\omega = \frac{1.745\,x10^{-2}\,radian}{7.292\,x10^{-2}\,radian/s} = 239 \text{ s}$$

! Mean Solar Day = $\boxed{P'' = 8.640\,x10^4\ s}$

Sidereal Day (Period of revolution of Earth) $\boxed{P' = 8.616\,x10^4\ s}$

Angular velocity of Earth, $\boxed{\omega = \frac{2\pi}{T} = 7.292\,x10^{-2}\ rad\ /\ s}$

6.5.1 **TENSION in a Rotating String**:

T_0 = Tension / unit area of string of density, ρ

$$T_{Max} = r\,\omega_{Max} = \sqrt{\frac{T_0}{\rho}}$$

6.5.2 **MOTION OF A PARTICLE, tied** to a string, in a vertical circle:

$$v_{min} = \sqrt{(5\,g\,R)}$$

6.5.3 **Motion of a particle on the outside** *of a vertical circle*:

$$v_{min} = \sqrt{(Cos\,\theta)\,g\,R}, \qquad Cos\,\theta = \tfrac{2}{3}\;.$$

6.5.4. Why Aircrtaft tyres have clearance between wheels.

The tension in an elastic band sufficient force for it to rotate with the disc only up to a critical velocity ω_c,

$$\omega_c = A - \frac{B}{r}; \; A = \frac{4\pi^2 k}{M}; \; k =\text{spring constant and M mass of the elastic band.}$$

At $\omega > \omega_c$, the elastic band will not stay on the elastic disc.

Thus aircraft tyres at high angular speeds increase in diameter, and a clearance of atleast 10 *mm* must be used always between the wheels and tyres.

+^+*+&+*+^+*+*+&*+*+

Chapter 7

STATICS – 1
ROTATION - 1 (PARALLEL LAW OF FORCES, MOMENT OF INERTIA, EQUILIBRIUM)

"Nature uses as little as possible as anything"   Johannes Kepler

7.1   PARALLEL LAW OF FORCES

When two or more forces act on the same point at the same time they are called **concurrent** forces. When two forces act concurrently in the same or in opposite directions, the **resultant** has a magnitude equal to the algebraic sum of the forces and acts in the direction of the greater force.

I) The graphic solution of the magnitude and direction of a resultant force consists of a diagram constructed to scale.

II) The trigonometric solution makes use of the facts that the opposite sides of a parallelogram are equal and that the diagonal of a parallelogram divides it into two congruent triangles

7.1.2   The Equilibrant Force

Equilibrium is the state of a body in which there is no change in its motion. A body in equilibrium is either at rest or moving at constant speed in a straight line.

A body at rest must be in both translational and rotational equilibrium. The first condition of equilibrium is that there are no unbalanced (net) forces acting on a body. The second condition of equilibrium deals with rotation.

7.2   PARALLELOGRAM LAW OF FORCES

If two forces, acting at a point, are represented in magnitude and direction by the two sides of a parallelogram drawn from one of its angular points, their resultant is represented both in magnitude and direction by the diagonal of the parallelogram passing through that angular point.

7.2.1   Magnitude and Direction of the Resultant of Two Forces:

Let OA and OB represent the forces $\vec{P}$ and $\vec{Q}$ acting at a point O and inclined to each other at an angle α then the resultant $\vec{R}$ and direction 'θ' will be given by

$$\vec{R} = \sqrt{P^2 + Q^2 + 2\,P\,Q\,Cos\,\theta}$$

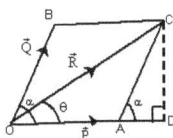

Case (i): If $\vec{P} = \vec{Q}$, then $\theta = \alpha / 2$

Case (ii): If the forces act at right angles, so that
$$\alpha = 90°,$$

then
$$\vec{R} = \sqrt{P^2 + Q^2}$$

and
$$\tan \theta = \frac{Q}{P}$$

7.2.2   LAMI'S THEOREM

According to this theorem, if resultant of three vectors $\vec{a}$, $\vec{b}$ and $\vec{c}$ is zero (null vector), then

$$\frac{\vec{a}}{Sin\ \alpha} = \frac{\vec{b}}{Sin\ \beta} = \frac{\vec{c}}{Sin\ \gamma}$$

7.2.3   The Law of Tangents

$$\left(\frac{a-b}{a+b}\right) = \frac{\tan\frac{(\alpha-\beta)}{2}}{\tan\frac{(\alpha+\beta)}{2}}$$

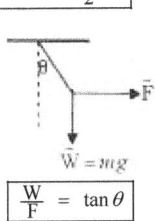

$$\frac{W}{F} = \tan \theta$$

7.3   ROTATION 1

A completely general motion f a body involves

(1) *Translational motion*: When a rigid body executes translational motion when each particle of the body has the same displacement in the same time interval.

(2) *Rotational motion*: When each particle of the rigid body travels in a circle, centered at the axis of rotation, excepting those lying in the axis.

7.3.1   ANGULAR COORDINATE: $2\pi r / P$.

$$\theta = \frac{s}{R} \quad \boxed{\text{Unit} \quad \text{rad (radian)} \quad (\text{Scalar}) \quad M^0 \ L^0 \ T^0}$$

$$1 \ rad = \frac{180^0}{\pi} = 57.3^0 \ . \ (\text{Dimensionless})$$

7.3.2 ANGULAR SPEED, ω, ANGULAR VELOCITY, $\vec{\omega}$, and **Right-Hand Rule**:
It is the rate of change of angular coordinate,

$$\omega = \frac{d}{dt}\theta \quad \boxed{\text{Unit} \quad \text{rad (radian)} \quad (\text{Scalar}) \quad M^0 L^0 T^{-1}}$$

7.3.2.1 For rotation **along z-axis**,

$$\vec{\omega} = \left(\frac{d}{dt}\theta\right)\hat{k} \quad \boxed{\text{Unit} \quad rad \ s^{-1} \quad (\text{vector}) \quad M^0 L^0 T^{-1}}$$

7.3.3 ANGULAR ACCELERATION, α

$$\alpha = \left(\frac{d}{dt}\vec{\omega}\right)\hat{k} \quad \boxed{\text{Unit} \quad rad \ s^{-2} \quad (\text{vector}) \quad M^0 L^0 T^{-2}}$$

7.3.4 ANGULAR MOMENTUM, $\vec{L}$ is moment of linear momentum $\vec{p}$.

$$\boxed{\vec{L} = \vec{r} \wedge \vec{p}} \quad \boxed{\text{Unit} \quad kg \ m^2 \ s^{-1} \quad (\text{vector}) \quad M^1 L^2 T^{-1}}$$

7.3.5 MOMENT OF A FORCE, $\vec{M}_F$

(Torque, $\vec{\tau}$ **Torque** $\vec{\tau}$ (or Moment) of a force $\vec{F}$ about a point O acting on a (rotating) body may be defined as its turning effect about that point and is measured by the product of the force $\vec{F}$ and the lever arm, *i.e.* d = r $\sin\theta$ perpendicular distance of the point OB from the line of action of the force.

Moment of Force $\vec{F}$ around point O: $\vec{M}_F$

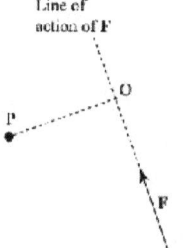

Line of
action of F

O

P

F

Moment of force $\vec{F}$ about OP

$$\boxed{\vec{M}_F = \vec{r} \wedge \vec{F} = \vec{F} \ \vec{r} \ \sin\theta} \quad \boxed{\text{Unit} \quad N \ m \quad (\text{vector}) \quad M^1 L^2 T^{-2}}$$

The true effect of the torque on a body can be seen in the space stations

$$\boxed{\vec{\tau} = \frac{d\vec{L}}{dt} = I \ \vec{\alpha}} \quad \boxed{\text{Unit} \quad N \ m \quad (\text{vector}) \quad M^1 L^2 T^{-2}}$$

where $\bar{\alpha}$ is angular acceleration,
If there are **several particles** relative to their CM (C-Frame),

$$\vec{\tau} = \frac{d}{dt}\sum_{i}^{n}\vec{L}_i \;.$$

Friction between the floor and your foot (or feet) can also generate a torque
The Wright brothers used the torque generated by aerodynamic surfaces to stabilize and control their aircraft. On an airplane, each control surfaces produces aerodynamic lift and drag. These forces are applied at some distance from the aircraft and therefore cause the aircraft to rotate. The elevators produce a pitching moment, the rudder produces a yawing moment, and the wing warping produced a rolling moment. The ability to vary the amount of the force and the moment allowed the pilot to maneuver the aircraft

7.4   **MOMENT OF INERTIA** $\overset{\smile}{I}$ (M.I.)

7.4.1   **Moment of inertia of a body about an axis** of rotation.

It takes the place of mass in the linear equations. $\overset{\smile}{I}$

$$\overset{\smile}{I} = (Mass) \bullet (\text{Distance of mass from Axis})^2 .$$

$$\overset{\smile}{I} = \sum_{i} m_i\, r_i^2 \qquad \text{Unit} \quad kg\ m^2 \ \text{(scalar)} \qquad M^1 L^2 T^0$$

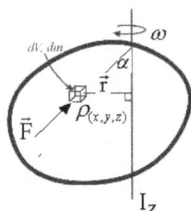

$$\overset{\smile}{I} = \int_{all} r^2\, dm$$

7.4.2   M.I of a body with respect to a plane

$$\overset{\smile}{I} = \int_{all} r^2\, dm$$

$$\overset{\smile}{I}_{YZ} = \int_{all} x^2\, dm$$

$$\overset{\smile}{I}_{XZ} = \int_{all} y^2\, dm$$

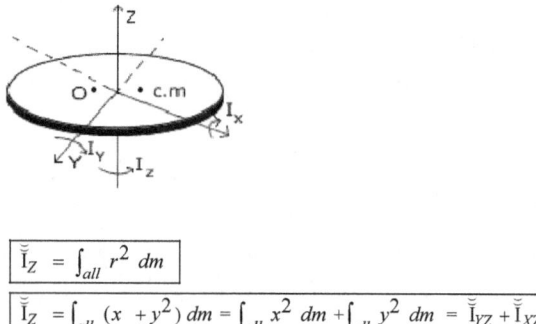

$$\bar{\bar{I}}_Z = \int_{all} r^2 \, dm$$

$$\bar{\bar{I}}_Z = \int_{all} (x + y^2) \, dm = \int_{all} x^2 \, dm + \int_{all} y^2 \, dm = \bar{\bar{I}}_{YZ} + \bar{\bar{I}}_{XZ}$$

r = distance of a differential element of mass dm to the Z-axis in the plane.

Stated in words: The sum of the moments of inertia of a mass with respect to two planes at right angles to each other is equal to the moment of inertia of the mass with respect to the axis formed by the intersection of the planes.

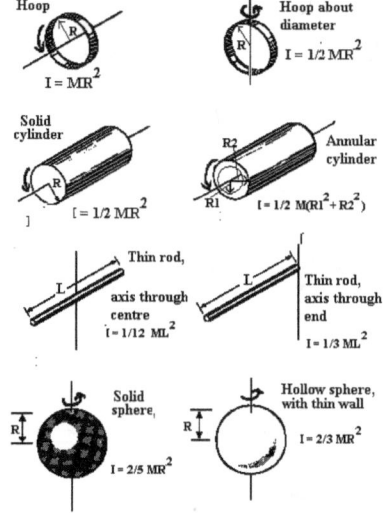

Moments of inertia of bodies about axes

7.4.3 Thin Lamina

Thin rectangular plate of height h and of width w and mass m (Axis of rotation at the end of the plate)

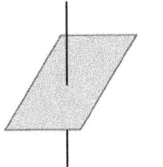

$$I_c = \frac{m(h^2 + w^2)}{12}$$

7.4.4 Rectangular body

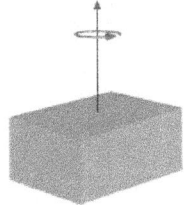

$I_h = \frac{1}{12} m (w^2 + d^2)$

$I_d = \frac{1}{12} m (h^2 + w^2)$

7.4.5. Right circular Cone

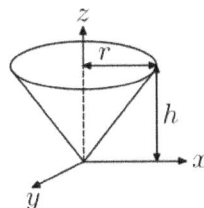

$I_z = \frac{3}{10} m\, r^2 \, ;$

$$\boxed{I_x = I_y = \frac{3}{5} m \left(\frac{r^2}{4} + h^2\right)}$$

7.4.6. Ellipsoid

b *and* c are semi-principal axes of the ellipsoid of mass m, I_a the M.I. along the semi-major axis a,

$$\boxed{I_a = \frac{m}{5} (b^2 + c^2)}$$

7.5.1 The Moment of Inertia of a composite object

I of a composite can be obtained by superposition of the moments of its constituent parts. The <u>Parallel axis theorem</u> is an important part of this process. For example, **a <u>spherical ball</u> on the end of a <u>rod</u>:**

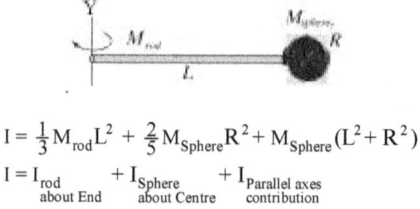

$$I = \tfrac{1}{3}M_{rod}L^2 + \tfrac{2}{5}M_{sphere}R^2 + M_{sphere}(L^2 + R^2)$$

$$I = I_{\substack{rod \\ about\ End}} + I_{\substack{Sphere \\ about\ Centre}} + I_{\substack{Parallel\ axes \\ contribution}}$$

7.5.2 Theorems of Moment of inertia $\vec{I}$:

Transfer of Axes

If the moment of inertia of a body is known about a centroidal axis, it may be determined easily about any parallel axis using the **Parallel-Axis theorem.**

7.6 ANALOGY BETWEEN ROTATION AND TRANSLATION:

	Quantity	Translational Motion of a particle	Rotational motion of a rigid body
1.	Inertia	Mass 'm'	M.I 'I'
2.	Displacement	$\vec{dr},\ \vec{s}$	$\vec{d\theta},\ \vec{\theta}$
3.	Velocity	$\vec{v} = \dfrac{\vec{dr}}{dt}$	$\vec{\omega} = \dfrac{\vec{d\theta}}{dt}$
4.	Acceleration	$\vec{a} = \dfrac{\vec{dv}}{dt}$	$\vec{\alpha} = \dfrac{\vec{d\omega}}{dt}$
5.	Momentum	$\vec{p} = m\vec{v}$	$\vec{L} = I\vec{\omega}$
6.	Cause of motion	$\vec{F}$	$\vec{\tau}$
7.	Law of motion	$\vec{F} = \dfrac{\vec{dp}}{dt} = m\vec{a}$	$\vec{\tau} = \dfrac{\vec{dL}}{dt} = I\vec{\alpha}$
8.	Equations of motion under uniform acceleration	$\vec{v_f} = \vec{v_i} + \vec{a}t$ $\vec{S} = \vec{v_i}t + 1/2\,\vec{a}\,t^2$ $v_f{}^2 = v_i{}^2 + 2\vec{a}.\vec{s}$	$\vec{\omega_f} = \vec{\omega_i} + \vec{\alpha}t$ $\vec{\theta} = \vec{\omega_i}t + 1/2\,\vec{\alpha}t^2$ $\omega_f{}^2 = \omega_i{}^2 + 2.\vec{\alpha}.\vec{v}$
9.	Work done	$\vec{F}.\vec{S}$	$\vec{\tau}.\vec{\theta}$

7.7.1 STEINER'S PARALLEL AXIS THEOREM:

Moment of Inertia of a rigid body of mass M about axis along $\breve{I}_S$ with respect to a line through its centroid separated d parallel to the axis through the CM is given by

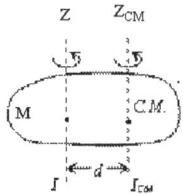

$$\boxed{\breve{I}_S = \breve{I}_{CM} + M\,d^2}$$

Z and Z_{CM} are two axes parallel to each other and one passing through the CM

7.7.2 Perpendicular Axes Theorem

M.I. of a body about an axis perpendicular to the plane of the body is equal to the sum of M.I. about two perpendicular axis in the plane of the body all three axis being mutually perpendicular and concurrent"

$$\boxed{I_Z = I_X + I_{CM}}$$

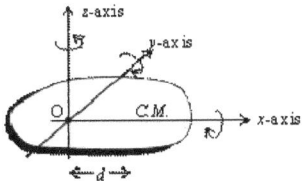

7.7.3 ROUTH'S RULE:

For Rectangular Lamina / Parallelepiped,

$$\boxed{\breve{I}_x = \tfrac{1}{2} M[(\text{Semi-}y-axis)^2 + (\text{Semi--}axis)^2]}$$

7.7.3.1 For circular / elliptical lamina

$$\boxed{\breve{I}_x = \tfrac{1}{4} M[(\text{Semi-}y-axis)^2 + (\text{Semi--}axis)^2]}$$

7.7.3.2 For sphere / spheroid

$$\boxed{\breve{I}_x = \tfrac{1}{5} M[(\text{Semi-}y-axis)^2 + (\text{Semi--}axis)^2]}$$

7.7.3.3 Product of inertia

In a few problems of advanced mechanics the integrals

$$\boxed{I_{xy} = \int xy\ dm, \qquad I_{yz} = \int yz\ dm, \qquad I_{xz} = \int xz\ dm}$$

are useful. These integrals are called the products of inertia of the mass m. They may be either positive or negative. In general, a three-dimensional body has three moments of inertia about the three mutually perpendicular axes and three products of inertia about the three coordinate planes. For an unsymmetrical body of any shape it is found that for a given origin of coordinates there is one orientation of axes for which the products of inertia vanish. These axes are called the **principal axes of inertia**. The corresponding moments of inertia about these axes are known as the **principal moments of inertia** and include the maximum possible value and the minimum possible value

7.8   Radius of gyration

The radius of gyration k of a mass m about some axis is defined by

$$k = \sqrt{\frac{I}{m}}$$

If the entire mass were concentrated at a point whose distance from the axis is equal to the radius of gyration k, the moment of inertia of the concentrated mass would be equal to that of the original mass

7.9   Rotational energy:

$$E_R = \tfrac{1}{2}\breve{I}\,\omega^2$$

| Unit | $N\,m = J$ | (scalar) | $M^1 L^2 T^{-2}$ |

7.9.1   Acceleration of a sphere (r) rolling down an Inclined Plane

$$\vec{a} = \frac{g\,\mathrm{Sin}\beta}{(1 + K^2/r^2)}$$

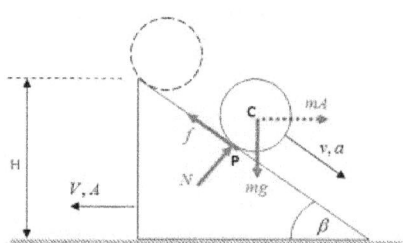

7.9.2   Torque acting on a Body
Relation between Torque and angular momentum of a body:

$$\sum \vec{\tau}_{ext} = \left(\frac{d}{dt}\vec{L}\right)$$

7.10.   EQUILIBRIUM:
A particle in equilibrium means there is no resultant force or Couple acting on it.
7.10.1   Conditions for Equilibrium:

i) Translational equilibrium: if $\sum_i \vec{F}_i = 0$

ii) Rotational equilibrium: if torques $\sum_i \tau_i = 0$

These two give algebraic equations:
$$\sum_i \vec{F}_{ix} = 0 \; ; \; \sum_i \vec{F}_{iy} = 0 \; ; \text{ and } \sum_i \tau_i = 0 \, .$$

7.10.2 Types of Equilibrium

7.10.2.1 Center of Gravity (CG) of a body is
$$CG = \text{Point } (\bar{x}, \bar{y}) = \left(\frac{\sum wx}{\sum w}, \frac{\sum wy}{\sum w} \right)$$

7.10.2.2 Centre of Mass (CM):
The point on the body where the entire mass of the body is can be thought to be concentrated.

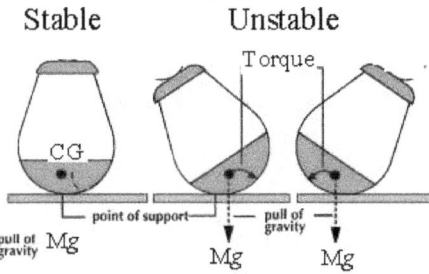

i) Stable Equilibrium: A body on rest must have its CG pass through its base.
ii) Unstable Equilibrium: If the vertical line passing through the CG does not pass through the base of the body.

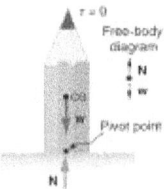

7.10.2.4 Applications
  (1)   Double Decker bus design
  (2)   How does a Marine Vehicle floats on Sea
  (3)   A tight rope walker seems to balance extreme precariously.

7.10.3 PRINCIPLE OF MOMENTS
Look at the ladder AB of length L, weight W, resting in equilibrium against a frictionless wall with angle θ to horizontal. What are the reactions of the ends of the ladder at the ground and the wall?
Taking Moments about point B
$$R_g = R_H + R_V$$

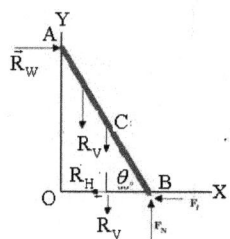

$$R_g = \sqrt{R_H{}^2 + R_V{}^2}$$

$$\tan^{-1}\theta = \frac{R_V}{R_H}$$

+*+&+*+&+*+&+*+&+

Chapter 8

STATICS - 2 :
ROTATION I1
COUPLE, CENTRE IOF GRAVITY, CENTRE OF MASS, EQUILIBRIUM, PRINCIPLE OF MOMENTS

"Intellectuals solve problems, geniuses prevent them"   Albert Einstein

8   **ROTATION II**

Law of Conservation of Angular Momentum $\vec{L}$

If $\vec{\tau}_{ext}$ is torque due to external forces (i.e. an isolated system), is zero then **its Total angular momentum is conserved**

i.e. $$\vec{\tau}_{ext} = \left(\frac{d}{dt}\vec{L}\right) = 0$$

and so $$\boxed{\vec{L} = \breve{I}\,\vec{\omega}}\ = \underline{\textbf{a constant of the motion}}$$

is **valid only when** $\vec{\tau}_{ext}$ and $\vec{L}$ are evaluated relative to a point fixed in an inertial frame of reference.

8.1.   **GYROSCOPE**

If $\theta_L = 0$; $\underline{\vec{L} = \breve{I}\,\vec{\omega}} = $ a constant, then the body will keep on rotating about an axis with constant $\vec{\omega}$ - principle of working of a gyroscope

i.e. $$\boxed{\left(\frac{d}{dt}\vec{L}\right) = \vec{\tau}_{ext}}\qquad \boxed{d\vec{\tau}_{ext} \neq 0},$$

$d\vec{L} = $ is always in the direction of $\vec{\tau}_{ext}$.

The motion of the axis of rotation about a fixed axis due to $\vec{\tau}_{ext}$ is called **Precession**.

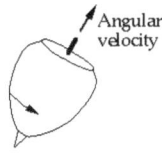

Angular velocity

8.2.   **COUPLE** *(C)*

Two unlike but equal parallel forces whose lines of action do not coincide constitute a Couple

A couple consists of two parallel forces that are equal in magnitude, opposite in sign and do not share a line of action. It does not produce any translation, only rotation. The resultant force of a couple is zero. But, the result of a couple is not zero; it is a pure moment

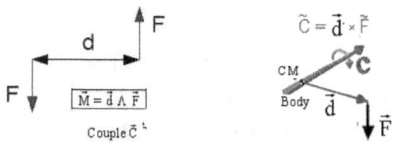

$$\boxed{\text{Moment of Couple } \tilde{C} \text{ (or Torque of a couple)} = \vec{F}\, d}$$

8.2.1 A wrench

A wrench is a force and couple system in which the force and couple are parallel

$$\vec{F}_{\substack{\text{on a body} \\ \text{to rotate}}} \equiv \vec{F}_{\substack{\text{through C.M} \\ \text{of the body}}} + \tilde{C}, \text{ couple}$$

$\vec{F}$ on a body may be replaced by $\boxed{\text{A Force } \vec{F} \text{ (through the CM)} + \text{Couple C}}$.

8.3. CENTRE OF GRAVITY *(CG)*:

CG of a body, $(\bar{x}, \bar{y})$, may be defined as that fixed point through which the line of action of the weight always passes for all the positions of the body.

$$\boxed{\text{CG of a body} \Rightarrow (x, y) = \left(\frac{\Sigma Wx}{\Sigma W}, \frac{\Sigma Wy}{\Sigma W} \right)}$$

$$\boxed{\text{Unit} \quad m \quad \text{(scalar)} \quad M^0 L^1 T^0}$$

8.3.1 C.G. of a **Triangular Lamina**:
$$\boxed{(x, y) = \text{at its CENTROID}}.$$

8.3.2 C.G. of an **Arc**
$$\boxed{(x, y) = \left(\frac{r \, \text{Sin } \alpha}{\alpha}, 0 \right)}$$

8.3.3 C.G. of a **Semi Circle**
$$\boxed{(x, y) = \left(\frac{2x}{y}, 0 \right)}$$

8.3.4 C.G. of a **Sector**
$$\boxed{(x, y) = \left(\frac{2}{5} \frac{r \, \text{Sin } \alpha}{\alpha}, 0 \right)}$$

8.35 C.G. of a **Solid Hemisphere**

$$(x, y) = \left(\frac{3r}{2}, 0\right)$$

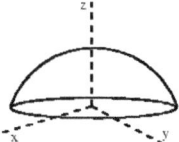

8.3.6 C.G. of a **Hollow Sphere**, Annular ring, Sphere

$$(x, y) = \left(\frac{1}{2}r, 0\right)$$ at its geometric centre

8.3.7 C.G. of a **Tetrahedron** (also Pyramid)

$$(\overline{x}, \overline{y}) = \left(\frac{1}{4}G, D\right)$$ on the line joining vertex D

8.3.8 C.G. of a **Right Solid Cone**

$$(x, y) = \left(\frac{3}{4}h, 0\right)$$ on the line joining vertex $= \left(\frac{3}{4}h, 0\right)$

$\frac{3}{4}h$ from the apex on the line from apex to the geometric centre of the base.

8.3.9 C.G. of a **Cube:**
At the body centre

8.3.10 C.G. of a **triangular body**

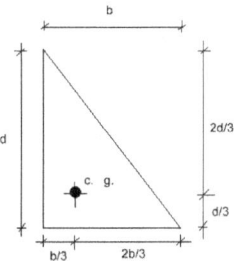

$(\overline{x}, \overline{y})$ = At the point of intersection of the cube diagonals.

8.4. **CM and CG**

The CM coincides with the CG in uniform gravitational fields such as those close to the surface of the Earth.

8.4.1 **Black hole**, the CM is not the same as CG.

8.4.2 **Fly Wheel**

* a wheel winds up through some system of gears and then delivers rotational energy until friction dissipates it
* stored energy = sum of kinetic energy of individual mass elements that comprise the flywheel
Tensile Strength is More important than density of material.

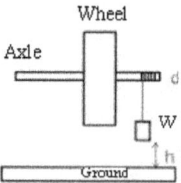

Height h.
time (t) it takes to hit the ground.
Number of revolutions (N_o) before the weight hits the ground
Number of revolutions (N_f) before the flywheel comes to rest.
Diameter of the axle d
Mass of the falling weight (m)

$$I = \frac{m\,d^2}{4}\frac{N_f}{N_f + N_o}\left(\frac{g\,t^2}{2h} - 1\right)$$

8.4.3   C.G. of a Human Body:

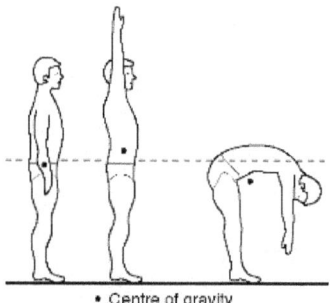

• Centre of gravity

8.5   EQUILIBRIUM OF A SOLID BODY

A body is in equilibrium if there is no resultant force or Couple C acting on it.
The stability of an object in equilibrium is determined by the Centre of Gravity concept.

8.5.1   Stable equilibrium

Stable equilibrium: is when resting on a surface, the vertical line passing through the object's C.G. must also pass through the base of the object.

8.5.2 Unstable equilibrium:

If the vertical line through the CG of the object does not pass through the base then unstable equilibrium prevails.

8.5.3 Conditions of Equilibrium:

(1)   For translational equilibrium: $\sum_i \vec{F}_i = 0$.

(2)   For rotational equilibrium: $\sum_i \vec{\tau}_i = 0$.

(3)   Forces are all should be in one plane.

8.6.   PRINCIPLE OF MOMENTS

When a body is in rotational equilibrium the algebraic sum of all the torques acting on the body about all its axes is zero. $\sum_i \vec{\tau}_i = 0$.

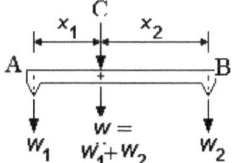

In other words, " When an object is in equilibrium the sum of the anticlockwise moments about a turning point must be equal to the sum of the clockwise moments."

$$\boxed{W_1 \bullet x_1 = W_2 \bullet x_2}$$

Thus the resultant divides AB in the reverse ratio.

8.6.1 Conditions of Equilibrium of Three Non-parallel Forces

1. The lines of action of the three forces must all pass through the same point.
2. The principle of moments: the sum of all the clock-wise moments about any point must have the same magnitude as the sum of all the anti-clockwise moments about the same point
3. a) The sum of all the forces acting vertically upwards must have the same magnitude as the sum of all the forces acting vertically downwards.
   b) The sum of all the forces acting horizontally to the right must have the same magnitude as the sum of all the forces acting horizontally to the left.

8.7   APPLICATIONS

8.7.1 To work out the strength of materials needed to construct bridges and building.

8.7.2 A uniform ladder resting on a wall rests in equilibrium

Consider no frictional force between the ladder and the wall. But

$\vec{F}_1$ = Frictional force between the ground and the ladder

m_2 = Mass of ladder of length L

m_1 = Mass of a man stands on the ladder at d from the ground (measured along the ladder),

μ_{min} = Minimum static friction coeff,. Required between the ladder and ground for no slip in equilibrium.

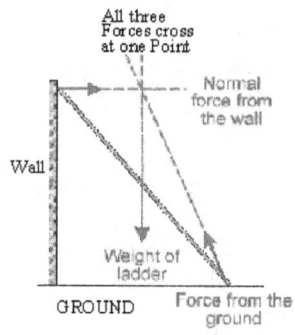

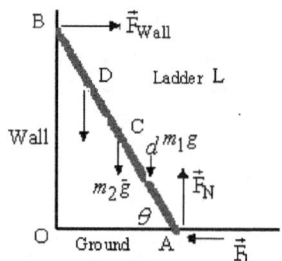

$$\mu_{min} = \left(\frac{d}{L} m_1 + \frac{1}{2} m_2 \right) \frac{\text{Cot } \theta}{m_1 + m_2}$$

$$\vec{F}_1 = \vec{F}_N = \left(\frac{m_2}{2} + m_1 \frac{d}{L} \right) g \text{ Cot} \theta \quad, \quad \mu_{actual} = \frac{3}{2} \mu_{min}$$

The first condition of equilibrium deals with only the forces
The **second condition** of equilibrium deals with torques):

$$\tan\theta = \frac{\vec{F}_N}{\vec{F}_1} \text{ at equilibrium.}$$

8.7.3 Action of a dancer

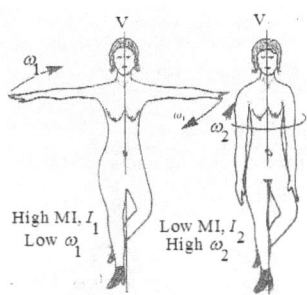

Conservation of Angular momentum

Turns in the air: no forces except gravity can act on the dancer. Angular momentum stays constant: changing $\widetilde{I}$ (by changing the orientation of arms, legs,...) will change $\vec{\omega}$ Correspondingly.

8.7.4   A **Tight Rope-Walker**
The walker seems to balance extremely precariously, but he is invariably less in danger than it appears, he is physically in **unstable equilibrium**. The long pliable pole lowers the C.G. of the walker. The **inertia** of the pole helps the walker to maintain equilibrium.

8.7.6   **A cyclist riding** a bicycle in equilibrium.

8.7.7   **Ever see a falling cat right itself?** The cat has zero angular momentum at all times, but somehow manages to turn over. It works like this:
i) **Upside-down cat** curves its back "the easy way."
ii) Cat straightens its back while bending around its middle to its right.
iii) Cat comes out of its bend-to-the-right while arching its back "the hard way."
iv) Cat straightens its back while bending around its middle to its left.
v) Cat comes out of its bend-to-the-left while curving its back "the easy way."

8.7.8   **Equal Arm Beam Balance**
A good balance has the three requisites, *viz.*, Truth, Sensitivity, and Stability.
Truth: A balance should have equals arms length, its pans of equal weight, its C.G. should pass through the fulcrum, and perpendicular beam.
Sensitiveness:

$$\text{Sensitivity} = \frac{\text{Angle of turn of beam, } \theta}{\text{weight difference in pans, } w}$$

$$\frac{\text{Tan } \theta}{w} = \frac{a}{[\,(2P + 2S + w)\,h + (h + k)\,W\,]} = \frac{\theta}{w} \Rightarrow \frac{\theta}{w} = \frac{a}{w\,k}$$

a = arm length,
W = weight of beam,
P = weight of each pan, S = weight added to each pan,
θ = angle through which the arm turns when a mass w is added to S,

Stability: For high stability the restoring moment should be more, *i.e.,* *h, k* and W should be large, Fulcrum and the CG of the beam must br far from the middle.. Hence a a balance cannot be both sensitive and stable.

Correction to buoyancy of air

Wt. of body in vacuum $= W(1 - \frac{\sigma}{\rho} + \frac{\sigma}{d})$

σ, ρ, d = density of air, material of weight, and od of the body, respectivrly.

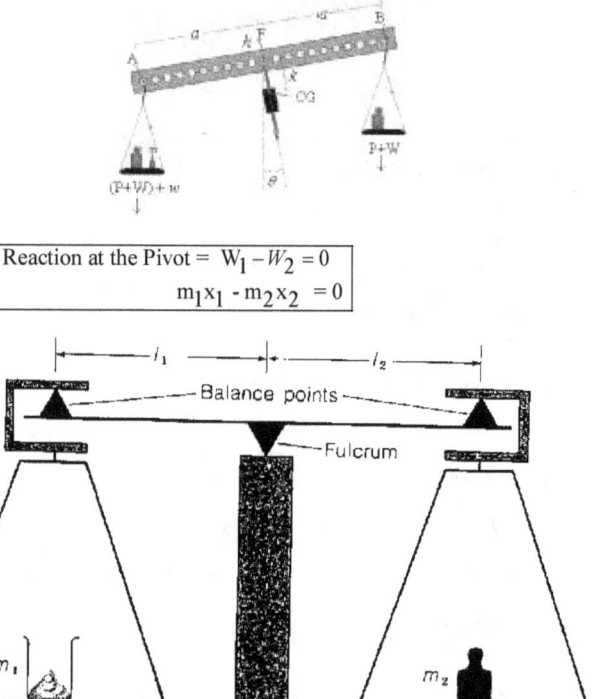

Reaction at the Pivot $= W_1 - W_2 = 0$

$$m_1 x_1 - m_2 x_2 = 0$$

8.7.9 What makes the Earth to rotate?

The Earth rotates simply because **it has not yet stopped moving**. Conservation of angular momentum meant that anybody formed from the gas would itself be rotating. It rotates around the Sun at $\sim 30\ km\ s^{-1}$

8.7.10 Spin of Earth:

Earth spins on its own axis at $\sim \sim 460\ ms^{-1}$ at the equator;

8.7.11. Physics in Sports

8.7.11.1. Magnus effect (Throw of a spinning cricket ball / base ball / curved ball)

When a spinning object is rapidly moving through air, it experiences a lateral force (one which deflects the object sideways from its normal path) (due to air resistance) (in aerodynamics) called Gustav Magnus effect (1853). The strength and direction of the Magnus force depends on how fast and the direction of spin.

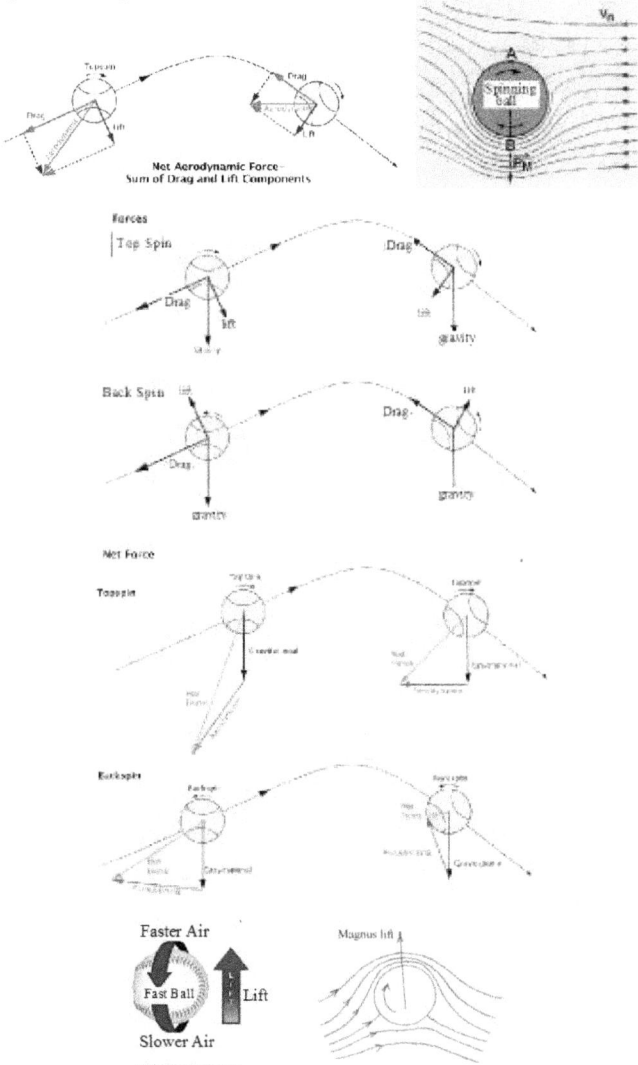

THE MAGNUS EFFECT

When the ball has 'no spin', the air flows around it and causes a 'wake' only behind the ball, and no Magnus force developed.

NONSPINNING BALL

8.7.11.2. Tennis Racquet

A tennis racquet, like a baseball or cricket bat, has a sweet spot, where the force transmitted to the hand is sufficiently small.

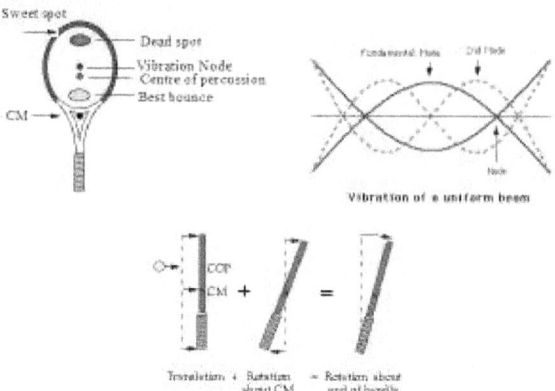

Forces on the hand arise from three independent motions of the handle, namely rotation, translation and vibration. The vibrational component is absent when a ball strikes the vibration node. The rotational component, arising from recoil of the racquet head, exerts a torque on the hand, causing it rotate about an axis through the wrist. As a result, a force is always exerted on the upper part of the hand, and a force in the opposite direction is always exerted on the lower part of the hand. These and some other significant spots on a racquet are shown in Figure.

The fundamental frequency of vibration for a typical tennis racquet is $\sim .30\ Hz$ (about 100 Hz for a relatively flexible frame or about 180 Hz for a stiff frame).

If a tennis ball hits the racquet anywhere on the strings, the impact will trigger the fundamental, the harmonics and the player will feel these vibrations in the hand.

By hitting the ball harder, players generate even more topspin to the ball.,

The COP is often regarded as a second sweet spot since the force on the hand should be zero for an impact at the COP. (Brody, H., Phys. Today, 48 1995).

When dropped from a height of say 1 m, onto a concrete slab, the ball will bounce to a height of about 0.70 m.

8.7.11.3. Bicycle (Motor bike)

"Trail" (shown below) is often an important contributor to bicycle stability. For the traditional bicycle design, if trail is positive, meaning the projection of the steering axis with the ground is in front of the contact point of front wheel and ground, then the bicycle is more stable when riding (i.e. it's less likely to fall over when riding it). If this projection is behind the contact point (negative trail) then the bicycle is less stable and the bicycle is more likely to fall down when riding it.

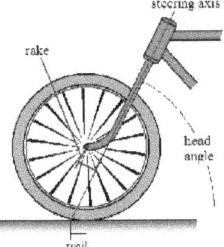

8.7.11.4.   Athlete jumper

Long Jump: Great jumpers seem to 'hang in the air', and this illusion is because of the principles of projectile motion.

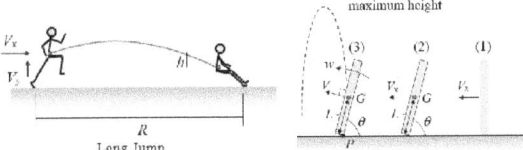

High Jump:, While watching sports on jump, remember the equation

$$u^2 = 2gH$$.

It explains why most of jumpers do the backward flip known as the Fosbury Flop. (John Barrow writes in his book *Mathletics: A Scientist Explains 100 Amazing Things about the World of Sports* (W. W. Norton, 2012), the Fosbury Flop keeps one's center of gravity low to the ground, and the lower one's center of gravity, the less energy is required to successfully jump over the bar. In the above equation, U is the speed of the jumper (and thus the energy required), g is the acceleration caused by gravity, and H is the height of the center of gravity. Surprisingly, it is possible for the high jumper's body to fly over the bar while his or her center of gravity passes below it.

8.7.11.5. Physics of Boomerang Flight

The wings of a boomerang are designed to generate lift as they spin through the air. But a phenomenon known as gyroscopic precession is the key to making a returning boomerang come back to its thrower. When the boomerang spins, one wing is actually moving through the air faster than the other [relative to the air] as the boomerang is moving forward as a whole, As the top wing is spinning forward, the lift force on that wing is greater and results in unbalanced forces that gradually turns the boomerang." The difference in lift force between the two sides of the boomerang produces a consistent torque that makes the boomerang turn. It soars through the air and gradually loops back around in a circle.

The returning trajectory of a Boomerang involves the 'aerodynamic lift' of its airfoil shape plus 'gyroscopic precession' associated with its rapid spin.

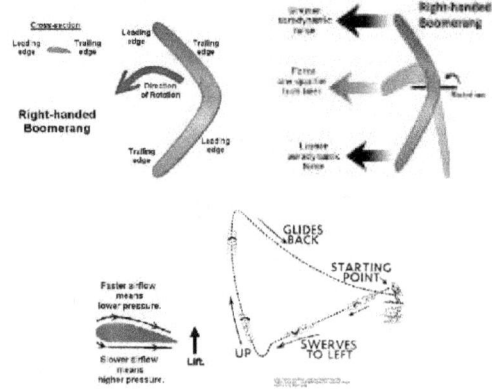

8.7.12 Kinetic energy of a Rolling Object:

$$E = (E_T + E_R) = \left(\frac{1}{2}MV^2 + \frac{1}{2}\overleftrightarrow{I}\omega^2 \right)$$

+*+*+*+*+*+*+*+*+*+

Chapter 9

STATICS - 3:
FRICTION
NON-CONSERVATIVE FORCES

"It is possible to fly without motors, but not without knowledge and skill"   Wilbur Wright

9.1   FRICTION

The opposition to the motion of an object moving over the surface of another object is called friction;

Two types:

The force arising due to friction is called the force of friction ($\vec{f}_S$).

9.1.1 $\vec{f}_S$ Opposing the motion of an object from rest is called **Static friction**;

9.1.2   The force of friction $\vec{f}_K$ opposing the moving object is called Sliding or **Kinetic friction**.

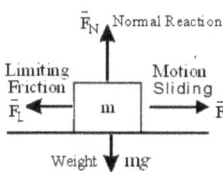

9.2   LAWS OF LIMITING FRICTION

(The forces of friction possess the following characteristics):

9.2.1

!)   The force of friction always act in the opposite direction to motion in which one body slides over the other.

2)   The force of friction increases in magnitude up to a certain maximum value, when there is equilibrium between the two bodies, is just enough to prevent the motion of one body with respect to the other. This maximum force is called '**limiting force of friction**' $\vec{F}_L$.

3)   The limiting force of friction is independent of the area of contact between the surfaces.

4)   The limiting force of friction $\vec{F}_L$ is proportional to the **normal reaction force**, $\vec{F}_N$.

$$\mu_S = \frac{\vec{F}_L}{\vec{F}_N} \;\; ; \text{and} \;\; \mu_k = \frac{\vec{F}_K}{\vec{F}_N}$$

μ_S = *Coefficient of sliding static friction,*

Coefficients of Friction		
Materials	Static Friction	Kinetic Friction
Steel on steel	0.74	0.57
Aluminum on steel	0.61	0.47
Wood on brick	0.60	0.45
Copper on steel	0.53	0.36
Rubber on concrete	1.0	0.80
Wood on wood	0.25 – 0.50	0.20
Glass on glass	0.94	0.40
Waxed wood on wet snow	0.14	0.10
Waxed wood on dry snow	—	0.040
Metal on metal (lubricated)	0.15	0.060
Ice on ice	0.10	0.030
Teflon on teflon	0.040	0.040
Synovial Joints in humans	0.010	0.0040

5)   For the same object, the limiting force of friction (and hence μ_S) is different for different surfaces. For the same object, the limiting force of friction is different for the same surface, depending upon the lubrication of the surface.

Kinetic friction $\vec{F}_K$ is slightly < static friction, $\vec{F}_L$.

$$\mu_S > \mu_k$$

9.2.2   **Angle of friction** λ is defined such that

$$\mu_S = \frac{\vec{F}_L}{\vec{F}_N} = \tan \theta$$

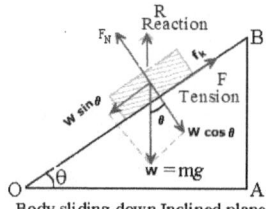

Body sliding down Inclined plane

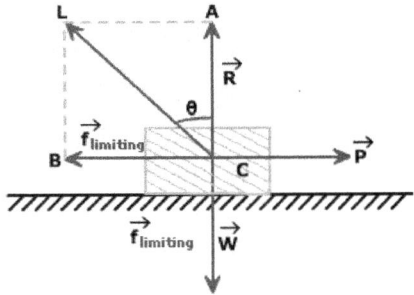

9.2.3. Static Friction and Kinetic Friction Plot

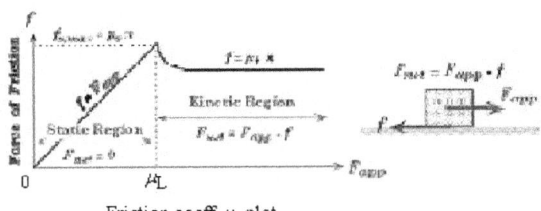

Friction coeff. μ plot

9.2.4. Determination of Coefficient of kinetic friction
        (Horizontal surface Method)

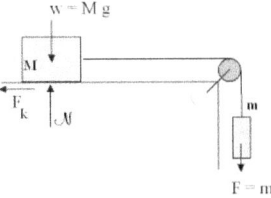

$\vec{F}$ is resisted by the friction force $\vec{F}_K$. If m is great enough to slide M to the right at constant velocity, then $\vec{F} = \vec{F}_K$. At a **constant velocity of M** to the right,

$$\mu_k = \frac{\vec{F}_K}{\vec{F}_N} = \frac{m}{M}$$

Graph of m *versus* M will also yield μ_k.

-o-0-o-0-o-0-o-0-o-0-o-

Chapter 10

STATICS – 4:
SIMPLE HARMONIC MOTION
& SIMPLE PENDULUM

"I have not failed, I've just found 10,000 ways that won't work"   Thomas Edison

10 1   **OSCILLATIONS [SIMPLE HARMONIC MOTION, Simple Pendulum]**

10.1.1 **Periodic Motion**:
Any motion that repeats after a certain period of time is called Periodic Oscillatory Motion.

10.1.2 **Simple Harmonic Motion**

A type of periodic oscillatory motion, in which the Restoring force ($\vec{F}$), and hence acceleration ($\vec{a}$), is directed towards the equilibrium (*i.e.* mean) position (**O**) and is directly proportional to its displacement (x) from the mean position, is called SIMPLE HARMONIC MOTION.

O : mean position

$\vec{k}$ = Spring (helical) constant (force constant)

$$\vec{k} = \frac{m\,g}{\ell}$$ | Unit $N\,m^{-1}$ (Vector) $M^1\,L^0\,T^{-2}$

10.2   **Elastic Potential energy**, V(x) of SHM

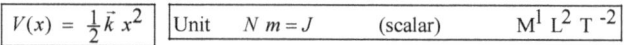

$$V(x) = \tfrac{1}{2}\vec{k}\,x^2$$ | Unit $N\,m = J$ (scalar) $M^1\,L^2\,T^{-2}$

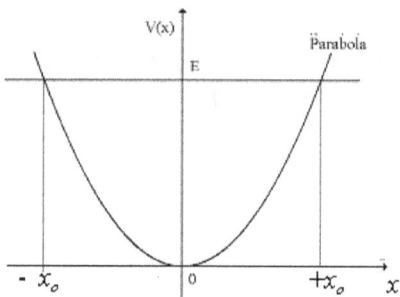

10.2.1 **Acceleration,** $\vec{a}$, *in SHM*

$$\boxed{\vec{a} = -\vec{k}\,x}\ ;$$

10.2.2 **Equation of motion** *of SHM in* **conventional form**

$$\boxed{m\frac{d^2x}{dt^2} = -\vec{k}\,x}\ ;$$

10.2.3 In **differential form**
Angular frequency of oscillation, $\vec{\omega}$

$$\boxed{\vec{a} = \frac{d^2x}{dt^2} = -\vec{\omega}^2\,x}\ ;$$

10.2.4 **Harmonic motion:** **displacement,** *x*

$$\boxed{\vec{x}(t) = x_O\ Sin\ (\alpha t)}$$

10.2.5 **Equilibrium (static) Mean position,** x_O

$$\boxed{\vec{x}(0) = x_O}$$

Any periodic motion represented in terms of Sine / Cosine. (Co sinusoid)

10.3. **GRAPH**
For $\quad \vec{x}(t) = x_O\ Sin\ (\alpha t)$

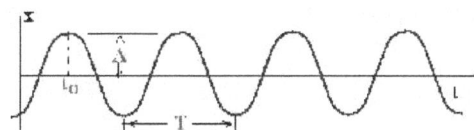

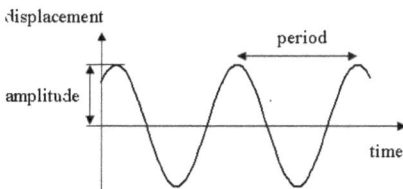

10.3.1 **GRAPHICAL DEFINITION OF SHM**

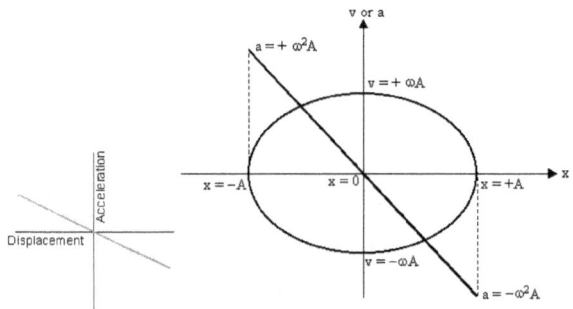

10.3.2 Illustration of Helical Spring & mass in SHM

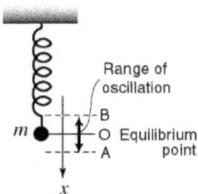

Position of Mass and its displacement in SHM and reference circle

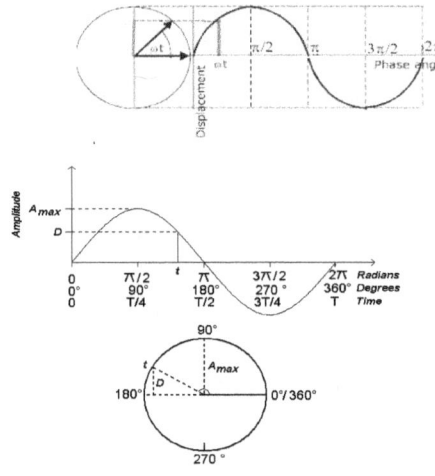

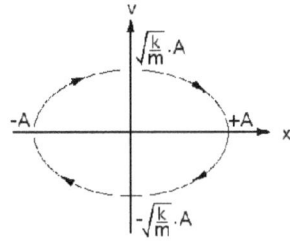

10.3.3 AMPLITUDE of SHM and reference circle:

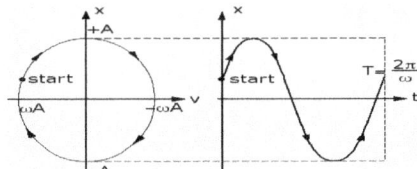

10.3.4 DISPLACEMENT,
General form:

$$\vec{y}(t) = r\ Sin\ (\omega t + \phi)$$

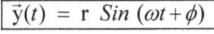

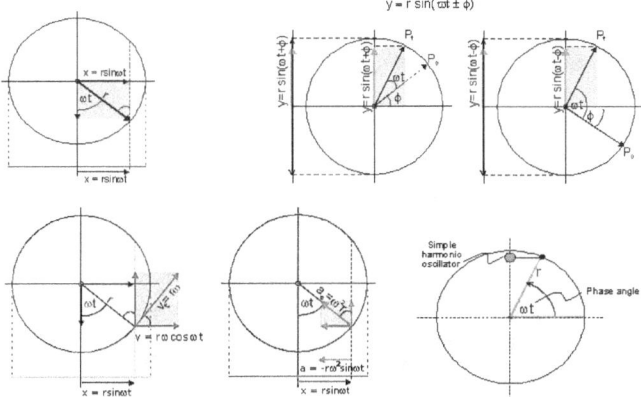

10.3.5 Illustrating two SHM waves with 180 out-of-phase

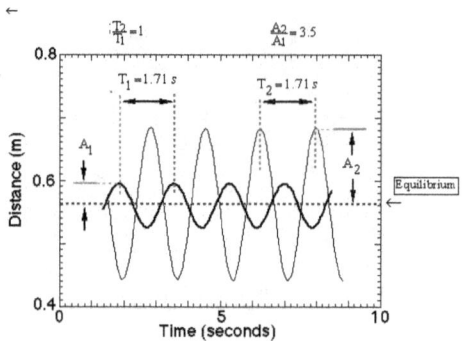

10.3.6 **DISPLACEMENT** *VERSUS* **TIME** curve in SHM

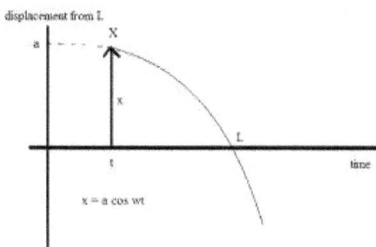

10.3.7 **DISPLACEMENT** *VERSUS* **SPEED**

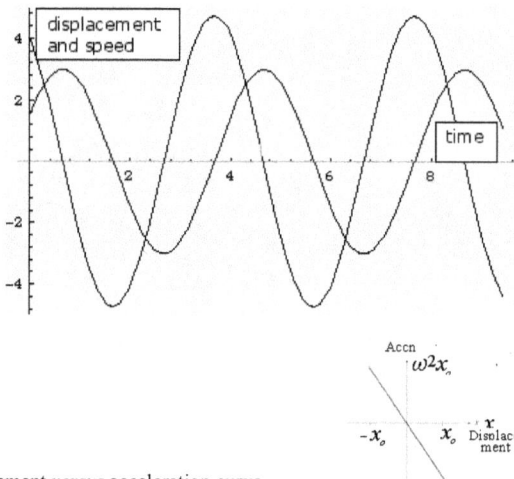

10.3.8 Displacement *versus* acceleration curve

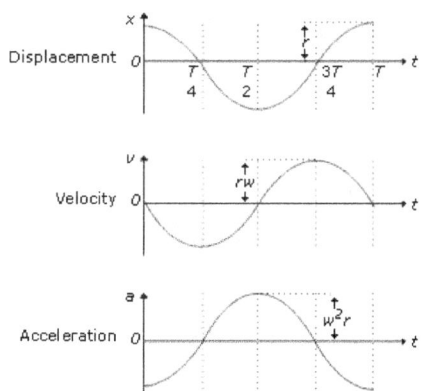

10.4 SIMPLE PENDULUM

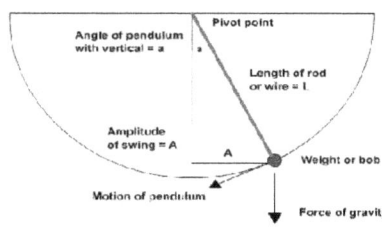

$$T = 2\pi \sqrt{\frac{\ell}{g}}$$

10.4.1 DISPLACEMENT and Reference circle

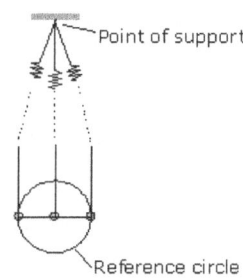

10.4.2 PARAMETERS of Simple Pendulum

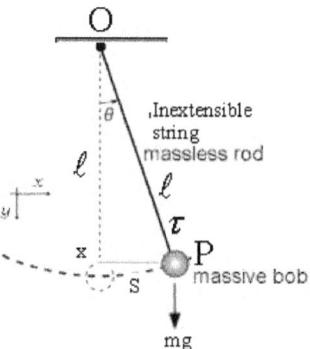

ℓ = length of string

ϑ = angle that the string making with the vertical at time t

      = small for the pendulum to execute linear SHM.

x_0 = Amplitude small,

10.4.2.1 Speed

$$v = \pm\omega\sqrt{(x_0 - x)^2}$$

| Unit | $m\ s^{-1}$ | (Vector) | $M^0L^1T^{-1}$ |

10.4.2.2 Maximum velocity,

$$v_{Max} = \pm r\omega$$

| Unit | $m\ s^{-1}$ | (Vector) | $M^0L^1T^{-1}$ |

10.4.2.3 Period

$$T = 2\sqrt{\frac{Displacement}{Acceleration}}$$

10.4.2.4 Angular frequency,

$$\omega = \sqrt{k/m} = \sqrt{g/\ell},\ \text{in linear SHM}$$

| $\omega = \sqrt{k/m}$ | Unit $rad\ s^{-1}$ | (Vector) | $(M^0L^1T^{-1})$ |

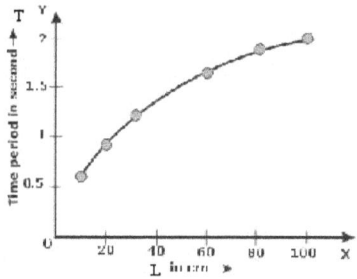

| Period | $T = 2\pi \sqrt{\dfrac{\ell}{g}}$ | Unit s (Vector) | $(M^0 L^0 T^1)$ |

When ϑ = large, and x_0 = Amplitude small

| **Period** | $T = 2\pi \sqrt{\dfrac{\ell}{g}} \left[1 + \dfrac{1}{4} Sin^2 (\vartheta_0 / 2) \right]$ |
| | $T = 2\pi \sqrt{\dfrac{\ell}{g}} \left[1 + (\vartheta_0{}^2 / 16) \right]$ |

Correction term $(\vartheta_0{}^2 / 16)$ in T < 1% for amplitude $< 23^0$ (*i.e.* 0.4^c)

$$T = T_0 \left[1 + (\dfrac{\rho_{air}}{2\, \rho_{bob}}) \right]$$

10.4.2.5. Potential Energy *versus* Time variation

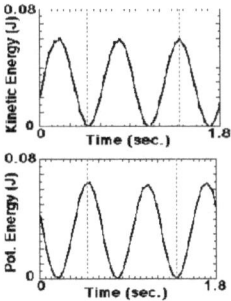

10.4.2.6 ENERGETICS *of* SHM

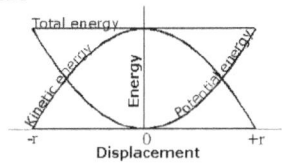

10.4.2.7 Variations of Displacement,
Total energy, Potential energy and Kinetic energy with time

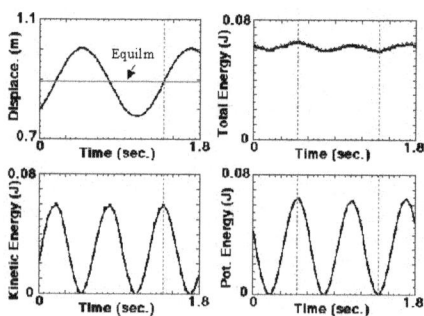

10.4.2.8 Variations of Displacement, Velocity and Acceleration with Time

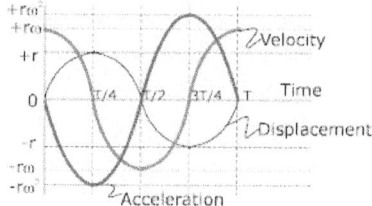

10.4.2.9 Period is independent of Mass

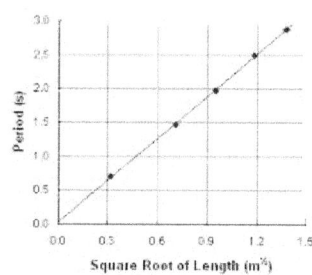

10.4.2.10 Period is dependent on k

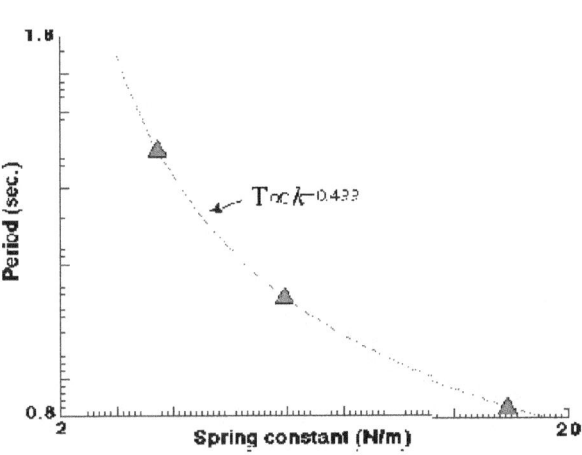

10.4.2.12. A simple pendulum of length 1 m is held in the horizontal position initially, and then it is released

        L: Lowest position

        H: Highest position

        P: Intermediate position

$$\boxed{T = 3\,m\,g\,Sin\,\vartheta}$$

$$\boxed{T_L = 3\,m\,g}$$

10.5. EXAMPLES of SHM:

10.5.1 A test-tube bobbing up and down in water (Floating cylinder)

        h = height of the bottom of the cylinder and water level

        r = radius of cylinder

        ρ = density of liquid

$$\boxed{\omega = \sqrt{\frac{\rho\,g}{r\ell}}}\,,$$

Period $\boxed{T = 2\pi\sqrt{\frac{h}{g}}}$

10.5.2 Compound pendulum

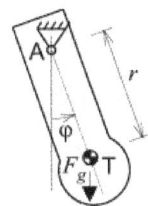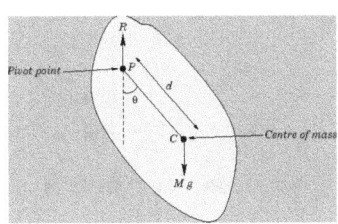

For a compound pendulum suspended about a horizontal axis through OX distant ℓ from the centroid, $\omega^2 = \dfrac{mg\ell}{I_{OX}}$.

$$t = 2\pi \sqrt{\dfrac{I}{m\,g\,h}} \quad \text{where } h = r$$

$$I = I_g + m\,h^2$$

$$t = 2\pi \sqrt{\dfrac{(k^2 + h^2)}{g\,h}}$$

k = radius of gyration of the body about the axis of rotation.

$\ell = \dfrac{(k^2 + h^2)}{h} = 2k$ (when T = minimum)..

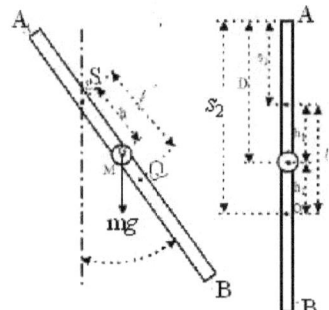

Compound Pendulum

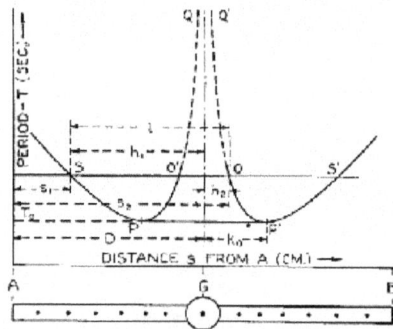

Compound Pendulum- Graphical Analysis

10.5.4 KATER'S (REVERSIBLE) PENDULUM

103

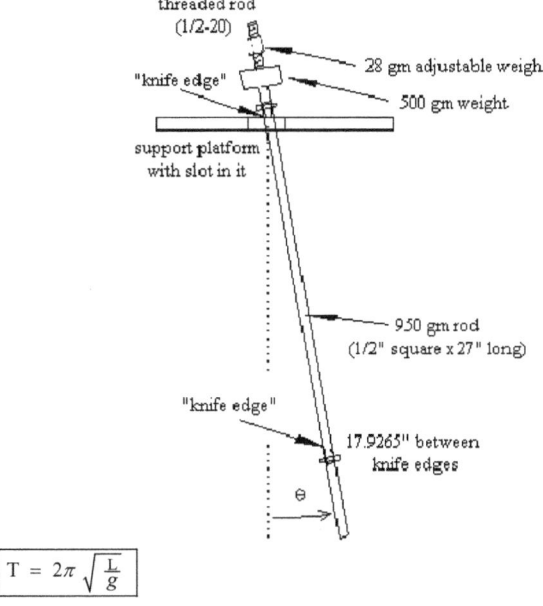

threaded rod
(1/2-20)

28 gm adjustable weight

"knife edge"

500 gm weight

support platform
with slot in it

950 gm rod
(1/2" square x 27" long)

"knife edge"

17.9265" between
knife edges

θ

$$T = 2\pi \sqrt{\frac{L}{g}}$$

10.5.5 Ballistic Pendulum

In the back courtyard of the munitions factory hung an old, scarred block of wood. As quality control for the cartridges coming off the assembly line, someone would regularly take a gun to the courtyard and fire a bullet into the block. Measuring the height

$$u = \frac{m+M}{m} v = \frac{m+M}{m} \sqrt{2gh}$$

$$v = \sqrt{2gh}$$

$$h = \frac{v^2}{2g}$$

$$v = \frac{m}{m+M} u$$

of the swing revealed the speed of the bullet, but since the block was increasing in mass with the added bullets, the mass of the block had to be checked as well as the mass of the bullet being fired.

In a perfectly inelastic collision, a bullet is fired into the stationary pendulum, which captures the bullet and absorbs its energy.

10.5.6. **Applications:**
  * This is one way to measure the speed of a bullet.
  * One can verify the law of conservation of momentum.

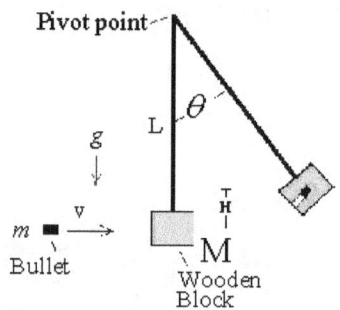

Ballistic Pendulum

$$mv = (M + m)V' \quad \text{and} \quad \frac{1}{2}(M + m)V'^2 = (m + m)gH$$

Speed of bullet, $v = \dfrac{M + m}{m}\sqrt{2gH}$

10.5.7 Vibrating helical spring
10.5.9   Atoms vibrating in a crystal lattice
10.5.10 Vibrating cantilever

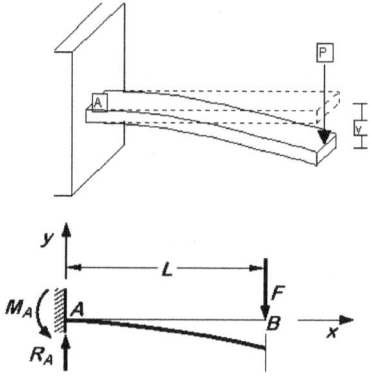

10.5.11 Mable on a concave surface

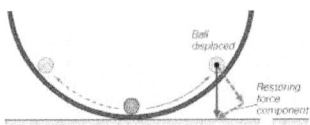

10.5.12 Torsional Pendulum

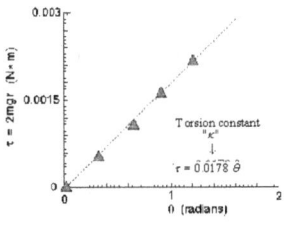

$$\omega = \sqrt{\frac{g}{L}}, \qquad L = \frac{I}{Md}$$

10.5.13 Liquid oscillating in a U-tube

$2h$ = length of the liquid in U-tube

Period $\omega = \sqrt{\rho\, g\, /\, r\, h}$, $T = 2\pi\sqrt{\frac{h}{g}}$

$$T = \pi\sqrt{\frac{2L}{g}}$$

L = length of the liquid column

10.5.14 Inertia balance.

10.5.15 Bifilar Pendulum

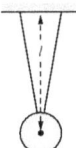

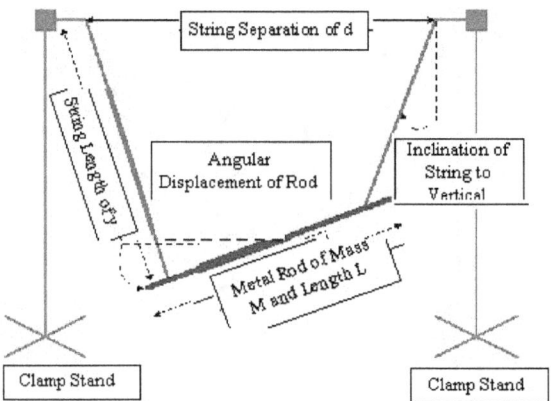

10.5.16 Piston in steam engine

10.5.17 *Oscillating disk*

10.5.18 **Second's pendulum**

A thin cotton string of length of $\ell = 50$ *cm*, tied to a small metallic sphere at one end and the other through a slit cork, clamped tightly to a vertical stand..

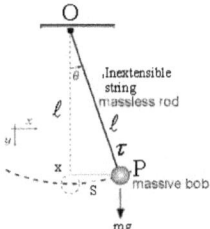

Plot a graph of T^2 *versus* ℓ. It will be a straight line. Second's pendulum is for which $T = 2s$. Find from the graph the length for which $T^2 = 4$ s.

10.5.19 **Helmholtz Resonator**

$$T = \frac{2\pi}{v_s} \sqrt{\frac{dV}{a}}$$

a = diameter of the neck of the resonator,
d = length of the neck of the resonator
V = volume of the spherical resonator

10.5.20 **HYDROMETER (Nicholson's)**

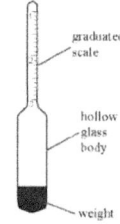

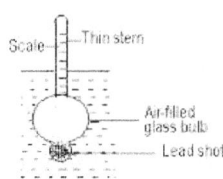

$$F = mg - B' = \pi r^2 \rho gx$$

$$\frac{d^2x}{dt^2} = \frac{-\pi r^2 \rho gx}{m}$$

Mass of hydrometer = m_H, ρ = density of liquid

$$\omega_0 = \sqrt{\frac{A\rho g}{m_H}}$$

Specific gravity of the body $= (\dfrac{\text{weight of body in air}}{\text{Loss of wt. of body in water}})$

Specific gravity of liquid $== (\dfrac{\text{weight of certain volume of liquid}}{\text{wt. of an equal volume of water}})$

$= (\dfrac{\text{Loss of weight of body in liquid}}{\text{Loss of wt. of body in water}})$

10.6   RIGID BODIES

The most general motion of a rigid body can always be considered as a combination of a rotation and a translation; Rotation is around an axis through the Centre of Mass (CM), translation being the displacement joining the two positions of CM.

Centre of Mass (CM):
The concept of CM is important in the analysis of composition of parallel forces and of a rigid body. If the body has centre of symmetry then the CM coincides with the Centre of symmetry.

Weight of the body = $W = \sum m_i g$
This sum extends over all the particles comprising the body is applied at a point called the CM, r_c :

$$r_c = (\sum m_i r_i) / (\sum m_i)   \text{Unit m . (Vector)}   (M^0 L^1 T^0)$$

10.7.1 DAMPED OSCILLATORY MOTION:

An oscillating system in which friction has an effect is said to be a damped system.

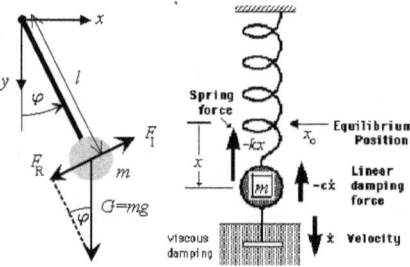

10.7.2 Undamped undriven pendulum

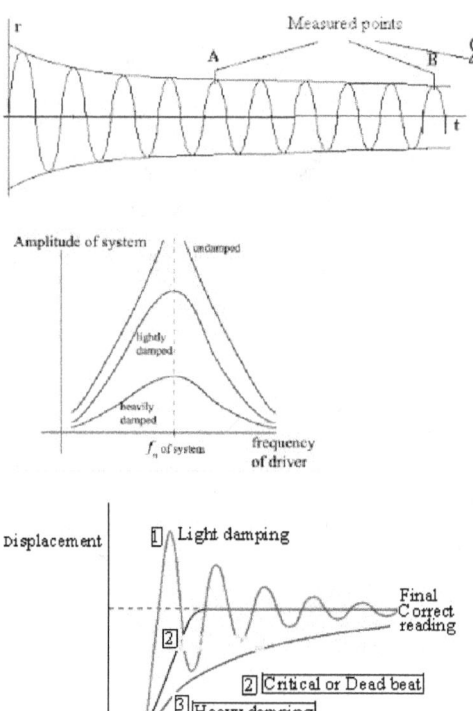

The condition for what is known as <u>critical damping</u> is that for more than critical damping the system ceases to oscillate.

10.7.3 **Car and suspension in it:**

The suspension is the link between the wheels and axles of a car and the body the passengers, and consists of a spring, which is damped by a shock absorber. A good suspension is one inch the damping is slightly under-critical damping for comfortable rides.

A **Skier's body** will have his thighs and calves act like damped spring.

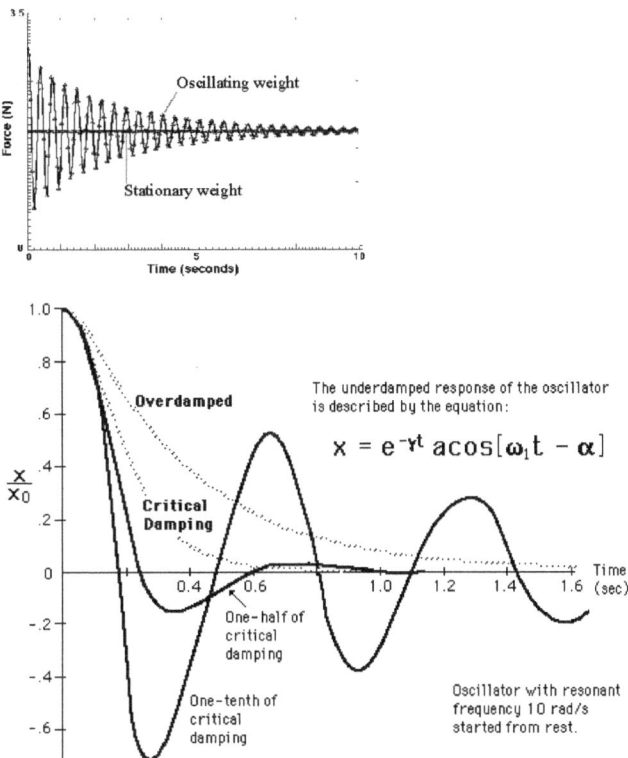

The underdamped response of the oscillator is described by the equation:

$$x = e^{-\gamma t} a \cos[\omega_1 t - \alpha]$$

Oscillator with resonant frequency 10 rad/s started from rest.

10.8. STATIC EQUILIBRIUM OF A RIGID BODY

10.8.1 Parallel Law of Forces:
10.8.3 Triangular Law of forces:
10.8.3 LAMI'S THEOREM
    If three forces P, Q, R keep a particle in equilibrium, then each force is proportional to the sin (angle α between the other two forces).

$$P / \sin \alpha = Q / \sin \beta = R / \sin \gamma$$

10.8.5   <u>TANGENT LAW:</u>

Moment = Force x Perpendicular Distance

$$M = \quad F \quad x \quad d$$

$$(Nm) = (N) \, x \, (m)$$

If a weight W suspended by a light inextensible string from a fixed point is drawn aside by horizontal force H, such that the string is inclined at an angle θ with the vertical, then

$$(W / H) = \tan \theta$$

10.8.6   *Parallel Forces and Moments*

$$P. \, AC = Q. \, BC$$

The resultant divides AB in the inverse ratio

10.8.7.   **VARIGNON'S THEOREM of Moments**

The algebraic sum of two forces about any point in their plane is equal to the moment of their resultant about that point.

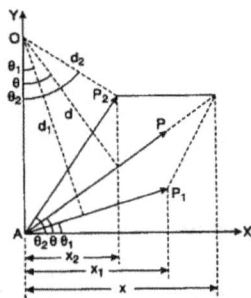

$$Pd = P_1 d_1 + P_2 d_2$$

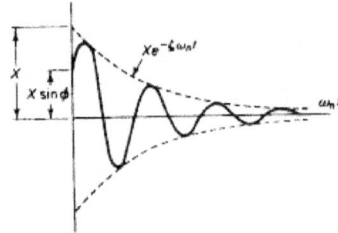

It states that the algebraic sum of the moments of two forces about any point in their plane is equal to the moment of their resultant about point. OR "The moment of a force about an axis is equal to the sum of the moments of its components about the same axis."

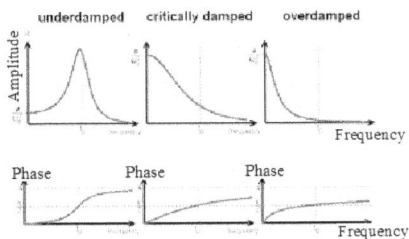

Damped Oscillations

(1) Equation: $\dfrac{d^2x}{dt^2} + 2\kappa\dfrac{dx}{dt} + \omega_0^{\,2}x = 0,$

ω_0 = undamped angular frequency,

κ = resistance coefficient.

(2) If $\kappa < \omega_0$, $\omega_d = \sqrt{\omega_0^{\,2} - \kappa^2}$

$$x = x_0\, e^{-i\kappa[Cos(\omega_d t) + \frac{\kappa}{\omega_d}Sin(\omega_d t)]}$$

(3) If $\kappa = \omega_0$, $x = x_0\,(1 + \omega_0\,t)\,e^{-(\omega_0 t)}$

(4) If $\kappa > \omega_0$, $\omega_d = \sqrt{\kappa^2 - \omega_0^{\,2}}$

$$x = x_0\, \frac{(\omega_d + \kappa)\, e^{(\omega_d - \kappa)\,t} + (\omega_d - \kappa)\, e^{-(\omega_d + \kappa)\,t}}{2\omega_d}$$

+*^+*^+*^+*%+*^

Chapter 11

STATICS - 5 –
SIMPLE MACHINES

"We must become the change we want to see"   Mahatma MK Gandhi

11.   **SIMPLE MACHINES**

11.1.   Input and output forces
A machine is basically a device for increasing force. Its output force is greater than its input force. For a small input force to produce a large output force, the principle of conservation of energy requires that the input force, $\vec{F}_{input}$, must move a greater distance than the output force , $\vec{F}_{output}$.

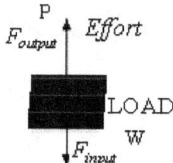

There are six basic simple machines..

They are Inclined plane, Lever, Wheel and axle, Pulley, and Screw.

11.1.1   **Mechanical Advantage (MA)** of a machine
Definition The ratio of its output force (F_{output}) (Load) to its input force (F_{input}) (Effort), when the machine is in *equilibrium*.

$$\boxed{Load \ \propto \ Effort}$$

$$\boxed{(M.A.) = \frac{\vec{F}_{output}}{\vec{F}_{input}} > 1 \qquad Unit \ none \quad (Scalar) \quad (M^0L^0T^0)}$$

11.1.2   Distance Ratio (DR) (or Velocity Ratio)

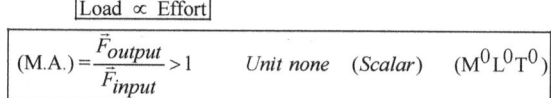
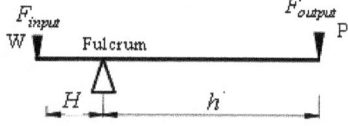

$d_{input} = H$, distance moved by the input force

d_{output} h, distance moved by the output force, in the same time,

$$(D.R.) = \frac{d_{input}}{d_{output}} > 1 \qquad Unit\ none \qquad (Scalar) \qquad (M^0 L^0 T^0)$$

11.1.3 Efficiency (η)

$$\eta = \frac{\vec{F}_{output}\, d_{output}}{\vec{F}_{input}\, d_{input}} = \frac{(M.A.)}{(D.R.)} < 1,\ always$$

For all machines $\eta < 100\%$ because some of the input energy is used to overcome frictional forces within the machine itself..

11.1.4 Theoretical Mechanical Advantage (TMA)

$$T.M.A. = \frac{W}{p} = \frac{2\pi\, r}{p}$$

r = length of handle,

p = pitch, or distance between corresponding points on successive threads..

TMA s the ratio of nth displacement H of the point of application of rth power to that of the load h.

$$V.R. = \frac{H}{h}$$

11.2.1 WHEEL & AXLE

In the wheel and axle, a small mass m just lifts a larger mass M. If the radii of the wheel and axle are R and r, the Mechanical Advantage (MA) and Distance Ratio are given by

$$M.A. = \frac{M}{m}\ ,$$

$$V.R = \frac{R}{r}$$

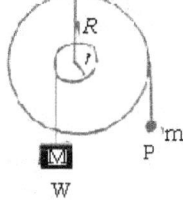

W

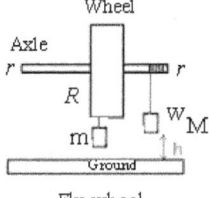

Fly wheel

11.2.2 INCLINED PLANE

A large mass M is placed at the bottom of the plane of angle θ and another mass m is connected to it over a pulley. If the mass m is just sufficient to pull the mass M up the plane at a slow, constant speed,

$$M.A. = \frac{M}{m}\ ,$$

$$D.R = \frac{L}{h} = \frac{1}{Sin\theta}\ .$$

114

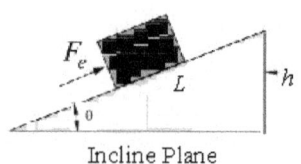

Incline Plane

11.2.3. P applied parallel to Plane

$$M.A. = \frac{1}{Sin\theta}$$

11.2.4 P applied horizontal

$$M.A. = Sin\theta$$

11.2.5 **SCREW AS AN INCLINED PLANE:**

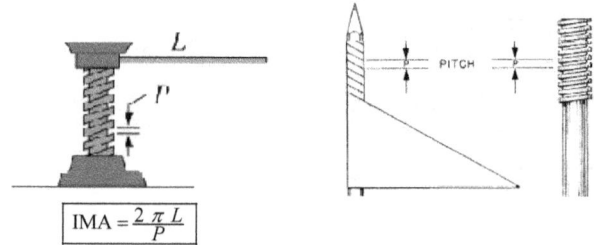

$$IMA = \frac{2\pi L}{P}$$

11.2.6 Wedge

L = Depth of Penetration, r = Separation of wedge surfaces

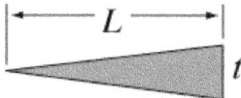

$$M.A. = \frac{L}{r}$$

11.3 **PULLEYS:**

Pulley systems are used to provide us with a mechanical advantage, where the amount of input effort is multiplied to exert greater forces on a load.

To lift heavy loads using forces much smaller than the weight of the load, pulley systems are used. Generally a force applied to a rope supplies the input work. This force usually acts in a direction parallel to the motion of the load.

Step #1   *First* calculate the theoretical mechanical advantage by doing a force analysis on the system.

Step #2 Measure the displacement ratio.

Step #3 Compute the efficiency of the system.

11.3.1 Single Pulley, FIXED PULLEY

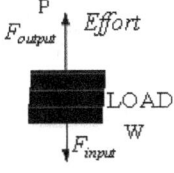

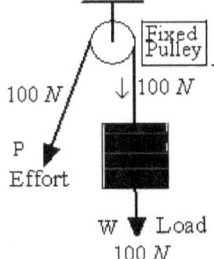

With no pulley - the effort force is similar to the load - in opposite direction.

I M.A. = 1

11.3.1.1 Lewis Carroll's Monkey

Consider an equilibrium realized when a "perfect" rope is passed over a *frictionless and massless* pulley with a weight M on one side and a monkey m on the other...If $M = m$, what happens when the monkey decides to climb up the rope?

Thus, if the monkey and the weight are initially motionless at the same height, they will always face each other no matter what the monkey does.

11.3.1.2. Elevator

W Empty elevator weight, W - Counter weight in the elevator system

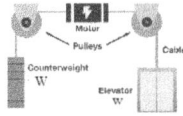

The motor has the power to pull with equal strength on either direction (Fig).

Ability of the motor to pull extra weight without its damage = $(W - w)$.

Being symmetric, the safe limit of the elevator to carry extra weight $= W + (W - w)]$. This means at least $[W + (W - w)] - w$ can be added the lift.

11.3.2. Movable Pulley:

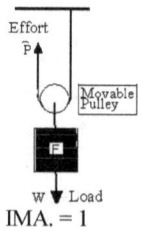

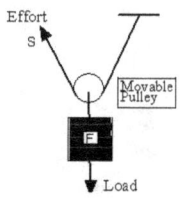

IMA. = 1

11.3.2.1 First System (**Combined Pulleys**)

If n = Number of movable pulleys

$$2^n \ P = W \left(2^n - 1 \right)$$

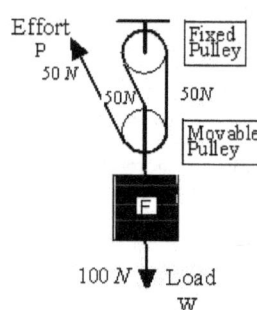

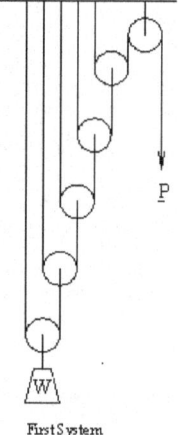

First System

11.3.2.2   FOUR different pulley systems with DRs of 1, 2, 3, and 4 are as shown.

- The MA in each case

  $M.A. = \dfrac{M}{m}$.

- The DR equals to the number of strings supporting the lower, movable, pulley block.

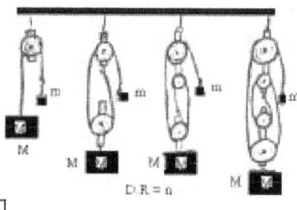

$$\boxed{D.R = n}$$

11.3.2.3 Second System (Block & Tackle or Pulley Block)

N = Total number of pulleys in the System.

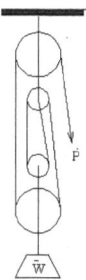

Second System
Block & Tackle

$$n\,P = w + W$$
$$M.A. = n - \frac{M}{m}$$
$$V.R. = \frac{H}{h} = n$$

11.3.2.4 <u>Third System</u>:

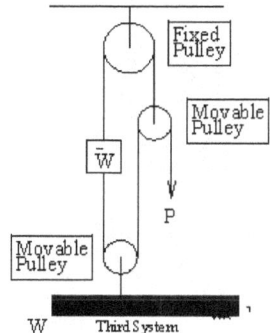

Fixed Pulley

Movable Pulley

$\bar{w}$

P

Movable Pulley

W   Third System

$$M.A. = \frac{W}{P} = [2^{n+1} - 1]$$

11.3.3 **DIFFERENTIAL PULLEY**

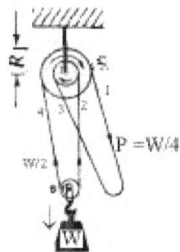

$P = W/4$

$W/2$

W

$$M.A. = \frac{W}{P} = \eta \; \frac{2R}{R-r}$$

11.4 LEVERS

A lever is one of the great number of machines for doing work.

11.4.1 The Law of Equilibrium

The law of equilibrium is: The effort (E) multiplied by its distance from the fulcrum (F) equals the load (or Resistance, R) multiplied by its distance from the fulcrum, and is true for all classes of levers.

(Effort, P) (Effort Arm, EA) = (Load, W) (Load Arm, RA)

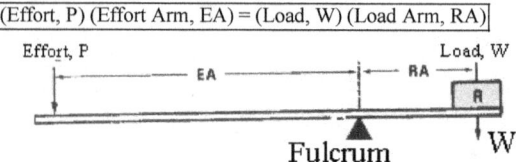

11.4..2 Class 1 F between P & W

$MA > 1, MA < 1; \; MA = 1$

Examples: Crow bar, Scissors, Water pump, Pliers, The arm and leg locks used in wrestling.

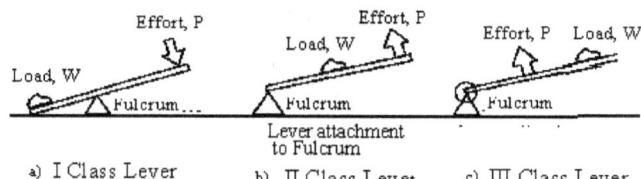

a) I Class Lever   b) II Class Lever   c) III Class Lever

11.4.3 **Class 2** W between F & P

$MA > 1$, always ;

Examples: Nut cracker, Wheel barrow, punch, Bottle lid opener, a runner on tiptoe in his foot,

11.4.4 **Class 3** P between F & W

$MA < 1$, always

Examples: Forceps, Fire tongs, the Jaw, the Forearm, the Fingers of the hand.

11.4.5 **SCREW JACK:**

h = pitch of screw,

ℓ = length of handle

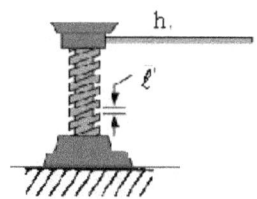

$$\eta = \frac{W}{P}\frac{h}{2\pi\ell}$$

$$\eta = \frac{W}{P}\frac{h}{2\ell}$$

$$M.A = \eta\,\frac{2\ell}{h}$$

$$V.R. = \frac{2\pi\ell}{h}$$

$$V.R.(Screw) = \frac{2\pi(\text{Radius of screw})}{\text{Pitch}}$$

$$V.R.(Screw) = \frac{2\pi\ell}{h}$$

11.4.6 GEAR WHEELS

$N_{\text{Driv Gear}}$ = Number of teeth on the driving gear (crank wheel, smaller size and attached to the wheel of vehicle)

$N_{\text{Free Gear}}$ = Number of teeth on the driven (free) gear (of bigger size).

$$V.R. = \frac{N_{\text{Driv Gear}}}{N_{\text{Free Gear}}}$$

The VR of a compound gear train is calculated by multiplying the velocity ratios for all pairs of meshing gears.

R = radius of the free driven wheel

ℓ = length of the pedal of the driving wheel

Number of revolutions made by the free wheel

$$= \frac{N_a}{N_b} < 1$$

$$V.R. = \frac{N_{\text{Driv Gear}}}{N_{\text{Free Gear}}} = \frac{\ell}{R}$$

11.5   SIMPLE BALANCE

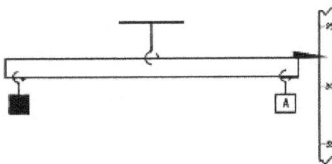

11.5.1 Common Balance

THREE requisites of a <u>good balance</u> are:

i)   Truth,
ii)   sensibility and
iii)   stability.

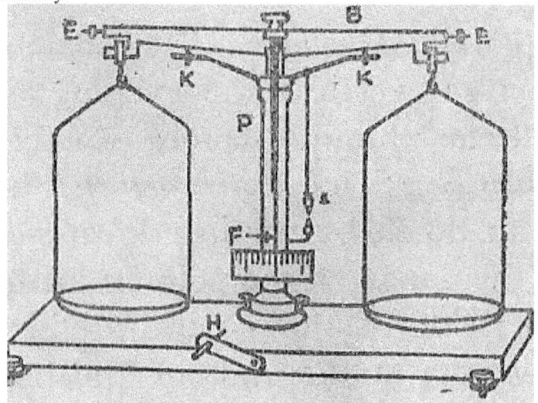

i) Truth: Both arms of equal length, both pans of equal weight, the Centre of gravity must pass through the fulcrum, and perpendicular beam.

a = length of each arm,.

$W = Weight\ of\ the\ beam,$

ii) Sensibility:

a)a should be large,

b) w should be small,

c)H and k (which determines θ) should be small.

$$\text{Sensibility} = \frac{\theta}{W}$$

θ = Angle through which the beam turns due to weight w.

a = Length of each arm;

W = Weight of beam

θ = Angle through which the beam turns due to a mass difference of $1\ m\ gm$ between the weights placed in the pans,

iii) Stability:

The beam should return to equilibrium, quickly as the beam is disturbed.

-o-0-o-0-o-0-o-0-o-

Chapter 12

FLUID MECHANICS
HYDROSTATICS, VISCOSITY, TURBULANT FLOW, FLOATATION, SURFACE TENSION, CAPILLARITY

"An equation for me has no meaning unless it expresses a thought of God"   Srinivasa Ramanujan

12.1.1 FOUR STATES OF MATTER:
Solid, Liquid, Gas, and Plasma (Fifth state **Super solid** by Bose-Einstein condensate is reported, 2017).

12.1.2.1 Volume, V

V	Unit m^3	(Scalar)	$(M^0 \, L^3 \, T^0)$

12.1.3 Density and Specific gravity:
For a matter with mass m and volume V, density ρ

$\rho = \frac{M}{V}$	Unit kgm^{-3}	(Scalar)	$(M^1 \, L^{-3} \, T^0)$

12.2 HYDROSTATICS

12.2.1 STATIC FLUIDS
(i) A static fluid can have **no shearing force** acting on it, and that
(ii) Any force between the fluid and the boundary must be acting at right angles to the boundary

12.2.2 PRESSURE $\vec{p}$

12.2.2.1 $\vec{p}$ at a point

Unit , Nm^{-2}	(vector)	$(M^1 L^{-1} T^{-2})$

$$\vec{p} = \lim_{A \to 0} \frac{\delta F}{\delta A} = \frac{dF}{dA} = \frac{Force}{Area}$$

$1 \, psi = 51.714 \, mm \, \text{Hg} = 2.0359 \, in \, \text{Hg}$
$= 27.680 \, in.\text{H}_2\text{O} = 6.8946 \, kPa$

$$1 \; bar = 14.504 \; psi$$

$$1 \; atm = 14.696 \; psi$$

12.2.2.2 PRESSURE ($\vec{p}$) acts normal to the surface.

$\hat{n}$ = normal vector

A = area

$\vec{F}$ = Forces on a body = (the vector sum of the pressure) (area around the entire body)

$$\vec{F} = \sum \vec{p}\hat{n}A = \oiint \vec{p}\hat{n} \; dA$$

12.2.3 THRUST

| Thrust = (Pressure) • (Area) | Unit   ,N | (vector) | $(M^1L^1T^2)$ |

A small element of fluid, in the form of a *triangular prism*, containing a point $\vec{P}$, the three pressures $\vec{P}_x$ in the x direction, $\vec{P}_y$ in the y direction and $\vec{P}_z$ in the direction normal to the sloping face, are

$$\vec{P}_x = \vec{P}_y = \vec{P}_z$$

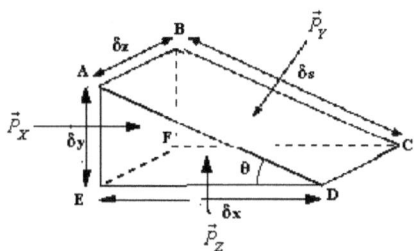

For a triangular prismatic element of fluid, when a fluid is at rest, there are
No shearing forces,
Acts perpendicular to surface ABCD,
Acts perpendicular to surface ABFE and
Acts perpendicular to surface CDEF'

12.2.4 PRESSURE AT A DEPTH IN LIQUIDS

Hydrostatic Pressure as a function of depth (vertical height is known as **head** of fluid), h in a static fluid. a = area of cross section of the vessel.

P.E. = $(a\rho)gh$

Gauge pressure, $\boxed{\vec{P}_{Gauge} = h\rho g}$

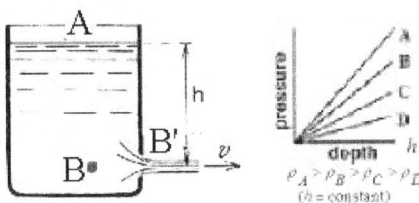

12.2.4.1 Liquid column and jet stream

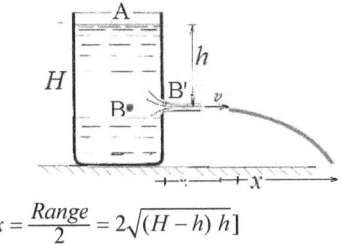

$$x = \frac{Range}{2} = 2\sqrt{(H-h)\,h}]$$

12.3.1 Incompressible Fluid:

12.3.1.1 When the density of a fluid ρ = constant, remains (the same everywhere), it is called *incompressible* fluid.

12.3.1.2 Static pressure in an incompressible fluid:
Evangelista Torricelli invented the mercury barometer, the Vacuum above the top oF the mercury surface in the tube Torricelli vacuum.

$$1\ torr = \frac{1}{\bar{P}_{Atm}} \approx 133.3\ Pa$$

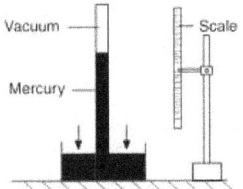

At sea level the barometric pressure is $14\ psi$.
This pressure is capable of supporting a 34 *foot* column of water

Atmospheric pressure, $P_{atm} = \rho_{Hg}\ g\ h$

Absolute pressure, $P_{absolute} = h\,\rho\,g + P_{atm}$

Static pressure at a certain point in a liquid does not depend on shape, total area, or surface area of the liquid.

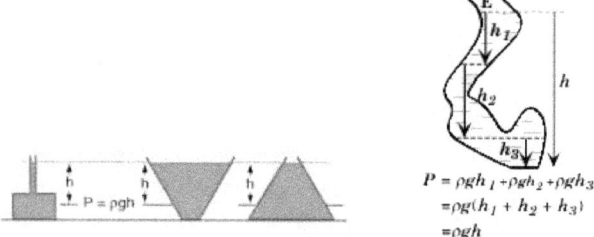

$P = \rho g h_1 + \rho g h_2 + \rho g h_3$
$= \rho g (h_1 + h_2 + h_3)$
$= \rho g h$

12.3.2 PASCAL'S LAW:

Definition: An increase in pressure P at any point exerted on a confined fluid is transmitted throughout the fluid increasing the pressure at every point in the fluid by the same amount P.

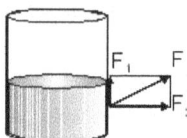

Pascal's Principle:

(i)   is used in hydraulic systems such as jacks and automobile brakes and
(ii)   is the fluidic equivalent of the principle of the lever, which allows large forces to be generated easily by trading large movement of a small piston for small movement of a large piston

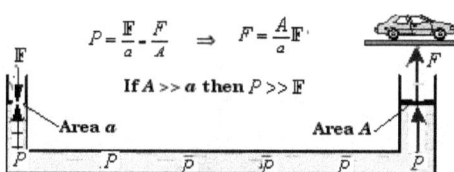

$$P = \frac{\mathbb{F}}{a} - \frac{F}{A} \quad \Rightarrow \quad F = \frac{A}{a}\mathbb{F}$$

If $A \gg a$ then $P \gg \mathbb{F}$

Area a Area A

12.3.3 **Compressible Fluids**:

Static pressure in a Compressible fluid (in which ρ varies)

P_h = static pressure at any point

$$\boxed{P_h = P_0 \, e^{-\{\rho_0 g / P_0\}h}}$$

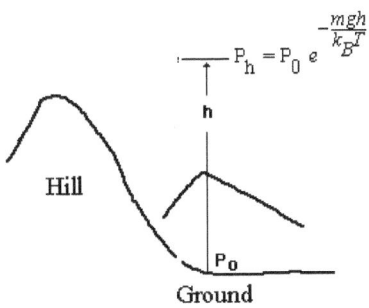

$$P_h = P_0 \, e^{-\frac{mgh}{k_B T}}$$

h

Hill

P_0

Ground

n = number of moles
N_A = Avogadro's number
m = mass of one molecule

$$\frac{R}{N_A} = k_B$$

In the case of troposphere, the pressure at an altitude point z

$$\rho = \frac{\text{Mass}}{\text{Volume}} = \frac{n \, N_A \, m}{n \, R \, T / P} \, ;$$

$$\frac{R}{N_A} = k_B \, ;$$

where
$\quad$ n = Number of moles
$\quad$ N_A = Avagadro's number
$\quad$ m = Mass of one molecule

$$\boxed{P_z = P_0 \, e^{-\{(z-z_0)g/RT\}}}$$

or
$$\boxed{P_z = P_0 \, e^{-\{mgh/k_B T\}}}$$

i.e., P_z decreases exponentially with altitude, h.

12.4 MEASUREMENT OF PRESSURES

12.4.1 BAROMETERS:

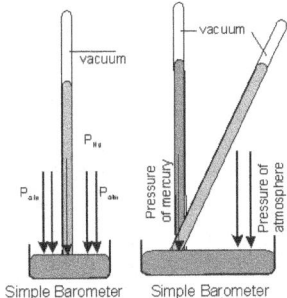

Simple Barometer $\qquad$ Simple Barometer

1) Bourdon-Tube gauge
2) Manometers
(i) Barometer
(ii) Piezometer
(iii) U-tube or Differential manometer
To measure the pressure exerted by the atmosphere

12.4.1.1 Barometric Formula

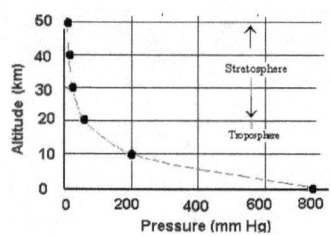

$$P_h = P_0 \, e^{-\frac{mgh}{k_B T}}$$

12.4.2 Aneroid Pressure Gauge

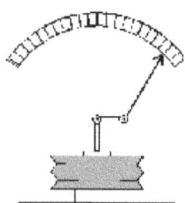

12.4.3 U-Tube Manometer

The pressure of a gas inside a vessel can be measure with a manometer, *i.e.*, used on

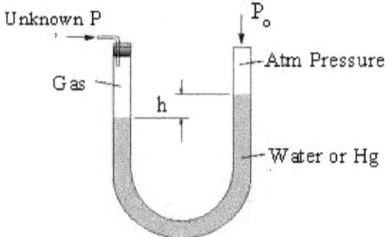

gauge, differential, and absolute sensors with a suitable reference. The difference between mercury column heights gives the pressure reading.

Gauge Pressure, $\boxed{\Delta P = (P - P_o) = h\rho g}$

<u>Water barometer</u> is impractical as h = 34 ft.

12.5.1 FLUID MOTION:

1) **Friction** leads to a need for <u>Poiseuille's law.</u>
2) Friction causes **viscosity**.
3) The motion of a fluid is said to be ***stationary*** when the motion pattern does not change with time. The path chosen by each fluid element in a stationary flow is called ***stream line***.(or laminar).

12.5.2 Laminar or Streamline Flow.

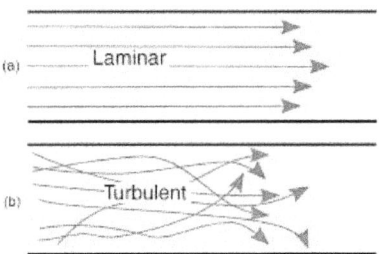

12.6. VISCOSITY, η

12.6.1 When a body moves at a relatively low velocity v through a fluid, the force of friction,

$$\boxed{\text{Dragging force F} = -\,\text{K}\ \eta\ \text{v} = \text{A}\ \eta\ \frac{dv}{dx}}$$

where η = coefficient of viscosity of the fluid

$$\boxed{\eta = \frac{\text{tangential Force}}{\text{Velocity gradient}}}$$

| Unit | ,Poise(P) or Nsm^{-2} | (scalar) | $(M^1\ L^{-1}\ T^{-1})$ |

= similar to shear modulus of a solid.

12.6.1.1 If a fluid has $\eta = 0$, it will flow through a level tube or pipe without an applied force.

12.6.1.2 Viscous forces dissipate energy.

12.6.1.3 Flow of a Liquid through a Tube
i) Expression for Capillary Flow
ii) Physics of Blood flow:

12.6.2 POISSEUILLE'S FORMULA (Laminar flow):

12.6.2.1 <u>Assumptions</u>:
1) The flow is steady and stream lined;
2) There is no radial flow,
3) The liquid in contact with the sides of the tube is at rest.
ℓ = length of capillary,
a = radius of capillary,
p = Difference in hydrostatic pressure,
Q = Volume of the liquid flowing through the whole capillary in time t.

12.6.2.2 For steady (laminar) flow:

$$Q_t = \pi \frac{(p_1 - p_2)\, a^4\, t}{8\, \ell \eta}$$

If for a liquid, $\eta = 0$, then it will flow through a level tube, without an applied force.

12.6.3 **STOKE'S FORMULA** for highly viscous liquid:

Viscous drag, $\boxed{V_f = 6\,\pi\,\eta\,a\,v}$

Effective gravitational force, $\boxed{F_{gravity} = \frac{4}{3}\pi\,a^3(\rho - \sigma)g}$

Determine $v = \frac{s}{t}t$, experimentally. (Millikan's oil drop experiment)

12.6.4 **Terminal (Limiting) velocity**:
At this velocity, Frictional drag due to viscous forces = Gravitational force.

Terminal velocity, $v = \boxed{v_{term} = \dfrac{2\,g\,a^2(\rho - \sigma)}{9\eta}}$

12.7 **TURBULENT FLOW** *and REYNOLDS NUMBER, R_e*

12.7.1 This is *chaotic fluid flow*. All vortices and swirls.
Turbulent flow is noisy and rough and is less efficient than Laminar flow

12.7.2 The Blood Circulatory System:
To measure blood pressure use a sphygmomanometer. Measurement made at the same level as the heart. Pressure is recorded as <u>systolic</u> and <u>or / diastolic</u> (120/80) In humans, blood pressure has 2 principal origins.
(1) The action of the heart and lungs.
(2) The gravitational factor. - $\rho g h$

12.7.3 The Reynolds's number, R_e .

This is ratio of the inertia forces and viscous forces in a fluid flow in a boundary layer.

Characterizes the onset of turbulence flow from laminar; is one of the most commonly used dimensionless quantity in fluid dynamics.

$$R_e = 2 v r \frac{\rho}{\eta}.$$

Unit	(A dimensionless number)	$(M^0 \ L^0 \ T^0)$

$R_e < 2000$, if the flow is laminar.

$$\boxed{R_e > 2200, \text{ if the flow is turbulent}}$$

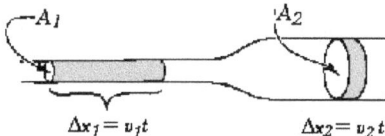

$$\Delta x_1 = v_1 t \qquad\qquad \Delta x_2 = v_2 t$$

Always at a single level, the <u>arterial pressure</u> is higher than the <u>venous pressure</u>. This serves to force the blood through the capillaries. In blood stream, as in all fluid types, the pressure is transmitted undiminished throughout the fluid - Pascal's Principle; so even the smallest capillaries feel the pressure.

12.7.4   Blood pressure at a man's foot

$$\boxed{P + P_{ground} = \rho \, g \, h},$$

the height of the hip above the foot, where C.G is located).

The pressure in the head (is lower than at the heart.) $= P - P_{ground}$.

12.8   EQUATION OF CONTINUITY

An important principle in fluid motion, expresses conservation of fluid (compressible) mass.

$$\boxed{\rho_1 \ A_1 \ v_1 = \rho_2 \ A_2 \ v_2}.$$

For incompressible fluid, $A_1 \ v_1 = A_2 \ v_2$.

12.8.1   BERNOULLI FLOW THEOREM:

The sum of the energies possessed by a flowing liquid at any point is constant, if the flow is steady and non-turbulent, the equation fundamental to much of hydrodynamics among p, kE and pE of fluid.

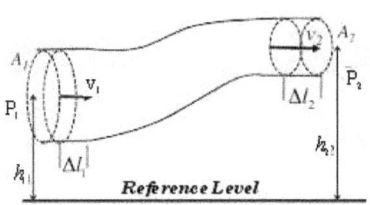

$$\frac{1}{2}\rho_1 \, v_1^2 + \rho_1 \, g \, h_1 + p_1 = \frac{1}{2}\rho_2 \, v_2^2 + \rho_2 \, g \, h_2 + p_2$$

$$\frac{1}{2}\rho \, v^2 + \rho \, g \, h + p = \text{a constant, for unit mass}$$

1st term = kE / unit volume;
2nd term = gravitational pE/ unit volume;
3rd term = pressure energy / unit volume.
v = velocity; p = fluid's pressure, h = change in height, g = acceleration due to gravity, ρ = density.

12.8.1.1   A fluid at rest or moving with constant velocity in a pipe.

$$\rho \, g \, h + p = p_o, \text{a constant}$$.

12.8.1.2. Pressure in an incompressible fluid in equilibrium is given by

$$p = p_o - \rho \, g \, h$$

Examples.   :(i) Pressure at the surface to the bottom of a lake increases.
                (ii) Atmospheric pressure decreases linearly with altitude. up to 10 km.

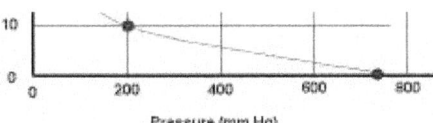

Pressure (mm Hg)

12.8.2   APPLICATIONS of Bernoulli's theorem:

12.8.2.1 Perfume Atomizer (Spray)

12.8.2.2 Airlift on an airplane wing:

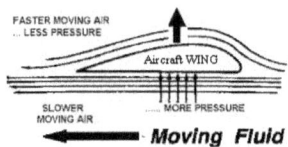

If the fluid is moving only horizontally,

$$\boxed{\tfrac{1}{2}\rho\, v^2 + p = \text{a constant,}}\; ..$$

In a horizontal pipe, the greater the velocity of flow, the lower is the pressure, and the vice versa. This produces the **lift** (a force generated by changing either speed or direction of a moving fluid) of an airplane due to design of the wing.

Lift ≡ upward force $\boxed{F_{\uparrow} = A\,\rho\, v\,(v_1 - v_2)}$.

Drag ≡ stream force = F_{Drag}.

12.8.2.3 During a storm passing through a house, closed window open.

12.8.2.4 Spinning Ball
A ball thrown with spin has two types of motion combined: 1) Translational and 2) Rotational. This causes the ball deflected up.

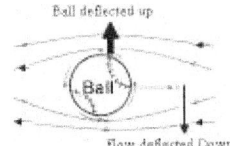

12.8.2.5 Siphon and Syringe

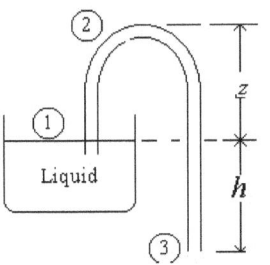
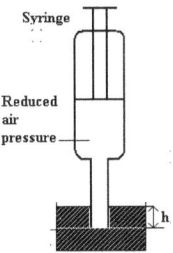

A siphon is a system of pipe or tubing with the fluid inlet

$\boxed{\text{P.E. of the fluid at the inlet of the siphon)} \gg \text{(P.E. at the outlet}}$

This difference in energy drives the fluid through the system.
How high can the siphon piping go above the inlet depends on
        (i)   the type of fluid ? and
        (ii)   the local barometric pressure Bernoulli equation for frictionless flow is

$$\frac{P}{\rho g} + \frac{v^2}{2g} + h = \text{constant}$$

Discharge Q = (Cross section area) (Velocity)

$$\boxed{z_{\text{liquid}}\,\rho_{\text{liquid}} = (76\ cm)\,(13.6\ g/cc)\,\rho_{\text{water}}}$$

This puts a limitation to the value of z of the empty tube above liquid level at point 2; h is the tip of the open end at point 3 of the tube below the liquid level, at point 1

12.8.2.6 VENTURI METER:

Principle: Lowering of the pressure due to the speed of the fluid flow.

It determines the velocity of a fluid in a pipe, flowing per unit time.

At the throat, the area is reduced from A_1 to A_2 and the velocity is increased from $(v=v_1)$ to $(V=v_2)$. Note that at the throat where the velocity is the greatest, the pressure is least. Bernoulli's equation gives:

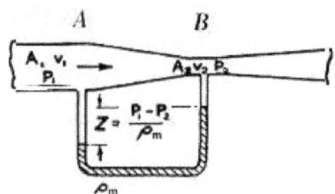

$$p_1 + \tfrac{1}{2}\rho v^2 + \rho g \tfrac{1}{2}h_1 = p_2 + \tfrac{1}{2}\rho v^2 + \rho g\, h_2.$$

satisfies the conservation of energy involving a fluid.

$Av = $ constant

$$v_1 = A_1 \sqrt{[\frac{2A_2{}^2(p_1-p_2)}{\rho A_2{}^2 - A_1{}^2}]}$$

Amount of fluid $= V = A_1 v_1 = K\, A_1 \sqrt{p_1 - p_2}$

$K = $ a quantity depends on the pipe and the density ρ_{Fluid},

A – Area of sections of pipe,

$v = $ velocity of fluid,

$$A_1 v_1 = A_2 v_2$$

If pipe is horizontal, $\qquad F = A_1 \rho v_1\, (v_1 - v_2)$

12.8.2.7 HOW DOES A BALLOON STAY AIRBORNE?

(i) A body in air

Upthrust, $U > $ Upthrust, $U_\Uparrow = $ Weight of Air replaced $\Downarrow$

(ii) For a balloon (of neoprene) filled with helium

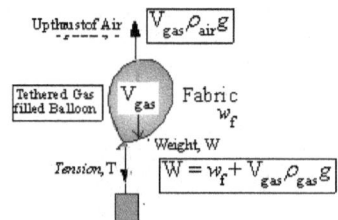

Upthrust, $U_\Uparrow$ > [Weight of Balloon (Neoprene)$_\Downarrow$ + Weight of gas (He)$_\Downarrow$]S the balloon rises in air and needs to be tethered by means of a string, which provides the necessary additional downward force $\downarrow$ to maintain equilibrium.

A balloon of volume V (and fabric of weight w) filled with a gas of density ρ_{gas} is anchored to the ground to prevent rising,

$$V = V_{balloon} = V_{gas}$$

Up thrust of surrounding air	=	Weight of balloon rest	+	Extra downward force supplied by the tension in the moving rope

Maximum weight of contents = $\boxed{V \rho_{Air} g - [w + V \rho_{gas}\ g]}$.

12.8.3. <u>Velocity of a Surface wave</u> of wavelength λ in an Incompressible inviscid Fluid
At depth h

$$c = \sqrt{gh}, \qquad (\lambda \gg h)$$

$$c = \sqrt{\frac{g\lambda}{2\pi}}, \qquad (h = \text{large})$$

$$c = \sqrt{\frac{2\pi\,T}{\lambda\rho} + \frac{g\lambda}{2\pi}}, \ (\lambda = \text{small})$$

T = surface tension

12.9. FLOATATION

12.9.1 ARCHIMEDES'S PRINCIPLE:
<u>Statement</u>: According to Archimedes's principle *"A body wholly or partially submerged in a liquid is buoyed up by a force equal to the weight of the liquid displaced."*

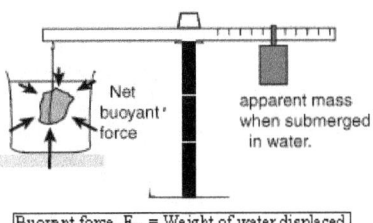

Buoyant force, F_B = Weight of water displaced

Buoyant force, F_B = weight of the liquid (density, ρ) displaced

$$= \rho\,g\,V$$

(Mass of body) – (Apparent mass when submerged) = $\boldsymbol{\rho}$ V
Apparent weight of a sphere of radius r and density ρ

$$W_{apparent} = \frac{4}{3}\pi\,r^3\,(\rho - \sigma)$$

12.9.2.1 **Centre of Gravity** (C. G.)

CG is that fixed point through which the line of action of the weight always passes for all the positions of the body.

12.9.2.2 Centre of Mass (CM)

CM of a body is the point through which any applied force produces translation of the body (not rotation).

Resultant Thrust

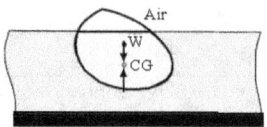

12.9.2.3 *Centre of Pressure* (CP)

of a rectangular lamina = (2/3) (length of vertical side) from the exposed side.

CP of triangular Lamina = (3/4)(altitude) on the medium from the surface

Centre of Buoyancy (CB), H, is the CG of the displaced liquid.

12.9.2.4 META CENTRE (M)

If a floating body is slightly disturbed whereby the vertical line through which the new CB (H) meets the line joining the C.G. of the body to the original CB, then the distance between M and the CG is called the *Meta-centric Height* (GM)

12.9.3.1 Ship Stability (EQUILIBTUM OF A BODY)

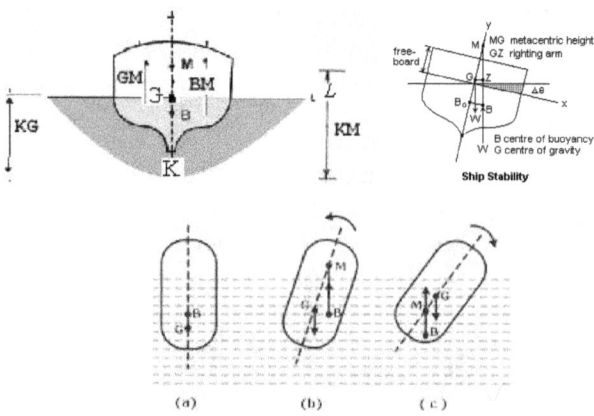

(i) **Stable:** M should be above CG. When resting on a surface the vertical line passing through the object's C.G. must also pass through the base of the object

(ii) **Unstable**: M should be below the CG, When resting on a surface the vertical line passing through the object's C.G. does not pass through the base of the object

(iii) **Neutral**: M and CG should coincide with a common point,

$$HM = \frac{AK^2}{V}$$

AK^2 = Moment of inertia of the plane of floatation about the axis of rotation,
V = volume of liquid displaced.

12.9.3.2 META CENTRIC HEIGHT OF A SHIP

$$GM = \frac{m\,a\,\ell}{M\,x}$$

M = Mass of the ship
m = weight of a a small mass placed at one end of the ship
ℓ = length of plumb line, making an angle θ with the line passing through Centre of Mass and CG.

$$x = \ell\theta$$

A = width of deck

12.9.4 HYDROMETER
12.9.4.1 Principle:
Law of hydrostatic pressure, *viz.*

$$\boxed{\Delta p = P_B - P_A = h\rho g}$$

Energy, $\quad E_A = g\,(h + x) + \dfrac{P_A}{\rho}$

$\qquad\qquad E_B = g\,x + \dfrac{P_B}{\rho}$

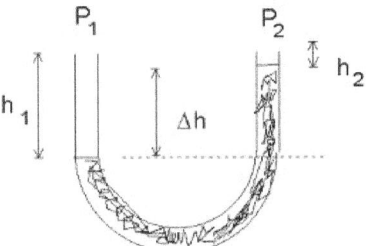

12.9.4..2 Nicholson's hydrometer
is based on the principle of constant immersion but of variable weight.

H = height of hydrometer immersed in water
w = Extra weight for the meter to sink up to a fixed mark In water

L = height extra in liquid to sink up to a fixed mark

At room temp,

$$\rho_{liq} = \frac{(H + L)}{(H + w)}$$

At temp, 4 °C,

$$\rho_{liq}(t = 4°C) = \frac{(H + L)}{(H + w)} \rho_{water}(t = 4°C).$$

12.9.4.3 Hydrostatic pressure due to Liquid column

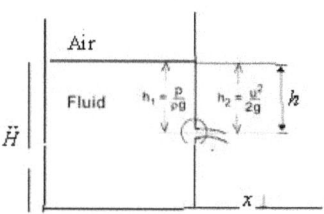

$$x = 2\sqrt{\frac{H - h}{h}}$$

$$\boxed{\begin{array}{l} P_x = P_{atm}(= 14.8\,psi) \\ \quad + 0.434 \text{ (wt. in } lbs \text{ of water } 1ft \text{ high and } 1in^2) \, D \, (depth \text{ in } ft) \end{array}}$$

12.9.5. PUMPs

    V = capacity of receiver tank
    v = volume of the barrel of pump
    n = Number of strokes
    P = initial pressure

12.9.5.1 Evacuation (Air) Pump

$$P_n = P\left(\frac{V}{(V + v)}\right)^n$$

12.9.5.2 Compression Pump (Foot ball, Bicycle Pump) (Inflation Pump)

$$P_n = P\left[1 + \frac{n\,v}{V}\right]$$

12.9.5.3 Water (Lift) Pump (Suction Pump)
        This pump can lift water to < 10 *m*.

12.9.5.4 Force (Lift) Pump
        This pump can lift water to > 10 *m*.

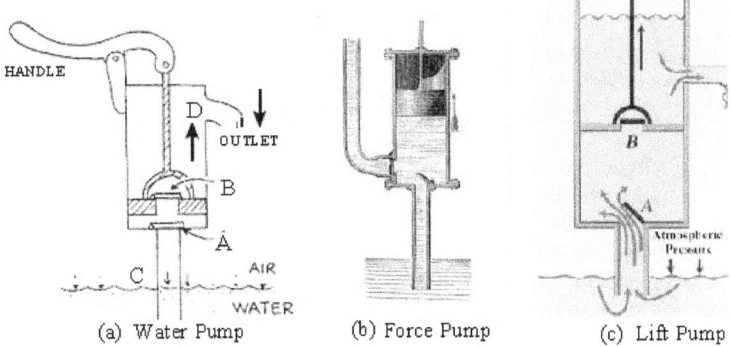

(a) Water Pump   (b) Force Pump   (c) Lift Pump

12.10. SURFACE TENSION: T

12.10.1 Definition: Surface tension is force, per unit length, acting on either side of a line drawn in the liquid surface in equilibrium, the direction of the force being tangential to the surface and perpendicular to the line.

$$T = \frac{\vec{F}}{\ell} \quad \text{Unit} \quad , Nm^{-1} \qquad \text{(vector)} \qquad (M^1 L^0 T^{-2})$$

12.10.2   Liquid assuming spherical surface:
Sphere has the least surface area; and cohesive forces of the liquid.

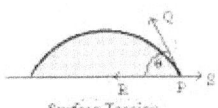

Surface Tension

12.11.1.3   Molecular range:
The maximum distance up to which cohesive force between two molecules can ac
$\rightarrow 10^{-7} cm = 1 nm$

12.10.3   Sphere of influence: Sphere whose radius is the molecular range.

12.10..4   Surface film: The layer of liquid at the surface whose thickness is molecular range.

> P.E. of the molecules lying within the surface film > that below it .

In equilibrium potential energy should be least. Since a sphere has the leas surface area, the surface of a liquid tends to be a sphere.

12.10.5   Angle of contact (θ):
The angle between the tangent to the liquid surface at a point of contact and the solid surface taken through inside liquid.

i) For a clear water-glass surface, $\theta = 0°$;

ii) For water in silver surface, $\theta = 90°$;

$$\text{Cos } \theta = 1 - \left(\tfrac{H}{h}\right)^2$$

or $\quad\quad \text{Sin } \tfrac{1}{2}\theta = \tfrac{1}{\sqrt{2}}\left(\tfrac{H}{h}\right)$

H *cm* = total depth of the liquid drop surface above the glass plate.

h *cm* = height of the drop from the upper surface to the bulged point (most protruding part)

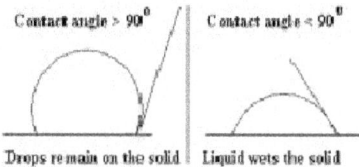

Drops remain on the solid   Liquid wets the solid

12.10.6 QUINCKE'S METHOD:

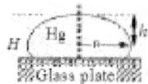

Surface tension, $T = \tfrac{1}{2}h^2 \rho g$ (*dynes / cm*)

Angle of contact, θ is given by $\text{Sin } \tfrac{1}{2}\theta = \tfrac{1}{\sqrt{2}}\left(\tfrac{H}{h}\right)$

Spherometer, vernier microscope, clean glass plate, small table resting on leveling screws, small quantity of clean mercury are required,.

12.10.7 **SURFACE TENSION by Method of Liquid Drops**

Surface tension determination by method of drops:

From the # of liquid drops dropping from the end of a glass tube having internal layer coated with paraffin wax (by dipping the end into molten wax). r = internal radius of the dropping end, = mean weight of a drop.

$$T = \frac{mg}{3.8\,r}$$

T of water provides the necessary wall tension for the formation of liquid drops with water. The tendency to minimize that wall tension pulls the bubbles into spherical shapes. Interfacial surface tension between two liquids of varying density (water ρ_1 and

Kerosene ρ_2) is

$$T = \frac{m\,g\{1-\frac{\rho_2}{\rho_1}\}}{3.8\,r}$$

12.20.7.1 **Table of Surface Tension of Materials**

Liquid	$T/\ N\ m^{-1}\ x\ 10^{-3}.$	
1) Methylated spirit	22.6	
2) Glycerol	63.4	
3) Water	72.7	
4) Benzene	28.9	
5) Mercury	472	
6) Olive oil	32	
7) Ether	17	
8) Gold		1102
9) Cocoanut oil		33
10) Kerosene	30	

12.10.7.2 Jaeger's Method of variation of Surface tension (T) with temperature

ρ_1 =density of Manometer liquid, h_1 = difference in Manometer levels, r = mean radius of the capillary tube

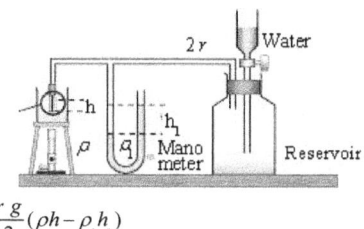

$$T = \frac{r\ g}{2}(\rho h - \rho_1 h_1)$$

12.10.7.3 LaPlace's Law:

For a bubble in equilibrium the relationship

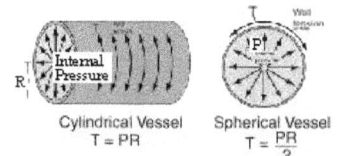

Cylindrical Vessel
T = PR

Spherical Vessel
$T = \frac{PR}{2}$

$$\Delta P = \frac{2\pi}{r}\ \text{holds at equilibrium.}$$

12.10.7.4 Relations between excess pressure p, T and r of a Bubble:

LIQUID BUBBLES:
   i) Synclastic surface

$$p = 2T\left(\frac{1}{r_1} + \frac{1}{r_2}\right)$$

   2) Anti-clastic surface:

$$p = 2T\left(\frac{1}{r_1} - \frac{1}{r_2}\right)$$

12.10.8. SPECIAL CASES:

12.10.8.1 Spherical Soap bubble:

$$r_1 = r_2 = r$$

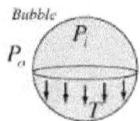

Excess pressure within a bubble

$$P_i - P_o = \frac{4T}{r}, \text{ for a bubble}$$

12.10.8.2 *Spherical drop: (A bubble of air within a liquid)*

$$p = 2T\left(\frac{1}{r_1}\right)$$

12.10.8.3 Cylindrical bubble: $r_2 = \infty, r_1 = r$

$$p = 2T\left(\frac{1}{r_1}\right)$$

12.10.8.4 Cylindrical drop:

$$p = T\left(\frac{1}{r_1}\right)$$

12.10.8.5 Force between two horizontal plates of glass with a small drop of liquid in between gap d:

$$r_1 = r ; \qquad r_2 = \frac{d}{2}; d \ll r$$

$$F = -2 \ T \ [\tfrac{1}{d}] \ (\text{Area } r^2)$$

12.10.8.6   Surface tension gives rise to the tendency of a volume of liquid to minimize it's surface tension.

e.g., Spherical droplets

$$\frac{\text{Cube Area}}{\text{Cube Volume}} = \frac{6}{r} \approx \frac{6r^2}{r^3}$$

$$\frac{\text{Sphere Area}}{\text{Sphere Volume}} = \frac{3}{r} \approx \frac{4\pi r^2}{(4/3)\pi r^3}$$

This tendency arises from molecular attractive forces.

12.10.8.7   **SURFACE TENSION EXAMPLES**

(1)   *Walking on water*

Small insects such as the water strider can walk on water because their weight is not enough to penetrate the surface.

(2)   *Don't touch the tent!*

Common tent materials are somewhat rainproof in that the surface tension of water will bridge the pores in the finely woven material. But if you touch the tent material with your finger, you break the surface tension and the rain will drip through.

(3) *Clinical test for jaundice*

Normal urine has a surface tension of about 66 dynes/cm but if bile is present (a test for jaundice), it drops to about 55. In the Hay test, powdered sulfur is sprinkled on the urine surface. It will float on normal urine, but sink if the S.T. is lowered by the bile.

(4) *Surface tension disinfectants*

Disinfectants are usually solutions of low surface tension. This allow them to spread out on the cell walls of bacteria and disrupt them. One such disinfectant, S.T.37, has a name which points to its low surface tension compared to the 72 dynes/cm for water.

(5) *Floating a needle*

If carefully placed on the surface, a small needle can be made to float on the surface of water even though it is several times as dense as water. If the surface is agitated to break up the surface tension, then needle will quickly sink.

(6) *Soaps and detergents*:

These help the cleaning of clothes by lowering the surface tension of the water so that it more readily soaks into pores and soiled areas. Why do we use soap? Surface tension plays a big role in many of our daily activities. Soaps and detergents include surfactants that reduce the surface tension of the liquid. This allows the liquid to have a good contact with the material and to remove the dirt from it efficiently

(7) *Washing with cold water*

The major reason for using hot water for washing is that its surface tension is lower and it is a better wetting agent. But if the detergent lowers the surface tension, the heating may be unnecessary.

(8) **Well fitness** Tooth paste, all Washing creams, Foot cream, Foot thorn,nails & coarse file, Mouth wash, *etc.*

The cream or paste used for various purposes of the type mentioned in the heading have to be prepared by the manufacturers by mixing **appropriate ingredients** having the best of the **basic physical properties**, such as wetting and soothing (capillarity), surface tension, least viscosity, required friction coefficient, which all act rapidly to throw out the unwanted dirt/ food particles / dead skin.

12.11. CAPILLARY ACTION

12.11.1 **Why?** Cohesive and adhesive forces give rise to **Capillary Action** *Capillary action* occurs when the adhesion to the walls is stronger than the cohesive forces between the liquid molecules.

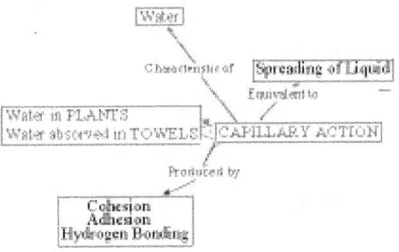

12.11.2 The height to which capillary action will take water in a uniform circular tube is limited by surface tension and circumference,
The height h to which capillary action will lift water depends upon the weight of water which the surface tension will lift:

$$(h + \tfrac{1}{2}r) = \frac{2T \cos\theta}{g\,\rho\,r}$$

$$\boxed{h = \frac{2T\,Cos\theta}{\rho\,r\,g}},$$

$h = \frac{2T}{\rho\,r\,g}$, approximately.

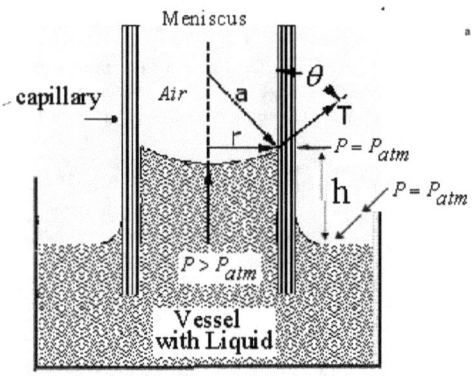

.12.11.2.1 CAPILLARY RISE OF WATER AND MERCURY DIFFER

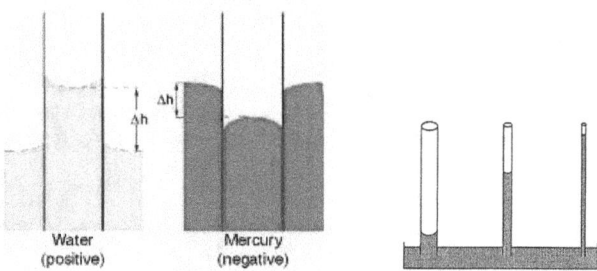

Liquid	$T / Nm^{-1} x\ 10^{-3}$
Methylated spirit	22.6
Glycerol	63.4
Water	72.7
Benzene	28.9
Mercury	472
Olive oil	32
Ether	17
Gold	1102

12.11.3 APPLICATIONS

12.11.3.1 BALL POINT PEN:
Ever sharp CA ballpoint (Invented in 1938 by Lazlo Biro)
It is primarily a gravity-reliant writing instrument. Typically the ballpoint pen requires a certain amount of pressure to be applied in order for it to work. It lacks the free-flowing quality found in some other pen designs.

12.11.3.2 STAY OF AIRBORNE BALLOONS
High-flying airman, if unprotected by high pressurized suit, is ejected from the aircraft (or for a space traveller), pressure will be reduced from 1 *atm.* to, say, 0.01 *atm.*, his body fluids will boil at pressures below about 25 *mm* Hg at $37\,^oC$. Secondly, for fear he holds his breath, punctures can occur in lung walls.

12.11.3.3 **DRINKING BIRD** This commercially available toy demonstrates the conversion of Thermal energy $\Rightarrow$ to Mechanical energy. As it dips its beak into a glass of water evaporative cooling induces the rise of volatile liquid from his tail toward his head.

12.11.3.4 FLOATING IN A JET STREAM
A nozzle projecting a jet of high velocity air can suspend various objects such as ping-pong balls, *etc*. The viscous force of air balance the weight and the low pressure in the jet keeps the object trapped in the air-stream

12.11.3.5 SPHERICAL OIL DROP
A layered mixture of water and alcohol has a region of density equal to that of vegetable oil. As oil is added to the mixture, it sinks to a stable depth and collects into a large spherical drop under the action of the oil's surface tension

12.11.3.6 Liquid jet:
If a vessel contains a liquid and if a small hole is present in the wall of the vessel at a point *h* below the liquid level, then p at the hole

$$p = \frac{F}{a} = h \rho g$$

H = length of the liquid column in the vessel,

x = position of the liquid jet striking at the ground = $\frac{1}{2}$ Range.

$$x = 2 \sqrt{[(H - h)\ h]}$$

12.11.3.7 Burning of an oil lamp or candle:
A solid block of wax (paraffin) in which embedded a wick, lighting the candle melts and vaporizes small amount of fuel. Once vaporized, the fuel combines with oxygen in the

atmosphere to form a flame. This flame provides sufficient heat to keep the candle burning *via* a self-sustaining chain of events: the heat of the flame melts the top of the mass of solid fuel, the liquefied fuel then moves upward through the wick *via* <u>capillary action</u>, and the liquefied fuel is then vaporized to burn within the candle's flame.

+*&+*&+*&+*&+*&+*&+*&+*&+

Chapter 13

ELASTICITY OF SOLIDS,
CANTILEVER, OSMOSIS, DIFFUSION

"Strength does not come from physical capacity. It comes from an indomitable will"
MK Gandhi

13.1 ELASTICITY OF A SOLID

13.1.1 Stress

$Stress = \dfrac{Force\ acting,\ \vec{F}}{Area,\ A}$	Unit	$, Nm^{-2}$	$, Pa$ (vector) $(M^1 L^{-1} T^{-2})$

13.1.2 Strain

$Strain = \dfrac{Change\ in\ Dimension,\ \Delta D}{Original\ Dimension,\ D}$	Unit	(*Dimensionless quantity*)	$(M^0 L^0 T^0)$

13.1.3 Modulus of Elasticity

Modulus of Elasticity, $\lambda = \dfrac{Stress,\ (F/A)}{Strain,\ (\Delta x/x)}$,	Unit Nm^{-2} Pa	(vector)	$(M^1 L^{-1} T^{-2})$

Young's modulus, $q = \dfrac{Stress\ (F/A)}{Longitudinal\ Strain\ (\Delta \ell / \ell)}$,	Unit Nm^{-2} Pa	(vector)	$(M^1 L^{-1} T^{-2})$

Rigidity (Shear) modulus, $n = \dfrac{Tangential\ Stress\ \Delta P = (F/A)}{Shearing\ Strain\ \theta = (x/y)}$,	Unit Nm^{-2} Pa	(vector)	$(M^1 L^{-1} T^{-2})$

13.1.4 ELASTIC LIMITS:

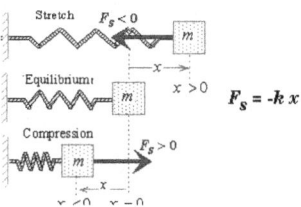

Elastic Modulus E

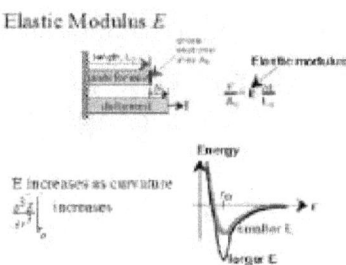

13.2.1 HOOKE'S LAW and Plastic flow:

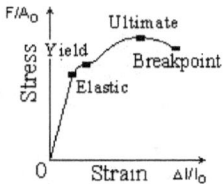

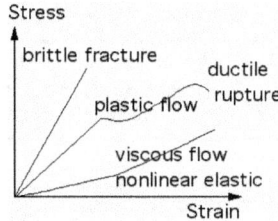

13.2.2 COMPRESSIBILITY:

$$\chi = 1 / K$$

Work done in stretching a wire of length ℓ by elongation x = Energy stored

$$W = \tfrac{1}{2}\vec{F}\,x$$

13.2.3 EXTENSION

$$x = \frac{\vec{F}\,\ell}{qA}$$

$$\boxed{Poisson's\ ratio,\ \sigma\ =\ \frac{lateral\ strain\ (\mu)}{longitudinal\ strain\ (\lambda)}}$$

$K = 1 / [3 (\lambda - 2 \mu)]$

$n = 1 / [2 (\lambda + \mu)]$

$q = 9 K n / (3 K + n)$

$\sigma = (3K - 2 n) / (6 K + 2 n)$

$\quad = [(q / 2 n) - 1]$

$K = q / [3 (1 - 2 \sigma)]$

$\quad \frac{1}{2} > \sigma > -1$

13.2.4 TORSIONAL RIGIDITY,

$$\boxed{c\ =\ \frac{Couple}{Unit\ twist}\ =\ \frac{\frac{1}{2}\Im\,n\,a^4}{\ell}}\ .$$

13.2.5 WORK DONE in twisting a twist angle θ,

$$\boxed{W\ =\ \tfrac{1}{2}c\ \theta^2}\ .$$

13.2.6 TORSIONAL PENDULUM:

Period, $\boxed{T = \sqrt{\Im / c}}$

13.2.7 BENDING:
R = radius of the beam when bent

Bending Moment $\boxed{C\ =\ \dfrac{q}{R}\left(AK^2\right)}$

$\left(AK^2\right)$ = geometric moment of inertia

13.2.7.1 For a **rectangular beam,**

$$\boxed{\left(AK^2\right) = \frac{bd^3}{12}}$$

13.2.7.2 For a rod,

$$\boxed{\left(AK^2\right) = \frac{\pi r^4}{4}}$$

13.2.8.1. **Searles's Method of measuring Young's modulus**
L– Length of the wire
l - Extension for a load M
r – Radius of the wire

g – Acceleration due to gravity
M - Mass added in the hanger

$$q = \frac{MgL}{\pi r^2 l}$$

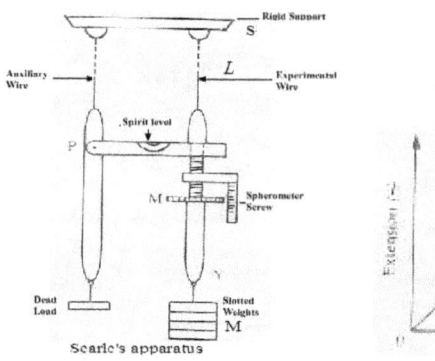

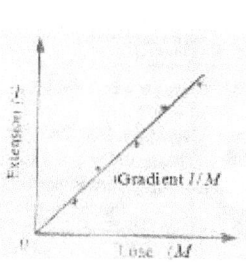

Scarle's apparatus

13.2.8 Cornu's method for Young's modulus by UNIFORM BENDING

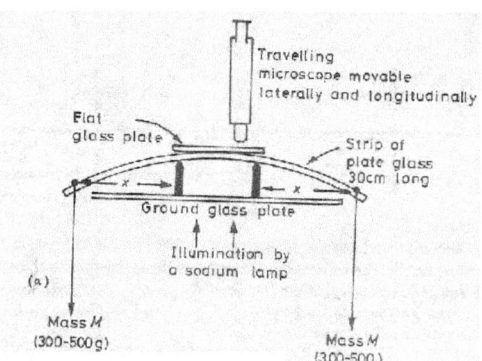

For two weights each W at either side of the support,
W = weight causing the bending
p = Separation of the suspension point and the weight.
h = shift of the centre of the bean from the centre when beam was not bent

$$q = \frac{WpL^2}{8hAK^2}$$

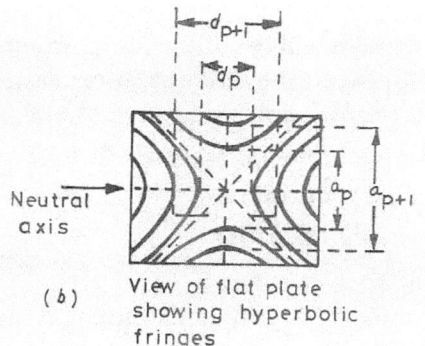

(b) View of flat plate showing hyperbolic fringes

13.2.9 CANTILEVER:

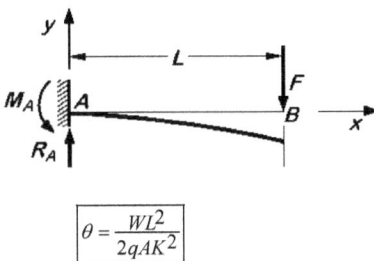

$$\theta = \frac{WL^2}{2qAK^2}$$

Total depression, δ

$$\delta = \frac{WL^3}{3qAK^2}$$

13.2.9.1 Young's Modulus by Non- **uniform Bending**

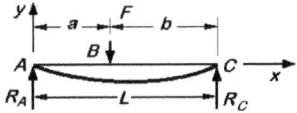

m - Mass loaded for depression.

g - Acceleration due to gravity.

l - Length between knife edges.

b - Breadth of the bar using vernier calipers.

d - Thickness of the bar using screw gauge.

δ - Depression of the bar.

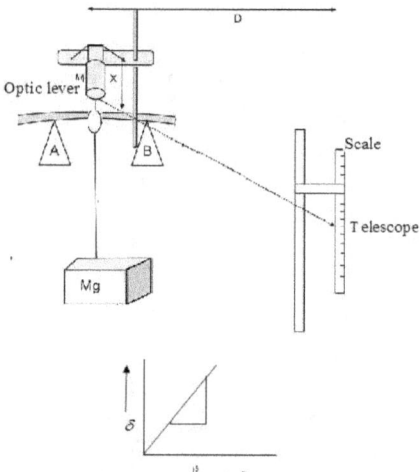

Graph between depression and length

Depression at the Centre of the bar,

$$\delta = \frac{WL^2}{48\,qAK^2} = \frac{WL^3}{4\,qbd^3} \quad \text{or} \quad q = \frac{mg}{4bd^3}\frac{L^3}{\delta}$$

13.3. PHOTOELASTICITY

Normally isotropic substances can become birefringent when under stress. This property called Photoelasticity. It can be used in stress analysis. It is also used for determining stress points in asymmetrical geometries. This detailed stress-analysis method is based on an opto-mechanical property called birefringence. Several transparent polymers and glass exhibit this property.

Photoelasticity is a very useful tool for engineers to see areas where a structure might break due to high concentrations of stress. Birefringent plastics will reveal areas of strain within a structure in the form of colorful light fringes when placed under polarized light.

13.3.1. Photoelastic basics:

The principles of photoelasticity can be succinctly elucidated by the following two laws:

a) For stress (or strain) induced birefringence, the normally incident polarized light is split into two components along the principal stress directions in a plane perpendicular to the direction of light propagation and are transmitted only along these planes through the model.

The velocities of light transmitted along these directions is directly proportional to the intensities of the respective principal stresses

13.4 **EFFUSION**

Effusion is the rate at which molecules pass through a hole

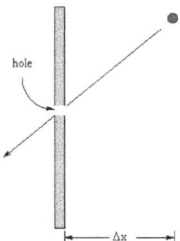

13.5 **DIFFUSION**:

Diffusion is a measure of how quickly a molecule can get from point A to point B.

13.5.1 **FICK'S LAW**:
The rate of diffusion Q/ t of a solute in any direction is directly proportional to the concentration c(x) gradient of the solute in that direction.

$$Q = -KA\frac{dc}{dx}$$ | Unit $\qquad m^2 s^{-1}$ (scalar) $\qquad (M^0 L^2 T^{-1})$

13.5.2. **OSMOSIS**:
Osmosis is the process of diffusion of one liquid is not the other through an semi-permeable membrane.
Osmotic pressure of a solution is the pressure that must be applied to the solution to prevent flow of solvent into it through a semi-permeable membrane.

13.5.2.1 PFEFFER'S LAWS of Osmotic pressure,
(i)   The temperature T remaining constant, the osmotic pressure p of a dilute solution of a given substance is proportional to its concentration, c.

$$p \propto c; \text{ but } c \propto \frac{1}{V},$$

So $\boxed{pV = \text{constant}}$.

(ii)   Concentration, c remaining constant, p of a dilute solution is directly proportional to its absolute temperature T *i.e.* $p \propto T$;

or $$\boxed{\frac{p}{T} = \text{constant}}$$.

Combining the two laws,

$$\boxed{\frac{pV}{T} = \text{constant}}$$.

Osmotic pressure and lowering of Vapor pressure:

$$\boxed{\frac{dp}{dp_1} = \frac{p}{H} \frac{\sigma_o}{\rho}}$$

ρ = density of solution at normal atm. pressure, H

σ_o = density of vapor of solvent

h = height of solution above solvent

p = osmotic pressure

dp = lowering of vapor pressure

13.5.3 ELEVATION OF BOILING POINT with vapour pressure.

$$\boxed{dT = \frac{pT}{\rho LJ}}$$

Elevation of the B.P. of water / gm molecule in 100 cc of it is known as the Molecular Elevation of the B.P. = 5.34 $^{\circ}C$.

If x g of a substance is dissolved in 100 cc of water, and molecular elevation its B.P. is dT , its molecular weight,

$$\boxed{M = (5.34) \frac{x}{dT}}$$

Depression of Freezing point with vapor pressure:

13.5.3.1 Molecular Depression of Freezing point,

$$\boxed{dT = \frac{pT}{\rho LJ} = 18.5^{\circ}C}$$.

13.5.3.2 Molecular weight

$$\boxed{M = (18.5) \frac{x}{dT}}$$

-o-0-o-0-o-0-o-0-o-0-o-

HEAT - 1
SOURCES OF HEAT, THERMOMETRY, THERMAL EXPANSION

"Ayam Brahmaasmi" (means Myself is Brahma) - Yajur Veda

14.1   **INTRODUCTION**
Four States of Matter:
They are Solid, Liquid, Gaseous and Plasma states.
14.1.1   **Sources of Heat** – Conventional Sources
14.1.1.1 **Flames**:

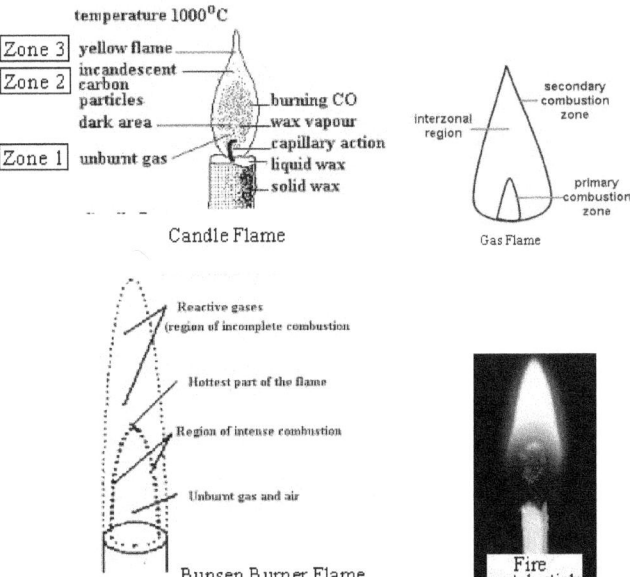

Candle Flame

Gas Flame

Bunsen Burner Flame

Fire match stick

14.1.1.2 Candle Flame{Tallow bee wax or Paraffin + Stearic acid 0.95, 60 °C }
Zone 1: The Inner and not Hot Non-luminous cone of vapourized fuel to the flame(cannot be seen as Oxygen is not available)
Zone 2: Middle one, highly heated Carbon particles, YELLOW, and the brightest part of burning carbon. It is waste of fuel.
Zone 3: Outer, invisible part of burning gas- BLUE, and the hottest part of the flame.

Beeswax is the major component of honeycomb. It is secreted in tiny flakes from the underside of the abdomens of worker bees, and molded into honeycomb.

$$H_{31}C_{15}-\overset{\overset{\displaystyle O}{\|}}{C}-O-C_{30}H_{61}$$

$$C_{46}H_{92}O_2$$

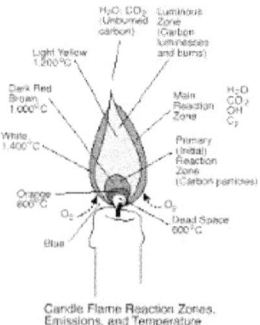

Beeswax is soft to brittle, with a specific gravity of about 0.95 and a melting point of over $60°C$, and consists of at least 284 different compounds, mainly a variety of long-chain alkanes, acids, esters, polyesters and hydroxy esters. These include free cerotic acid (hexacosanoic acid, $CH_3(CH_2)_{14}COOH$), the ester of cerotic acid and triacontanol $\left(CH_3(CH_2)_{29}OH\right)$, myricin (myricyl palmitate, $CH_3(CH_2)_{14}COO(CH_2)_{12}CH_3$), and hentriacontane, $CH_3(CH_2)_{29}CH_3$. Hentriacontane.

Paraffin is used synonymously with <u>alkane,</u> indicating hydrocarbons with the general formula C_nH_{2n+}.

Candle Flame Reaction Zones, Emissions, and Temperature

14.1.1.3   Nernst Glower
            Infrared source
14.1.1.4 Globars
            Made of silicon carbide, used as source of IR radiation.
            It has three Zones.
14.1.1.5. LPG stow
14.1.1.6. Oxy-Acetylene flame
14.1.1.7. Microwave Oven (vide Chap 28)
14.1.1.8   Induction Heating (Vide Chap 28)

14.1.2.1. **Pure Substance**:

Fixed chemical composition, throughout H_2O, N_2, CO_2, Air (even a mixture of ice and water is pure)

14.1.2.2. Compressed Liquid:

NOT about to vapourize (Sub-cooled liquid), *e.g.*, water at 20 °C and 1 *atm*.

14.1.2.3 Saturated Liquid:

About to vapourize.

e.g., water at 100 °C and 1 *atm*.

14.1.2.4. Saturated Vapour:

About to condense.

e.g., water vapour (steam) at 100 °C and 1 *atm*.

14.1.2.5 Superheated Vapour:

NOT about to condense

e.g., water vapour (steam) at >100 °C and 1 *atm*.

14.1.2.6. Critical Point

The saturated liquid and saturated vapour states are identical. No saturated mixture exists - the substance changes directly from the liquid to vapour states.

14.1.2.7. Gaseous System

(1)   A microscopic description of a gas is at the molecular level.

(2)   Thermodynamics is the larger scale, *i.e.,* macroscopic, description of a gaseous system.

Variables of thermodynamic state of a gas include pressure (P), volume (V), and amount of gas as number of moles (n), Temperature (T), internal energy (U) and entropy (S).

In the equilibrium state the variables of state are constant.

14.1.3. The Zeroth Law of Thermodynamics:

Theorem: "Two systems in thermal equilibrium with a third system are in thermal equilibrium with one another".

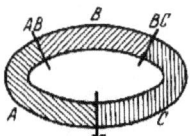

Thermal equilibrium of
three substances *A. B. C.*

14.2.   **THERMOMETRY**

14.2.1 Temperature, T

Concept of temperature is at the very heart of thermodynamics. Zeroth law of thermodynamics allows determining the condition of thermal equilibrium.

Two systems in thermal equilibrium have the same temperature (T). It is the degree of hotness of a body.

An equilibrium state is one in which all bulk properties of the system are uniform throughout the system at all time. Thermodynamic variables (P, V & T)(coordinates or state variables) specify the equilibrium state of a simple system.

Equation of state is $\boxed{f(P,V,T)=0}$

Galileo constructed the earliest thermometer in 1593.

14.2.2 Temperature Scales:

Before 1954, temperature scales were based on the relation,

$X = cT_x + d$,

where c and d are fixed by <u>two fixed points</u>, X is a thermodynamic variable.

i)   Ideal gas scale: $\boxed{T_{Gas} = 273.16 \lim_{P_{TP} \to 0} (P/T_P)K}$

ii)   The **Celsius scale** of temperature by Andres Celsius in 1742.

$$\boxed{t(°C) = T(K) - 273.15}$$

By definition the temperature of the fixed point, the <u>Triple point of water</u>, is $0.01°C$.
Two fixed points are:
a) Ice point (273.15 K) and
b) Steam point (373.15K).
The **Fahrenheit scale** by Gabriel Fahrenheit in1720.
The **Kelvin (Absolute) scale** by William Thomson (later Lord Kelvin).
The **Reumer scale**, and **Rankine scale** by William J.M. Rankine.

The Absolute Zero is the lowest temperature

iii)   Modern method of adopting a Scale of temperature using a <u>Single Fixed Point</u>.
$X_{Steam} = 100\,c + d, \quad X_{Ice} = d$.

$$\boxed{\theta_X(°C) = 100\left[\frac{X - X_{Ice}}{X_{Steam} - X_{Ice}}\right]}$$

The Triple point water (273.16 K) is precisely reproducible: and the Absolute Zero is precisely determined as the limiting temperature at which the pressure.
$$P \xrightarrow{\text{in an ideal gas thermometer}} 0 .$$

14.2.3 **Thermodynamic Scale** of Temperature:
It is one that is used for scientific measurement. It is measured in units called Kelvin (K).
It is defined using one fixed point in the Triple point of water.
<u>Triple point</u> temperature:

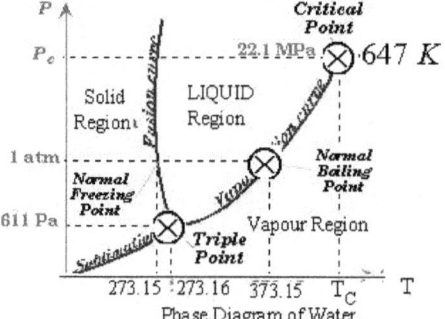

Phase Diagram of Water

This is the temperature where saturated water vapour, pure water and ice are all in equilibrium at a temperature of $273.15K$.
Triple point temperature, $T_3 = 273.15K$.

14.2.3.1 <u>Ideal Gas Temperature</u> of the substance,

$$T = (273.16) \lim_{P_{Tp} \to 0} (P/P_{Tp})$$

14.2.3.2. Triple point of water, $t(0^o C) = T_p = 273.16$

P = Pressure of boiling liquid.

P_{Tp} = Triple point pressure

Scale		Temp of Melting point of ice (Lower Fixed Point)	Temperature of Steam (Upper Fixed Point)
(1)	C	0°	100°
(2)	F	32°	212°
(3)	Re	0°	80°
(4)	K	273	373
(5)	R	492°	672°

14.2.3.3 Relations

$$\frac{C}{100} + \frac{F - 32}{180} = \frac{Re}{80}$$
$$= \frac{K - 273}{100} = \frac{R - 492}{180}$$

$$C = \frac{5}{9}(F - 32)$$

$$F = [(\frac{K \, 9}{5}) - 459.67 \, K] = \{\frac{C \, 9}{5} + 32\}$$

14.2.3.4 (a) Mercury-in-glass thermometer is the most familiar one.

(b) Platinum Resistance thermometer: First designed by Siemen in 1871;
Improved by Callendar and Griffiths.

Resistance R_t of platinum at a given temperature t^o C

$$R_t = R_0(1 + \alpha t + \beta t^2)$$

where α and β are constants.

$$t(^o C) = \{(R_T - R_0)/(R_{100} - R_0)\} \, 100$$

14.2.3.5 Correction

If θ = temperature on the gas scale (thermodynamic temperature)and
t = temperature on the Platinum scale

$$(\theta - t) - \delta\left[(\frac{\theta}{100})^2 - (\frac{\theta}{100})\right]$$

$$\delta = -\frac{(1000)^2 \beta}{\alpha + 100\beta}$$

14.2.4 Ranges of Measurements

1.	Liquid thermometer	Mercury	down to - 30° C up to 600° C
		Alcohol	- 100° C
2.	Gas thermometer	H_2.	- 250° C up to 1100° C
		He	-268° C
		N_2.	up to 1500° C
3.	Resistance	Pt	-190° C up to 1200° C
4.	Thermo-electric	Cu - Constantan	-250° C up to 300° C
		Pt - Silver	
5.	Vapor pressure	He vapour	-272° C

14.2.5 STP (Saturated Temperature and Pressure)

STP is a set of conditions which is usually applied is defined as a T of 273 K, and a pressure of 760 mm of Hg $(1.013 \times 10^5 \ Pa)$, At STP, one *mole* of any gas has a volume of 22.4 l $(22.4 \times 10^{-3} \ m^3)$ '

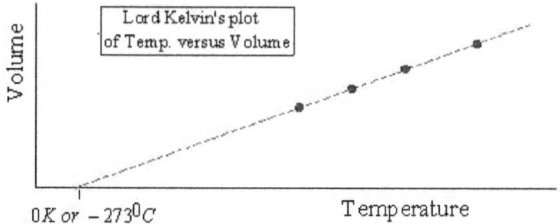

14.3. **HEAT TRANSFER**

Why does it take place? The quantity of heat energy Q transferred between a system and its surroundings solely because of a temperature difference.

14.3.1 Three categories of Heat Transfer Processes

The heat energy may be transmitted from a body to another by

1. Conduction: Transfer of heat by contact, energy transfer from molecule to molecule in the medium. Metals have high thermal conductivity – 1000 times greater than non-metals and 10,000 times than gases. Rate of heat transfer by conduction depends on a) type of material, b) area of cross section of material, c) temperature gradient.
2. Convection: Transfer of heat by movement of a hot fluid (air, oil, or water). Convection currents cook food more rapidly than conduction. Effectiveness of convection depends on the viscosity of the fluid (medium).
3. Irradiation: Emission of radiation in the form of IR waves from hot objects. It can travel in vacuum.

14.3.2 THERMAL EXPANSION of SOLIDS

Reason: Anharmonicity in the potential energy of crystalline lattice causes thermal expansion of a **solid**.

ℓ_1 = the original length at temperature θ_1,

$(\theta_2 - \theta_1)$ = degree rise in temperature

Linear thermal expansion coefficient, α

$$\alpha = \frac{(\ell_2 - \ell_1)}{\ell_1 (\theta_2 - \theta_1)} \, , \quad \boxed{\text{Unit} \quad K^{-1} \quad (\text{tensor}) \quad (M^0 \, L^0 \, T^0)}$$

$$\ell_\theta = \ell_0 (1 + \alpha \, \theta)$$

14.3.2.1 For Isotropic (cubic) solids,

Coefficient of *areal expansion* of an isotropic solid, β

$$\beta = 2\alpha \qquad \boxed{\text{Unit} \quad K^{-1} \quad (\text{tensor}) \quad (M^0 \, L^0 \, T^0)}$$

Coefficient of *Volume Expansion* of an isotropic solid, γ

$$\gamma = 3\alpha \qquad \boxed{\text{Unit} \quad K^{-1} \quad (\text{tensor}) \quad (M^0 \, L^0 \, T^0)}$$

14.3.3 THERMAL EXPANSION OF LIQUIDS:

Coefficient of *apparent* expansion, ℓ

Coefficient of *absolute* expansion, m

$$m = \frac{(V_2 - V_1)}{V_1 (\theta_2 - \theta_1)} \quad \boxed{\text{Unit} \quad K^{-1} \quad (\text{tensor}) \quad (M^0 \, L^0 \, T^0)}$$

$$\rho_0 = \rho_\theta (1 + m \, \theta)$$

Coeffcient of Absolute T of a liquid	Coeffcient of Apparent T of the liquid	Coeffcient of Absolute T of Container
=		+

$$m = \ell + \gamma_{Solid}$$

14.3.3.1 Principal applications of Expansion of liquids:

14.3.3.2 Toluene thermostat

Barometric scale correction

H = observed height of barometer at $\theta^\circ C$

H_0 = actual height of barometer at $0^\circ C$;

$\rho_0 \, \& \, \rho_\theta$ = densities of Hg at $0^\circ C$ and $\theta^\circ C$,

α = Linear thermal expansion coefficient of material of scale

$$H = (1 + \alpha \, \theta) = H_0 (1 + m \, \theta)$$

+*+*+*+*+*+*+*+*+

Chapter 15

HEAT – 2
IDEAL GASES, SPECIFIC HEATS

"The profound study of nature is the most fertile source of mathematical discoveries"   Joseph Fourier

15.1   GAS LAWS

15.1.1 BOYLE'S LAW

Theorem: Robert Boyle stated that temperature (T K) remaining constant; the volume (V) of a given mass of gas varies inversely as its pressure (P).

Mathematically,

$$P \propto \frac{1}{V},$$

$$\boxed{PV = \text{Constant}} \quad , \left(n = \frac{m}{M} \text{ and T both fixed}\right)$$

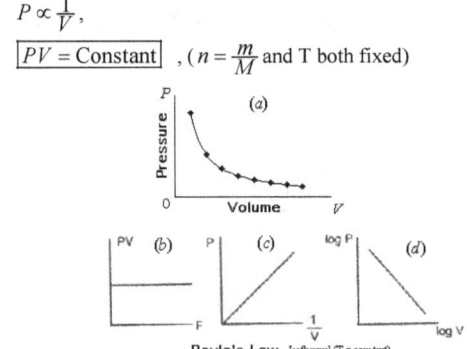

Boyle's Law Isothermal (T = constant)

15.1.2 CHARLES' LAW

Theorem: According to J.A. Cesar Charles, P remaining constant, V is proportional to T;

$$\boxed{V \propto T}$$

$$\boxed{V / T = \text{constant}} \quad . \quad \left[n = \frac{m}{M}, P \text{ fixed}\right]$$

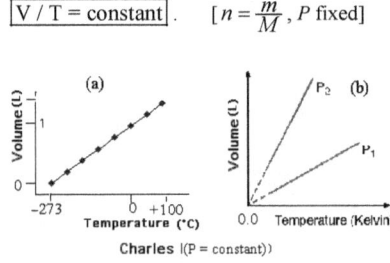

Charles I(P = constant))

15.1.3. GAY-LUSSAC LAW:

$$\boxed{P \propto T}$$

$$\boxed{P \; / \; T = \text{constant}} \qquad [\, n = \frac{m}{M}, \; V \text{ fixed}]$$

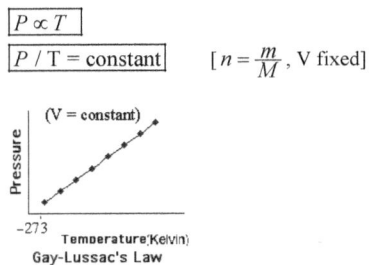

Gay-Lussac's Law

15.1.4 EQUATION OF STATE: Perfect (or Ideal) Gas:

The equation of state is an equation that relates the variables of the (gaseous) state. Ideal gas molecules do not exert any appreciable attraction to each of the molecules.
Mathematically, the <u>ideal gas equation of state</u>

$$\boxed{P\,V = nR\,T}$$

for an ideal gas exactly; and approximately for a real gas (*i.e.*, dilute gas, whose p is not too high and T is not too low), unless it is at high pressure or at $T \approx T_C$.
where R = *Universal Gas Constant*

$$R = 8.3145 \times 10^7 \;\boxed{\text{Unit } J \; K^{-1} \; Mol^{-1} \qquad \text{(Scalar)} \quad (M^0 \; L^2 \; T^{-2})}$$

$$\boxed{n = N \, / \, N_A} = \text{number of gram molecules}$$

N = number of gas molecules of the gas

N_A = Avogadro's constant, $N_A = 6.0225 \times 10^{23} \, mol^{-1}$.

V = 22.4 *l* / *gm* molecule
Boltzmann constant

$$\boxed{k_B = \frac{R}{N_A}, \;\text{Unit} \quad 1.3805 \times 10^{-23} \; JK^{-1} \qquad \text{(scalar)} \quad (M^1 \; L^2 \; T^{-2})}$$

15.1.5 The $p - V$ diagram and **Isotherm***:

A point on a $p - V$ diagram represents a state of a system. There is a set of states on the $p - V$ diagram that have same T; called an isotherm.

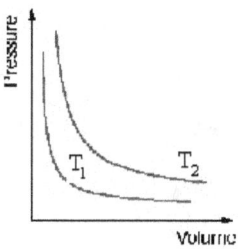

15.1.6 HENRY'S LAW

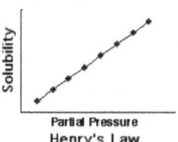

Partial Pressure
Henry's Law

THE AMOUNT OF GAS
DISSOLVED IN SOLUTION IS
DIRECTLY PROPORTIONAL
TO THE PRESSURE OF THE
GAS OVER THE SOLUTION.

PV against T plot extrapolated until PV = 0.

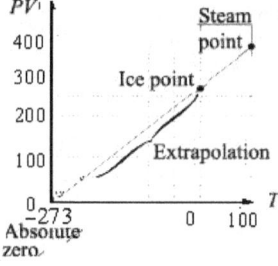

If the plot is extrapolated, then the point at which it crosses the T-axis (PV = 0) occurs when T = 273.15 °C . Kelvin defined this point as *absolute zero* and it represents the theoretical lowest temperature attainable. All temperatures can now be defined with respect to Absolute Zero.

15.2. THERMOMETERS

15.2.1 Callendar's compensated Constant Pressure Air Thermometer (Useful up to 600 °C)

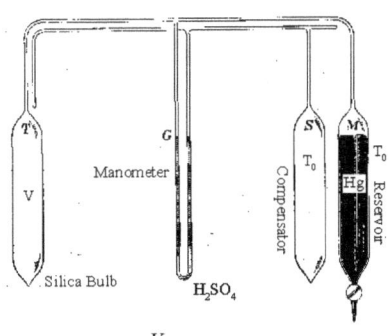

$$T = [\frac{V}{V - V_o}] \, 273 \, K$$

$$t \, (^oC) = (T - 273) = [\frac{V_o}{V - V_o}] \, 273$$

15.2.2 International Standard Thermometer (Constant Volume H_2 Thermometer)

(Harker & Chappuis) (Useful down to - 250 ° C)

$$t \, (^oC) = [\frac{P_t - P_0}{P_{100} - P_0}] \, 100$$

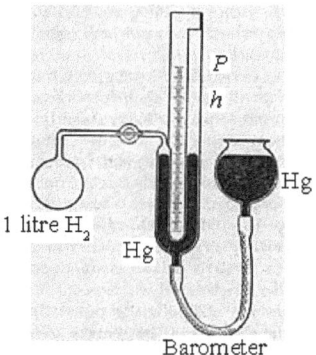

Barometer

15.2.3 Philipp Jolly's Law (Pressure Law of Ideal Gases)

The pressure coefficient of expansion of a gas at constant volume is defined as the fraction of its pressure at 0°C by which the pressure of a fixed mass of gas increases per °C rise in temperature.

Jolly's Constant Volume Air Thermometer.

$$t \ (^{o}C) = [\frac{h_t - h_0}{h_{100} - h_0}] \ 100$$

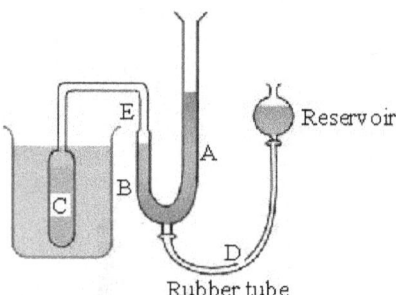

15.3. DALTON'S LAW of Partial Pressures

P = pressure in a container
p_i = partial pressure
For any two gases 1 and 2

$$\frac{P_1}{P_2} = \frac{N_{A1}}{N_{A2}}$$

Every gas in a mixture of gases exerts its own partial pressure independently of the others. The pressure of a single gas in a mixture is known as the 'partial pressure'.

Formula

$$P_t = P_1 + P_2 + P_3 +$$

As a rule, a pressure in excess of 760 *mm* is stated in atmospheres (atm), not in *mm* mercury.

Pressure   Unit $1Pa = 1 \ Nm^{-2}$ (Vector) $(M^1 \ L^{-1} \ T^{-2})$

$1 \ atm = 101 \ kPa = 1.01 \ x \ 10^6 \ Nm^{-2}$ $= 760 \ torr = 760 \ mm$ of Hg

15.4 AVOGADRO'S LAW

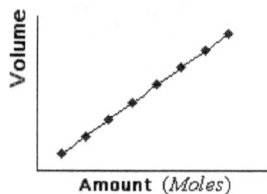

Avogadro's Law

Theorem: An equal volume, V of all gases at the same temperature T and pressure p, contains equal number of molecules

$$. P_t = P_2 \,; \; V_t = V_2 \,; \; T_t = T_2 \,; \; n_t = n_2 \,;$$
$$(N_A)_1 = (N_A)_2$$

15.5. SPECIFIC HEATS

15.5.1 **Definition:** Specific heat is the amount of heat that must be absorbed or lost to change in temperature of a substance. C at constant pressure (C_P) is the limiting case of infinite temperature changes (a function of T) as

$$dQ = m \; C_P \; dT \,, \qquad n = \frac{m}{M}$$

$C_P = \frac{1}{m} \frac{dQ}{dT}$	Unit $J \; K^{-1} \; Mol^{-1}$	(Scalar)	$(M^0 \; L^2 \; T^{-2})$

15.5.2 The Calorie (*cal*):

1 Calorie	Unit $1 \; cal = 4.186 \; J$	(Scalar)	$(M^1 \; L^2 \; T^{-2})$

15.5.3 SPECIFIC HEATS OF SOLIDS

15.5.3.1 Bunsen's Ice Calorimeter

$$V(1 \; gm \; Ice) = 1.0908 \; cm^3$$

$$V(1 \; gm \; Ice) - V(1 \; gm \; melted \; Ice) = 0.0907 \; cm^3$$

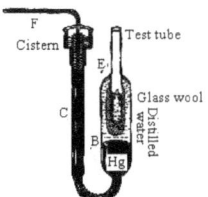

Ice calorimeter (Bunsen)

Calibrating F with water

$$m \; \theta = q \; y \,, \quad q = \text{quantity of heat for recession of 1 scale.}$$
$$M \; C \; \theta = q \; x$$

15.5.3.2 Jolly's Steam Calorimeter

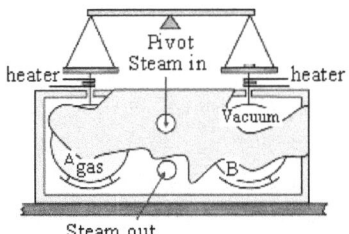

$$(MC + wC)(\theta_2 - \theta_1) = m(L + 100 - \theta_2)$$

15.5.4. The DULONG-PETIT'S LAW:

Since $U = nRT$

$$C_V = 3R = 24.9 \ JK^{-1}mol^{-1}.$$

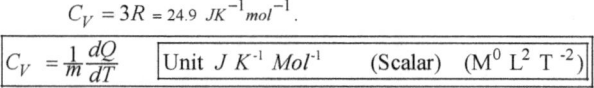

| $C_V = \frac{1}{m}\frac{dQ}{dT}$ | Unit $J \ K^{-1} \ Mol^{-1}$ | (Scalar) | $(M^0 \ L^2 \ T^{-2})$ |

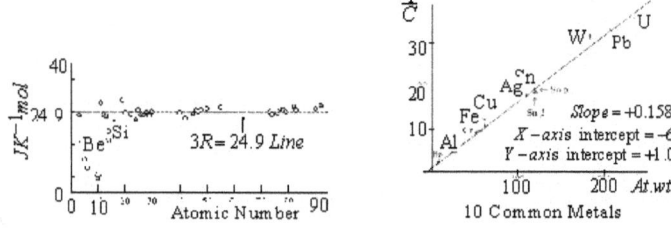

| Atomic heat $=$ (Atomic weight) (Specific Heat) $=$ constant, 6.4 |

Specific heat of Ice $= 2.09 \ J \ g^{-1}K^{-1}$

Specific heat of Water $= 1.00 \ cal \ g^{-1}C^{-1} = 4.186 \ J \ g^{-1}K^{-1}$

Water has a high value of specific heat

In order for molecules to move faster, hydrogen bonds must be broken. This requires energy.

15.6. **SPECIFIC HEATS OF GASES** C_P **and** C_V :

15.6.1 Two molar specific heats of gases: C_P and C_V .

C_P .is the amount of heat required raising the temperature of 1 *mole* of a gas keeping pressure constant through 1 °C.

C_V .is the amount of heat required raising the temperature of 1 *mole* of a gas at constant volume through 1 °C.

| $C_P - C_V = \frac{R}{J}$ |

= Thermal energy equivalent of the work done in expansion of the gas against external pressure

15.6.2 Regnault's method for C_P .

$$m \ C_P[t - \tfrac{1}{2}(\theta_2 + \theta_1)] = w \ (\theta_2 - \theta_1)$$

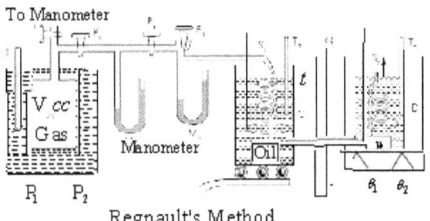

Regnault's Method

15.6.3 Jolly's Differential Steam Calorimeter for C_V.

$$w\,(\theta_2 - \theta_1)\,C_V = w\,[L + 100 - \theta_2]$$

15.6.4 Callendar & Swann's Electrical Continuous Flow for C_P.

$$\frac{e\,i\,t}{J} = m\,(\theta_2 - \theta_1)\,C_P$$

Specific heat of Water Vapor: $= 1.85\ J\,g^{-1}K^{-1}$

15.6.5 <u>Newton's Law of Cooling</u>
<u>Theorem</u>: For natural convection, the ratio of loss of heat of a body by cooling in a steady stream of air is $\propto$ to excess of temperature $(\theta_2 - \theta_1)$ of the body above the surroundings.

i.e., $\quad -\dfrac{dH}{dt} = k\,(\theta_2 - \theta_1)^{5/4}$

15.6.6. **SPECIFIC HEATS OF LIQUIDS by Method of Cooling C_P and C_V**

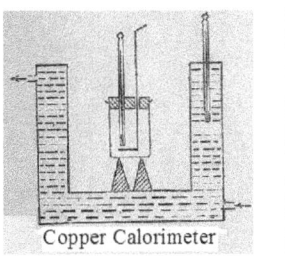

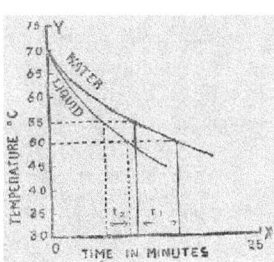

Copper Calorimeter

w_1 = weight empty calorimeter with stirrer,

$w_2 = w_1 +$ water

$w_3 = w_1 +$ liquid

t_1 = time of w_2 to cool from, say 55 °C to 50 °C

t_2 = time w_3 to cool from, 55 °C to 50 °C

Specific heat of liquid

$$C = -\frac{[(w_1 \times \text{Sp.heat water, } c) + (w_2 - w_1)]\,\{(t_2/t_1) - (w_1 \times c)\}}{w_3 - w_1}$$

168

16.6.7. <u>Newton's Law of Cooling</u> states that

The rate of cooling (rate of loss of heat) of a hot body is ∝ excess of temperature of the body over the surroundings

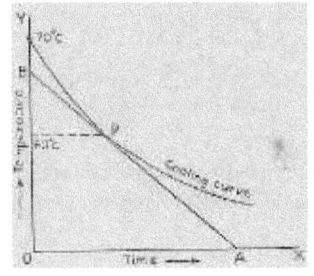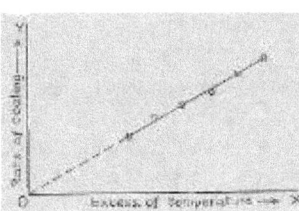

15.6.8. **Mechanical Equivalent of Heat** J

It is the amount of work that must be expended to produce unit quantity of heat.
<u>Searle' Friction-Cone apparatus</u> may be used to determine J .
c = specific heat of water
w_1 = weight empty calorimeter with stirrer,

$w_2 = w_1$ + water
θ_1 & θ_2 initial and final temperatures of water,
n = number of revolutions made by the counter,
$2\pi r$ = circumference of the disc,
m = mass attached to the end of spring

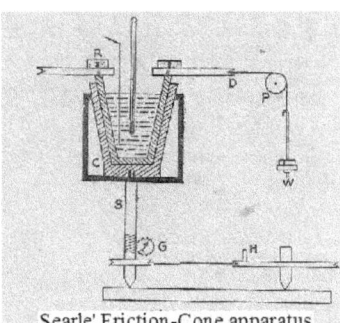
Searle' Friction-Cone apparatus

$$J = \frac{m\ 2\pi\ r\ n}{\pi\ r^2\ (w_1\ c + w_2 - w_1)\ (t_2 - t_1)}\ ergs\,/\,Cal.$$

15.7. MELTING

15.7.1 LATENT HEAT, L

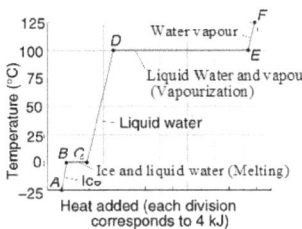

BC and DE represent Latent heats, L_f and L_v.

The amount of heat added to or removed from a substance undergoing a phase change.

$$L = \frac{Q}{m} \quad \boxed{\text{Unit } J\ kg^{-1} \quad \text{(Scalar)} \quad (\text{M}^0\ \text{L}^2\ \text{T}^{-2})}$$

15.7.2 TROUTON'S RULE -1

$$\boxed{\frac{L_f}{T_M} = \text{Constant}}$$

A = Atomic weight
L_f = Latent heat of fusion
T_M = Melting Point in K.

15.7.2.1 Don't get stuck!

Why ice trays are made of plastic? The moisture on the hand comes in contact with the cold surface of the tray the moisture freezes, cementing the fingers to the tray, causing the loss of some skin.

15.7.2.2 Role of Salt

* Pure water freezes at $0^o C$.
* 23% of NaCl (by weight) + pure water freezes at $-21\,^\circ C$.
* Above or below this critical concentration freezing temperature becomes higher than $-21\,^o C$.
* In winter in high latitude countries salt is added to melt the unwanted snow on the roads, when atmospheric T $> -21\,^o C$; at T $< -21\,^o C$, salt water starts to crystallize.

15.7.2.3. Physics of Ice Skating

At $0^o C$ the specific volume3s of ice and water are $1.09\ cm^3 g^{-1}$ and $1.00\ cm^3 g^{-1}$, respectively. The latent heat of fusion = $335\ Jg^{-1}$. From the Clausius-Clapeyron equation, the slope of the fusion curve is $\frac{dP}{dT} = -134\ atm\ \text{K}^{-1}$.

The bottom of an ice skate hollow ground and so enormous pressure is built up under the sharp edge, of the order of 100 *atm.* or more. (Fig.).

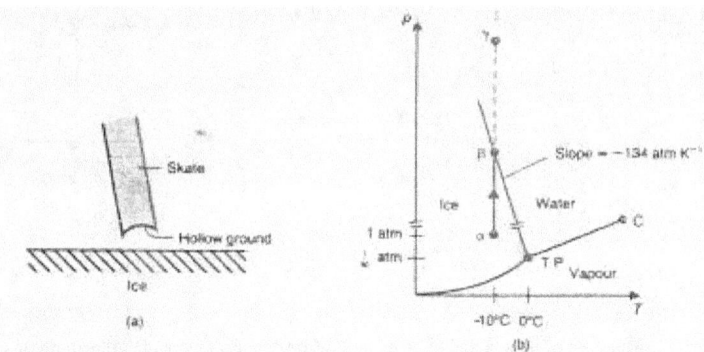

The physics of skating. (a) Hollow-ground ice-skate. (b) $P-T$ projection for pure water.

Skating is not a pressure melting effect!

15.7.2.4 EVAPORATION and BOILING

Evaporation, $\underline{Liquid \rightarrow Gas}$, occurs at the surface of a liquid and can occur at any temperature. Volatile liquids (alcohol, petrol, perfume) chane the state easily at normal temperatures. Factors affecting evaporation are a) Temperature, b) Surface area, c) A current of air above the surface, *i.e.*, atmospheric humidity.

Saturated Vapour Pressure - Boiling Point

Definition: Boiling Point (B.P.) of a liquid is the temperature at which its saturated vapour pressure (s v p) = the external pressure acting on it, *i.e.*, the atmospheric pressure of the place.

Specific volume of water at $100\,^{\circ}C = 1.043\ cm^3 g^{-1}$

Specific volume of steam $100\,^{\circ}C = 1673\ cm^3 g^{-1}$

Latent heat of vaporization $= 2257\ Jg^{-1}$

Clausius-Clapeyron equation $\frac{dP}{dT} = \frac{L}{T(V_2 - V_2)}$ gives $\left[\frac{dP}{dT} = +\frac{1}{28} atm\ K^{-1}\right]$.

i.e., on the top of Mt Everest, $(0.35\ atm)$ BP of water is $82\,^{\circ}C$.

- Water boils at $100\,^{\circ}C$ at which its s v p = $1\ atm$ = $76\ cm$ of Hg;
- Water boils at $90\,^{\circ}C$ at which its s v p = $3\ km$ above sea level;
- Water boils at $80\,^{\circ}C$ at which its s v p = $5\ km$ above sea level.

15.7.2.5 Variation of L_V with θ.

Q_θ = total heat of the saturated vapor of a liquid at any temp $\theta\,^{\circ}C$ is the quantity of heat required to raise $1\ gm$ of liquid from $0\,^{\circ}C$ to $\theta\,^{\circ}C$ and convert it into saturated vapour at $\theta\,^{\circ}C$.

$$Q_\theta = (L_\theta + \theta) = (606.5 + 0.305 \ \theta) \ ;$$
$$L_\theta = (606.5 - 0.695 \ \theta) \ ;$$

- L_f (water, fusion) $= 33 \ x \ 10^4 J \ kg^{-1}$.

- L_v (water, vaporization) $= 23 \ x \ 10^5 J \ kg^{-1}$.

15.7.3   TROUTON'S RULE - 2

$$\boxed{M \frac{L_v}{T_b} = \text{Constant}, \approx 20}$$

M = Molecular weight of liquid
L_v = Latent heat of vaporization
T_b = Boiling Point in K.

15.8.   VAPOUR PRESSURE
15.8.1   Vapour Pressure

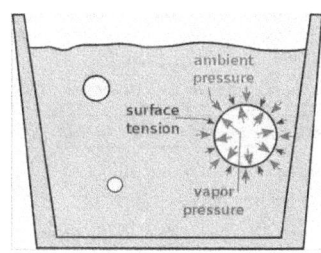

$$\boxed{\text{Log } p = A + \frac{B}{T} - C \ \log T}$$

$A, B \ \& \ C$ are constants, and
$p = $ S v p, at temperature T.

$$\boxed{S \ v \ p \propto T}$$

- S v p is independent of change in volume, V, *i.e.*,
- S v p does not obey Boyle's Law.

Vapour pressure of water at $100\,^oC = 1 \ atm. = 0.101325$ M*Pa*.

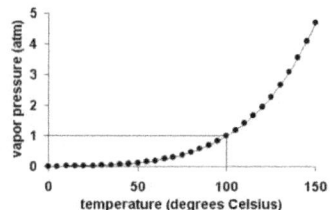

Table showing variation of vapour pressure of water with pressure

Temperature ($^{\circ}C$)	Vapor Pressure (MPa)
0.01	0.000 612
25	0.003 17
50	0.012 35
75	0.0386
100	0.1014
150	0.4762
200	1.555
250	3.976
300	8.588
350	16.529
373.946	22.064

15.8.2 **Unsaturated vapours behave approximately like** <u>gases</u>.
   * Dalton's Law (of partial pressures) holds good for both saturated vapours and unsaturated vapours, but with a difference that the

   $\boxed{pressure\ of\ saturated\ vapour\ =\ constant}$,

- $\boxed{pressure\ of\ \text{un}saturated\ vapour\ \neq\ constant}$.

- At a 760 *mm* height the Boiling Point $= 100\,^{\circ}C$,

   * At 1520 *mm* height the Boiling Point $= 120.5\,^{\circ}C$,

   i.e., saturated steam at $120.5\,^{\circ}C$ balances a 1520 *mm* column of mercury, that is, it exerts a pressure of 2 *atm*. As a rule, a pressure in excess of 760 *mm* is stated in atmospheres (*atm*), not in *mm* mercury.

15.8.3 Beyond 374 $^{\circ}C$, even at the highest pressure, <u>water cannot exist as a fluid</u>, (no Boiling Point for it is above its <u>critical temperature</u>, $374\,^{\circ}C$).

15.8.4 Boiling and Saturated vapor pressure:
   Definition: Boiling Point: where the vapour pressure of a liquid = the external pressure. Mercury is toxic at room temperature; <u>all liquids have some vapour pressure</u>.

15.8.5. <u>Evaporation of a liquid takes place only at its surface</u>.
   1) Boiling takes place throughout the whole volume of liquid
   2) Vapor pressure inside a bubble = S v p of the surrounding liquid at the temperature concerned.
   3) At $T < T_b$, SVP of liquid at T [$\equiv$ Vapour pressure (VP) inside the bubble 'h' below the liquid. Surface] < [atm. pressure, $(p + h\rho g)$] = external pressure on the bubble.
   So the <u>bubble cannot grow in size</u>.
   Water has
   4) $T_b = 100\,^{\circ}C$ at which its SVP = 1 *atm*, at sea level
   5) $T_b = 90\,^{\circ}C$ = 3 *km* above sea level;
   6) $T_b = 80\,^{\circ}C$ = 5 *km* above sea level.

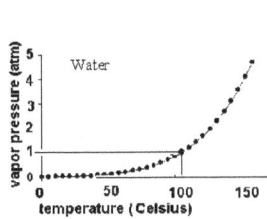

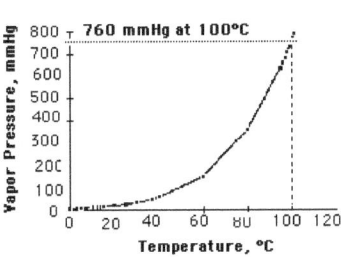

7) **Surface *Area***: the surface area of the solid or liquid in contact with the gas has no effect on the vapor pressure.

8) If sunflower oil has a density of 0.919 g cm^{-3} am3 at 20 ^{o}C, the vapour pressure of ethanol at 20 ^{o}C $(5.58\ kPa)$ will produce 650 mm of height difference. (Huge!!!).

9) When mercury is used as a liquid for manometer, the vapour pressure of water at 20 ^{o}C $(2.33\ kPa)$ will produce 17.5 mm of height difference.

10) Elevation of B.P with vapour pressure of water per *gm molecule* = 5.34 ^{o}C

11) The molecular elevation of B.P. $\boxed{dT = \dfrac{pT}{\rho L_v J}}$

12) The molecular weight $\boxed{M = \dfrac{534x}{dT}}$

13) A liquid can be made to boil.
    (i)   by increasing its temperature to the boiling point under environmental pressure, or
    (ii)   by reducing the environmental pressure until boiling point equals the temperature of the liquid.

14) Bubbles form during boiling contain vapour of the liquid (Nucleation) at small imperfections in the walls of the container or grains of solid material

15) Cavity formation is the formation of bubbles in a liquid by mechanical means; typically, rapid rotation or vibration of an immersed solid surface

16) Liquids can exist at negative pressures by keeping them under tension.

17) The sublimation temperature of water ice is 198 K under Martian atmospheric conditions.

18) The sublimation temperature of ice in a vacuum is 152 K.

19) Water from deep sea thermal vents can be as hot as 700 °F and yet not boil.

20) Boiling point of water decreases 1 °F for every 500 foot increase in altitude .

21) Sublimation, freeze drying, condensed milk, freezer burns on food stored in the freezer for awhile; freeze drying is a controlled form of freezer burn.

22) In 1856 Gail Borden received first patent on condensed milk from the United States and England.

23) Vapours can be condensed by compression alone, gases must also be cooled.

24) Anomalous behaviour of water, expands upon freezing.

25) Frozen carbon dioxide is also known as dry ice since it cannot exist as a liquid under normal pressures. Dry ice doesn't melt, it sublimates.

15.8.6 Pressure Cooker (How does it do that thing it does?)

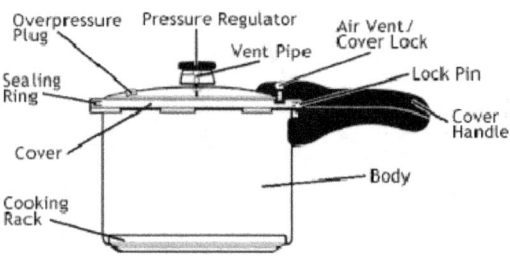

Dennis Papin (1679) introduced first to produce steam, then in 1939 to cook food. Water boils at $100\,^{o}C$ at NTP; evaporates and becomes steam at the same temperature, $100\,^{o}C$. A pressure cooker works on a simple principle: Steam pressure. A sealed container, with a lot of steam inside, builds up high pressure, which helps food cook faster; as steam has six times the heat potential when it condenses on a cool food product. Steaming is the way to retain nutrients and provide a healthy meal. Physics behind the working of pressure cooker is shown with Fig. below (TV diagram for heating water with pressure) (vide Section 15.8.5).

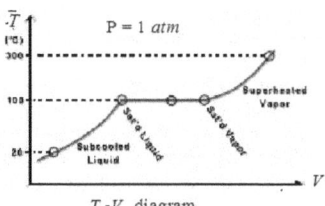

T-V diagram

Consider example below:

- The pressure gauge has weight at top (Pressure Regulator) = $0.1\,N\ mm^2$, usually.

- The pressure (SVP) inside the cooker vessel = $0.2\,N\ mm^2$,

- The external pressure on the Cooker vessel = $0.1\,N\ mm^2$,

- Then food in a container is immersed in water at $120\,^{o}C$.

Pressure (P) inside the Cooker $1\ atm = 14.696\ psi \approx 1Pa$ $=1\ Nm^{-2}$ or $kg{\cdot}m^{-1}{\cdot}s^{-2}$	Cooking T (C)
0 psi	100
5 psi	104
10 psi	113
15 psi	121
* Higher the P shorter the cooking time	

----o-0-o-0-o-0-o---

Chapter 16

HEAT – 3 - REAL GASES,
KINETC THEORY AND
LIQUEFACTION OF GASES

When a man sits with a pretty girl for an hour, it seems like a minute.
But let him sit on a hot stove for a minute it's longer than an hour. That's relativity!"
-   Albert Einstein

16.1   **REAL GAS**

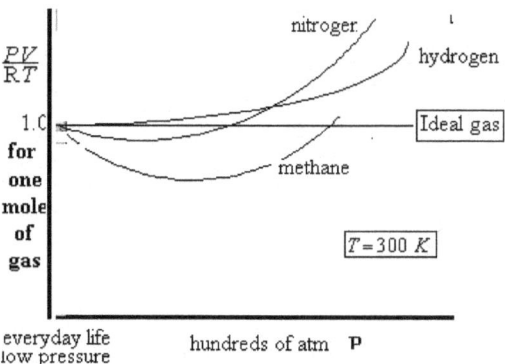

16.1.1   Ratio of Specific heats of gases, Gruneisen γ

1) For an ideal **monatomic** gas,

$$\boxed{C_P - C_V = (5/2)\,R - (3/2)\,R = R}$$

$$\gamma = \frac{C_P}{C_V} = 5/3 = 1.66$$

2) For a **diatomic** molecule, degree of freedom,

d.f. = $(3N - 1)$

$n = (3 \times 2 - 1) = 5$

Total kE $= \frac{5}{2} RT$

$$\gamma = \frac{C_P}{C_V} = \frac{7}{5} = 1.40$$

3) For a **complex** molecule,

$$\gamma = \frac{C_p}{C_v} = (1 + \tfrac{2}{n})$$

16.1.2 Table of γ

Gas	γ
1) Air	1.410
2) Ammonia	1.31
3) Argon	1.66
4) CO_2	1.30
5) CO	**1.40**
6) He	**1.66**
7) H_2	1.41
8) O_2	1.40

16.1.3 Variation of γ with T for air

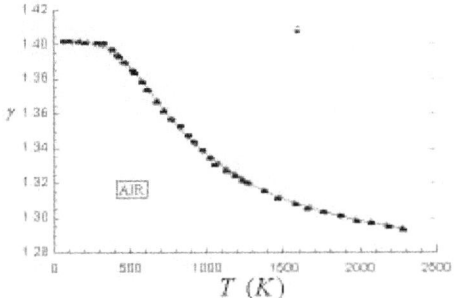

$$T \ (K)$$

16.2. IDEAL GAS
16.2.1 <u>Adiabatic Transformation</u> of an ideal Gas (Infinitesimal Transformation)
When work is due to change in volume,

$$dW = p \, dV$$

$$dU = -dW = -p \, dV$$

16.2.1.1 Equation for an adiabatic process for an ideal gas

$$U = (3/2)\, n\, R\, T = N\, C_v dT$$

$$pV^\gamma = \text{constant}$$

$$TV^{\gamma-1} = \text{constant,}$$
$$Tp^{(1/\gamma)-1} = \text{constant,}$$
$$\gamma-1 = n\, R/C_V$$

16.2.1.2 (1) For a <u>monatomic</u> gas at STP,

$$\gamma = \frac{C_p}{C_v} = \frac{5}{3} = 1.67$$

(2) for a diatomic gas,

$$\gamma = \frac{C_P}{C_V} = \frac{7}{5} = 1.4$$

16.2.1.3 Using the ideal gas law

For all gases, though, the following is true.

$$\boxed{C_P - C_V = R}$$

16.3. KINETIC THEORY OF GASES

16.3.1 The distribution P(v) of molecular speeds at a temperature T, for an ideal gas, is given by Maxwell-Boltzmann (Classical), are given equivalently as

$$P(v) = \sqrt{\frac{2}{\pi}} \left(\frac{M}{k_B T}\right)^{3/2} v^2 e^{(-Mv^2/2k_B T)}$$

$$P(v) = 4\pi \left(\frac{M}{2\pi RT}\right)^{3/2} v^2 e^{\left(-Mv^2/2RT\right)}$$

$$\boxed{dN = 4\pi N \left(\frac{M}{2\pi k_B T}\right)^{3/2} e^{\left(-Mv^2/2k_B T\right)} v^2 dv}$$

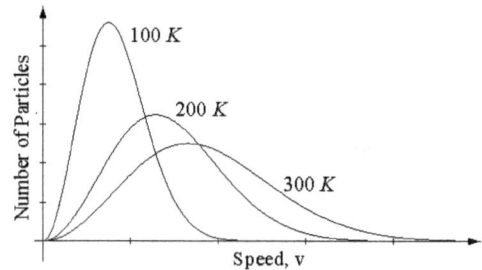

The distribution is broader for a gas with smaller mass than for a gas with larger molecular mass at the same T.

16.3.1.2 Molecular model: Postulates:

1) Every Gas composed of molecules; which are <u>alike</u>. A molecule of one gas differs from those of another.

2) The molecules of a gas are <u>rigid, perfectly elastic</u>, solid spheres, <u>identical</u> in all respects (such as mass, form, *etc.*), point masses when compared to the separation between them.

3) Molecules are in a state of incessant <u>random motion</u>, in all directions, with different all possible velocities, *i.e.*, gas is in a state of molecular chaos.

4) The gas molecules collide with one another and with the walls of the container, yet in the <u>steady state collisions</u> do not affect the molecular density.

5) Gas's molecular collisions are perfectly <u>elastic</u>; there are no forces of attraction or repulsion between them, i.e., the energy of a gas is completely <u>kinetic</u>.

6) The time spent in a collision is negligible as compared to the tat during which they move independently.

16.3.2 Pressure of a Gas according to Kinetic Theory

16.3.2.1 <u>The Kinetic Theory Equation</u>

$$P = \frac{1}{3}\frac{M}{V}c^2 = \frac{1}{3}\rho\, c^2$$

16.3.2.2 Mean speed,

$$c = \sqrt{\frac{3P}{\rho}} = \sqrt{\frac{8\, k_B T}{\pi\, m}}$$

16.3.2.3 Total mass of gas,

$$M = m\, N_A$$

16.3.2.4. Density

$$\rho = \frac{m\, N_A}{V}$$

16.3.2.5. Total kinetic energy of all the molecules = U of an ideal gas

$$U = \frac{1}{2} n\, RT$$

16.3.3.1 Mean free Path, λ

The average distance traversed by a molecule between successive collisions is called "mean-free path".

r = radius of molecules
ℓ = distance a molecule travels/ unit time
n = # of molecules / unit volume
N = # of collisions made

$$\lambda = (\frac{1}{4\pi r^2 \ell n}) \quad Unit\ m \quad (scalar)\ (M^0\, L^1\, T^0)$$

16.3.3.2 Example:

(1) At sea level $\lambda_{Air} = 7x\,10^{-8}\,m$

(2) At the altitude 300 km, pressure is $10^{-10}\,atm$ $\lambda_{Air} \approx 3\,km$

$$\lambda = (\frac{r^2}{\ell})$$

$$n = \pi\,\lambda\,r^2.$$

16.3.4 Thermal resistance,

$$Thermal\ resistance\ \ Unit\ tog. \quad (scalar)\ (M^{-1}\, L^0\, T^3)$$

$$1\ tog = 1\ m^2\ kW$$

Example: A medium quality blanket has a thermal resistance of ~ 1 tog.

16.3.4.1 Kinetic Energy and Pressure Relation

$$P = (\frac{2}{3})E$$

16.3.4.2 Kinetic Interpretation of Temperature

$$E = \frac{1}{2}Mc^2 = \frac{3}{2}RT$$

16.3.4.3 Kinetic Energy per molecule,

$$\varepsilon = \frac{3}{2}k_B T$$

16.3.4.4 RMS speed

$$c = c_0 \sqrt{\frac{T}{T_o}} = c_0 \sqrt{\frac{(273 + \theta)}{273}}$$

$$= \sqrt{\frac{3P}{\rho_{NTP}} \frac{(273 + \theta)}{273}}$$

RMS speed, $\quad c = \sqrt{\frac{3k_B T}{M}} \quad$ Unit $ms^{-1} \quad (scalar) \quad (M^0 \, L^1 \, T^{-1})$

16.3.4.5 Most probable speed,

$$v_m = \sqrt{\frac{2k_B T}{M}}$$

16.3.4.6 Graham's Law of Diffusion

For two gases of densities ρ_1 and ρ_2, and RMS speeds, c_1 and c_2,

Ratio of diffusion

$$\frac{r_1}{r_2} = \frac{c_1}{c_2} = \sqrt{\frac{\rho_2}{\rho_1}}$$

16.3.4.7 Viscosity coefficient η of a gas molecule,

Viscosity coef $\eta = \frac{1}{3} \rho \, c \, \lambda \quad$ Unit $Poise \; (P)(or) \quad (scalar) \quad (M^1 \, L^1 \, T^{-1})$

16.3.5.1 Degree of Freedom (d.f.) of Motion

A d.f. is the motion of the molecule in a direction

For a polyatomic molecule with N atoms,

i)   for a linear molecule

Number of df, n = 3N

ii)   for non-linear molecule,

Number of df, n = (3N -1)

N = # of atoms in the molecule.

16.3.5.2 Boltzmann Law of Equi-partition of Energy:

E per df $= \frac{1}{2} RT$

16.3.5.3 A gas heated from T to $(T + \Delta T)$, when volume V = constant,

Increase in energy input ΔE = kinetic energy of the molecule

$$\Delta E = C_V \Delta T = \frac{3}{2} R \, \Delta T$$

16.3.5.4 An idealized **graph of the heat capacity of hydrogen with temperature**.

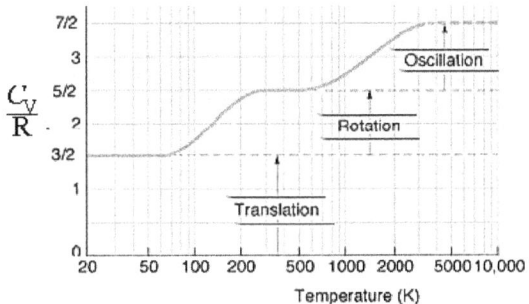

As the moment of inertia for H_2 molecule is small, the temperature by which the Equipartition Law holds for rotational modes is actually quite high.
(i)   **Isentropic** At constant entropy.
(ii)   **Isobaric** At constant pressure.
(iii)   **Isochoric** At constant volume.
(iv)   **Isothermal** At constant temperature.
(v)   **Reversible** A reversible process is one, which would change direction with an infinitesimal change in external conditions.

16.3.6.1   Deviations from Ideal Behavior of a Gas
     * All real gasses fail to obey the ideal gas law to varying degrees
     * The deviation from ideal behavior is large at high pressure
     * The deviation varies from gas to gas
     * At lower pressures (<10 *atm*) the deviation from ideal behavior is typically small,
       and the ideal gas law can be used to predict behavior with little error
     • As temperature increases the deviation from ideal behavior decreases
     *As temperature decreases the deviation increases, with a maximum deviation near the temperature at which the gas becomes a liquid

16.3.6.2   **Two of the characteristics of ideal gases**
     *The gas molecules themselves occupy no appreciable volume
     *The gas molecules have no attraction or repulsion for each other
     *Real molecules, however, do have a finite volume and do attract one another
     *At high pressures, and low volumes, the intermolecular distances can become quite short, and attractive forces between molecules becomes significant
Neighboring molecules exert an attractive force, which will minimize the interaction of molecules with the container walls. And the apparent pressure will be less than ideal (PV / RT will thus be less than ideal).

16.3.6.3   **Real gas curve**
       As pressures increase, and volume decreases, the volume of the gas molecules becomes significant in relationship to the container volume.
       In an extreme example, the volume can decrease below the molecular volume, thus (PV/RT) will be higher than ideal (V is higher).

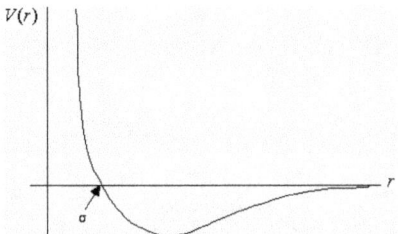

At high temperatures, the kinetic energy of the molecules can overcome the attractive influence and the gasses behave more ideal.

At higher pressures, and lower volumes, the volume of the molecules influences (PV/RT) and its value, again, is higher than ideal.

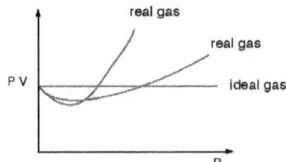

16.4. **LIQUEFACTION OF GASES** (PREPARATION OF CRYOGENTS):
March towards the Absolute Zero of Temperature

Van Marum (1800) examined if the gaseous Ammonia obeyed Boyle's law as air does. His experiment proved that around $P = 7$ *atm*, the gas violated the Boyle's Law, but on the other hand was found to liquefy!!

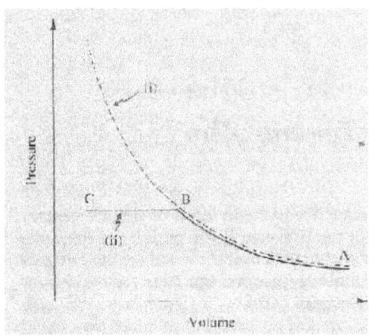

Michael Faraday and Paris performed an experiment with Chlorine in a sealed and bent glass tube as in Fig. With chlorine at one end and made hot to find the other kept cold with liquid chlorine.

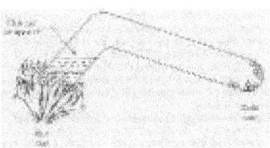

16.4.1 <u>Andrew's experiments</u> on CO_2.

Thomas Andrew made very careful experiments with Carbon dioxide.

16.4.1.1 Isothermals, *i.e.*, pV diagram for CO_2

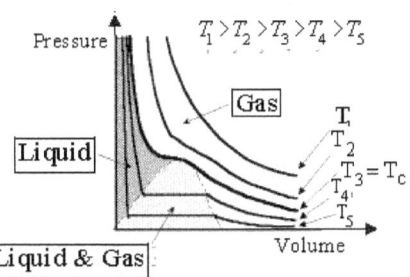

16.4.1.2 <u>Equation of State</u> -- relates P, V and T. Typically, plot P *vs.* V for constant T.

For $T > T_c$ (<u>Critical temperature</u>) can have gas and liquid in equilibrium. As decrease in volume system develops, more liquid, pressure is called the vapour pressure, only a function of T not V. At small volume, no vapor left, system is all liquid and then to change V must apply very large pressure. Liquids (and solids) are much less <u>compressible.</u>

16.4.1.3 **Phase diagram for Carbon dioxide; X is Triple Point and Z is T_C**

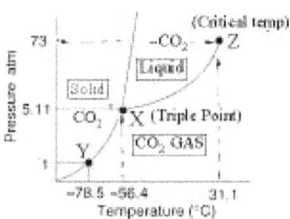

16.4.1.4 **Gas**

Gas is the term applied to a substance, which is in the gaseous phase and T above its critical temperature, T_C.

16.4.1.5 **Vapour**

is the term applied to a substance, which is in the vapour phase at T below its critical temperature, T_C.

16.4.1.6 Saturated Vapour

is the term applied to a substance, which is in the vapor phase at T in equilibrium with its own liquid.

16.4.1.7 Unsaturated vapours obey gas laws to the same extent as real gases.

The deviation from the gas laws is intimately connected with the process of liquefaction.

16.4.1.8 To every gas there is a critical point, T_C,

below which alone the gas can be liquefied by means of pressure alone at $P_{Liq} = P_{Gas}$;

pressure and volume corresponding to this is P_C and V_C.

16.4.1.9 There is a continuity of state between the liquid and gaseous phases.

16.4.2 Amagat's Experiments on CO_2 (PV versus P curves for CO_2, H_2, N_2. gases)

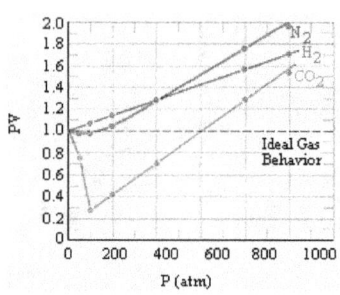

16.4.2.1 Dewar Flask

Sir James Dewar was first to **liquefy hydrogen**. To store liquefied gases for subsequent use in experiments Dewar invented (1898) the Dewar flask. Kamerlingh Onnes was the first to **liquefy a noble gas, Helium**, in 1908, and soon discovered **superconductivity**, received the Nobel Prize in 1915.

16.4.2.2 The van der Waals equation

This contains two constants, a and b, that are characteristic properties of a particular gas. The first of these constants corrects for the force of attraction between gas particles. The other van der Waals constant, b, is a rough measure of the size of a gas particle. Empirically, for a gas

$$pV = A + Bp + Cp^2 + Dp^3 +$$

A, B, C, D are Virial coefficients depending on T.

16.4.2.3 Boyle Temperature, T_B.

At T_B,   B = 0.

(1) and the gas approximates to an ideal gas.

$$T_B = 3T_C$$.

(2) When $T < T_B$, the gas is highly compressible and intermolecular forces are significant.

(3) When $T > T_B$, the gas obeys the Boyle's Law, and the gas approximates to an ideal gas.

16.4.3 Permanent Gases

O_2, N_2, H_2, He, and CO_2 are those gases at room temperature

$$\boxed{T > T_c}.$$

16.4.3.1 Van der Waal's Equation of State *for Gases*

$$\boxed{P = \frac{RT}{V-b} - \frac{a}{V^2}}$$

$$\boxed{\left(P + \frac{an^2}{V^2}\right)(V - nb) = n\,RT}$$

a and *b* are van der Waal's constants.

16.4.3.2 *Van der Waals Constants for Various Gases*

Gas	a (L^2-atm mol^{-2})	b (L mol^{-1})
He	0.03412	0.02370
Ne	0.2107	0.01709
H_2	0.2444	0.02661
Ar	1.345	0.03219
O_2	1.360	0.03803
N_2	1.390	0.03913
CO	1.485	0.03985
CH_4	2.253	0.04278
CO_2	3.592	0.04267
NH_3	4.170	

16.4.3.3 At the Critical point (P_c, V, T_c).

$$\boxed{V_c = 3b}. \quad \boxed{T_c = \frac{8a}{27\,b\,R}}. \quad \boxed{P_c = \frac{a}{27\,b^2}}.$$

16.4.3.4 The predicted constraint on the critical parameters is now seen to be

$$\boxed{\frac{RT_c}{P_c V_c} = \frac{8a/27b}{a/9b} = \frac{8}{3} \approx 2.67} = 3 \text{ (calculated)}$$

16.4.3.5 Inversion temperature, T_i

$$\boxed{dT = \frac{\left(\frac{P_1 - P_2}{C_P}\right)}{\left(\frac{2a}{RT} - b\right)}}$$

Three cases:

1) If dT = +ve, cooling is $T < T_i$,

2) If dT = -ve, heating is $T < T_i$,

3) If dT = 0, null effect, and $T = T_i$, where

$$T_B = \frac{a}{R\,b},$$
$$T_i = \frac{2a}{R\,b},$$
$$T_C = \frac{8}{27}T_B = \frac{4}{27}T_i$$
$$\frac{T_i}{T_C} = 6.75$$

16.4.3.6 T_i and T_C for typical Gases

Gas	BP	T_i	T_C
Oxygen	90 K	> Room T	155 K (-119°C)
Nitrogen	77 K	> Room T	126 K (-147°C)
Hydrogen	20 K	143 K	33 K (-240°C)
Helium	4.2 K	30 K	5.3 K (-268°C)
Carbonic acid	31°C		
Sulphurous acid	-10°C		157°C
Water steam		374°C	224.2

* Beyond 374 °C, even at the highest pressure, water cannot exist as a fluid, whence there is no boiling point for it above 374°C (its critical temperature).

16.4.3.7 <u>Corresponding State Equation</u> (Reduced van der Waal's Equation)

Reduced Isothermal

Put $P = \pi P_C$, $V = \varphi V_C$, $T = \eta T_C$.

$$\left(\frac{\pi}{27} + \frac{1}{9\varphi^2}\right)(3\varphi - 1) = \frac{8\eta}{27}$$

$$\left(\pi + \frac{3}{\varphi^2}\right)(3\varphi - 1) = 8\eta$$

16.5.   CRYOGENICS

16.5. 1 Summary

Generally the science of cryogenics is when the temperature goes below that which we can reach with conventional refrigeration equipment, around $250°F$ below zero. Many gases are liquid at these low temperatures. They can be colder, but the following list is the temperature at which these gases boil. Before their temperature can get any higher all the liquid must boil away and turn back into a gas.

Fluid	BP (°C)	BP (°F)
Oxygen	-183°	-297°
Air($70\%N_2 + 21\%O_2$)	-195°	-319°
Nitrogen	-196°	-320°
Neon	-246°	-411°
Hydrogen	-253°	-423°
Helium	- 270°	-452°

Liquid air has a density of approximately 870 kgm^{-3} (0.87 gcm^{-3}). The BP of liquid air is approximately 78 K (-195 $^\circ$C)(-319 $^\circ$F).

16.5.2 **Liquefied Petroleum Gas (LPG, GPL, LP Gas)**

Liquid petroleum gas or simply Propane or Butane, is a flammable mixture of hydrocarbon gases used as a fuel in heating appliances (cooking food) and vehicles. It is increasingly used as an aerosol propellant and a refrigerant, replacing chlorofluorocarbons in an effort to reduce damage to the ozone layer. When specifically used as a vehicle fuel it is often referred to as *autogas*.

Varieties of LPG bought and sold include mixes that are primarily propane (C_3H_8), primarily butane (C_4H_{10}) and, most commonly, mixes including both propane and butane, depending on the season — in winter more propane, in summer more butane.

16.5.3 **Liquefied Natural Gas (LNG)**

Natural gas (predominantly Methane, CH_4) that has been converted to liquid form for ease of storage or transport.

Liquefied natural gas takes up about $\frac{1}{600}$ th the volume of natural gas in the gaseous state. It is odorless, colourless, non-toxic and non-corrosive Hazards include flammability after vaporization into a gaseous state, and freezing.

The liquefaction process involves removal of certain components, such as dust, acid gas, He, water, and heavy hydrocarbons which could cause difficulty downstream. The natural gas is then condensed into a liquid at close to atmospheric pressure (maximum transport pressure set at around 25 kPa (4 psi)) by cooling it to approximately -162 $^\circ$C (-260 $^\circ$F).

LNG achieves a higher reduction in volume than Compressed Natural Gas (CNG) so that the /volumetric / energy density of LNG is 2.4 times greater than that of CNG or 60 % of that of diesel. This makes LNG cost efficient to transport over long distances.

&%&%&%&%&%&%&%&%&

Chapter 17

THERMODYNAMICS
TYPES OF PROCESSES, LAWS, HEAT ENGINES, PHASE TRANSITIONS, THERMAL CONDUCTIVITY, THERMO-ELECTRICITY

There is something fascinating about science.
One gets such wholesale returns of conjecture out of such a trifling investment of fact.
~Mark Twain, Life on the Mississippi, 1883

17.1   THERMODYNAMICS

17.1.1   Thermodynamics is the study of systems involving energy in the form of heat and work. *Thermal equilibrium* is an important concept in thermodynamics.
Example: The gas confined by a piston in a cylinder, as shown in the diagram.

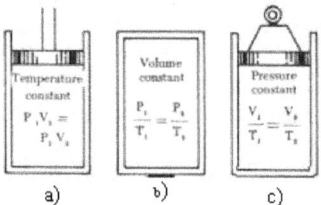

17.1.2 Types of Thermodynamic Processes:
Four Types of p-V changes for Ideal Gases:
Work depends not on the end points but also on the path direction.

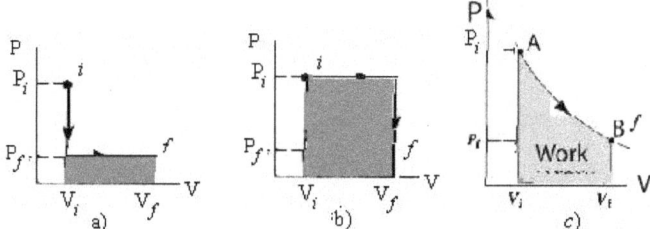

There are a number of different thermodynamic processes that can change the pressure p and/or the volume V and/or the temperature T of a system.

17.1.2.1 **Isobaric**

The pressure $\boxed{P = \text{kept constant}}$.

An isobaric system is a gas, being slowly heated or cooled, confined by a piston in a cylinder, and the *P-V* graph looks like:

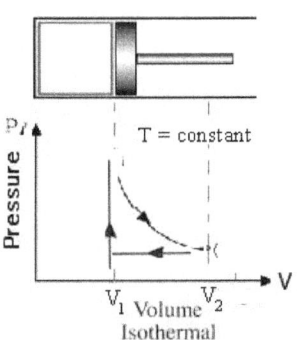

isobaric

$$\boxed{W_{isobaric} = p(V_2 - V_1)}$$

17.1.2.2 **Isochoric**

The volume $\boxed{V = \text{kept constant}}$.

An example of this system is a gas in a box with fixed walls. The work done is zero in an isochoric process, and the P-V graph looks like:

$$\boxed{W_{isochoric} = 0}$$

17.1.2.3 **Isothermal**

The temperature $\boxed{T = \text{kept constant}}$.

A gas confined by a piston in a cylinder is again an example of this, only this time the gas is not heated or cooled, but the piston is slowly moved so that the gas expands or is compressed.

$$\boxed{W_{isother} = P_1 V_1 \ln\frac{V_2}{V_1} = n\,R\,T\,\ln\frac{V_2}{V_1}}$$

If the volume increases while the temperature is constant, the pressure must decrease, and if the volume decreases the pressure must increase.

17.1.2.4 **Adiabatic**

In an adiabatic process, no heat is added or removed from the system
$\boxed{Entropy = \text{kept constant}}$.

Example is a gas expanding so quickly that no heat can be transferred. The expansion does work, and the temperature drops.

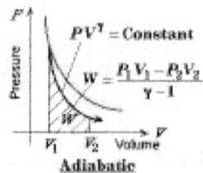

$$\boxed{W_{adiabatic} = \frac{P_1V_1 - P_2V_2}{\gamma-1}}$$

Example: CO_2 Fire Extinguisher: The gas coming out at high pressure and cooling as it expands at atmospheric pressure.

17.1.2.5 The isothermal and adiabatic processes examined in a little more detail.

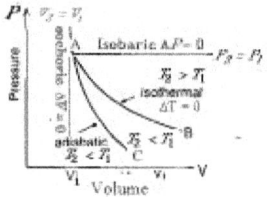

17.2. LAWS of Thermodynamics

Thermodynamics is based on THREE main laws. Like many scientific laws, **the laws of thermodynamics are <u>universal</u> for everything and have yet to be proven wrong in nature.**

17.2.1 ZEROTH LAW:

It states that <u>of the three systems A, B & C, if A and B are separately in thermal equilibrium with C, then A and B are also in thermal equilibrium with one another.</u>

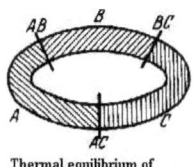

Thermal equilibrium of three substances A. B. C.

Zeroth law helps to define the term thermodynamic temperature, T (measured by a gas thermometer):

$$T \propto \lim_{p \to 0} (pV)$$

17.2.1 Equation of State of the Fluid:

$$\Phi(p,V) = T = \text{Constant}$$

a) **Heat** (Q) = the energy transferred by conduction, convection or radiation from one body to another, because one body is at a temperature T_2 higher than the other T_1.

b) **Work** (W): energy, which is transferred from one system to another by a force moving through a distance.

c) **Internal Energy** (U): the energy in a system.

17.2.2 FIRST LAW

It states that <u>the increase in internal energy, dU equals the heat received by the substance, dQ less the work done by the substance dW</u>,

$$dW = pdV$$

$$\text{For a Gas, } dQ = dU + pdV$$

i.e., Energy is conserved, if heat is taken into account that the amount of energy in the universe does not change.

17.2.2.1 **Clausius** statement:

"The energy of the Universe remains constant".

17.2.2.2 Indicator (p-V) curve

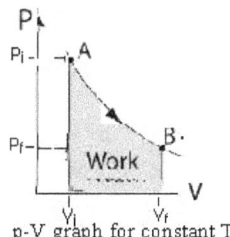

p-V graph for constant T

$$W = \int_{V_1}^{V_2} p(V)dV$$

17.2.2.3 Determination of Mechanical equivalent of heat, J
(i) Searle's friction cone apparatus,
(ii) Callendar & Barnes Continuous Flow Method:

17.2.3 SECOND LAW

17.2.3.1 Clausius' statement:

Heat Engine: <u>It is impossible for a self-acting machine, unaided by any external agency, to transfer heat from a body at lower temperature to a body at a higher temperature</u>.

17.2.3.2 Kelvin's statement:

It is impossible to derive a continuum supply of work by cooling a body to a temperature lower than that of the coldest of the surroundings

17.2.3.3. CARNOT CYCLE (Ideal heat engine)

The cycle is represented on the P- V (indicator) diagram
Work done is path dependent.

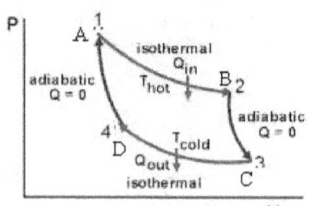 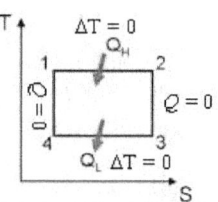

The Carnot cycle or .   T-S diagrams of Carnot Cycle

(i) Isothermal, AB: Heat absorbed = $\boxed{Q_1 = W_{isother} = P_1 V_1 \ln\frac{V_2}{V_1} = n\,R\,T\,\ln\frac{V_2}{V_1}}$

(ii) Adiabatic, BC: Heat absorbed = 0, and $\boxed{Q_2 = W_{adiabatic} = \frac{P_1 V_1 - P_2 V_2}{\gamma - 1}}$

(iii) Isothermal, CD;

Heat transferred to Sink = $\boxed{Q_3 = W_{isother} = P_2 V_2 \ln\frac{V_2}{V_1} = n\,R\,T\,\ln\frac{V_2}{V_1}}$

(iv) Adiabatic, DA; Heat transferred to sink = 0, and $\boxed{Q_4 = W_{adiabatic} = \frac{P_1 V_1 - P_2 V_2}{\gamma - 1}}$

$$\boxed{\gamma - 1 = \frac{n\,R}{C_V}}$$

$$\boxed{TV^{\gamma-1} = \text{Constant}}.$$

17.2.3.4 Net amount of heat absorbed, in units of work, by gas

$$\boxed{Q = W_{Net} = R \ln\frac{V_1}{V_2}(T_1 - T_2)}$$

17.2.3.5 Efficiency, η_{Th}

$$\boxed{\eta_{Th} = \frac{\text{Useful work}}{\text{Total Heat absorbed}} = \frac{(T_1 - T_2)}{T_1}}$$

17.2.3.6 Entropy, S

Entropy is that thermal property of a body which remains constant during an adiabatic process, when no heat is given to or removed from it.

17.2.3.7 Clausius'statement of the II Law in terms of entropy S

$\boxed{\text{Entropy of the Universe tends to a maximum}}$.

$$\boxed{dQ = T\,dS}$$

17.2.3.8 ICE

The change in entropy dS when temperature of m gm of solid ice from 0 °C to - 10 °C
(C_V of ice = 0.5)

$$dS = m\,C_V \int_{T_1}^{T_2} \frac{dT}{T} = m\,C_V \ln\frac{T_2}{T_1} = m\,C_V\,(2.3026)\,lg\frac{263}{273}\,.$$

17.2.4 THIRD LAW:

It will never be possible to reach the Absolute Zero of Temperature (- 273.1 °C).
An example of why this law is thought to be true is the constant expansion of the
universe

17.3 HEAT ENGINES

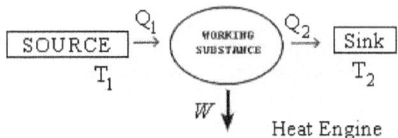

Heat Engine

Technical conversion of heat into work is achieved by thermodynamic machines (piston
steam engines, steam turbines) and combustion machines (gas engines, oil engines). The
heat is always that of highly heated gases, into which fuels (carbon, gas, oil) convert, if
they combust in the oxygen of the atmosphere

17.3.1. Ideal Heat engine Gas cycles:

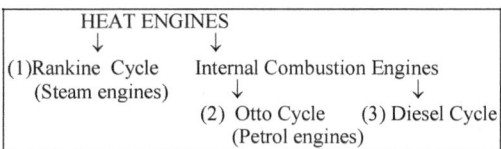

17.3.2. Rankine Cycle (Steam Engines)

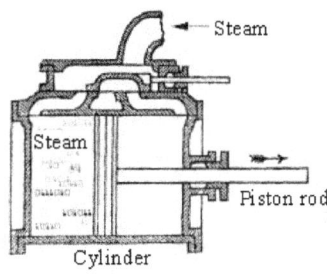

Rankine engine cycle had been very popular before 1850s in India. There primary source of heat was coal. Examples are the passenger bus, Lorries, and coal engine driven railway trains.

17.3.3. Otto Cycle (Constant volume Ignition)
  * Working substance: Air; Fuel is Petrol
  * There are 4 strokes. Spark plug for igniting,

  * Point 2 is at 600 $^{\circ}$C, 5 *atm*; Point 3 is at 2000 $^{\circ}$C, 15 *atm*;
  * Adiabatic, Compression Stroke of gas fuel mixture in the cylinder
  *Ignition of gas fuel mixture at top of the compression stroke while the volume $V = $ Constant.
  * Adiabatic (isentropic) expansion of gases in the cylinder after fuel mixture is ignited, the cycle that does partly positive work.
  * Exhaust of the spent gases and the intake of a new fuel mixture into the cylinder

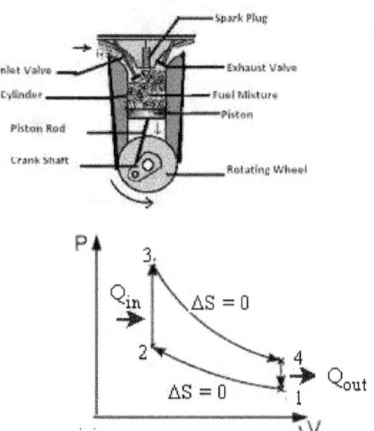

Thermal Efficiency, $\eta = \eta_{Th}$

$$\eta_{Th\ Otto} = \frac{\text{Useful work}}{\text{Total Heat absorbed}} = 1 - \left(\frac{V_2}{V_1}\right)^{\gamma-1} = 1 - \left(\frac{1}{\rho}\right)^{\gamma-1}$$

Adiabatic compression ratio, ρ

$$\rho = \left(\frac{V_1}{V_2}\right).$$

$$\rho = 1 - \left(\frac{T_4-T_1}{T_3-T_2}\right) = 1 - \left(\frac{V_1}{V_2}\right)^{1-\gamma}$$

$$Q_{in} = m\ C_V(T_3 - T_2)$$

$$\boxed{Q_{out} = m\,C_V\,(T_4 - T_1)}$$

17.3.4 Diesel Cycle :(Constant Pressure Ignition)
Working substance: Air; Fuel is heavy crude oil
There are 4 strokes (as shown)

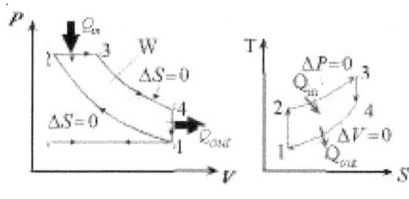

$P_1 = 1\,atm$, V_1;

$$V_2 = \tfrac{1}{17}V_1;\ \ T_2 = 2000\ ^{\circ}C, P_2 = 35\,atm.$$

$$T_3 = 2000\ ^{\circ}C$$

Efficiency $\boxed{\eta_{\text{Th } Diesel} = \left[1 - \tfrac{1}{\gamma}\left(\tfrac{V_2}{V_1}\right)^{\gamma-1}\right]\left\{\dfrac{\left(\tfrac{T_3}{T_2}\right)^{\gamma-1} - 1}{\left(\tfrac{T_3}{T_2}\right)}\right\}} = 53\ \%$

$$\boxed{\rho = \left(\dfrac{V_1}{V_2}\right)},$$

$$\boxed{\rho = 1 - \tfrac{1}{\gamma}\left(\dfrac{T_4 - T_1}{T_3 - T_2}\right)},$$

$$\boxed{Q_{in} = m\,C_p(T_3 - T_2)},$$

$$\boxed{Q_{out} = m\,C_V\,(T_4 - T_1)}$$

17.3.5 Stirling Cycle

The cycle is reversible, meaning that if supplied with mechanical power, it can function as a heat pump for heating or cooling, and even for cryogenic cooling. The cycle is defined as a closed regenerative cycle with a gaseous working fluid.

Closed cycle means the working fluid is permanently contained within the thermodynamic system. This also categorizes the engine device as an external heat engine. Regenerative refers to the use of an internal heat exchanger called a regenerator which increases the device's thermal efficiency.

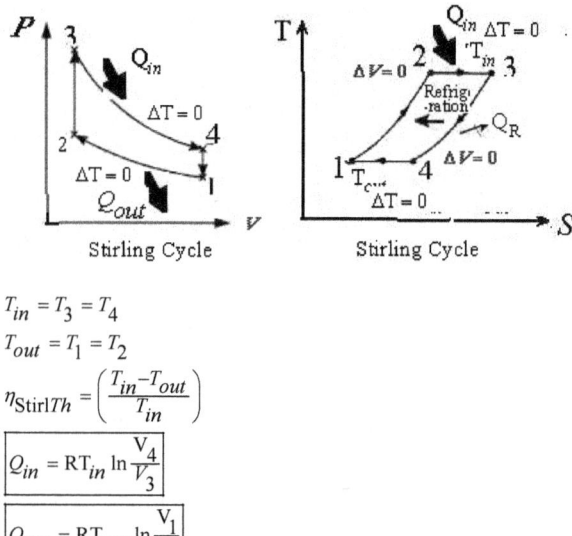

Stirling Cycle   Stirling Cycle

$$T_{in} = T_3 = T_4$$

$$T_{out} = T_1 = T_2$$

$$\eta_{StirlTh} = \left(\frac{T_{in} - T_{out}}{T_{in}} \right)$$

$$\boxed{Q_{in} = RT_{in} \ln \frac{V_4}{V_3}}$$

$$\boxed{Q_{out} = RT_{out} \ln \frac{V_1}{V_2}}$$

17.4   PHASE CHANGES (MELTING, VAPOURIZATION AND SUBLIMATION)

17.4.1   FIRST ORDER transitions

Common phase changes (melting / freezing, vaporizing / condensing, and subliming, and in certain crystalline structural change) are called "first-order transitions" because the first-order derivative of the Gibbs Free energy of finite changes during the transition. The transition is normally to be isobaric processes.

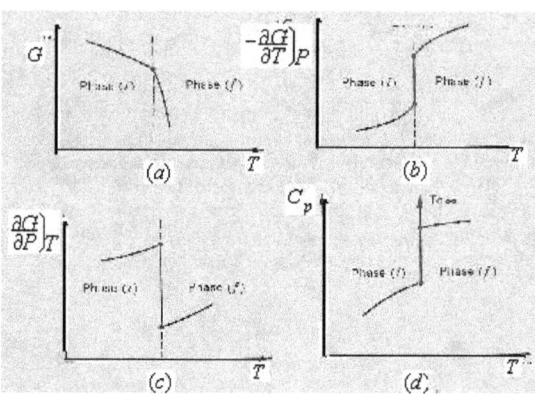

$\boxed{T \text{ and } P \text{ are constants}}$

$\boxed{S \text{ and } V \text{ are constants}}$

If the transition is reversible, latent heat involved L, whose existence shows change in S.

$$L = T(V_2 - V_1)$$

$$\boxed{S = -\frac{\partial G}{\partial T}\Big)_P \; ; \; V = \frac{\partial G}{\partial P}\Big)_T}$$

Characteristics of First-order phase change are:
1) S and V both change,
2) First order derivatives of Gibbs energy <u>change discontinuously</u>.

$$\boxed{C_p = T\frac{\partial S}{\partial T}\Big)_P \; ; \; \beta = \frac{1}{V}\frac{\partial V}{\partial T}\Big)_P \; ; \kappa = -\frac{1}{V}\frac{\partial V}{\partial P}\Big)_T}$$

17.4.2 Higher order transitions
are those in which G and its first derivatives remain constant during the transition, but higher order derivatives undergo finite transitions. Examples are normal-superconducting transitions at H= 0, and lambda transitions.

17.4.3 <u>First Latent Heat Equation</u> (Clapeyron's Equation) (First order Equation)
There is change in both S and V.
(i) A Carnot engine working with a liquid and a sink at temp T K The two isothermals one of and the other at $(T + \Delta T)$ K The liquid has boiling point at $(T + \Delta T)$ K corresponding to pressure p .L= Latent heat
(ii) The vapour adiabatically expands
(iii) The vapor is compressed isothermally
(iv) The vapour is compressed adiabatically
The second TdS equation is

$$\boxed{TdS \; = C_V\, dT + T\frac{\partial P}{\partial T}\Big)_V \, dV}$$

$$\frac{Q_1}{Q_2} = \frac{T_1}{T_2}; \quad i.e., \frac{L+dL}{L} = \frac{T+\Delta T}{T_2},$$

$$\partial Q = L \; ; \quad \partial V = V_2 - V_1$$

$$\frac{\partial Q}{\partial V}\Big)_T = T\frac{\partial p}{\partial T}\Big)_V$$

$$\boxed{\frac{dP}{dT} = \frac{L}{T(V_2 - V_1)}}$$

First order phase transitions: There is change in S and V
Second order phase transitions: There is no change in S and V

17.4.3.1 <u>Ehrenfest's Equations</u>

$$\boxed{\frac{dp}{dT} = \frac{C_{p1} - C_{p2}}{TV(\alpha_2 - \alpha_1)} = \frac{\alpha_2 - \alpha_1}{(K_2 - K_1)}}$$

C_{p1} = Specific heat of liquid in contact with its own vapour,

C_{p2} = Specific heat of saturated vapour in contact with liquid.

Second Latent Heat Equation (Clausius' Equation)

  There is no change in both S and V.

  (I) Adiabatic, $C_{p1}dT$

  (II) Isothermal, $L + dL$,

  (III) Adiabatic, $C_{p2}dT$,

  (IV) Condensation , L

$$\boxed{C_{p2} - C_{p1} = \frac{dL}{dT} = \frac{L}{T}}$$

17.5 THERMAL CONDUCTIVITY

17.5.1 Thermal Conduction

  When two bodies at different temperatures come into direct contact, in each point of their contact surface heat s from the body at a greater temperature to the one at a smaller temperature, till is reached the thermal equilibrium condition, that is till in all the points of both the bodies there is the same temperature.

  The thermal equilibrium corresponds to equal amplitudes of the harmonic oscillations and then to equal thermal kinetic energies of all the atoms of both the bodies. If instead between two bodies are placed one or several layers of other materials, the heat transfer between the bodies happens indirectly, by means of the atomic harmonic oscillations of the mediate materials.

  <u>Metals</u> are the best thermal conductors, by means of free electrons, whose thermal agitation energy is added to the one of the atoms in the crystalline structure. <u>Gases</u> have values of the thermal conductivity from about 10000 to 100000 times smaller in comparison with metals.

<u>Example:</u> The thermal flux across a plate of copper with the surface $A = 1\ m^2$ and thickness $\Delta x = 2\ cm$, among whose sides there is a temperature difference $\Delta T = -50°C$, is

$$Q_{th} = \frac{\Delta Q}{\Delta T} = -KA\frac{\Delta T}{\Delta x} = 230000\ Cal\ s^{-1}.$$

If it is considered instead a plate of cement with the same surface and the same thickness, the thermal flux is reduced to

$$Q_{th} = \frac{\Delta Q}{\Delta T} = 0,\ \ (002)\,(10000)\,(-50/2) = 500\ Cal\ s^{-1}.$$

17.5.2.1 Fourier's Law

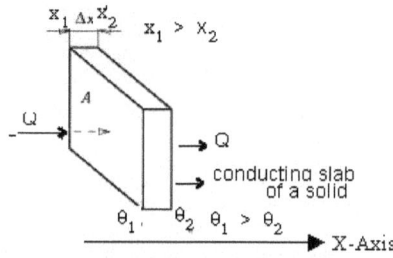

17.5.2.2 Quantity of heat flowing per second

$$Q = -KA\frac{(\theta_2-\theta_1)\,t}{x} = -KA\frac{d\theta}{dx} \quad \text{unit } Jm^{-2}s^{-1}K^{-1}$$

K = Thermal conductivity of the material, J m^{-2} s^{-1} K^{-1}

$\frac{d\theta}{dx}$ = Temperature gradient, A = area of surface, Thermal resistance = $\frac{1}{K}$

17.5.2.3. Searle's apparatus for a Metal Rod / slab

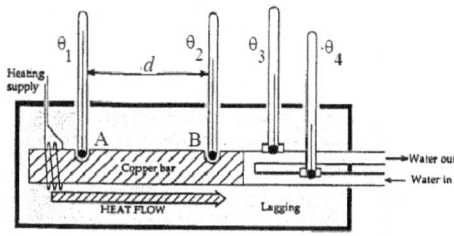

Searle's apparatus

In the steady state, heat conducted across the solid , $Q = KA\,\frac{\theta_2 - \theta_1}{d}t\ Cal$

Mean radius of the solid bar = r

d = distance between thermometers at T$_1$ and T$_2$.

m = mass of water,

$$K = \frac{m\,(\theta_4 - \theta_3)\,d}{\pi\,r^2\,(\theta_4 - \theta_3)\,t}$$

17.5.2.4. Lee's Disc Method for Insulators

A = Cross sectional area of the sample disc

x = thickness of the sample insulator

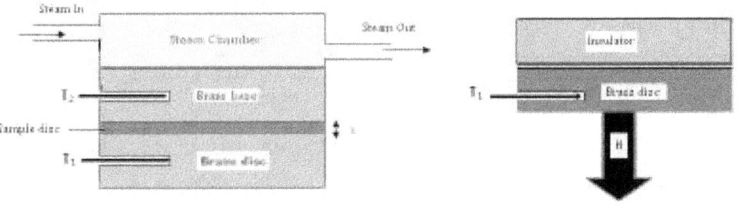

Heat loss, $H = mc\frac{dT}{dt}$

$$K_T = \{mc\frac{dT}{dt}\}/\{A\frac{T_2-T_1}{x}\}$$

17.5.3 The Fundamental Thermodynamic Relation, WHICH CONTAINS THE PHYSICS

The I and II Laws when combined gives

$$dE = TdS - pdV$$

implies S and V $E = E(S,V)$

$$T = \left(\frac{\partial E}{\partial S}\right)_V \;;\; p = -\left(\frac{\partial E}{\partial V}\right)_S$$

$$\left(\frac{\partial T}{\partial V}\right)_S = -\left(\frac{\partial P}{\partial S}\right)_V$$

Problem	Maxwell's Thermodynamic Relation
(1) Clausius-Claperon LH Eqn →	$\left(\frac{\partial Q}{\partial V}\right)_T = T\left(\frac{\partial P}{\partial T}\right)_V$
(2) Effect of P on substances →	$\left(\frac{\partial Q}{\partial P}\right)_T = -T\left(\frac{\partial V}{\partial T}\right)_P$
(3) Variation in intrinsic energy with V →	$\left(\frac{\partial Q}{\partial V}\right)_P = T\left(\frac{\partial P}{\partial T}\right)_P$
(4) Joule-Kelvin Effect →	$\left(\frac{\partial S}{\partial P}\right)_T = -\left(\frac{\partial V}{\partial T}\right)_P$ & $U + PV =$ constant
(5) Adiabatic change →	$\left(\frac{\partial T}{\partial V}\right)_S = -T\left(\frac{\partial P}{\partial Q}\right)_V$
(6) Stefan-Boltzmann law of radiation →	$\left(\frac{\partial Q}{\partial V}\right)_T = T\left(\frac{\partial P}{\partial T}\right)_V$
(7) Helmholtz Function →	$F = U - TS$
(8) Thermodynamic potential (Gibb's Function) →	$G = U - TS + PV$
(9) First TdS Equation →	$TdS = C_V dT + T\left(\frac{\partial P}{\partial T}\right)_V dV$
(10) Second TdS Equation →	$TdS = C_V dT - T\left(\frac{\partial V}{\partial T}\right)_P dP$
(11) $C_P/C_V = \gamma$ →	All the 4 Maxwell's Equations
(12) For Perfect Gas, → $\left(\frac{\partial U}{\partial V}\right)_T = 0$ →	$\left(\frac{\partial S}{\partial V}\right)_T = \left(\frac{\partial P}{\partial T}\right)_V$, & $dQ = dU + PdV$, & $PV = RT$
(13) For a Perfect Gas →	$C_P - C_V = R(1+\frac{2a}{RTV})$
(14) Homogeneous Fluid →	$C_P - C_V = T\left(\frac{\partial P}{\partial T}\right)_V \frac{\partial V}{\partial T})_P$
(15) Any Substance →	$C_P/C_V = \beta_S/\beta_T = \gamma$

By starting with F, H and G, we can get three more relations.

Maxwell's Thermodynamic Relations consists of SIX Fundamental relations, from

$$\frac{dT}{dy} \cdot \frac{dS}{dx} - \frac{dp}{dy} \cdot \frac{dV}{dx} = \frac{dT}{dx} \cdot \frac{dS}{dy} - \frac{dp}{dx} \cdot \frac{dV}{dy}$$

x and $y \rightarrow$ any two of quantities, $p, V,$

T and $S,$ keeping the other two constants

$$C_V = T\frac{\partial S}{\partial T}\Big)_V = \frac{\partial E}{\partial T}\Big)_V ;$$

$$C_P = \frac{\partial H}{\partial T}\Big)_P = T\frac{\partial S}{\partial T}\Big)_P$$

$$C_P - C_V = -V\,T\frac{\alpha^2}{\kappa_T}$$

$$C_P - C_V = R\left(1 - \frac{2\alpha(V-b)}{RTV^3}\right)$$

17.5.3.1 Useful listing of various matching Problems and Thermodynamic Relations.

17.6.1 FREE EXPANSION OF A GAS

A gas confined within an insulated container is initially confined to a volume V_1 at pressure P_1 and temperature T_1. The gas then is allowed to expand into another insulated chamber with volume V_2 that is initially evacuated. What happens?

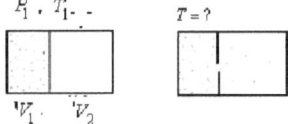

17.6.2 The Joule-Thomson (Joule-Kelvin) Effect experiment
(Cooling by Van der Waals equation). For Permanent Gases

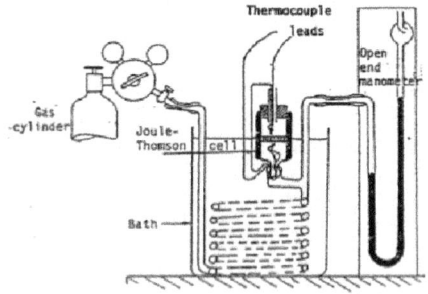

Joule-Thomson Apparatus

17.6.2.1 The Porous-Plug experiment

202

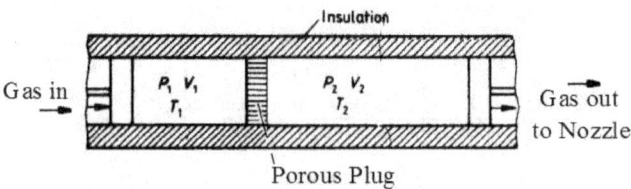

Insulation

Gas in
P_1, V_1, T_1
P_2, V_2, T_2
Gas out
to Nozzle

Porous Plug

Experiment was done with air, O_2,, N_2 and CO_2 between 4 °C and 100°C, $P_{initial}=$ 4.5 *atm* and $P_{final}=1$ *atm*. All gases except H_2 and He showed cooling effect. Greater the $\Delta P = P_{final} - P_{initial}$ higher is the cooling.

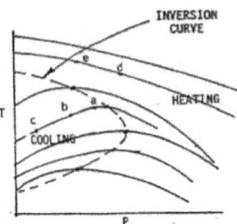

For $T > T_i$: All gases show <u>heating</u> effect, and the *vice-versa*.

$$W = R(T_2 - T_1) + 2a\left(\frac{1}{V_1} - \frac{1}{V_2}\right) - b(P_1 - P_2)$$

$$dT = (T_2 - T_1) = \left(\frac{P_1 - P_2}{C_P}\right)\left[\frac{2a}{RT} - b\right]$$

$dT = (T_2 - T_1)$	Effect
(*i*) Positive	→ Cooling or $T < T_i$
(*ii*) Negative	→ Heating or $T > T_i$
(*iii*) Zero	→ No effect or $T = T_i$

Gas	BP	T_B(K)	T_C (K)	T_i K
(*i*) O_2	90	155	$T > RT$	
(*ii*) N_2	77	126	$T > RT$	
(*iii*) H_2	20	33	143	
(*iv*) He	4.2	5.3	30	

$$T_B = \frac{a}{Rb}$$ $$T_i = \frac{2a}{Rb}$$; $$T_C = \frac{8a}{27Rb}$$

$$T_C = \frac{8}{27} T_B = \frac{4}{27} T_i \; ;$$

17.6.2.2 Throttling

Throttling is of great technical importance for real gases, Initial cooling of the gas should be

Initial cooling $T < T_C$

Two processes, (1) external, (2) internal work, are superimposed. The throttle valve is insulated so that no heat is transferred during the process. The gas initially has a pressure P_1, temperature T_1 and volume V_1. After is passes through the valve, its pressure is P_2 and the volume is V_2.

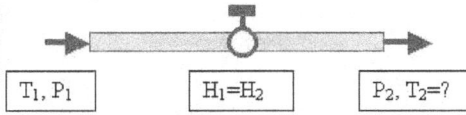

| T_1, P_1 | $H_1 = H_2$ | $P_2, T_2 = ?$ |

17.7.1 **Joule-Thomson coefficient** for the van der Waals gas

This can be approximated as

$$\mu = \frac{2}{5R}\left[\frac{2a}{RT} - b\right]$$

implies that at low temperatures, $\mu > 0$ and a gas should cool upon expansion.

17.7.2. Two primary methods for liquefying gases on a commercial basis are.

i)   Cascade and
ii)   Linde processes.

17.7.3 Cascade Process

By expanding compressed gas in a turbine and extracting work (thereby lowering the temperature).

17.7.3.1 Liquefaction of Oxygen:

K. Onnes (following Pictet, 1878) used three compression pumps and oxygen is initially cooled using cold water, methyl chloride (T_C = 143 $^\circ$C) and liquid ethylene (T_C = 10 $^\circ$C); which was allowed to boil at reduced pressure by means of a pump, T = -160 $^\circ$C was reached. So Oxygen (T_C = -119 $^\circ$C) is compressed to 25 *atm*, liquefaction takes place.

17.7.3.2 Liquefaction of Nitrogen:

Using liquid Oxygen to boil (T_{BP} = -183 $^\circ$C) under reduced pressure a T = -218 $^\circ$C) was reached;

Nitrogen (T_C = -146 $^\circ$C) gets liquefied using the cascade process.

Cascade process cannot be used to liquefy hydrogen, (T_C = -240 $^\circ$C); Neon (T_C = -229 $^\circ$C), and helium (T_C = -268 $^\circ$C).

17.7.3.3. Regenerative Cooling:

In cooling of a gas by The Joule-Thomson expansion depends on the difference of pressure between the two ends of the porous oplug and the initial temperature. If, on the other hand, the Joule-Thomson cooled gas is allowed to undergo a second Joule_thomson cooling before reaching the nozzle, the incoming gas is cooled further. This commutative process used to cool a gas continuously is called Regenerative cooling. Linde used this method.

17.7.4 **ADIABATIC COOLING** Linde Process

The graph above compares the actual inversion curves for hydrogen and nitrogen with the van der Waals prediction, all in reduced coordinates.

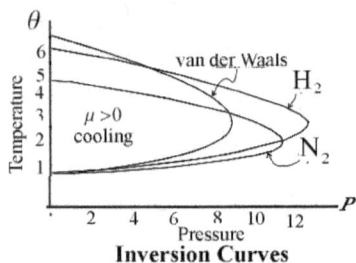

Inversion Curves

17.7.4.1 Liquefaction of air:

Linde (1896) liquefied air using the Joule-Thomson Effect. Air is freed of water vapour and CO_2); compressed to 200 *atm* and to T = -20 °C ; using Joule-Kelvin Effect air is liquefied. It is then stored in a Dewar Flask.

17.7.4.2 Liquefaction of hydrogen:

Dewar's (1898) modified apparatus; hydrogen (T_i = -83 °C, T_C = -240 °C) under 200*atm* passed through solid CO_2 and alcohol; passed through a chamber when liquid air is boiled at reduced pressure (*10 mm* Hg) so that hydrogen is cooled to T = -200 °C ; due to Joule-Kelvin Effect it is liquefied.

17.7.5 Magnetic Cooling, or **Adiabatic Demagnetization** for $T < 1K$.

A magnetic field H from certain materials serves to lower their T (Peter Debye, 1926 and William Francis Giauque, 1927), provides a means for cooling an already cold material (at about 1 K) to a small fraction of 1 K. The process involves:

1) The sample to be cooled (typically a gas) is allowed to touch a cold reservoir (which has a constant temperature T of around 3 - 4 K, and is often liquid He, and н is induced in the region of the sample.

2) Once the sample is in thermal equilibrium with the cold reservoir, the H strength is increased; entropy S of the sample decreases; because the system becomes more ordered as the particles align with the H . The T of the sample remains the same.

3) Then the sample is isolated from the cold reservoir, and the H strength is reduced. The sample has ΔS =0., but its temperature T drops in reaction to the reduction in the magnetic field strength. If the sample was already at a fairly low T, this temperature decrease can be ten-fold or greater.

4) This process can be repeated, permitting the sample to be cooled to very low T .

By magnetizing and demagnetizing a **paramagnetic sample** while controlling the heat flow, one can lower its temperature.

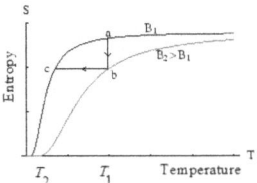

Start with the sample in field B_1 at a fairly low temperature T_1.

$a \rightarrow b$: With sample at T_1 increase the field to B_2

$b \rightarrow c$: Isolate the sample, slowly return to B_1, *i.e., slowly* adiabatic process thereby S is unchanged.

These steps in a T_S plot shows that the $b \rightarrow c$ step lowers the temperature; is a function of B/T only (not B or T separately).

The following Fig shows what is happening to the spins.

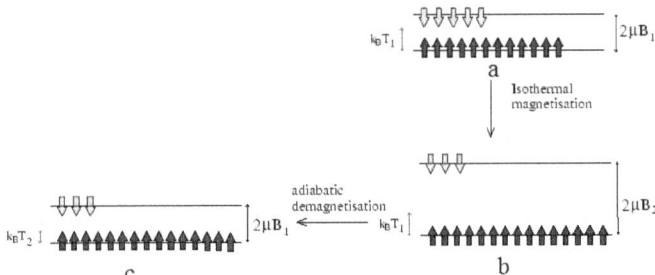

This is an important and general result. There is always a minimum excitation energy ε of the system, and once $k_B T \ll \varepsilon$, there is no further way to reduce T; *i.e.,* the Absolute Zero is never attainable!

17.7.5.1 NEGATIVE ABSOLUTE TEMPERATURES

Casimir and Du Pre in 1939 asked an important question: "What will happen to a system of particles (electrons) with spin as heat energy is supplied to it?"

It is known that

Application of Heat to a system leads to
$\begin{cases} \rightarrow \text{ increase of internal energy, U} \\ \rightarrow \text{ increase of T} \\ \rightarrow \text{ increase of Energy, Q} \end{cases}$

According to quantum mechanics a non-relativistic particle of spin s can have (2s+1) levels called degeneracy. They have magnetic substates in the presence of a magnetic field, having lowest and highest state. For N atoms system, the total energy of the system would ne N times the lowest energy state. So the spin system has two limits of energy state, upper and lower, giving the concept of negative Absolute Temperature. Temperature is derived from

$dS = \left(\frac{dQ}{T}\right)$, and so $dS \propto dQ$, One can write therefore,

$dS = -\theta . dQ$, where θ is a new temperature function.

This leads to concept of negative temperatures are above positive temperatures.

This is against the Kelvin's Law, which requires modification as per Norman Ramsay. Experiments (Purcell and Pound, with LiF crystal in NMR) have proved it!! Thus negative temperatures are not encountered in normal experience, but the concept is a valid one.

17.7.5.2 Cryogenic Storage Dewar Flask

A complete line of Dewar Flasks in sizes from 150 ml to 5000 ml designed for use in laboratory and research for handling Liquid Nitrogen. All Dewars are fully silvered, borosilicate glass in models from open vessels for use with glass flasks, wide mouth with mesh or aluminum housings, with handles or narrow necks

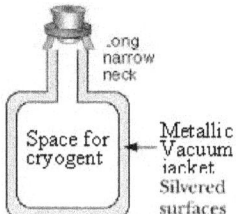

17.7.6 Refrigerator (Cooling by Adiabatic Expansion)

Freon is the refrigerant and can be liquefied at room temperature. Similarly compressing it (T_C > room T).

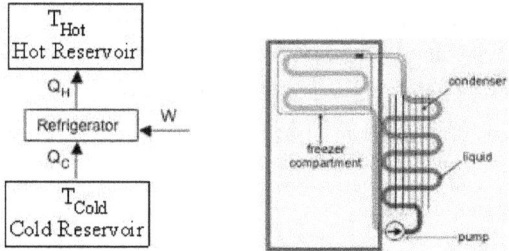

There are two things that need to be known for refrigeration.

1) A gas cools on expansion.

When you have two things that are different temperatures that touch or are near each other, the hotter surface cools and the colder surface warms up. This is a law of physics called the Second Law of Thermodynamics.

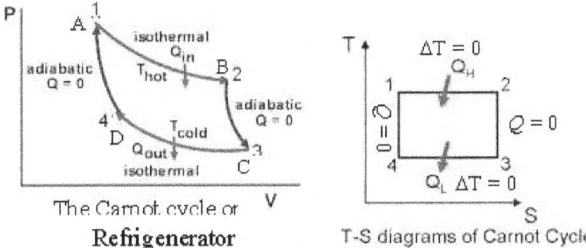

The Carnot cycle or **Refrigenerator**

T-S diagrams of Carnot Cycle

Modern refrigerators don't use CFC (Chloro-fluoro-carbon) as CFCs are harmful to the atmosphere if released. Instead they use another type of gas called HFC-134a (Tetra-fluoro-Ethane). HFC turns into a liquid when it is cooled -26.6 °C (to -15.9 °F). A motor and compressor squeezes the HFC.

17.7.8 STEFAN'S Radiation Law
Total energy emitted by a Black Body at temp T K per unit area surface per second

$$E = \sigma T^4$$

If the body is surrounded by an enclosure at T_0,

$$E = \sigma (T - T_0)^4$$

For a body of surface area A

$$E = \sigma A \ (T - T_0)^4$$

If the body is not a Black Body,

$$E = \varepsilon \ \sigma A \ (T - T_0)^4$$

$\varepsilon < 1$ always, is the emissivity of the body
Stefan's Law applies to loss of energy by radiation; while Newton's Law of Cooling applies to loss of energy by convection and conduction.

17.8 **THERMO-ELECTRCITY:**

17.8.1 Joule Heating
Joule Effect is the heat produced Q when an electric current i passes through a resistance R for a time t is a IRREVERSIBLE process.

$$Q = i^2 R \ \frac{t}{J} \quad \text{unit } kCal$$

17.8.2 REVERSIBLE phenomena are:
1. Seebeck Effect,
2. Peltier Effect,
3. Thomson Effect.

17.8.3 Seebeck Effect (Thomas Johann Seebeck, 1823) Thermocouple

Two dissimilar metals, say Copper and Iron, in a closed circuit, ΔT is established between these junctions, forms <u>thermocouple</u>.

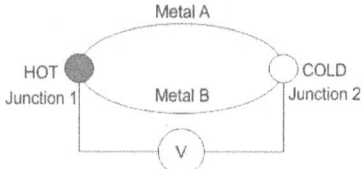

For a given pair of junction, the emf of a circuit composed of metals A and B,

$$\varepsilon_{AB} = \varepsilon_{AC} - \varepsilon_{BC}.$$

a and **are characteristics of the metals A and B**

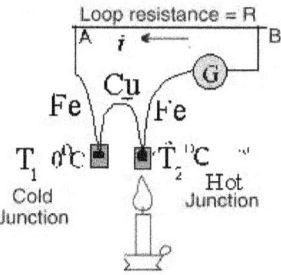

Thermo-electric current $\quad \boxed{i = \dfrac{(T_2 - T_1)(\kappa_A - \kappa_B)}{R}}$

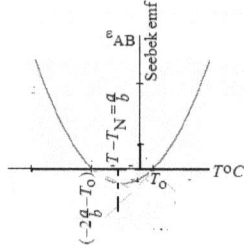

17.8.4 Peltier Effect:

It is the converse of Seebeck Effect.

Neutral Temperature, T_N,

$$T_N = -\frac{a_{AB}}{b_{AB}}$$

At $T = T_o$, $\qquad \varepsilon_{AB} = 0$.

$$T = T_i = (-2\frac{a_{AB}}{b_{AB}}) - T_o$$

where T_i = Inversion temperature

The two junctions of the thermocouple are at the same temperature, $\Delta T = 0$. Then one junction is heated and the other cooled.

Thermo-emf, $\quad \boxed{e_{AB} = \pi_1 - \pi_2 = \frac{\pi_1(T_2 - T_1)}{T_1}}$

π = Peltier emf.

17.8.5 Thomson Effect

When a temperature gradient is maintained between different points in the same metal there exists a variation of potential Δe along the metal.

e_{AB} = Thomson coefficient of a metal;

$$\boxed{e_{AB} = (\pi_1 - \pi_2) - \int(\sigma_A - \sigma_B)dT}$$

17.8.6 Thermo-electric Power

For most metals or alloys, commonly used in thermocouple circuits, the operating temperature range is sufficiently far removed from the T_N so that the rate of change of thermo-emf of the hot junction $\frac{de}{dT}$ of a thermocouple at a particular T, is **linear**.

$$\boxed{\frac{de}{dT} = a + 2b\,T}$$

17.8.7. The thermoelectric diagram is as shown.

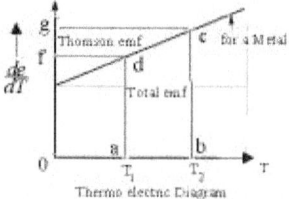

Thermo electric Diagram

a and *b* are constants for a given thermo-couple.

$$\pi_{AB} = T \frac{de}{dT} K$$

$$(\sigma_A - \sigma_B) = -T \frac{d^2 e}{dT^2} K = \frac{d\,[a + b\,T]}{dT} = -b\,T\,K$$

$$T_N = -\frac{a}{2b} \; ^\circ C$$

$$T_i = -\frac{a}{b} \; ^\circ C$$

$$T_N = \frac{273 + T_i}{2} \; ^\circ C$$

17.8.8. Thermoelectric coefficients a and b, reference metal Lead (Pb), when ε_{AB} =+ve, the conventional current flows from metal A to metal B.

TABLE : Thermoelectric coefficients a and b; Reference metal -LEAD

Substance	$a = a_{A\text{-Pb}}$ (μV)	$b = b_{A\text{-Pb}}$ (μV)
1) Antimony (Sb)	+35.6	+0.145
2) Bismuth (Bi)	-74.4	+0.032
3) Constantan (60%Cu +40% Ni)	-38.1	-0.0888
4) Copper (Cu)	+2.71	+0.0079
5) Iron (Fe)	+16.7	-0.0297
6) Nickel (Ni)	-19.1	-3.02
7) Platinum (Pt)	-3.03	-3.25
8) Silver (Ag)	+3.34	+0.008

17.8.9. Useful Thermocouples

Mullite type insulating material is recommended as a tight sheath up to $1500\,^\circ C$.

1) Chromel-Alumel

Chromel (90% Cr, 10% Ni); Alumel (95% Ni, 5% Al, Si, Mn)

Range: $-200\,^\circ C$ to $700\,^\circ C$. It is robust and gives a relatively large emf of $\sim 40\,\mu V / K$.

2) Tungsten-Molybdenum

Range: $1250\,^\circ C$ to $2600\,^\circ C$. ($< 1250\,^\circ C$, the emf of this couple changes sign).

3) 40% Rh – Pt against 20% Rh – Pt: up to $1900\,^\circ C$.

4) Pt – 10% Rh-Pt: Range: $0\,^{\circ}C$ to $1700\,^{\circ}C$.

5) Pt – 13% Rh-Pt: Range: $0\,^{\circ}C$ to $1700\,^{\circ}C$.

6) Iron – Constantan: : Range $-200\,^{\circ}C$ to $1000\,^{\circ}C$.

7) Copper – Constantan: Range $-200\,^{\circ}C$ to $400\,^{\circ}C$.

8) Fe/Cu: $T_N = 240\,^{\circ}C$.

12.8.10. Calibration of a Thermocouple

A slide-wire potentiometer (or a commercial Pye type Precision potentiometer) three resistance boxes, $10{,}000\,\Omega$ each, 3 on-off keys, a BG, Weston Cadmium Standard cell, 2V accumulator, pure metals like Tin (MP $232\,^{\circ}C$), Lead (MP $327\,^{\circ}C$), Zinc (MP $419\,^{\circ}C$), Ice, electric oven, A Hg thermometer up to $400\,^{\circ}C$, beaker, hypsometer (steam bath) and mercury bath (up to $360\,^{\circ}C$) are required.

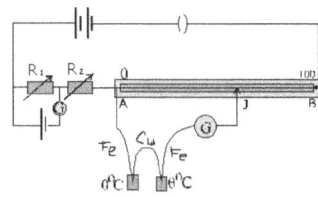

+*+*+*+*+*+*+*+*

Chapter 18

GRAVITATION –
Universe,
Equations of Circular Motion

"We apprehend time only when have marked motion, .. we measure movement by time,
but also time by movement"   Aristotle

18.1.   THE UNIVERSE

18.1.1   Basics
(i)   Gravity acts between all particles that have mass. Mass will attract other mass with a force that gets weaker as the distance between them gets larger. Gravity is responsible for the large scale structure of the Universe..
(ii)   Although gravity appears to be a very powerful force, when it comes to things on smaller scales, like tiny particles, can be ignored because of its weakness. The carrier of the gravitational force is the *graviton*. Although it has never been observed in experiment, it is strongly believed to exist.
(iii)   **Goldilocks Zone**
There will be a hospitable zone suitable for life, where liquid water can exist.. The development of intelligent life necessitates that planetary temperatures are "just right". This zone is tiny, and fortunately, our Earth fell within it in the Solar system. Large orbital eccentricities are not conducive to the existence of life. It is an apparent miracle that this factor is only ~ 2 % for the Earth.
(iv)   Enceladus, a small (500 *km* diameter), icy moon of Saturn (a billion miles farther from the Sun than Earth) is reported (NASA, 14 Apr 2017) to have all the ingredients for supporting life. This is the Cassini Mission Spacecraft.

18.1.2   Mach's Principle (1893):
"The inertia of any system is the result of the interaction of that system with the rest of the Universe. In other words, every particle in the universe ultimately has an effect on every other particle."

18.1.3   THREE MODELS of the Universe:

18.1.3.1 Both the Hindu and Aristotle Models of the Universe – Earth centred.

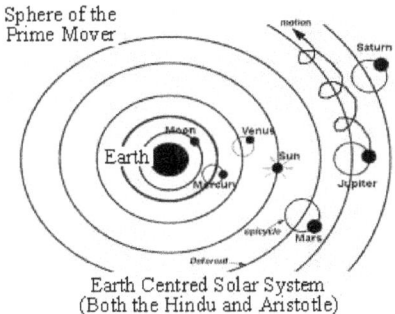

Earth Centred Solar System
(Both the Hindu and Aristotle)

18.1.3.2 Ptolemaic Model of the Solar System: <u>Earth centred</u>. (about 100 AD)

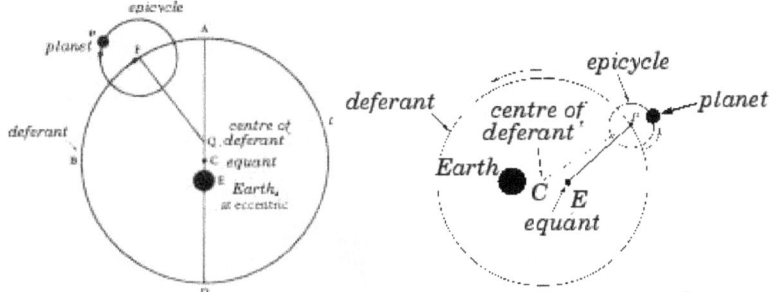

While the Ptolemaic model was very good at predicting the positions of the planets, it wasn't precise, and over the centuries its predictions got worse and worse. Ptolemaic model had big epicycles to explain the retrograde motions of the planets.

18.1.3.3 Nicolaus Copernicus Model of the Solar System <u>Helio-centred</u>.(1543).

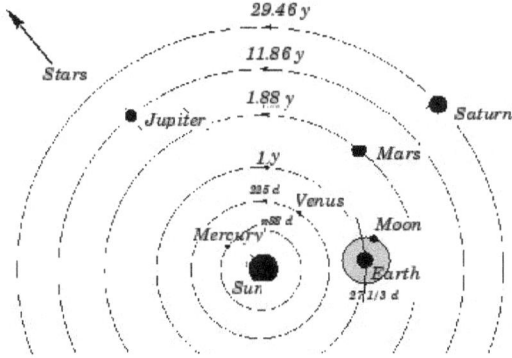

The main simplification of the Copernican model was that the retrograde loops of the planets as seen from the Earth occur naturally as a result of the Earth's motion combined with the motions of the planets.

18.1.3.4. Tycho Brahe (1546 – 1601) model

Brahe model had all the planets (except Earth) orbiting around the Sun, but then the Sun orbited around the Earth.

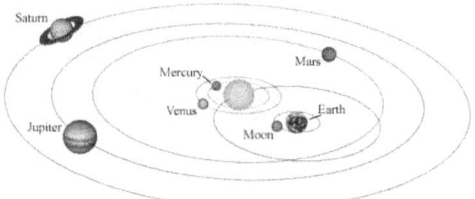

Tycho Brahe's Model

18.1.3.5   Johannes Kepler Model

Kepler used very precise data provided by Tycho Brahe. Today, we remember Kepler's insight as 3 laws:

1. The orbits of the planets are ellipses, with the Sun at one focus.

This tells us that the motion is not uniform circular motion. Not only is the shape of the orbit no longer a circle, but also the Sun is not at the center. Remember the "eccentric" of Ptolemy? This is now something that we expect, and only one position is allowed.

2. The line joining the planet to the Sun sweeps out equal areas in equal intervals of time.

This tells us that the planet sometimes moves quickly (when it is closer to the Sun) and sometimes more slowly (when it is farther away).

3. The squares of the sidereal periods (P) are proportional to the cubes of the semi-major axes.

What does this mean? It tells us that the planets are all obeying some common rules. It lets us figure out how long a new planet will take to go around the Sun if we know the size of its orbit. For example: The Earth takes 1 *yr* to go around, and it is 1 AU from the Sun. So if there were a planet at 4 AUs, it would take 8 years to go around the Sun once: P squared = a cubed = 4x4x4 = 8x8 so P = 8.

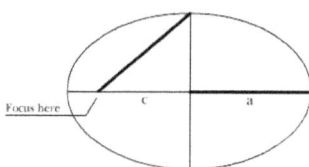

18.1.3.6. Galileo Galilei

He was the first astronomer to use a telescope to study the heavens. Galileo made a number of observations that finally helped convince people that the Sun-centered solar system model (the heliocentric model), as proposed by Copernicus, was correct.

18.2 EQUATIONS OF CIRCULAR MOTION

18.2.1 Relation between linear speed v and angular speed ω:
If the body goes from C to D in time t

$$s = r\theta$$

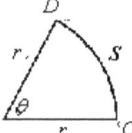

$$\boxed{v = r\omega}$$

18.2.2 Formulae for centripetal acceleration:
The instantaneous acceleration, a

$$\boxed{a = \frac{v^2}{r}}$$

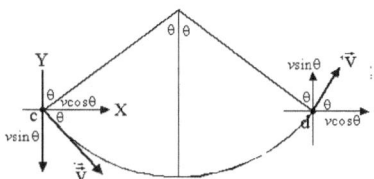

The instantaneous acceleration acts in towards the centre of the circle.

$$\boxed{a = r\,\omega^2}$$

18.2.3 Formula for centripetal force
By Newton's Second Law of motion

$$\boxed{F = m\,a \rightarrow m\frac{v^2}{r} = m\,r\omega^2}$$

18.2.4. Formulae for periodic time and frequency, ν

Periodic time T $\boxed{T = \dfrac{\text{Length of one complete orbit}}{\text{Speed}} = \dfrac{2\pi r}{v} = \dfrac{2\pi}{v}}$

Frequency, ν $\boxed{\nu = \dfrac{1}{T} = \dfrac{v}{2\pi r} = \dfrac{\omega}{2\pi}}$

= Number of orbits in one second

18.3 KEPLER'S Laws of Planetary Motion:
Johannes Kepler (1609) made observations of the Dutch astronomer Tycho Braho and deduced the laws to describe planetary motion.

18.3.1 Law I
The orbits of the planets around the Sun are elliptical, with the Sun at one of the foci. Either observe the change in the diameter of the Sun as the Earth makes one orbit around it, or plot the orbit of one component of a double star.

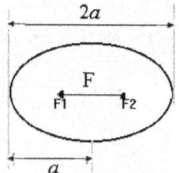

<u>Eccentricity</u> $e = \frac{F}{2a} < 1$

(When F = 0, then e = 0, a CIRCLE)
(Eccentricity is a maximum at e = 1)

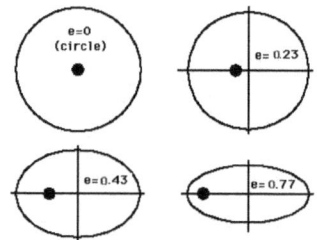

18.3.2 **Law II**

*An imaginary line connecting a planet and the sun **sweeps out equal areas during equal time intervals**.*

The Earth's orbital speed varies at different times of the year;
It moves fastest when closest to the sun; slowest when farthest away.

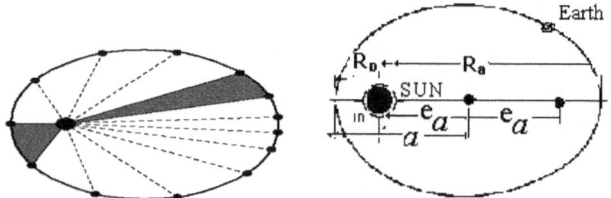

Terms to know:
1.   PERIHELION = where a planet is closest to the Sun
2.   APHELION = where a planet is farthest from the Sun

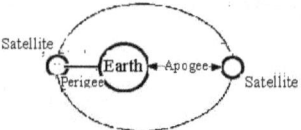

3. Kepler's Second Law was calculated for Earth, then the hypothesis was tested using data for Mars, and it worked!
    A planet's speed changes with its distance from the Sun.
Observe the orbit of one component of a double star.
A line, drawn from the Sun to a planet, sweeps equal areas in equal time.

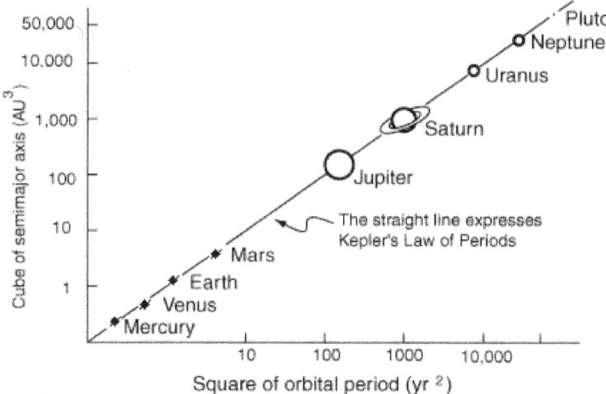

The angular momentum, $I\omega$ of the planet is also conserved,

$$\boxed{\bar{L} = I\omega = \text{constant}},$$

since it moves fastest when closest to the Sun and slowest when at its greatest distance .

18.3.3 **Law III**
        The square of the time of revolution about the Sun is directly proportional to the cube of the mean radius of the planet's orbit.
It showed the relationship between the **size a of a planet's orbit** and its **orbital period T**,

$$\boxed{a^3 = T^2}$$

which is the same for all planets.
Constancy of ratio of Planetary Motion showed the relationship between the **size of a planet's orbit** r from the Sun and its **orbital period, T**.

$$\boxed{T^2 = \frac{4\pi^2}{GM_\odot} r^3}$$

$\dfrac{a^2}{T^2}$ = same for all planets.

The distance of a planet from the Sun varies.
        Its orbital size a = the semi-major axis.
IF THE PERIOD OF A PLANET IS KNOWN IN EARTH YEARS, ITS SEMI-MAJOR AXIS CAN BE CALCULATED IN units of AU. (and *vice versa*).
        Kepler's three laws replaced the cumbersome epicycles to explain planetary motion with three mathematical laws that allowed the positions of the planets to be predicted with accuracies ten times better than Ptolemaic or Copernican models.

18.4   NEWTON'S LAW OF GRAVITATION

18.4.1   Newton's Law of Gravitation

Everybody in the Universe attracts every other body with a force $\vec{F}$ that varies directly as the product of the masses m_1 and m_2 of the two bodies and inversely as the square of r the distance between them.

$$\vec{F} = G\frac{m_1 m_2}{d^2} \quad \text{unit } N \quad \text{(vector)} \quad (M^{-1}L^3T^{-2})$$

Gravitation constant,

$$G = \frac{\vec{F}d^2}{m_1 m_2} = 6.6 \times 10^{-8} \ CGS \ units \quad \text{(vector)} \quad (M^{-1}L^3T^{-2})$$

18.4.1.1 Gravitational Field Intensity at a point distant r from a point mass, m

$$F = G\frac{m}{r^2}$$

$$F = -\nabla V = -\frac{dV}{dr}$$

18.4.1.2   Gravitational Potential $\boxed{V = G\frac{m}{r}}$

18.4.2   Acceleration due to gravity, g :

$$g = 10 \ N \ kg^{-1} \ \text{(Vector)} \quad (M^0 L^1 T^{-2})$$

18.4.3 Variation of g with altitude, h

g_h = value of g at height h above earth's surface.

$$g_h = g \ [1 - \frac{2h}{R}]$$

r = radius of earth

18.4.4 Variation of g with latitude, θ :

$$g' = g_0 - R \ \omega^2 \cos^2 \vartheta$$

$R_E = R$ = radius of Earth, ω angular velocity of Earth's spin,

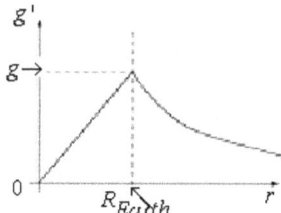

18.4.5 Variation of g with depth, d below the surface of Earth

g_d = value of g at depth d below earth's surface. R = radius of Earth.

$$g_d = g \ (1 - \frac{d}{R})$$

$$g_{Equator} = 9.78 \ ms^{-2}$$

$$g_{\vartheta=45^o} = 9.81 \ ms^{-2}$$

$$g_{Poles} = 9.83 \ ms^{-2}.$$

18.4.6 **Relation between g and G:**

Gravitational field strength,

$$g = \frac{GM_E}{R_E^2} \quad units \ Nkg^{-1} \ \text{(Vector)} \quad (M^0L^1T^{-2})$$

18.4.6.1 Mean Density of Earth

$$\rho = \frac{M}{V} = \frac{(gR^2/G)}{\frac{4}{3}\pi R^3} = \frac{3g}{4\pi GR} = 5510 \ kg.m^{-3}.$$

18.4.6.2. Mass of the Sun

Angular velocity of the Earth around the Sun, $\omega = 1.9 \times 10^{-7} \ rad.s^{-1}$.

$$M_S = \frac{r^3 \omega^2}{G} = 2.00 \times 10^{30} \ kg$$

18.4.6.2.1   Potential at r outside due to a spherical shell of radius a

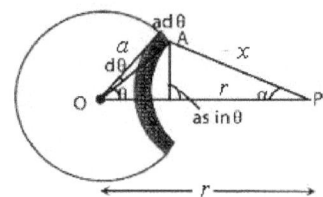

$$V = \frac{GM}{r}$$

(i)   V inside a point:

$$V = \frac{GM}{a}$$

(ii)   Force outside the shell:

$$F = \frac{d}{dr}\left(\frac{GM}{r}\right) = -\frac{GM}{r^2}$$

(iii)   Force inside the shell:

$$F = \frac{d}{dr}\left(\frac{GM}{a}\right) = 0$$

18.4.6.3   Potential at x due to a spherical shell of radius a bound by a sphere of radius b :

$$V = \frac{4}{3}\frac{G\pi\rho}{x}(b^3 - a^3)$$

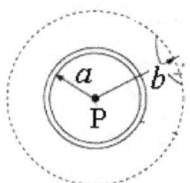

18.4.7 Weightlessness (Free fall)

This is a condition either in space where gravity due to a planet on the object is zero, or if the object travels in an acceleration against gravity, and acceleration equal to the gravity.

Weightlessness = Free fall

Such a situation occurs when an artificial satellite is launched at a speed equal to the escape velocity.

Micro-gravity circumstances also occur in geo-synchronous satellites.

18.4.7 NEWTON, Apple and the Moon

Isaac Newton (1666) proposed Law of Universal Gravitation, for a planet of mass m orbiting Earth of mass M,

$$F = m\omega^2 r = m\left(\frac{2\pi}{T}\right)^2 r \text{ , Also } F = k\frac{m}{r^2}$$

Giving

$$\boxed{T^2 = \frac{4\pi^2}{k} r^3} = \text{Kepler's Law III.}$$

$$\boxed{\vec{a}_{Apple} = 9.8 \ ms^{-2}}$$

$$\boxed{\vec{a}_{Moon} = 0.0272 \ ms^{-2}}$$

$$\frac{\text{Earth -Moon Distance}}{\text{Radius of Earth}} = 60$$

$$\frac{\vec{a}_{Apple}}{\vec{a}_{Moon}} = 3600 = \left(\frac{\text{Earth -Moon Distance}}{\text{Radius of Earth}}\right)^2$$

18.4.8 **For the Moon:**

$$\left(\frac{4\pi^2 m \ r}{T^2}\right) : mg = \left(\frac{1}{r^2}\right) : \left(\frac{1}{R^2}\right)$$

$$\boxed{g_0 = \frac{4\pi^2}{R^2} \frac{r^3}{T^2}}$$

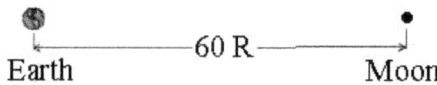

←————— 60 R —————→
Earth   Moon

18.4.9 **Determination of the value of G**:
1. Cavendish's Method,
2. Boys' Method (More accurate).

18.4.9.1. Measurement of G – by Cavendish (1798)

R is a fine metal ribbon, and the masses m_1, m_2 connected through a rigid rod is suspended. Angular deflections are detected by mirror M. L is light beam and S is the scale. A small metal plate P dipped in oil dampens unwanted vibrations.

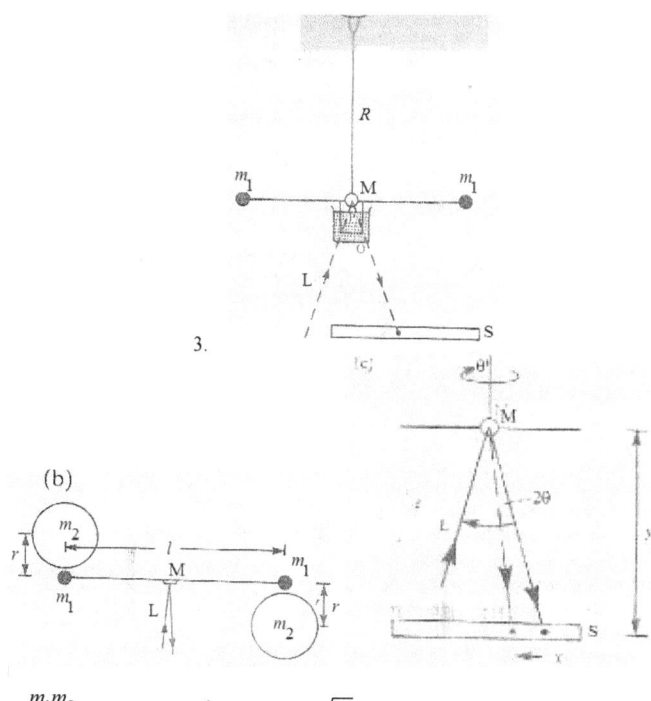

$$G\frac{m_1 m_2}{r^2}.l = C\theta \; ; \; \theta = \frac{1}{2}\frac{x}{y} \; ; \; T = 2\pi\sqrt{\frac{I}{C}}$$

where C = suspension constant of the metal ribbon.

18.5. **ARTIFICIAL SATELLITES**

18.5.1 Synchronous Satellites

$$g_0 = \frac{GM}{R^2}$$

18.5.1.1 Period of satellite, $$T^2 = \frac{4\pi^2}{g_0 R^2} r^3$$

18.5.1.2   For **synchronous (Geo-stationary)** satellites,

$$T = 1 \text{ day} = 86400 \text{ seconds}$$

$$\boxed{T_{\text{Geo sta}} = 1d = T_{\text{Eart spin}}}$$

For a satellite Circular orbit occurs at 27,000 kms^{-1}.

$T_{\text{Sat}} = 88.4\ m$;

$$r_{\text{sat}} = [\frac{T_{\text{Sat}}}{T_{\text{mean earth}}}]^{2/3}\, r_{\text{m e}} = r_{\text{m e}}[\frac{1d}{27d}]^{2/3}$$

i.e., r_{sat} = Altitude of <u>42,400 km</u> from the centre of Earth

$\approx$ Circumference of Earth
$\approx$ at an altitude of $(42400\ km - 6400\ km)$
$= 36000\ km.$

18.5.2 Velocity and acceleration of a body on Earth's surface:

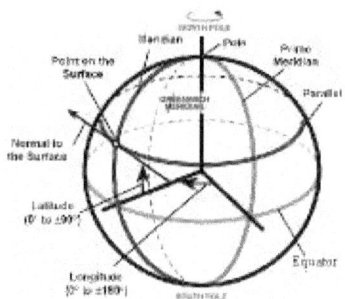

λ = latitude of point A (of radius r) with Equatorial radius R
$\quad$ R= $r \cos \lambda$
ω = angular velocity for Earth
$\omega = 7.292 \times 10^{-5}\ s$
$a = \omega^2\ r \cos \lambda = 3.39 \times 10^2 \cos \lambda\ ms^{-2}$
v = R $\omega = 464 \cos \lambda ms^{-1} = 1672 \cos \lambda\ km\ hr^{-2}$
$\lambda = 0$, at a point on the Equator.

18.5.3 Total Gravitational potential of a satellite (Total Energy in orbit)

$$\boxed{V = -\frac{G\ m\ M}{2\ r}}$$

18.6 Principle of Equivalence:
The inertial and gravitational masses are the same, for all bodies.
All bodies at the same place in gravitational field experience the same acceleration; ie all bodies fall on Earth with the same value of g.

18.6.1 Which will hit the Ground First? A feather or a Coin?
$\quad$ A lighter object falling at the same rate as a heavier object when air resistance is removed from the medium by setting up vacuum. Remember Galileo's experiment proved <u>uniform</u> *gravitational acceleration*, rejecting Aristotle's view.

18.6.2 Terminal velocity
Falling Body in Air *(Gravity and Drag)*

v = velocity

ρ = gas density; A = Frontal area; c_d = Diffusion coefficient

Net Force = Drag – Weight; $F = D - W$

$$D = c_d \frac{\rho v^2}{2} A$$

If $F = 0$, $\qquad c_d \frac{\rho v^2}{2} A = W$

Terminal velocity, $\boxed{v_T = \sqrt{\frac{2W}{c_d \, \rho A}}}$

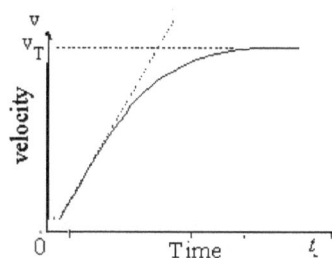

18.6.2. Escape velocity, v_e :

Escape velocity, v_e , is the vertical velocity that a body must be given at the surface of a planet so that it will just escape from the gravitational attraction of that planet.

Kinetic Energy,

$$\frac{1}{2} m v^2 = \text{Work done}, \ \frac{GmM}{R}$$

i.e., $\qquad \boxed{v_e = \sqrt{\frac{2GM}{R}} = \sqrt{2Rg_0}}$

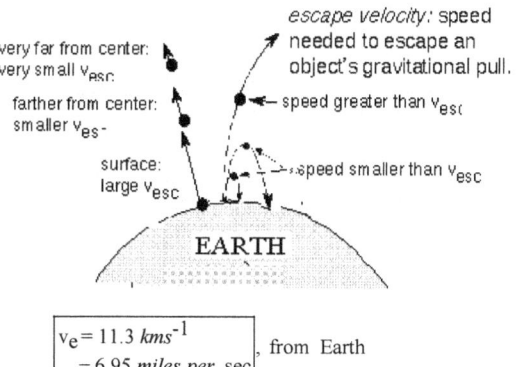

$\boxed{\begin{array}{l} v_e = 11.3 \ kms^{-1} \\ = 6.95 \ miles \ per \ sec \end{array}}$, from Earth

Condition		Orbit type
i) $v = v_e = \sqrt{2Rg_0}$		Parabolic
ii) $v < v_e$		Elliptical
iii) $v (< v_e) = \sqrt{Rg_0}$		Circular
iv) $v > v_e$		Hyperbola

$$\frac{T_{Sat}}{T_{Moon}} = \left(\frac{r_{Sat}}{r_{Moon}}\right)^{3/2}$$

18.6.3 **Applications**:
18.6.3.1 Launching of Space Vehicles, Astronauts' travel.
18.6.3.2 Loss of atmosphere from the Earth:
   d = diameter of a molecule

   No. of molecules $\quad n\ (m^{-3}) = \dfrac{1}{\sqrt{2}\ \ell\ d^2}$

   $$v = \sqrt{\frac{3k_B T}{m}}\ ,$$

Mean free path, $\ell = \dfrac{v\ t}{\text{\# of collisions}}$

Molecule	Velocity ($m\ s^{-1}$)
(1) He	1350
(2) N_2	510
(3) O_2	477
(4) CO_2	407
(5) H_2	1908

18.6.3.3 Is any Oxygen molecule would be lost from the Earth's upper atmosphere?
18.6.3.4 Would any Hydrogen molecules be lost from Earth's atmosphere?

18.6.4 Rocket with variable mass:

$$v_f = v_0 + v_e\ \ln\left(\frac{m_0}{m_f}\right) - g\ t$$

t = time required to burn all fuel.

m_f = final mass, when velocity is v_f ;

m_0 = mass of rocket when velocity is v_0 .

v_e = velocity of exhaust gases – velocity of rocket.

18.7 **INDIAN SATELLITES**:

18.7.1 FATHER OF INDIAN SPACE PROGRAMME:
   Vikram Ambalal SARABHAI
18.7.2 Abbreviations of Launch Vehicle all built at VSSC, Thiruvananthapuram:

i)   ROHINI RH-75 (Sounding rocket) (first launch 1967).
Weight: 32 *kg*; height: 1.5 *m*; payload: 1 *kg*; Propulsion: solid.

ii)   ASLV – Augumented Satellite Launch vehicle (1987).
Weight: 39 tonnes; height: 23.5 *m*; payload: 150 *kg*; Propulsion: solid.

iii)   PSLV - Polar Satellite Laucnh Vehicle (1993).
Weight: 2.95 tonnes; height: 44 *m*; payload: 1750 *kg*; Propulsion: solid & liquid.

iv)   GSLV –Mk3– Geo-stationary Satellite Launch Vehicle (2017).
Weight: 640 tonnes; height: 44 *m*; payload: 4000 *kg*; Propulsion: solid, liquid & cryogent.

v)   ULV (expected from 2020)'

18.7.3   List of names of Satellite Launches by ISRO

Name	Date	Weigght(kg);	Launch Vehicle	
1). ARYABHATA	19.04.1975	358	C-1 INTERCOSMOS,	USSR
2). BHASKARA-1	07.06.1979	444	C-1 INTERCOSMOS,	USSR
3). RTP	10.08.1979	35	SLV-3,	INDIA
4). RS-1	18.07.1980	35	SLV-3,	INDIA
5). RS-D1	31.05.1981	38	SLV-3,	INDIA
6). APPLE 19.	06.1981 670		ARIANE,	EUROPE
7). BHASKARA-2	20.11.1981	436	C-1 INTERCOSMOS,	USSR
8). INSAT-1A	10.04.1982	1150	PROCURED: DELTA,	USA
9). RS-D2	17.04.1983	41.5	SLV-3,	INDIA
10). INSAT-1B	30.08.1983	1194	PROCURED: SPACE SHUTTLE, USA	
11). SROSS-1	24.03.1987	150	ASLV,	INDIA
12). IRS-1A	17.03.1988	980	PROCURED: VOSTOK, USSR	
13). SROSS-2	13.07.1988	150	ASLV,	INDIA
14). INSAT-1C	22.07.1988	1190	PROCURED: ARIANE,	EUROPE
15). INSAT-1D	12.06.1990	1293	PROCURED: DELTA,	USA
16). IRS-1B	29.08.1991	990	PROCURED: VOSTOK,	USSR
17). SROSS-C	20.05.1992	106	ASLV, INDIA	
18). INSAT-2A	10.07.1992	1906	PROCURED: ARIANE,	EUROPE
19). INSAT-2B	23.07.1993	1932	PROCURED: ARIANE,	EUROPE
20). IRS-1E	20.09.1993	845	PSLV,	INDIA
21). SROSS-C2	04.05.1994	113	ASLV,	INDIA

Name of Satellite	Date	Weight(kg)	Launch vehicle	
22) IRS-P2	15.10.1994		PSLV-D2,	INDIA
23) INSAT-2C	07.12.1995		ARIANE-44L H10-3	EUROPE
24) IRS-1C	29.12.1995		MOLNIYA Baikanur	USSR
25) IRS-P3	21.03.1996	920	PSLV-D3,	INDIA
26) INSAT-2D	04.06.1997	2070	PROCURED: ARIANE, EUROPE	
27) IRS-1D	29.09.1997	1200	PSLV-C1,	INDIA
28) INSAT-2DT	7.01.1998		PROCURED IN ORBIT FROM ARABSAT	
29) INSAT-2E	03.04.1999	2550	PROCURED: ARIANE, EUROPE	
30) IRS-N	26.05.1999	1050	PSLV-C2,	INDIA
31) INSAT-3B	22.03.2000	2070	PROCURED: ARIANE, EUROPE	
32) GSAT-1	18.04.2001	1530	GSLV-D1,	INDIA
33) IRS	22.10.2001	1108	PSLV-C3,	INDIA
34) INSAT-3C	24.01.2002	2750	PROCURED: ARIANE, EUROPE	
35) KALPANA-1	12.09.2002	1060	PSLV-C4,	INDIA
36) INSAT-3A	10.04.2003	2950	PROCURED: ARIANE, EUROPE	
37) GSAT-2	08.05.2003	1825	GSLV-D2,	INDIA
38) INSAT-3E	28.09.2003	2775	PROCURED: ARIANE, EUROPE	
39) RESOURCESAT-1	17.10.2003	1360	PSLV-C5,	INDIA
40) Edusat	20.09.2004	1950	GSLV-F01,	INDIA
41) CARTOSAT-1	05.05.2005	1560	PSLV-C6,	INDIA
42) HAMSAT	05.05.2005	42	PSLV-C6,	INDIA
43) INSAT-4A	22.12.2005	3080	PROCURED: ARIANE, EUROPE	
44) INSAT-4C	10.07.2006	2168	GSLV-F02,	INDIA
45) CARTOSAT-2	10.01.2007	650	PSLV-C7,	INDIA
46) SRE-1	10.01.2007	550	PSLV-C7,	INDIA
47) INSAT-4B	12.03.2007	3025	PROCURED: ARIANE, EUROPE	
48) INSAT-4CR	02.09.2007	2140	GSLV-F04,	INDIA
49) CARTOSAT-2A	28.04.2008	690	PSLV-C9,	INDIA
50) IMS-1	28.04.2008	83	PSLV-C9,	INDIA
51) Chandrayaan-1	22.10.2008	1380	pslv-c11,	INDIA
52) RISAT-2	20.04.2009	300	PSLV-C12,	INDIA
53) OCEANSAT-2	23.09.2009	960	PSLV-C14,	INDIA
54) GSAT-4	15.04.2010	2215	GSLV-D3,	INDIA
55) Cartosat-2B	12.07.2010	694	pslv-c15,	INDIA
56) GSAT-5P	25.12.2010	2310	GSLV-F06,	INDIA
57) RESOURCESAT-2	20.04.2011	1206	PSLV-C16,	INDIA
59) YOUTHSAT	20.04.2011	92	PSLV-C16,	INDIA
59) GSAT	21.05.2011	3093	PROCURED: ARIANE, EUROPE	
60) GSAT-12	15.07.2011	1410	PSLV-C17,	INDIA
61) Megha-Tropiques	12.10.2011	1000	PSLV-C18,	INDIA
62) RISAT-1	26.04.2012	1858	PSLV-C19,	INDIA
63) GSAT-10	29.09.2012	3400	ARIANE	EUROPE
64) SARAL	25.02.2013	407	PSLV-C20,	INDIA
65) IRNSS-1A	01.07.2013	1425	PSLV-C22,	INDIA
66) INSAT-3D	26.07.2013	2060	ARIANE	EUROPE
67) GSAT-7 (for Navy)	30-08-2013	2650	ARIANE	EUROPE
68) MARS ORBITER (MANGALYAAN) MOM	07.11.2013	1,350	PSLV C25	SHAR, INDIA
69) GSAT-14	05.01.2014		GSLV-D5 LiqH2 Cryogenic SHAR,	INDIA

70) IRNSS-1B, 4 April 2014, PSLV-C24

71) IRNSS-1C, 10 November 2014, PSLV-C26

72) GSAT-16, 7 December 2014, Ariane -5

73) IRNSS-1D, 28 March 2015, PSLV-C27,

74) GSAT-6, 27 August 2015, GSLV-D6

75) ASTROSAT, 28 September 2015, PSLV-C30

76) Feb 15, 2017; <u>Milestone Mission</u>: Scripting history, PSLV-C37 injected India successfully launched from Satish Dawan Centre a record of 104 satellites. Excepting the CARTOSAT-2 series(664 *kg*), INS -1A, INS-1B and 101 foreign nano-satellites into precise orbits, So far Russia in 2014 had launched 37 satellites in a single launch.

* Till April 2001, INSATs were launched by Ariane launched at Kororou.

* SROSS (Stretched Rohini Satellite Service): 20-05-1992; ASLV; 100 *kg*; SHAR centre.

* GSAT: by GSLA from Sriharikota, 1540 *kg*; on 18 – 04 – 2000;

*77) May 05, 2017; India's Geosynchronous Satellite **Launch** Vehicle (GSLV-F09) successfully ... Geosynchronous Transfer Orbit (GTO).

* 78) June 05, 2017: India scripted history as the 'heavy lift rocket' GSLV MkIII, D1 launched the communication satellite GSAT 19 (3,136 kg) into orbit and from the Sriharikota Centre.

* 79) June 23, 2017: Launch of Cartostat-2 along with 30 passenger satellites, using PSLV-C38 to reach Sun synchronous Orbit from Sriharikotta Centre.

18.7.4. NavIC (Indian GPS) (Global Positioning System)

The Indian Regional Navigation Satellite System (IRNSS) named NavIC (Navigation with Indian Constellation) is going to be operational from 2018. It will have a position accuracy of 5 m.

18.7.5. Satellites Launched till July 1, 2017

For the USA: 123

For Germany: 11

For UK: 9

For Singapore 8

For Japan 4

Total foreign satellites, during 1999 -2016: $\boxed{209}$.

-o-0-o-0-o-0-o-0-

Chapter 19

WAVES & SOUND –

Simple Harmonic Motion , Superposition, Lissajous Figures, Stretched String, Longitudinal Waves in Fluids, Gases and Solids, Beats, Resonance, Closed and open Pipes, Doppler Effect

"No human inquiry can be called science unless it pursues its path through mathematical exposition and demonstration" Leonardo da Vinci

19.1. SIMPLE HARMONIC MOTION (SHM)

19.1.1 Sound Waves
Sound waves are <u>Longitudinal Progressive (Travelling) Waves</u> having compressions & rarefactions in the medium as shown. These waves are propagated through a medium. The individual particle in the medium oscillates about its equilibrium position.
This wave motion is represented by an equation of simple harmonic motion.

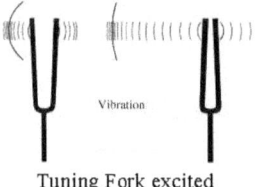

Vibration

Tuning Fork excited

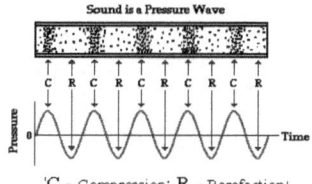

Sound is a Pressure Wave

'C - Compression; R - Rarefaction'

19.1.2 Simple Harmonic Motion (SHM)
SHM is the resolved, parallel to a fixed straight line, of uniform circular motion.

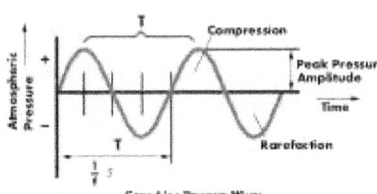

Sound is a Pressure Wave

19.1.3 Displacement, x

$$x = x_0 \operatorname{Sin} \omega t$$

Since the periodic motion is represented in terms of Sine / Cosine it is called harmonic motion.
Period, T

$$T = 2\sqrt{\frac{h \text{ (displacement)}}{g \text{ (Acceleration)}}}$$

19.1.4. Equation of simple harmonic motion can be rewritten as

$$x = x_0 \operatorname{Sin} (\omega t + \varphi)$$

(i)   **Amplitude,** x_0

(ii)   Phase $(\omega t + \varphi)$

(iii)   **Epoch** $= \varphi$

(iv)   **Angular frequency,** ω

$$\omega = 2\pi v$$

(v)   **Time frequency** $= v$

19.1.5 Differential form, acceleration

$$a = \frac{d^2 x}{dt^2} = -\omega^2 x$$

When a sound wave travels from one medium to another its frequency does not change.

19.1.6 The maximum value of KE

$$KE_{Max} = \tfrac{1}{2} m a^2 \omega^2$$

19.1.7. Potential energy PE = V(x)

$$V(x) = \tfrac{1}{2} m a^2 \omega^2 \operatorname{Sin}^2 (\omega t + \varphi)$$

19.1.8 Wave Velocity of propagation, v of the wave

$$\text{Wave velocity, } v = \frac{\omega}{k} = v\lambda$$

Wavelength of the wave, λ.

19.2.1 General expression for Displacement $y(x,t)$ of a particle in the medium,

$$y(x,t) = x_0 \operatorname{Sin} \left[\frac{2\pi}{\lambda}(v t - x + \varphi)\right]$$

$$\text{Particle velocity} = \text{(wave velocity)} \cdot \text{(Slope of the wave curve)}$$

$$\frac{dy}{dt} = -v \left(\frac{dy}{dx}\right).$$

19.2.2 Speed of sound, or the ultra-sound, is very much smaller than the speed of EM radiation,

Speed of sound in some solids			
Solid	Speed ($m\ s^{-1}$) Rod waves	Speed ($m\ s^{-1}$) compressiona waves	Speed ($m\ s^{-1}$) shear waves
(1) Aluminium	5102	6374	3111
(2) Brass	3451	4372	2100
(3) Crown glass	5342	5660	3420
(4) Perspex	2177	2700	1330

Speed of sound in some materials	
Gases and Liquids	Speed ($m\ s^{-1}$.)
(1) Air	331.46
(2) Hydrogen	1286
(3) Helium	971.9
(4) Nitrogen	337
(5) Neon	434
(6) Carbon dioxide	259
(7) Carbon tetrachloride	940
(8) Distilled water	1482
(9) Acetone	1190
(10) Ethanol	1162
(11) Glycerol	1860
(12) Sea water	1521
(13) Acetic acid	1173

viz., $\boxed{c = 3\ x\ 10^8\ m\ s^{-1}}$;

This leads to an interesting use of ultrasonic waves in Radar, in television sets, and in digital computers, namely to delay an EM wave.

19.2.3 **Differential Equation** of a SHM.

$$\boxed{\frac{d^2 y}{dt^2} = -v^2 \left(\frac{d^2 y}{dx^2} \right)}$$

19.2.4 **Energy per unit volume** of the affected medium

$$\boxed{\frac{E}{m^3} = \tfrac{1}{2} \rho\, a^2 \omega^2}$$

Density of the medium $= \rho$.

19.3 **RESULTANT OF TWO SHMs SUPERPOSED**

19.3.1 When waves are superposed resultant waves are produced

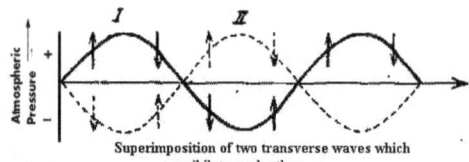

Superimposition of two transverse waves which
annihilate each other

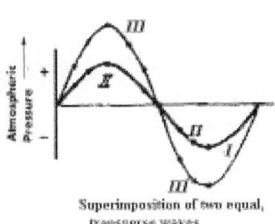

Superimposition of two equal,
transverse waves

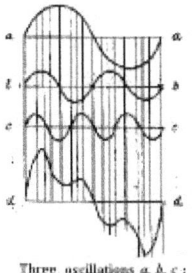

Three oscillations *a, b, c :*
superimposed into an
oscillation of arbitrary type

19.3.2 Lissajous Figures
19.3.2.1. Superposition of two waves of the type

$$x = a \, Sin \, \omega t \quad \text{and}$$

$$y = b \, Sin \, (\omega t + \varphi)$$

give the resultant

$$\frac{x^2}{a^2} + \frac{y^2}{b^2} - \frac{2xy}{ab} Cos\varphi = Sin^2 \varphi$$

which is the equation of an ellipse with major and minor axes $2a$ and $2b$.

19.3.2.2 Special cases: Lissajous' Figures
When a particle is acted upon simultaneously by two SHMs at right angles to each other, the curious resultant path traced out by the particle (displayed in a CRO) is called Lissajous' Figures. The nature of these figures depends on

(i)   Amplitudes,
(ii)   Frequencies, and
(iii)   Phase difference between the two component vibrations.

19.3.2.3 **Horizontal to vertical frequency ratios**: 1:1; 1:2; 2:1; 3:1

19.3.2.4 Phase difference:

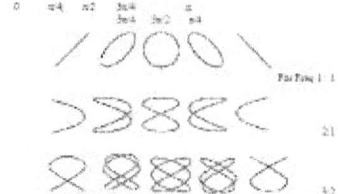

19.3.2.5 Fundamental frequency (Base tone), ω_o,

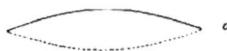

It is the natural frequency, ω_o, of a system in vibration.

19.3.2.6 First overtone (2nd harmonic), $2\omega_o$,

19.3.2.7 Second overtone (3rd harmonic), $3\omega_o$

19.3.2.8 Third overtone (4th harmonic) $4\omega_o$

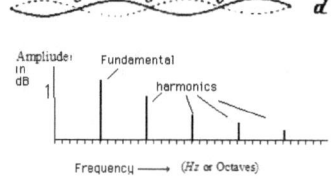

19.3.2.9 What is Middle C = 226 Hz tone ?

A musical instrument playing a fundamental 262 Hz tone sounds different because it produces
262 Hz + many harmonics (overtones) at the same time it produces pure 262 Hz tone.

What is an <u>Octave</u>?

(1) One octave above middle C = 262 x 2 = 524 Hz;	
(2) Three octaves below middle C = (262) (½) (½) (½) = 33 Hz;	
(3) Three octaves above 1 MHz = (1 MHz) (2) (2) (2) = 8 MHz.	
(4) Octaves increase as	1, 2, 4, 8,.. rate,
i.e.	$2^0, 2^1, 2^2, 2^3$, &.;

Harmonics:	1,	2.	—,	4, ..., in steps of fundamental
	Fundamental			
i.e.	3,	5,	6,	*and* 7 harmonics are not involved in the Octave.

An octave is a ratio of 2:1 and, in equal temperament; an octave comprises 12 equal semitones. Each semitone therefore has a ratio of $2^{1/12}$ (approximately 1.059). By convention, A4 is often set at 440 *Hz*. These data were used to calculate the frequency of any standard keyboard note or MIDI note number. To convert from any frequency to pitch (*i.e.*, to the nearest note and how far it is out of tune), how to do the calculation? Suppose that two notes have frequencies f_1 and f_2, and a frequency ratio of f_2/f_1. An octave is a ratio of 2:1, so the number of octaves between f_2 and f_1 is

$$n_o = log_2\left(f_2 / f_1\right).$$ The relations between physical and perceptual properties of sound are

shown below in Table.

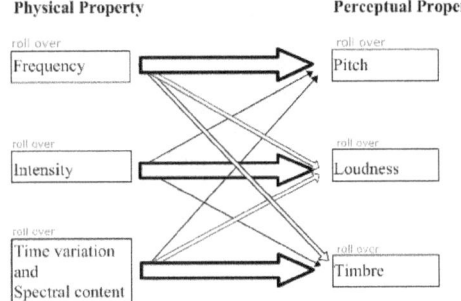

19.4. STRETCHED STRING
A string of infinite length having mass per unit length ρ, kept under tension T.

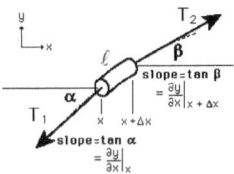

P = mass per unit length

19.4.1.1 Velocity of Transverse waves along a Stretched String:

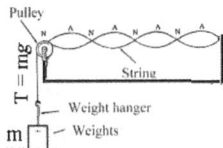

Velocity, $\boxed{v = \sqrt{\frac{T}{m}}}$

m = mass per unit length of wire string.

T = Tension along the string.

19.4.1.2 Frequency of vibrations of a stretched wire

Frequency, v $v = \frac{p}{2\ell}\sqrt{\frac{T}{m}}$

$2\ell = p\lambda$

ℓ = distance between two successive nodes in the stretched wire

p = Number of vibrations = 1, 2, 3,

19.4.1.3 Laws of transverse vibrations in a Stretched string

(i) $v \propto \frac{1}{\ell}$,

(ii) $v \propto \sqrt{T}$,

(iii) $v \propto \sqrt{\frac{1}{m}}$, (i.e., $v \propto \sqrt{\frac{1}{\rho}}$)

19.4.2.1 <u>Sonometer</u>

A Sonometer is a device for demonstrating the relationship between the frequency of the <u>transverse wave</u> produced by a plucked string, and the tension T, length l and mass per unit length m of the string. These relationships are usually called Mersenne's laws after Marin Mersenne (1588-1648), who investigated and codified them. For small amplitude vibration, the frequency v is proportional to:

a. the square root of the tension of the string,

b. the reciprocal of the square root of the linear density of the string,

c. the reciprocal of the length of the string.

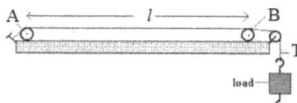

$\boxed{v = \frac{1}{2l}\sqrt{\frac{T}{m}}}$

In the sonometer

$\boxed{(Natural\ frequency\ of\ the\ wire) \neq (frequency\ of\ the\ tuning)}$.

Eg., If a sonometer wire ($\rho = 0.075\ g.cm^{-1}$) vibrates in unison at tuning fork at 256 *Hz*, with

vibrating segment 10 *cm*, tension T = $\frac{(256^2)(20^2)(0.075)}{(980)(1000)} = 2 - 006\ kg.wt$.

19.4.2.2 <u>Melde's String</u>

(i)   <u>Transverse mode</u> of vibration:

(Natural frequency of the wire) = (frequency of the tuning fork)

μ = linear density (mass per unit length) of the string

$\boxed{v = \frac{p}{2\ell}\sqrt{\frac{T}{m}} = \frac{1}{\lambda}\sqrt{\frac{T}{m}}}$, or $\boxed{v = \sqrt{\frac{g\,M}{4\mu\,\ell^2}}}$

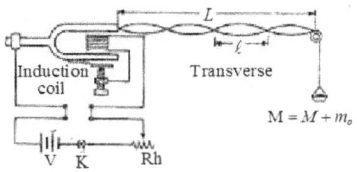

Longitudinal mode of vibration:
(Natural frequency of the wire) = ½ (frequency of the tuning fork)

$$v = \frac{p}{2\ell}\sqrt{\frac{T}{m}} = \frac{2}{\lambda}\sqrt{\frac{T}{m}} \quad \text{or} \quad v = \sqrt{\frac{g\,M}{4\mu\,\ell^2}}$$

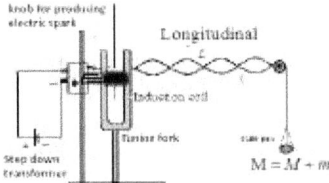

Applications
1. Tuning of instruments like guitar.
2. Standing waves in air column, soprano saxophone, *etc.*
3. Human speech analysis.

19.5   LONGITUDINAL VIBRATIONS
19.5.1   In Fluid medium
Resonance method to determine speed of sound in Air.

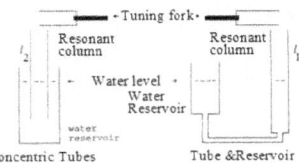

$$v = n\lambda = 2(l_2 - l_1)$$

19.5.1.1 Newton's Formula

Velocity, $v = \sqrt{\frac{E}{\rho}}$

E = elasticity of the medium

19.5.1.2 Laplace's correction

Velocity, $v = \sqrt{\frac{\gamma P}{\rho}}$

γ = Ratio of specific heats

19.5.1.3 Effect of Pressure

Velocity, $\boxed{v = \sqrt{\dfrac{P}{\rho}} = \sqrt{\dfrac{1}{m}}}$

19.5.1.4 Effect of Temperature, θ

Velocity, $\boxed{\dfrac{v_\theta}{v_0} = \sqrt{\dfrac{\rho_0}{\rho_\theta}} = \sqrt{\dfrac{(\theta+273)}{273}}}$

From Kinetic Theory of gases, $P = \frac{1}{3}\rho\bar{c}^2$

$\boxed{v = \bar{c}\sqrt{\dfrac{\gamma}{3}}}$

19.5.2 In Solids

Velocity of sound $\boxed{v = \sqrt{\dfrac{q}{\rho}} = \sqrt{\dfrac{K+(n/3)}{\rho}}}$

19.5.2.1 Kundt's Tube

The apparatus consists of a long transparent horizontal pipe G, which contains a fine powder such as cork dust or talc or lyco-podium powder.

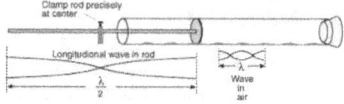

$\boxed{v = \dfrac{p}{2\ell}\sqrt{\dfrac{E}{\rho}}}$

19.5.3 Intensity (I) of Sound

Intensity $\qquad I = \dfrac{Power}{Area} = \dfrac{W \text{ or } Js^{-1}}{m^2}$

DECIBEL $\qquad \boxed{I(dB) = 10\log\dfrac{I}{I_0}}$

$\boxed{I_0 = 0 \ dB, \text{ the minimum augible sound intensity} = 10^{-12} \ Wm^{-2}}$

A voltage gain A_v of 100 gives,

$A_v = 20\log 100 = 40 \ dB$

A power gain A_p of 100 gives,

$A_p = 10\log 100 = 20 \ dB$

19.6 BEATS

19.6.1 The phenomenon in which two wave trains, of **nearly the same frequency**, v_1 and v_2 , travel along the **same straight line** in the **same direction** travel, the resultant displacement at a point constituted by a minimum and maximum sound, as the difference in their frequencies.

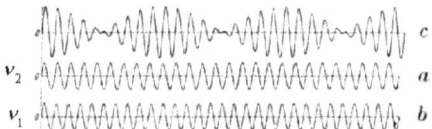

**Generation of beats (c) by two sounds (a and b)
with slightly different Hz**

Time interval between two consecutive maximum $\boxed{\Delta t = \dfrac{1}{\nu_1 - \nu_2}}$

$\boxed{\text{Beat frequency} = 2\left(\text{Amplitude frequency}\right)}$

$\boxed{\text{Number of beats} = \text{Beat frequency} = (\nu_1 - \nu_2)}$

Resultant frequency

$\boxed{\text{Resultant frequency} = \text{Arithmetic Mean} = \tfrac{1}{2}(\nu_1 + \nu_2)}$

If there are *m* beats in *t seconds*,

$$\boxed{\left(\frac{1}{\lambda_1} - \frac{1}{\lambda_2}\right) = \frac{|\nu_1 - \nu_2|}{v} = \frac{m}{t}}$$

Velocity, $\boxed{v = \nu\,\lambda}$

19.6.2 Stationary (Standing) Wave

$$y(t) = [2x_0\,Cos(\tfrac{2\pi}{\lambda})]\,Cos\,(\tfrac{2\pi}{\lambda})(v\,t)$$

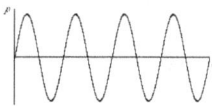

$$\frac{dy}{dt} = -\,v\left(\frac{dy}{dx}\right)$$

19.6.3 Progressive (Travelling) Wave

$$y(x,\,t) = x_0\,Sin\,(\tfrac{2\pi}{\lambda})(v\,t\,-\,x\,+\,\varphi)$$

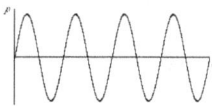

$$\frac{dy}{dt} = -\,v\left(\frac{dy}{dx}\right)$$

19.6.4 Free Vibrations

19.6.5 Damped Vibrations
Damping

19.7 **RESONANCE**

19.7.1 The Helmholtz Resonator and its mechanical correlate

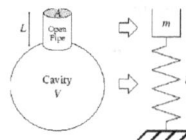

Frequency

$$\omega_0 = \bar{c} \sqrt{A/(LV)}$$

where $\bar{c}$ = the speed of sound in air, A is the cross-sectional area of the tube, L is the
length of the tube, and V is the volume of the cavity.

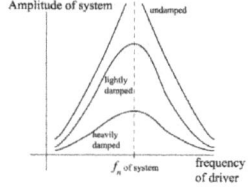 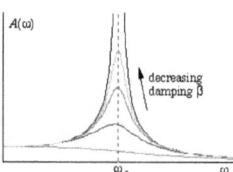

19.7.2 Standing Waves in Pipes
* One-dimensional sound waves propagate well in cylindrical pipes, just as mechanical waves
   travel along a string.
* Discontinuities in a pipe, like open and closed ends, cause wave reflections that result in the
formation of standing waves patterns. These boundary conditions impose limitations on the
particular frequency components that can be made to ``resonate''.
* Open end: acoustic pressure = ambient room pressure = 0.
* Closed end: particle velocity (and volume velocity) = 0.
* The acoustic length of a cylindrical pipe is slightly greater than its physical length by an
   additional $0.61\, r$, where r is the radius of the pipe.

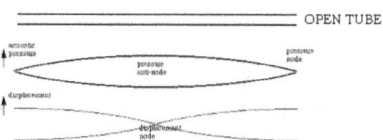

19.7.3 **Harmonics of Closed Pipe Resonator**

The closed pipe is $\ell = \frac{\lambda}{4}$; resonates odd harmonics and sounds hollow.

$$\boxed{\lambda = \ell + 0.4d}$$

ℓ = length
d = diameter
Anti-node is at one end and node at the other.

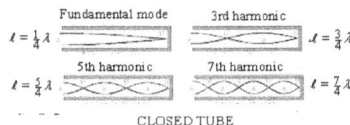

CLOSED TUBE

Progressive ([Travelling]) Waves	Standing ([Stationary]) waves
1) Disturbance trvels forwards. Along with Particle transformation.	Disturbance is fixed. No particle transforms its motion to next any time.
2) Each particle has the same constant amplitude. Phase varies along the wave.	Amplitude of each particle is not same, Maximum at antipode, decreases from antinode to node as per . $A = 2a\, Sin\, (\frac{2\pi}{\lambda})\, x$
3) No particle is permanently at rest. Every particle is at rest at the extreme, positions of its displacements. Different particles reach this position at different times.	Particles at nodes are permanently at rest. Other particles are also momentarily at the extrema of their position of displacements. All the particles reach the same position at the same time. The condition is repeated after half period
4) When passing through their mean position one after the other all particles have the same maximum velocity.	The velocity at the nodes is always zero. increases as it passes to antinodes and maximum at antinodes. All particles have maximum velocity when they reach mean positions at the same time.
5) Every region passes successfully through conditions of compression, and density, and rarefaction, and these conditions travel forward	The condensation, regions of normal density rarefaction is fixed. In any region the normal same condition appears and disappears alternatively.
6) $y(x, t) = x_0\, Sin\, (\frac{2\pi}{\lambda})(v\, t - x + \varphi)$ $\frac{dy}{dt} = -v\,(\frac{dy}{dx})$	$y(t) = [2x_0\, Cos(\frac{2\pi}{\lambda})]\, Cos\,(\frac{2\pi}{\lambda})(v\, t)$ $\frac{dy}{dt} = -v\,(\frac{dy}{dx})$
7) There is transmission of energy across any plane.	There is no flow of energy across any plane. Because condensation and velocity curves differs in phase by $\frac{1}{4}\pi$.

19.7.4 Harmonics of Open pipe Resonators

The open pipe is $\ell = \frac{\lambda}{2}$ resonates even harmonics and has a bright sound.

$$\boxed{\lambda = 2(\ell + 0.8d)}$$

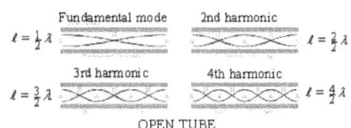

OPEN TUBE

ℓ = length

d = diameter
<u>Anti-nodes at each end</u>.

Eg., The Organ Pipe is extensively used and is of open pipe.

19.8   Travelling versus Standing Wave

19.9   **MUSICAL SOUND** PRODUCED BY STRINGED AND WIND INSTRUMENTS:
Hearing is one of the most important senses a man can have. It allows a person to hear sounds through the mechanical waves in harmony that are transmitted and which stimulate the hearing organs. Music is a form of art with sound as a medium, and the musical note is a combination of pitch and duration as its foundation. When these frequencies become inharmonious, they produce noise instead of music. Music has pleasing effect on the listener, while noise is unpleasant.
Indian music has therapeutic significance! Drone has psychic importance. According to an Indian text, '*Swara Saastra*' the 72 *melakarta* ragas control the 72 important nerves in a human body.

19.9.1 GUITAR
Standing waves in Guitar a freely vibrated string simultaneously in a number of different modes, and in a harmonic series, 1, 2, 3, 4, 5, . The relative amplitudes of each mode which would have frequencies of, say 440 *Hz*, 880 *Hz*, 1320 *Hz*, 1760 *Hz*, 2200 *Hz*, and so on (*i.e.*, a single pitch with a 1st harmonic 440 *Hz*), determines the sound quality.

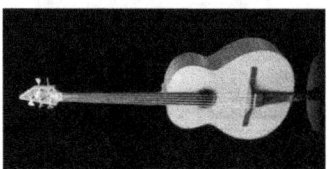

Lightly touching the centre of the string tends to extinguish the odd-numbered modes, which have anti-nodes there. Thus only even modes remain, having frequencies 880 *Hz*, 1760 *Hz*, 2640 *Hz*, *etc.* This combination of modes is a new pitch with 1st harmonic of 880 *Hz*. This new tone, or harmonic, as it is called by musicians, is one octave higher in pitch and radically different in quality. Similarly touching the string $\frac{1}{3}$ of the way along its length tends to extinguish all modes except those that are multiples of 3, and these predominate producing a new tone with 1st harmonic at 1320 *Hz*.
Sounds are distinguished from one another by (1) pitch (frequency), (2) loudness (intensity), and (3) quality (depends n the relative intensities of the harmonics (overtones) produced).

19.9.2 **VIOLIN**
Violin has a full vibrant sound because of its tones are characteristically rich in overtones. The idea is that by plucking, bowing, or hitting a string, a violinist can make it vibrate. The strings are wound around the peg, and so the tension in the string can be changed when the pegs are loosened or tightened. The shorter the length of the string causes higher the frequency (obtained by putting one's fingers down makes the string length effectively shorter). The "mass per unit length" is just how thick the strings are, the E-string is thinner than the A-string; the thinner the string, the higher the frequency. The violin resonates in two main ways: The top and

back plates radiate most of the sound; air circulating within the violin. When a string is plucked in the middle, all of the even modes will be still, while all of the odd modes will oscillate furiously.

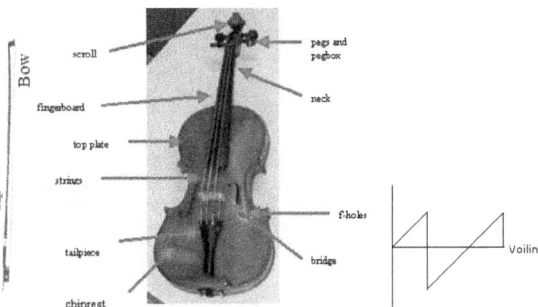

In violin the string is played close to one end; so some modes will be more excited than others, giving the standing waves. When a string is bowed, a force is created in the direction of the bow's motion, as the bow pulls the string along the force on the bridge increases in the direction of bowing. When the string slips, the force reverses direction to be opposite from the bow's motion. Thus the force transmitted to the bridge takes the form of a sawtooth wave. (Ref: Cremer, Lothar, "The Physics of the Violin", trans. Cambridge: MIT, 1984).

19.9.3 **Human voice:**

When air passes from the lungs to the mouth it may vibrate the vocal cords. All vowels are produced by vocal cords, which the human equivalent of the double-reed mouth piece of an oboe or bassoon. Many consonants are produced without the aid of vocal cord, for example, is produced by the vibration of the lips. Pitch of a human vowel sound depends on (1) length of the vocal cord, (2) tension of the vocal cord.

Quality of human vowel sound depends on the shape of the resonant cavities, the mouth and the throat, in which it is produced.

19.9.5 Galton Whistle

The maximum upper range of human hearing is about 20 kHz for children, declining to 15–17 kHz for middle-age adults. The top end of a dog's hearing range is about 45 kHz, while a cat's is 64 kHz. To human ears, a dog whistle makes only a quiet hissing sound. It was invented by Francis Galton.

19.9.6 VEENA

It is an Indian plucked stringed instrument used mainly in Indian classical music. It derives its distinctive timbre and resonance from sympathetic strings, bridge design, a long hollow neck and a gourd resonating chamber.

In the words of Sir CV Raman, the sound coming out on playing a Veena is almost equal to human voice, so rich in harmonics that the real music comes out of it.

The Body of '*Saraswathy* Veena' is composed of three parts: The main **Resonator**, the **Dandi** (neck, which is hollow), and the **Pegbox** (called *Yaali*). The Resonator is made from specifically chosen Jack wood without flaws and parallel grains (from the broader portion closer to the root). Rosewood is used for Pegbox, Sound board and the Bridge (Figure).

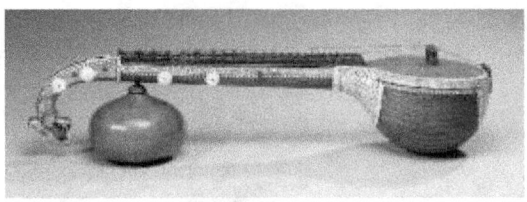

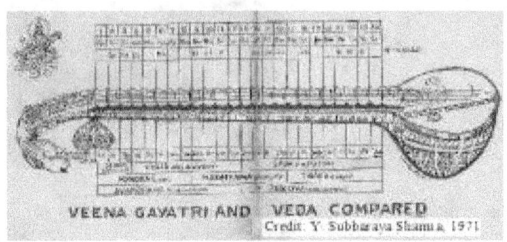

VEENA GAYATRI AND VEDA COMPARED
Credit: Y. Subbaraya Sharma 1971

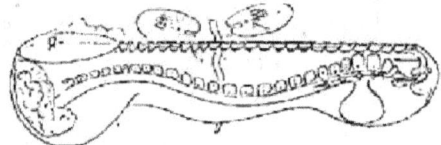

Veena and Spinal cord of human superposed

The top board or sound board of the main resonator plays a very important part on the quality of timbre of the Veena (Figure). There are four and twenty frets (apartments, based on '*swara*' modulation and control in acoustics).

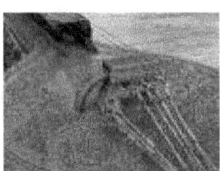 

Of strings, there are four *melody* (*Satva guna*) strings on top (symbolize the 4 Vedas) and three on one side as *subsidiaries* (minor / Drone, standing for *satva, rajo & tamo* gunas). Two highest melody strings (1st & 2nd) ('*Sarani*' for Rig Veda, and '*Panchama*' for Yajur Veda) are of steel and the lowest two strings (3rd & 4th) ('*Manthra*' for Saama Veda and '*Anumandra*' for Atharva Veda) are made of brass. They are tuned to provide the tonic, fifth, the octave and the fourth.

The characteristics of the bridges are responsible for the timbre of the Veena. Three Drone rhythmic *taana* (*Taala*) strings are known as '*Anusarani*', '*Anupanchama*' and '*Anumandra*'. The four knobs on which the Taala strings pass though are made out of antler and are fixed on the side of the Dandi (Ref: '*Eternality of the Veda (knowledge) and the Veena (Sound)*', Y Subbaraya Sharma, Bangalore, 1971).

Veena is played with three fingers on right palm struck towards or outwards with a plectrum. The musical notes produced on stringed instruments like the veena, are based on the physical and mathematical formulations associated with the vibrations of stretched strings. Such instruments operate over a vibrational frequency range of perhaps 60 to 1000 Hz in the fundamental, spanning over about four octaves. The frequency span of 60 to 1000 Hz is covered with four main strings. Pressing the wires against any of the frets and plucking them on the resonator side produces a distinct musical note (Rao, PP, *Current Sci.*, **81**, (6), 25 Sep 2001).

19.9.7 MRUDANGAM

This *percussion* instrument is made of jack tree wood in an angular barrel shape, having an outline like an elongated hexagon. Thong hoops around each end of the drum, leather thong lacing, and small wooden dowels slipped under the lacings control the skin tension. Goat skin smaller aperture is smeared in the central part made of rice flour, starch and ferric oxide and gets the colour black. It is important to note that both the layers are different in size, which helps in production of both bass and treble sounds from the Mridangam. The 'thoppi' or 'edabhaaga' is low-pitched aperture whereas the 'valanthalai' or 'valabhaaga' is high-pitched aperture. A removable patch of tuning paste is affixed to each end, giving the drum a definite pitch. The left head is usually tuned an octave lower than the right. The drum is held across the lap and played on both ends with the hands and fingers.

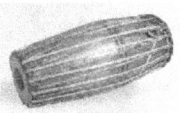

19.9.8 FLUTE

A flute is an aerophone or reedless wind instrument that produces its sound from the flow of air across an opening. It produces sound when a stream of air directed across a hole in the instrument creates a vibration of air at the hole. The air stream across this hole creates a Bernoulli, or siphon. This excites the air contained in the usually cylindrical resonant within the flute (say at 450 Hz).

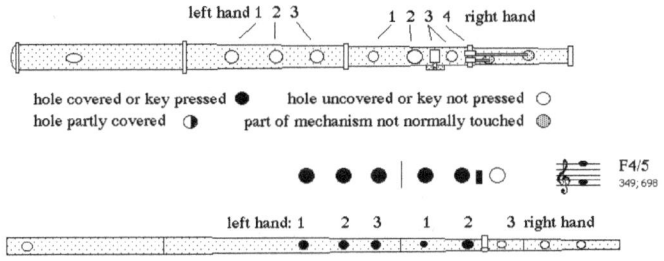

Classical flute fingering

The player changes the pitch of the sound produced by opening and closing holes in the body of the instrument, thus changing the effective length of the resonator and its corresponding resonant frequency.

By varying the air pressure, a flute player can also change the pitch of a note by causing the air in the flute to resonate at a harmonic rather than the fundamental without opening or closing any holes.

19.9.9. Harmonium

Harmonium (also called **Reed Organ)** is a free-reed keyboard instrument that produces sound when wind sent by hand / foot-operated bellows through a pressure-equalizing air reservoir causes metal reeds screwed over slots in metal frames to vibrate through the frames with close tolerance. There are no pipes; pitch is determined by the size of the reed. Separate sets of reeds provide different tone colours, the quality of the

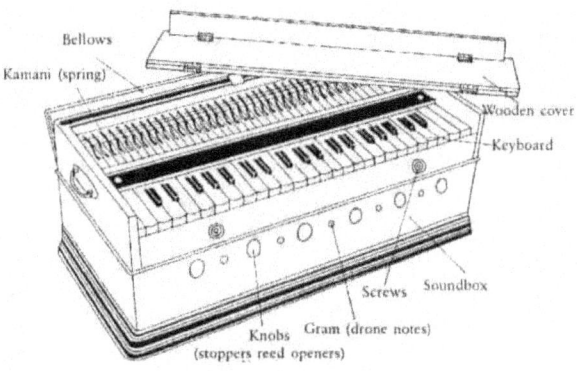

Harmonium

Scale Changer Harmonium parts and functions - DMS

sound being determined by the characteristic size and shape of the tone chamber surrounding each reed of a given set; constricted chambers, for instance, induce powerful vibration and incisive tone. Volume is controlled by a air valve or directly. The instrument's compass is normally four to five octaves.

19.9.10. Piano

Piano differs from harmonium. Piano generates sound by striking hammers against strings, unlike harmonium where air induce vibrations in those strings. Otherwise the two instruments are quite different.

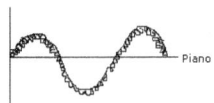

19.10.1 **Sound Pressure Level**

* Threshold of audibility" or the minimum pressure fluctuation detected by the ear is less than 10^{-9} of atmospheric pressure or about $2 \ x \ 10^{-5} \ Nm^{-2}$ at 1000 Hz.

* `Threshold of pain" corresponds to a pressure 10^6 times greater, but still less than $\frac{1}{1000}$ of atmospheric pressure.

* Because of the wide range, sound pressure measurements are made on a logarithmic scale (decibel scale).

* Sound Pressure Level

$$\left(SPL\right) = 20\log(P/P_0) = 10\log(P/P_0)^2,$$

where $\qquad P_0 = 2 \ x \ 10^{-5} \ Nm^{-2}$.

*SPL is proportional to the average squared amplitude.

Sound Power

    * Total sound power emitted by a source in all directions.

    * Measured in watts (joules / second).

    * Sound Power Level (PWL)

$$\left(PWL\right) == 10 \ \log(W/W_0),$$

where $W_0 = 2 x 10^{-12} W$.

Sound Intensity

    * Rate of energy flow across a unit area.

    * Sound Intensity Level

$$\left(IL\right) = 10\log(I/I_0)$$

where $I_0 = 10^{-12} Wm^{-2}$ ..

19.10.2 The Doppler Effect

19.10.2.1 Observer (*i.e.*, listener) moving toward source:

    $f' = `Apparent` frequency$ or pitch of sound

$$f' = f_S \left(\frac{V + v_0}{V} \right)$$

Increased wavelength $\quad \Delta\lambda = (\frac{V + v_0}{f_S})$

where $\quad f_S$ = the frequency of the source,

v_0 = the speed of the observer, and

V = the speed of sound.

19.10.2.2 Source moving toward observer (frequency of the approaching source)

$$f' = \frac{V}{(V - v_S)} f_S$$

$$\Delta\lambda = (\frac{V - v_0}{f_S})$$

*Frequency of the approaching source > frequency f_S of the same source at rest, where

v_S = the speed of the source.

* Direct determination of the velocity of a moving object by ultrasonics (say, in blood-   flow rates, study of heart movements).

19.10.2.3 Observer is moving at velocity v_0, away from a stationary source:

The observed frequency will be

$$f' = (\frac{V - v_0}{V}) f_S$$

$$\Delta\lambda = (\frac{V - v_0}{f_S})$$

19.10.2.4 Both the source and the observer are moving:

Using the sign convention that $v_0 = +$ ve, if it is in the same direction as V and

$v_0 = -$ ve, if it is in the opposite direction as V

$$f' = \frac{V - v_0}{V - v_S} f_S$$

(i) *Sign convention for Doppler Effect formula*, **moving source**:

Sound, V $\qquad\qquad$ Sound, V

$\rightarrow$ $\qquad\qquad\qquad\quad$ $\rightarrow$

$\rightarrow$ $\qquad\qquad\qquad\quad$ $\leftarrow$

Source, $v_S = +$ ve $\qquad$ Source, $v_S = -$ ve

(ii) *Sign convention for Doppler Effect formula*, **moving observer**:

Sound, V $\qquad\qquad$ Sound, V

$\rightarrow$ $\qquad\qquad\qquad\quad$ $\rightarrow$

$\rightarrow$ $\qquad\qquad\qquad\quad$ $\leftarrow$

Source, $v_0 = +$ ve $\qquad$ Source, $v_0 = -$ ve

(iii) **Change in frequency**

$$\Delta f = f' - f = [\frac{(v - v_0)}{v_0} v] - v = [\frac{(-v_0)}{v} v]$$

Here, for a moving source or observer, it is the velocity which is unaltered and both the frequency and wavelength that change.

19.10.2.5 A stationary observer receiving ultrasound from a stationary source, after the ultrasound has been reflected by a mirror moving with a velocity v_M ,

Change in frequency

$$\Delta f = f' - f = \frac{(-2\ v_0)v}{v_0}$$

19.10.2.6 Effect of motion of the medium:

$$f' = \left\{ \frac{V \pm v_M - v_0}{V \pm v_M - v_S} \right\} f_S$$

where v_M = speed of the wind (medium) moving in the direction of the source (or in the opposite direction)

19.10.11 REVERBERATION

The time taken for sound's intensity to diminish to 10^{-6} of its initial value is called the time of reverberation. It varies from $10\ s$ in an acoustically controlled auditorium to $< \frac{1}{2}\ s$ in a room crowded with people. Richness to music is added by a relatively long reverberation time (characteristic of all concert halls). In a lecture hall, a short reverberation time gives clarity to a speaker's voice. In Concert Halls, it is common to control sound by placing large, angled reflecting surfaces behind the performers, and absorbent surfaces behind the audience. Erecting false ceilings and walls with convex surfaces avoid focusing of sound waves in unwanted places. Concave surfaces such as domes inside buildings must be avoided.

19.10.12 Rhythm and Beat

A beat is like the tempo and never changes, the rhythm is the pattern of how the melody sounds the difference is the beat is steady like a a a a as in rhythm would be like l o f s the rhythm changes as beat stays steady.

19.11. Determination of the speed of Ultrasound in a Liquid by Diffraction

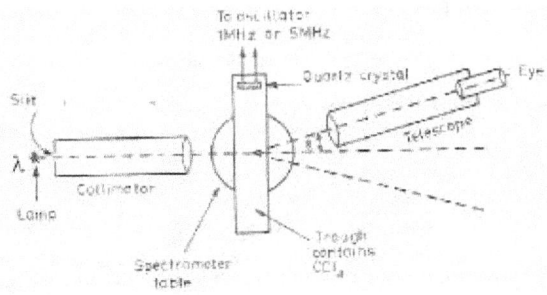

A specially constructed rectangular trough having two plane glass windows, one in each long side. This is filled with carbon tetrachloride CCl_4, and ultrasonic generator (quartz plate, f = 5 *MHz*) is mounted within the trough at one end. Longitudinal waves of ultrasonic frequency are set up in the liquid. This acts as a diffratometer when light of λ = 589.3 *nm* pass through it..

$d \sin \theta = n\lambda$.

Speed v of sound in the liquid, $v = fd$.

+*+*+*+*+*+*+*

Chapter 20

EM WAVES & LASERS

Maxwell's Equations, Poynting Vector, Optical Fibres, Holography

To be confused about what is different and what is not, is to be confused about everything"

- David Bohm

20.1 INTRODUCTION

In 1864 James Clerk Maxwell published a paper entitled "Dynamical Theory of the Electromagnetic Field".

20.1.1 Maxwell's Equations

(1) $$\oiint \vec{E}.d\vec{S} = \iiint \frac{\rho_j}{\varepsilon} \, dV \qquad \text{Gauss's Law}$$

it is equivalent to Coulomb's Law for static E fields

(2) $$\oiint \vec{B}.d\vec{S} = 0 \qquad \text{Gauss's Law for Magnetic field},$$

(3) $$\oint \vec{E}.d\vec{\ell} = -\frac{d}{dt} \oiint \vec{B}.d\vec{S} \qquad \text{Faraday's Law},$$

(4) $$\oint \vec{B}.d\vec{\ell} = \mu \iint \vec{j}_f.d\vec{S} + \mu\varepsilon \frac{d}{dt} \iint \vec{E}.d\vec{S} \qquad \text{Ampere's Law (modified form)}$$

20.1.2 The **flux of electric field intensity** Φ_E is defined as

$$\Phi_E = \oiint \vec{E}.d\vec{S}$$

where $d\vec{S}$ is the vector outwardly normal to the surface.
The **permittivity of free space,**

$$\varepsilon_0 = 8.8542x10^{-12} \, Fm^{-1}.$$

$$\varepsilon = K_e\varepsilon_0.$$

20.1.3 The **magnetic flux is**

$$\Phi_B = \iint \vec{B}.d\vec{S},$$

$$emf = -\frac{d\Phi_B}{dt}.$$

$$\boxed{emf = \oint \vec{E}.d\vec{\ell}}$$

$$\boxed{\oint \vec{B}.d\vec{\ell} = \mu \iint \vec{j}.d\vec{S}} \,,$$

where j = the **current density**.

20.1.4 Permeability of free space

$$\mu_0 = 4\pi x10^{-7} TmA^{-1} \; ; \quad \mu_0 = 4\pi x10^{-7} NA^{-2} \,.$$

$$\mu = K_B \mu_0,$$

20.1.5 Maxwell hypothesized the existence of an additional current, the **displacement current,**

$$\boxed{i_d = \varepsilon \iint \frac{\partial \vec{E}}{\partial t}.d\vec{S}} \,.$$

When this is combined with Ampere's law in a region with no physical currents, we get

$$\boxed{\oint \vec{B}.d\vec{\ell} = \mu\varepsilon \frac{d\varphi_E}{dt}} \,.$$

20.2.1 Differential Form of Maxwell's Equation.

Maxwell's equations, in differential form, will be necessary for discussing the wave nature of light.

$$\boxed{\vec{\nabla}.\vec{E} = \frac{\rho}{\varepsilon}} \qquad\qquad \text{Differential Form}$$

and $\qquad \boxed{\vec{\nabla}.\vec{B} = 0} \,. \qquad\qquad \text{Differential Form}$

$$\boxed{\vec{\nabla} \wedge \vec{E} = -\frac{\partial \vec{B}}{\partial t}} \qquad\qquad \text{Differential Form}$$

and $\qquad \boxed{\vec{\nabla} \wedge \vec{B} = \mu\left(\vec{j} + \varepsilon\frac{\partial \vec{E}}{\partial t}\right)} \,. \qquad \text{Differential Form}$

20.2.2 Maxwell's Equations in a Charge free Region:

How are Maxwell's equations used to show wave motion? In a charge free vacuum region, the Maxwell's equations become

$$\vec{\nabla}.\vec{E} = 0$$
$$\vec{\nabla}.\vec{B} = 0$$
$$\vec{\nabla} \wedge \vec{E} = -\frac{\partial \vec{B}}{\partial t}$$
$$\vec{\nabla} \wedge \vec{B} = \mu_0 \varepsilon_0 \frac{\partial \vec{E}}{\partial t}$$

20.2.3 The EM WAVE EQUATION:

$$\boxed{\nabla^2 \vec{E} = \mu_0 \varepsilon_0 \frac{\partial^2 \vec{E}}{\partial t^2}}$$

Velocity of EM Wave, $\boxed{\vec{v} = c}$

$$\vec{v}_s = (\vec{v}_0 - \vec{v}_i)$$

$$\vec{v} = \frac{1}{\sqrt{\mu_o \varepsilon_o}}$$

$$= \frac{1}{\sqrt{\left(4\pi x 10^{-7} \frac{m\ kg}{c^2}\right)\left(8.85 x 10^{-12} \frac{c^2}{J\ m}\right)}}$$

$$= 3.00 x 10^8 \frac{m}{t}$$

20.3.1 Light as Transverse Waves

Longitudinal waves oscillate in the same direction as the direction of propagation, while transverse waves oscillate in a direction perpendicular to the direction of propagation (the x direction), $E = E(x,t)$. The flux is through the faces in the y-z planes, so Gauss's law becomes

$$\frac{\partial \vec{E}_x}{\partial x} = 0$$

$$\left(\vec{\nabla} \wedge \vec{E} = -\frac{\partial \vec{B}}{\partial t}\right) \Rightarrow \left(\frac{\partial \vec{E}_y}{\partial x} = -\frac{\partial \vec{B}_z}{\partial t}\right)$$

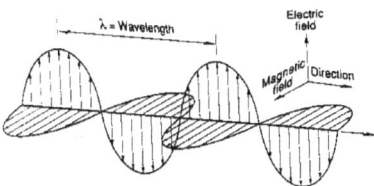

From these it is seen that, **in free space, the plane electromagnetic wave is transverse**. *Orthogonally of the Electric and Magnetic Fields*

1) $E_y = E_{yo} \mathrm{Cos}\left[\omega(t - \frac{x}{c}) + \phi\right]$

2) $\frac{\partial \vec{E}_y}{\partial x} = \frac{\omega}{c} E_{yo} \mathrm{Sin}\left[\omega(t - \frac{x}{c}) + \phi\right]$
$= -\frac{\partial \vec{B}_z}{\partial t}$
$\Rightarrow B_z = \frac{1}{c} E_{yo} \mathrm{Cos}\left[\omega(t - \frac{x}{c}) + \phi\right]$

3) $E_y = c\,B_z$

20.3.2 Energy in an Electromagnetic Wave

The energy density u_ε stored in the E-field

$$u_E = \frac{1}{2}\varepsilon_o E^2$$

- The energy density stored in the B-field:

$$u_B = \frac{1}{2\mu_o} B^2$$

* Velocity of light in vacuum, $c = 1/\sqrt{\mu_o \varepsilon_o}$,

$$c = 2.997925 \times 10^8 \, m \, s^{-1}$$

$$c = \lambda \nu$$

where frequency is symbolized by ν

- **Total energy density,** $U = u_E + u_B$.

20.3.3 The **Poynting vector** $\vec{S}$

$\vec{S}$ = the transport of energy per unit area. For an isotropic media, the energy flows in the direction of propagation of the wave, and the corresponding vector S is

$$\vec{S} = \frac{1}{\mu_o} \vec{E} \wedge \vec{B}$$.

20.3.4 Irradiance, I.

The time averaged value of the magnitude of the Poynting vector

$$\langle S \rangle = \frac{c^2 \varepsilon_o}{2} |\vec{E}_o \wedge \vec{B}_o|$$,

$$I = \langle S \rangle = \frac{c \varepsilon_o}{2} E_o^2$$
$$= \mu_o \varepsilon_o \langle E^2 \rangle$$

20.3.5 The **Optical field**. E:

Since the electric field is considerably more effective at exerting forces and doing work on charges than the magnetic field, the electric field E is referred to as the **optical field**.

20.4.1 Hertz's Experiment:

Heinrich Hertz in 1887 demonstrated that EM waves do exist.

An accelerating electric charge produces an EM Wave.

20.4.1.1. Lecher wires Experiment

The Lecher lines (a balanced transmission line) (Ernst Lecher, 1888) is made up of two parallel wires or rods, of 1.5 m, which are connected at one end. An Oscillator (200 MHz)is connected to the fixed end. The wires are spaced no more than a foot apart. The length is selected based on what frequency one wishes to put on the line. A metal bar(having two terminals connected to a neon/ bulb) is put across the width of the two lines, shorting them. As the bar is slid along the wires, the bulb stops glowing at the nodes (positions of $\frac{1}{2}\lambda$ or λ) of the waves.

$$\lambda = \frac{c}{\nu} = \frac{\text{Speed of light}}{\text{Frequency of oscillator}}$$.

20.4.2 Michelson-Morley Experiment (1887)

Michelson and Morley interferometer essentially consists of a light source A, a half-silvered glass plate P, two mirrors A and B and a screen / telescope C. The mirrors are placed at right angles to each other and at equal distance from the glass plate, which is obliquely oriented at an angle of 45° relative to the two mirrors. In the original device, the mirrors were mounted on a rigid base that rotates freely on a basin filled with liquid mercury in order to reduce friction.

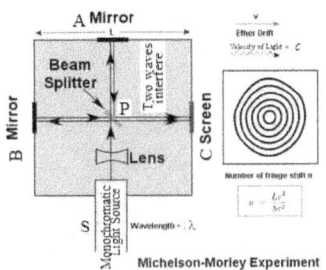

Michelson-Morley Experiment

Albert Michelson and Edward Morley were able to measure the speed of light by looking for interference fringes between the light which had passed through the two perpendicular arms of their apparatus.

In 1895, HA Lorentz concluded that the "null" result obtained by Michelson and Morley was caused by a effect of contraction made by the ether on their apparatus and introduced the length contraction equation,

$$L = L_0 \sqrt{\{1 - \frac{v^2}{c^2}\}}$$

where L is the contracted length, L_0 is the rest length, v is the velocity of the frame of reference, and c is the speed of light. The Michelson-Morley measurements were the first with sufficient accuracy to challenge the existence of the ether. The explanation of their null result awaited the insights provided by Einstein's theory of special relativity in 1905.

20.4.3 Bunsen Grease Spot Photometer

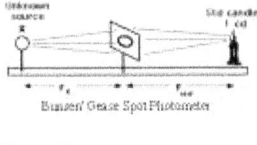

Bunsen' Grease Spot Photometer

$$E_x = E_{std},$$
$$\frac{I_x}{r_x^2} = \frac{I_{std}}{r_{std}^2}$$

20.5.1 **EM SPECTRUM**:

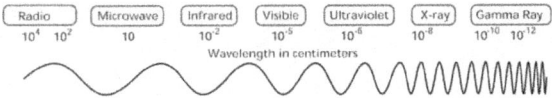

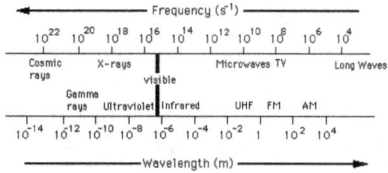

20.5.1.1 SPONTANEOUS EMISSION:

Einstein introduced the coefficient of spontaneous emission, A_{km} and induced emission B_{km}.

The rate of emission of EM radiation is Spontaneous emission (Downward transition rate):
$R_{k \rightarrow m} = A_{km}$
Radiation density of states $g(\omega)$,
At equilibrium temperature, T, according to the BOLTZMANN CRITERION,

$$\frac{N_m}{N_k} = e^{\hbar \omega_{km} / k_B T}$$

$$g(\omega_{km}) = \frac{A_{km}}{B_{km}} \left[e^{\hbar \omega_{km} / k_B T} - 1 \right]^{-1}$$

$$A_{km} \propto \omega_{km}^3.$$

20.5.2 LASERS

Difference between two Light Sources

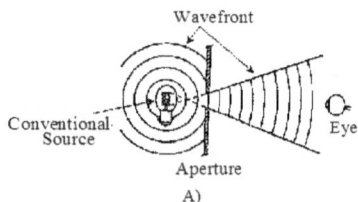

A)

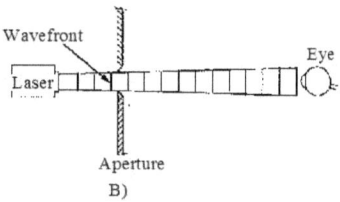

B)

MASER

Gordon, Zeiger and Townes (1954) made the two-level pumping scheme using Ammonia molecules as the active medium, in the MW region.

20.5.2.1 Acronym LASER stands for "Light Amplification by Stimulated Emission of Radiation".

T. Maiman invented the first LASER; CH. Townes and A. Schawlow (in 1957) developed the first Maser (Nobel Prize in 1981),

20.5.2.2 Characteristics of a Laser beam:

Characteristics	
Laser beam	Light beam
1. EMwaves, Photons	EMwaves, Photons
2. Liht Amplification byStimulated (induced) Emissionof Radiation	Spontaneous Emission
3. Highly monochromatic (single λ) light, Single colour	Polychromatic (white) All colours
4. Very high degree of coherency, (all waves in a beam are in phase with one another)	Incoherent
5. Well collimated (Parallel) Plane wave front always	Highly Divergent Spherical wave, mostly, Planewaves in lab
6. Highly directionsl (60 *mm* diameter *beam* can illuminate an object at distance of 1 *km*	Not directional
7. Plane polarized	Unpolaeized
8. Very intense beam	Lowpower

20.5.2.3 GAS lasers;

The first laser invented by Charles Townes was the <u>Ammonia Maser</u>. It involves two-levels of energy,

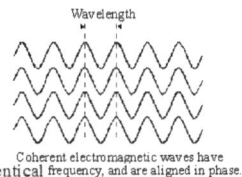

Wavelength

Coherent electromagnetic waves have Identical frequency, and are aligned in phase.

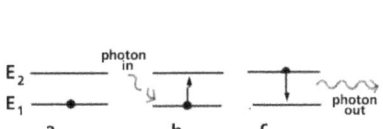

20.5.2.4 STIMULATED (Induced) Emission

(A. Einstein in 1917 suggested its existence):

$|\, k \,\rangle \rightarrow |\, m \,\rangle$ corresponds to **emission.**

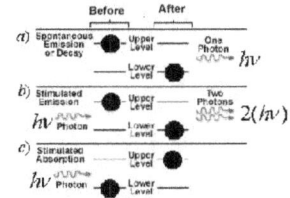

Spontaneous and Stimulated Processes

The rate of emission of EM radiation is Stimulated (induced) emission

$$R_{k \to m} = B_{km} \, g(\omega_{km}).$$

20.5.2.5 POPULATION INVERSION (Negative Temperature)

A condition for stimulated emission to dominate the spontaneous emission, or for lasing to occur, one requires the system to have more atoms in the excited state than in the ground state

$$\boxed{\frac{N_2}{N_1} = e^{-[-(E_2-E_1)/k_B T]}}$$

Is population inversion (negative temperature) under normal conditions of thermo-dynamic equilibrium, between the atoms in two levels.

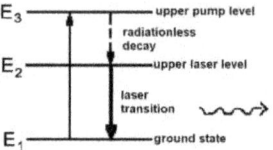

20.5.2.6. Laser Components

All lasers have three primary Components
1) Core Medium
2) Population inversion Pump
3) Lasing medium in Resonant cavity

Most lasers use EM pumping, meaning collisions via electrons or ions. Dye lasers and solid state lasers use optical pumping using flash lamp type light source. Chemically pumped lasers use chemical reactions as energy source.

20.5.2.7. Laser Resonator

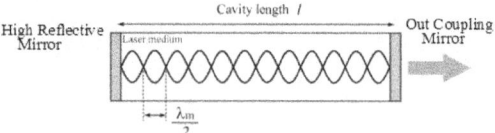

20.5.3 Types of Lasers:

20.5.3.1 **Gas lasers** (Free ion)

20.5.3.1.1. Nitrogen N_2 laser gives in UV $337.1 \, nm$.

The component of N_2-laser are: (i) A container for liquid nitrogen with pressure of 0.5 - 1.0 bar. (ii) A vacuum pump containing heavy oil. (iii) Cooling system (kerosin), for the transformer and spark gap. (iv) Power supply that gives a voltage of $12 \, kV$.

20.5.3.1.2. He-Ne laser (4 - level laser)

He-Ne laser emits mainly at $\lambda = 632.8 \, nm$ $\lambda = 1152.3 \, nm$, and $\lambda = 3390 \, nm$.

Helium–neon (He–Ne) lasers are a frequently used type of continuously operating gas laser, at a power level of a few mW and with excellent beam quality. The gain medium is a mixture of helium and neon gas in a glass tube, which normally has a length of the order of 15–50 cm.

257

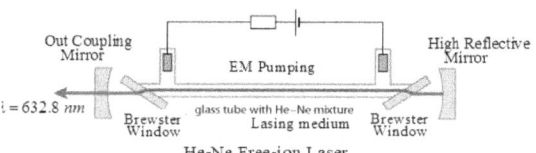

He-Ne Free-ion Laser

20.5.3.1.3. CO_2 laser

This is a molecular gas laser, consisting of a mixture of CO_2, He, N_2, some H_2, and Xe. It emits at $\lambda = 10.6 \ \mu m$ and $9.6 \ \mu m$.

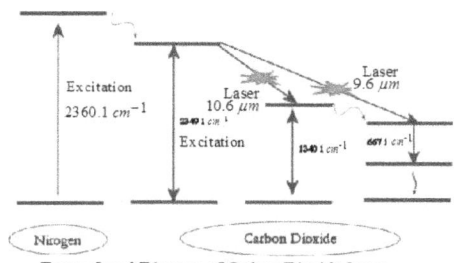

Energy Level Diagram of Carbon Dioxide Laser

20.5.3.1.4 Ar-ion laser

The core component of an argon ion laser is an argon-filled tube, made *e.g.*, of beryllium oxide ceramics. It is having mainly $\lambda = 514.5 \ nm$.

20.5.2. **Solid State Lasers**;

Ruby laser (Maiman invented) (3- level laser)

It is made of synthetic sapphire, Al_2O_3; doped approximately to 0.05% by weight by Cr_2O_3 working at $\lambda = 693.4 \ nm$; Pulses, 3-level laser. Method is optical pumping using a flash light.

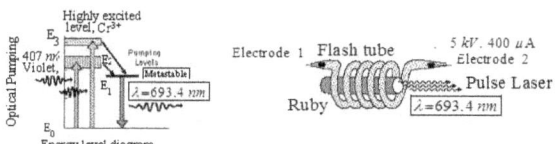

20.5.3 YAG Laser (4-level laser)

The core component is 1% Nd^{3+} in host material of Yttrium Aluminum Garnet, $Y_3Al_5O_{12}$ or $CaWO_4$, working at $\lambda = 1065 \ nm$. Optical pumping is used.

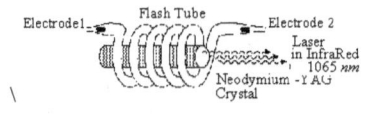

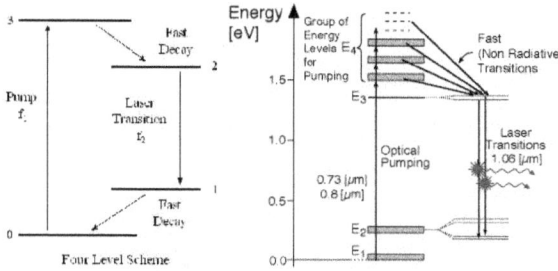

20.5.4. Semiconductor <u>diode lasers</u> (injection lasers)

$$\lambda(nm) = 1240 / E_g(eV)$$

GaAlAs, InGaAsP, AlGaInP or Lead salt materials are used..

20.5.5 <u>Dye Lasers</u>. Coumadin family of laser dyes.

20.5.6 Laser Gain

GAIN = amount of stimulated emission of a photon can generate as it travels a given distance.

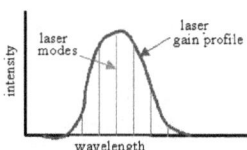

A Gain of 0.05 / *cm* means an amplification $= \dfrac{Output}{Input} = (1.05)$, raised to the power of length measured in *cm. i.e.*, A =1.63 for 20 *cm*, and 2.65 for 20 *cm*, 11.5 for 50 *cm*.

20.5.7. <u>Resonance</u>

Light waves are amplified strongly if they meet a resonance condition, and satisfy

$$\boxed{N\lambda = \text{(Cavity length)}}, \text{ N = integer.}$$

Each resonance value of N is said to be 'Longitudinal' Modes and so lasers oscillates in different <u>Longitudinal modes</u>.

Lasers oscillate in different <u>Transverse modes</u>, which are manifested as different beam patterns.

First order Transverse Electric Magnetic Mode $\boxed{\text{TEM}_{mn}}$, *m* and *n* indicate the number of walls in the E and M directions, and $\vec{E} \perp \vec{M}$.

TEM$_{00}$ TEM$_{10}$ TEM$_{11}$

$$\text{Beam Divergence} = \text{arcSin}\frac{K\lambda}{D}$$

D = beam diameter, K = constant.
L = distance of laser spot from laser.

$$\text{Beam Radius} = L\frac{K\lambda}{D}$$

$$\text{Sot Radius} = \frac{\lambda L}{\pi w_s}$$

w_s = radius of beam waist.

$$\text{Coherence length} = \frac{\lambda^2}{2\,\Delta\lambda} = \frac{c}{2\,\Delta\nu}, \text{ for range } \Delta\lambda.$$

20.5.8. APPLICATIONS OF LASERS

1) Cutting

CO_2 laser operating continuously cuts almost everything when focussed, and is the high speed and easy control of laser beam.

2) Medical Applications:

A number of medical applications had been suggested for the laser, *e.g.* - Skin disease treatment. - The remedy for detached retina (ruby laser). - Dentistry (Nd:YAG- CO_2 laser) - Treatment for leukemia patients. - In general surgery. - For bladder cancer (Nd:YAG laser - In urology (Dye laser).

3) Communications

Light -wave communications, using optical fibre as a transmission medium, is an important configuration in both public and private telecommunications networks.

4) Radar

Shorter laser pulses (ruby and Nd:YAG) can be used as signals for distance measuring by the radar principle. Giant- pulses from crystal lasers are suitable as signals for short distances, pulsed gas or semiconductor - lasers. Laser radar is interesting for extreme resolutions required for short measurement paths *e.g.*, air craft landing systems.

5) Spectroscopy Applications

The use of synchrotron radiation and lasers in the experiments of atomic and ion spectroscopy, and the advanced techniques in the photoelectrons spectroscopy and fluorescence measurements, make possible strong development in the theoretical investigations of resonant or auto-ionizing states of atoms and ions.

6) Industrial Applications

The efficiency of continuous wave CO_2 laser is about 14% . The range of output power allows using it, in processing of plastic, ceramics and some other non-conducting materials.

7) Environmental Applications:

Laser can detect the pollution in the air e.g. the existence of SO_2 , and also it detects the percentage of chlorophyll in the trees, depths of minerals in the earth, and depths of rivers and seas (geological uses).

8) Holography:

It is a new type of photography which needs no image forming lenses and gives pictures in a three dimensional view of objects. He-Ne laser is used as light source for holography. The unique properties of holograms give rise to numerous potential applications and great deal of work, *e.g.,* - Holographic interferometry. - Particle analysis. - Character recognition. - Acoustic holography. - Data storage by holography

20.6   OPTICAL FIBRES

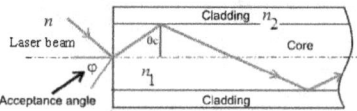

Light is transmitted by the core (of refractive index, n_1) and is surrounded by a cladding layer (n_2), $n_2 < n_1$.

The numerical aperture (NA) is the indicator of the fibre's light acceptance angle.

$$NA = n \, Sin \, \varphi = \sqrt{ \{ n_1^2 - n_2^2 \} } \, .$$

NA is also a measure of the dispersion of the fibre.

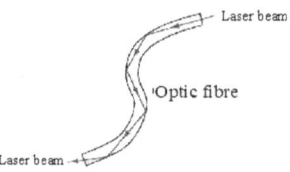

Light in the core striking the interface with the cladding at glancing angle s reflected back into the core.

In fibre-optic communication systems, i) Step index multimode fibre, ii) graded-index fibre, iii) single mode fibre are the three types used.

PIN photodiode and avalanche photodiode (APD) are the solid state detectors usedv as receivers in fibre optic communication.

LEDs and laser diodes (LD) are used as transmitters.

In order to build a complete fibre optic communication system, the transmitter and receiver   must be interfaced with the optical fibre. Connecters, Splices and Couplers are used as these components

20.7.   HOLOGRAPHY

20.7.1. Holography

Discovered by Dennis Gabor in 1947, is a technique for producing three-dimensional photography of an object using interference of light, and the image is obtained by the method known as "*wave front reconstruction*". The special screen in which the image is obtained is called the "hologram".

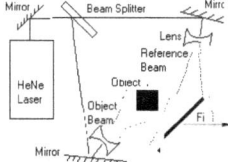

20.7.2 Recording a transmission hologram.

Light from a laser is divided into two beams. One beam goes directly to the photographic plate. The other beam reflects off the object before hitting the photographic plate. The two beams combine to produce a pattern on the plate which contains information about the 3-D shape of the object.

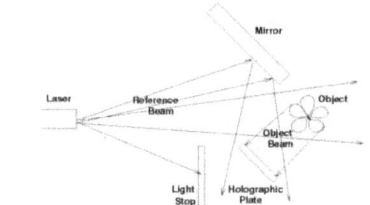

Transmission Hologram

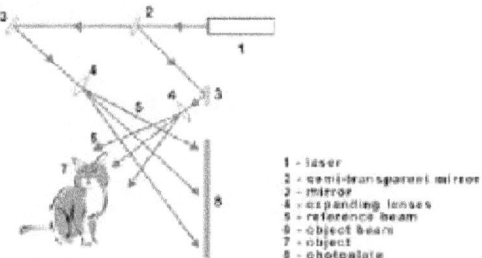

1 - laser
2 - semi-transparent mirror
3 - mirror
4 - expanding lenses
5 - reference beam
6 - object beam
7 - object
8 - photoplate

Recording of a Transmission Hologram

If the exposed and developed plate is illuminated by laser light, the pattern can be seen as a 3-D picture of the object.

+&+&+&+&+&##+*+*+*+*

Chapter 21

OPTICS -1

Nature of Light, Reflection, Refraction, Snell's Laws,
Dispersion, Achromatic prism, Cauchy's Relation, Rainbow,
Colours and Mixing, Concave and Convex Mirrors

"Mathematic seem to ends one with something like a new sense"   Charles Darwin

21.1 NATURE OF LIGHT:

The basic principle of light is its <u>reversibility</u> in path. What would the World be without light? Anything that gives light is called a <u>source</u> of light.

1. Natural light Sources
    Luminous objects such as The Sun, S tars, Fire, Candle flame, Fireflies are earliest sources of light.
2. Artificial light sources

21.1.1 Artificial Sources of light

Burning wood, candle, gas or oil are the earliest forms

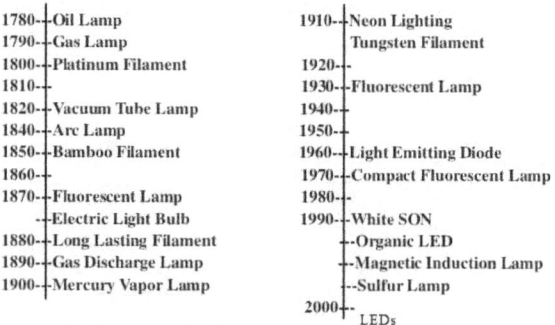

1780-- Oil Lamp	1910-- Neon Lighting
1790-- Gas Lamp	Tungsten Filament
1800-- Platinum Filament	1920--
1810--	1930-- Fluorescent Lamp
1820-- Vacuum Tube Lamp	1940--
1840-- Arc Lamp	1950--
1850-- Bamboo Filament	1960-- Light Emitting Diode
1860--	1970-- Compact Fluorescent Lamp
1870-- Fluorescent Lamp	1980--
-- Electric Light Bulb	1990-- White SON
1880-- Long Lasting Filament	-- Organic LED
1890-- Gas Discharge Lamp	-- Magnetic Induction Lamp
1900-- Mercury Vapor Lamp	-- Sulfur Lamp
	2000-- LEDs

Lighting Technology Timeline

21.1.2 Candle Flame {Tallow bee wax, (*viz.*, animal fat)
or Paraffin + Stearic acid 0.95, 60 °C }, as detailed in Chap 14.

21.1.3 Oil Lamps: Use plant or animal oil and a wick to burn.

21.1.4. Petromax Lamp (Gas Light)

One of the first and most successful Kerosene fueled pressure lantern was invented by Max Graetz (1910) is the world famous Petromax.

The mantle of this gas lamp is made up of cotton fabric loosely woven and saturated with salts of two rare minerals–Thorium (99%) and Cerium (1%). This lamp is 6 times as bright as the open flame lamp.

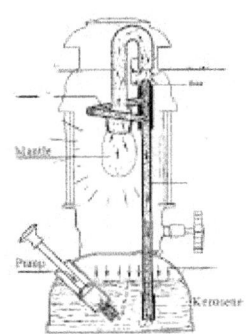

21.1.5 Electric Light:
Thomas Edison (1879) and separately Joseph Swan (1879) invented electric light bulb having oxygen, using a carbon filament what glowed when an electric current flowed through it. Then tungsten filament bulbs were made lasting 1000 hrs with yellow-white light. Service life is between 750 – 1000 *hrs*.

21.1.5.1. Carbon Electric Arc
When two adjustable arms (scissors shaped) having at their ends two Carbon rods and connected to electricity, around 40 VDC. & A arc will give 1000 candle-power.

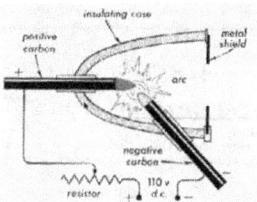

21.1.5.2 Neon Light: High voltage electricity is passed through neon gas enclosure.
By changing the flow of current up to five different colours of light would be produced in the same lamp.

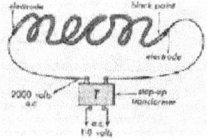

21.1.5.3. Fluorescent Lamp:

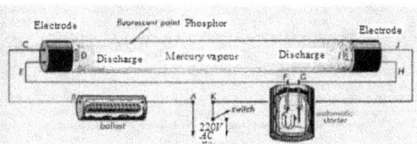

Glass tubes containing mercury vapour through which as current passes produces UV light. The phosphor powder that is coated inside the glass tube absorbs the UV light to give blue-white light.

21.1.6. Lasers: Lasers, invented in 1960, emits powerful monochromatic light.

The working principle is amplification &stimulated emission of radiation. Lasers find applications in astronomy, entertainment, image storage, medicine, *etc*. Lasers ushered a new era. (Vide Chap 20, Section 5.2).

21.1.7 Invisible Light sources (Both UV and IR.) Chemo-luminescence.

21.1.8 CFL bulb:

Compact Fluorescent Lamp is a mini version of standard fluorescent lamp, and is designed to replace incandescent lamp. Spiral CFL was invented by Edward E Hammer in 1976. Like fluorescent lamps CFLs contain toxic mercury and become hazardous waste on disposal. Service life is 6000 - 15000 *hrs*.

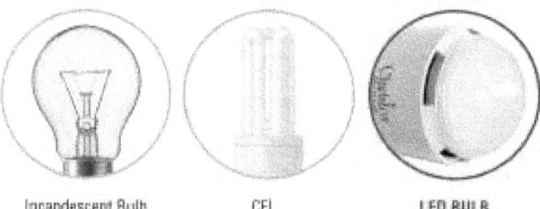

Incandescent Bulb   CFL   LED BULB

21.1.9 LEDs light the wave to sustainability.

The long-lasting, energy efficient LEDs that provide today light for scientific equipment, consumer electronics, general solid-state lighting and many other technologies for many future

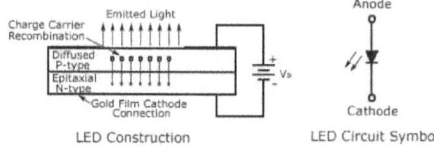

LED Construction   LED Circuit Symbol

decades. The emergence new semiconductor materials in 1970s made green, orange, yellow LEDs possible. For their development of efficient blue LEDs Amano, Akasaki and Nakamura received 2014 Nobel Prize in Physics (Physics Today, Dec 2014, page 14).

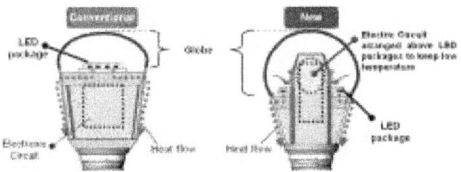

21.2. **Light is EM wave in nature**. Quantum mechanically light is a photon.
21.2.1 Theories of Light:
    (1) The Corpuscular Theory (Sir Isaac Newton, 1690),
    (3) The Wave Theory,(Christian Huygens, 1679),
    (3) The EM Theory,(James Clerk Maxwell, 1865),
    (4) Theory of Photons (Quantum Theory) (Max Planck, 1900 and Albert Eistein1905).

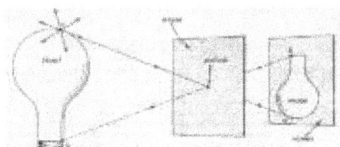

    Geometrical optics is an approximation to the results of Maxwell's equations that is valid when the dimensions of the physical system are much greater than the wavelengths λ of the light used. The laws of geometrical optics are:
(1) Light travels in straight lines in any homogeneous medium (rectilinear propagation). Image of an object will get inverted.
(2) The angle of reflection = the angle of incidence,
(3) Snell's law of refraction holds good.

21.2.2 **Mathematical expression for a Light wave**
21.2.2.1 **Spherical wave**

$$\phi = \frac{a}{r} Sin(\omega\, t - kr)$$

ϕ =Displacement of the particle distant r from the Source.
a = Amplitude of the wave at unit length from the Source,

$$\omega = 2\pi v = k\; c$$

$\omega = 2\pi v$, $k = 2\pi / \lambda$
Here the wave front is spherical in shape.

21.2.2.2 **Plane Wave**
When the direction of propagation of a light wave is perpendicular to the plane wavefront, the wave is plane wave.

$$\phi = a\, Sin(\omega\, t - kr)$$

$$\text{Phase difference} = (k)(\text{Optical path difference})$$

21.3 REFLECTION
21.3.1 Laws (Willebrord Snellius, 1621)

      *The angle of incidence θ_i = angle of reflection θ_r.

      *The incident ray, the reflected ray and the normal all lie in the same plane.

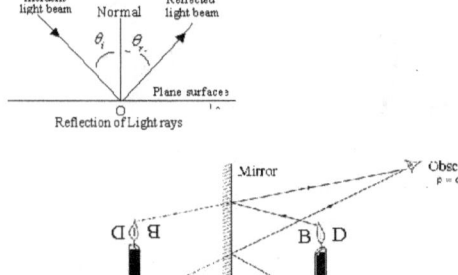

Virtual image shows lateral inversion of the object.

21.3.2 Images

    Real image is one through which the rays of light actually pass and which can be formed on a screen.

    Virtual image is one through which the rays do not actually pass.

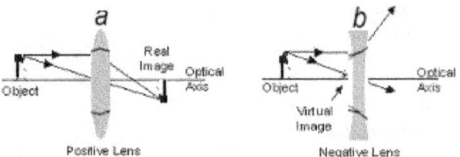

21.3.3 Reflection properties:

    Intensity of light I transmitted through a material of thickness x,

$$\boxed{I = I_0 \, e^{-\mu x}}.$$

    μ = constant for the material,

    $\mu = 4 \; cm^{-1}$ at $\lambda = 600 \; nm$

    $\mu = 1000 \; cm^{-1}$ at $\lambda = 250 \; nm$

21.3.3.1 Inclined Mirrors

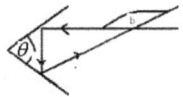

If θ = angle between the two mirrors,

Number of images of the object seen,

$$\boxed{\text{\# of Images, N} = [\frac{360}{\theta} - 1]}, \text{ if } (\frac{360}{\theta}) = \text{even,}$$

$$\boxed{\text{\# of Images, N} = [\frac{360}{\theta}]}, \text{ if } (\frac{360}{\theta}) = \text{Odd}$$

21.3.3.2 Two Mirrors at right angle

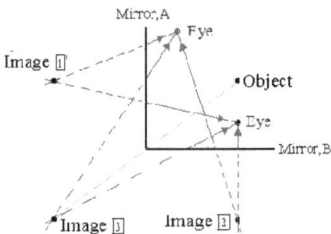

21.3.3.3 Minimum height of mirror for Full vertical image

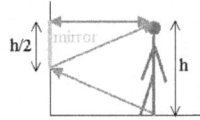

21.4 **REFRACTION**

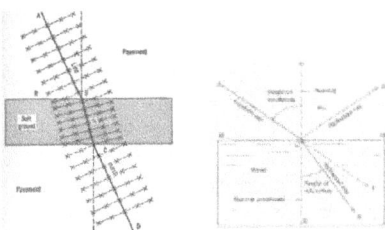

21.4.1.1 Snell's Law
(i) The incident ray and the refracted ray lie in the same plane, as the normal to the refracting surface, at the point of incidence O and on opposite sides of it.
(ii) Light is refracted toward normal when it passes to a denser medium.
The diagrams illustrate the effect of refraction of an object to an observer, when in different media.

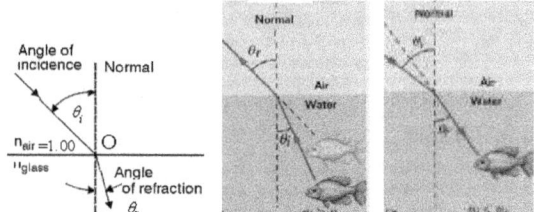

21.4.1.2 Refractive index *versus* Index of refraction: Unknown refractive index of medium

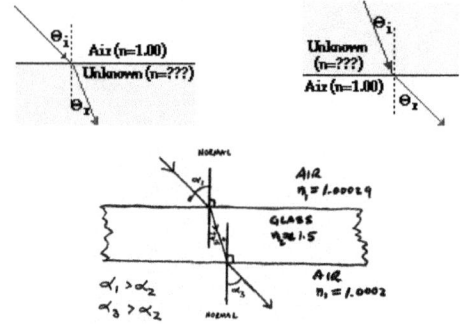

(a) For a pair of media, a and b,

Index of Refraction, $_a n_b$ $\boxed{_a n_b = (\dfrac{\text{Angle of incidence in Air medium, a}}{\text{Angle of Refraction in medium b}})}$

i.e., $\boxed{_a n_b = (\dfrac{\text{Speed of Light in Air medium, a}}{\text{Speed of light in medium b}}) = \dfrac{c}{v}}$

Index of refraction $\boxed{_a n_b = \dfrac{Sin\theta_i}{Sin\theta_r} = \dfrac{c}{v}}$

$\boxed{c = \text{Speed of light in vacuum}}$

v = Speed of light in the medium

of medium 1 to medium 2., and is denoted by $\boxed{_1 n_2 \equiv _1 \mu_2}$

and of medium 2 to medium 1, and is denoted by $\dfrac{1}{_2 \mu_1}$.

$\boxed{_1 \mu_2 \equiv \dfrac{1}{_2 \mu_1}}$

Indices of Refraction Medium	$air^{n}med$
(1)Vacuum	1.00
(2) Air	1.0003
(3) CO_2	1.00045
(4) Water (20°C)	1.333
(7) Ethanol	1.36
(6) Acetone	1.36
(8) Sugar Solution (30%)	1.38
(9) Fused Quartz	1.46
(10) Gycenine	1.4729
(11) Sugar solution (80%)	1.49
(12) Crown Glass	1.52
(13) Glass (typical)	1.5 to 1.9
(14) Cubic Zirconia	2.25 to 2.18
(15) Quartz	1.64
(16) Flint glass	1.61
(17) Sapphire	1.77
(18) Diamond	2.419
(19) Silicon	4.01

(b) If the medium 1 is air, then the ratio $\dfrac{Sin\theta_i}{Sin\theta_r}$ = generally, refractive index of medium 2 is μ_2.

$$a^{\mu_b} \equiv \frac{c^{\mu_b}}{c^{\mu_a}}$$

Refractive index of water means index of refraction from air into water.

$$\frac{1}{1^{\mu_2}} = \frac{Actual\ depth}{Apparent\ depth}$$

21.4.2. Abbe Refractometer

Abbé refractometer working principle is based on critical angle. Sample is put between two prisms - measuring and illuminating. Light enters sample from the illuminating prism, gets refracted at critical angle at the bottom surface of measuring prism, and then the telescope is used to measure position of the border between bright and light areas. Telescope reverts the image, so the dark area is at the bottom, even if we expect it to be in the upper part of the field of view. Knowing the angle and refractive index of the measuring prism it is not difficult to calculate refractive index of the sample. Surface of the illuminating prism is matted, so that the light enters the sample at all possible angles, including those almost parallel to the surface. The most widely used wavelength of light for refractometry is the sodium D line at 589 nm.

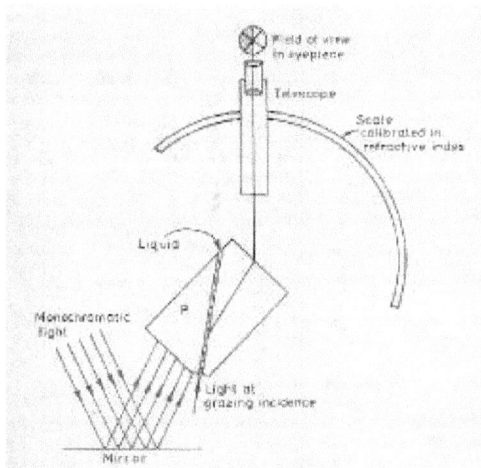

21.4.3.1 Mirages

Different densities of medium (air or water) due to different temperatures cause refraction by the atmosphere.

21.4.3.2 **Refraction of Light by Earth's Atmosphere** and Sun and Moon below Horizon

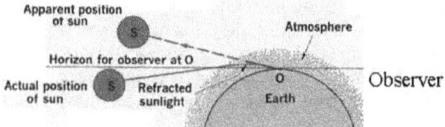

21.4.4.1 Critical Angle, θ_C:

The critical angle θ_C is the angle of incidence where the refracted ray lies on the surface., *i.e.*, $\theta_r = 90°$.

$$Sin\ \theta_C = \frac{1}{\text{Index of the denser medium}}$$

21.3.2.2 Total Reflection

When the incident angle exceeds the critical angle θ_C one gets Total Reflection

21.3.3.1 **Principle of Reversibility:**

If a reflected or refracted ray be reversed in direction, it will retrace its original path.

21.4.3.2 Optical path, *d*

In order to state a more general principle which will include both the law of reflection and that of refraction, one requires defining "optical path". When light travels a distance '*t*' in a medium of refractive index *n* the optical path, *d* is

$$d = n\,t$$

21.4.3.3. Fermat's Principle of Least Time:

$$\partial \int n ds = 0$$

21.5   WAVE OPTICS

21.5.1 Huygens' Principle

It tells us that every point on a wave can be considered as a source of *tiny wavelets that spread out at the speed of the wave*, and the new *wave front is the envelop* that is tangent to all of them.

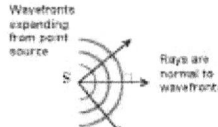

21.5.1.1 Huygens' Principle explains refraction:

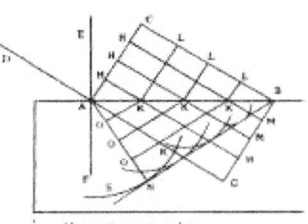

Huygen's construction

Wave front at point, say A, travels at speed

$$v_2 = \frac{c}{n_2}$$

while wave front at point, say B, travels at speed

$$v_1 = \frac{c}{n_1}.$$

Therefore the envelope bends, and the ray perpendicular to the envelope bends.
Snell's law of refraction (originally obtained by experimentation)
Highway mirages explained by using wave fronts.

21.6. NEWTON'S VISIBLE SPECTRUM

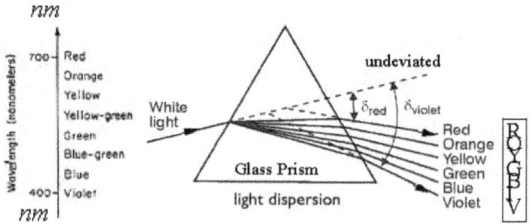

light dispersion

21.6.1 Spectra:
Various types:
- a. Emission spectra of Ne, Hg and Na
- b. Absorption spectra
- c. Continuous spectra
- d. Line spectra
- e. Band spectra
- f. Fraunhofer lines.

Each element in the Periodic Table has its own unique Emission Line Spectrum

21.6.2 DEVIATION and DISPERSION
21.6.2.1 Prism
A wedge shaped transparent medium bounded by two plane faces inclined at the angle A at the edge of the prism, Planes at right angle to the two plane faces are called principal planes of the prism.

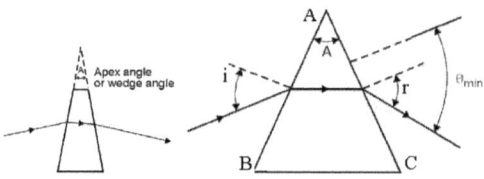

ABC Glass Prism –Transverse Section
A – Refracting angle of prism

θ_i Angle of incidence

θ_e Angle of emergence

θ_{r1}, θ_{r2} Angles of refraction

δ Angle of deviation,

μ Refractive index of glass with respect to air

$$A = \theta_{r1} + \theta_{r2}$$
$$\delta = \theta_i + \theta_e - A$$

21.6.2.2 Stoke's Formula:

$$\tan(\theta_{r1} - \theta_{r2}) = \left(\tan\frac{1}{2}A \cdot \frac{\tan\frac{1}{2}(\theta_i - \theta_e)}{\tan\frac{1}{2}(\theta_i + \theta_e)} \right)$$

21.6.2.3 Angle of Minimum Deviation, θ_D

The smallest angle, $\delta = \theta_D$, occurring at $\boxed{\theta_{r1} = \theta_{r2}}$; and $\boxed{\theta_i = \theta_e}$.

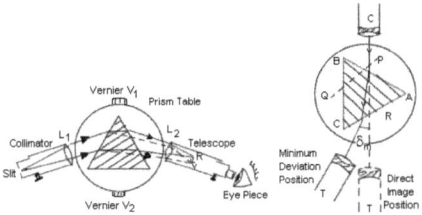

$$\boxed{\theta_i = \tfrac{1}{2}(A + \theta_D)}$$ and

$$\boxed{\theta_r = \tfrac{1}{2}A}$$

$$\boxed{n = \frac{Sin\ \theta_i}{Sin\ \theta_r} = \frac{Sin\ \frac{1}{2}(A + \theta_D)}{Sin\ \frac{1}{2}A}}$$

21.6.2.4 The Seven colours of Newton's spectrum

A White light ray incident on one side of a prism <u>dispersion</u> takes place and there emerge into 7 colours. The <u>Violet will be the more deviated one than the Red.</u>

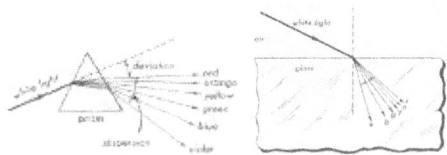

A prism separates white light into a rainbow of colors because the index of refraction of a material depends on the wavelength, λ.

The index of refraction is smaller for the larger wavelength

$$\boxed{n_{Red} < n_{Violet}}$$

So Red component moves faster than Violet ray in a material $\boxed{v_n = \frac{c}{n}}$**, although same** c **in vacuum.**

21.6.2.5 Variation of Refractive Index of a material with λ

Variation of the Index of Refraction		
Colour λ nm	Crown glass	Flint glass
Red	1.515	1.622
Yellow	1.517	1.627
Blue	1.523	1.639
Violet	1.533	1.663

$$\boxed{Sin\theta_i = n_\lambda Sin\theta_\lambda}$$

i.e., $$\boxed{Sin\theta_i = n_{Violet} Sin\theta_{Violet}}$$, *etc.*

21.6.2.6 For a thin prism:

A = small so that

The Mean deviation, $$\boxed{\delta \equiv D = (n-1)A}$$

δ is independent of *i*.

$$\boxed{\delta_V - \delta_R = (n_V - n_R)A}$$

21.6.2.7 Dispersive Power, ω of the prism

$$\boxed{\omega = \frac{\delta_V - \delta_R}{\delta} = \frac{\delta_F - \delta_C}{(n-1)A} = \frac{n_F - n_C}{(n_D - 1)}}$$

21.6.2.8 Angular dispersion,

$$\boxed{\varphi = (n_F - n_C)A}$$

21.5.2.9 Refractive Indices of glass for the C, D, and F-lines:

Notations used for standard wavelengths :		
1) *Red* :	the C line of Hydrogen with λ_C	= 656 nm.
2) *Yellow*	the D – line of Sodium with λ_D	= 589 nm.
3) *Blue* :	the F – line of Hydrogen, λ_F	= 486 nm.

Material	μ_C	μ_D	μ_F
1) *Crown*	1.5150	1.5175	1.5233
2) *Flint*	1.6434	1.6550	1.6648

21.6.2.10 Achromatic Prism:

a) Dispersive powers of prisms of different materials a and b are different:

$\omega_V > \omega_C$

$$\boxed{A_a(n_{Fa} - n_{Ca}) = -A_b(n_{Fb} - n_{Cb})}$$

$$\boxed{A_a(n_{Da} - 1) = -A_b(n_{Db} - 1)}$$

b) The **deviation** of the compound system of two prisms for Red light is

$$\boxed{\text{Residual Deviation} = A_a(n_{Da} - 1) - A_b(n_{Db} - 1)}$$

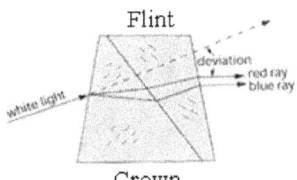

Flint

Crown

Note: '- ve' sign, when an achromatic is formed, represents that the two prisms must be combined such that the apex of prism 'a' is joined to the base of prism 'b', and the vice-versa, with the two sides joined together.

c) Deviation Produced by a Sphere:

$$D_m = 2 (\theta_i - \theta_r) + m (180° - 2\theta_r)$$

d) Minimum deviation occurs when $\dfrac{dD_m}{dt} = 0$

i.e.,

$$Cos\theta_i = \sqrt{\frac{(n^2 - 1)}{(m^2 + 2m)}}$$

where m = number of reflections inside the sphere.

21.6.2.11 Direct Vision Spectroscope

A combination of m crown and m Flint prisms when joined as shown provides spectrum of sources of light viewed the other side.

21.6.2.12 Cauchy's Dispersion Law (Empirical Equation (Augustin-Louis Cauchy, 1836)

One can find the index of refraction of light of a particular wavelength once we know its angle of minimum deviation through a prism. How do we relate that to the wavelength itself? That's the whole point of the experiment:

TABLE   Mercury Arc Spectrum

Spectral Line	Wavelength (nm)
1) Red	623.435
2) Yellow 1	579.066
3) Yellow 2	576.960
4) Green	546.073
5) Greenish Blue	491.640
6) Blue	435.844
7) Violet 1	407.810
8) Violet 2	404.656

$$n_\lambda = A + \frac{B}{\lambda^2} + \frac{C}{\lambda^4} + ..,$$

A, B and C are Cauchy's constants for a material.

21.6.2.12 Single Prism spectrometer

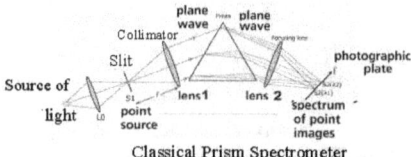

Classical Prism Spectrometer

<u>Schuster's method</u> of obtaining the image of slit in the field of view of a prism spectrometer

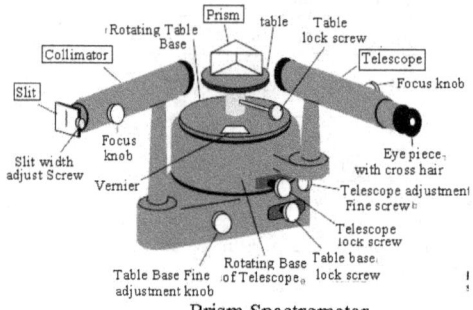

Prism Spectrometer

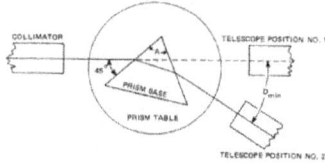

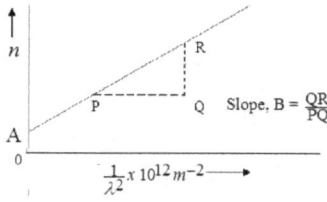

As the 'prism table' is rotated slowly, the slit image, which is moving in one direction, suddenly reverses its direction of motion. This position is the minimum deviation location.
A and B (Cauchy's constants) are also determined from the graph

25.6.2.13. Various types of prisms

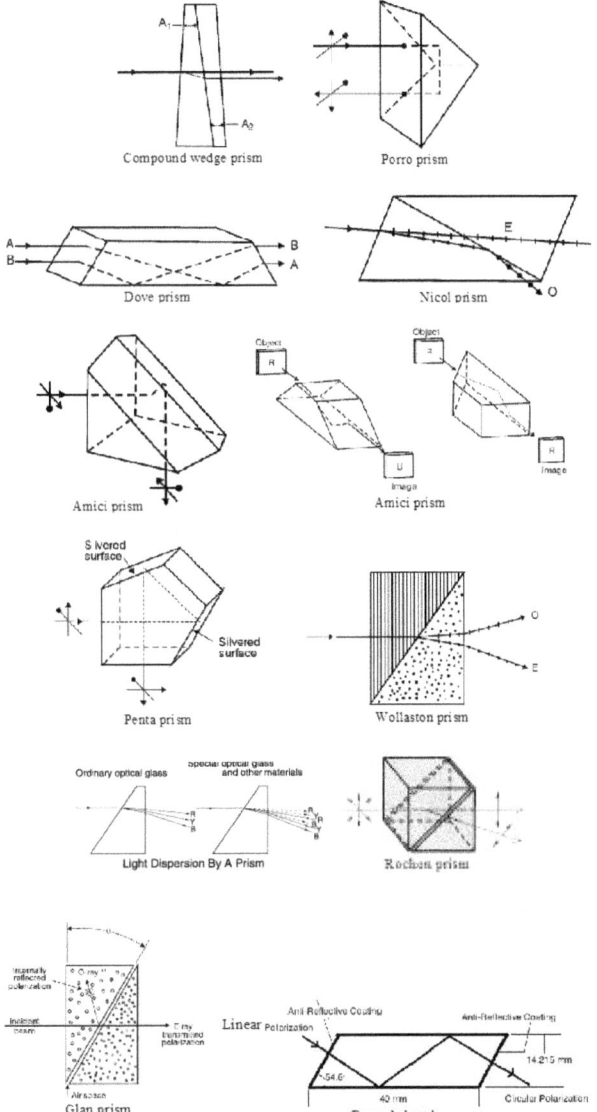

Compound wedge prism

Porro prism

Dove prism

Nicol prism

Amici prism

Amici prism

Penta prism

Wollaston prism

Light Dispersion By A Prism

Rochon prism

Glan prism

Fresnel rhomb

21.6.2.14 Rainbow

Rainbows are spherical arc pattern of spectral colours hanging in the sky seen by an observer standing with his back towards the Sun. These are produced when sunlight is incident on rain drops in the atmosphere is dispersed after suffering refraction and internal reflection.

Primary bow:

Light suffers one internal reflection, characterized by Red on the outer edge and Violet in the inner edge of concentric circles of different radii, with the centre of the circles lies below the horizon. It is the brightest and its radius subtends a **mean angle of about 41^0** at the observer's eye.

The emergent rays suffering <u>one</u> <u>internal</u> <u>reflection</u> will all be contained in a right circular cone with '**half the vertical angle**' $= (180^0 - D_r) = (180^0 - 137.3^0) = 42.7^0$ for the Red range and

$$= (180^0 - 139.2^0) = 40.8^0 \text{ for the Violet range.}$$

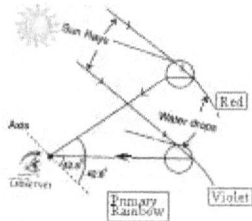

Secondary bow:

Due to <u>two internal reflections</u>; Fainter bow, its radius subtends a mean angle of about 50° at the observer's eye, has Red on its inner edge and Violet at the outer edge.

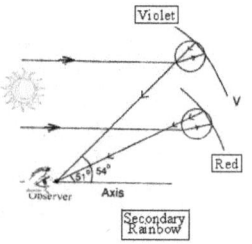

The emergent rays suffer two internal reflections,

The maximum deviation $= 360^\circ$,

Minimum deviation for the Red $= 230.8^\circ = 180^\circ + 50.8^\circ$,

for the Violet $= 234.52^\circ = 180^\circ + 54.52^\circ$,

The rays are packed in a cone **of half the external angle** $= 50.8^\circ$ for Red.

No two observers see the same rain bow! Because the two cannot subtend the angles of 41° and 52.5° for the primary and secondary rainbows from the same two drops.

21.6.2.15 Diamond Action

Diamond is one of the hardest materials, an allotrope of carbon. A diamond (crystalline in nature) has strong covalent bonds. The refractive index of diamond is pretty high (2.417) and is also dispersive (coefficient is 0.044). A jewel cut diamond is one whose faces are given selective cutting to sparkle as light undergoes total internal reflection followed by dispersion. .

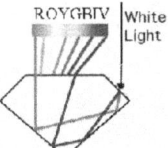

21.6.2.16 PRIMARY COLOURS OF LIGHT

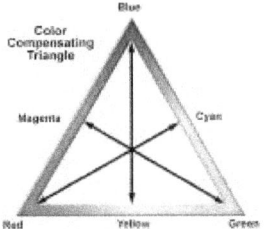

a) | The Primary Colours are: BLUE, GREEN & RED |

b) | The Primary Colours of Pigments: CYAN & MAGENTA | .

c) | Complementary colors are: TWO that add to give WHITE | .

d) Primary colour Mixing

	Additive Primaries
	-------------------------------
i.	Red + Green = Yellow Green + Blue = Cyan Blue + Red = Magenta

	Subtractive Primaries:
	-----------------------------
ii.	Magenta - Yellow = Red Cyan + Yellow = Green Magenta + Cyan = Blue

21.6.3. <u>Constant Deviation Spectrometer</u> (say, Hilger and Watts D186)

21.6.3.1. Emission spectra of a carbon arc source (50 – 240 V) can be viewed or photographed. For Hg spectrum exposure time of 5 *s* is enough; for Copper arc around 7 *mts*, and brass 8 *mts* may be required (ORWO film NP 27). Film processed by Develop for 6 *mts* and fix for 10 *mts*. in a **dark room**.

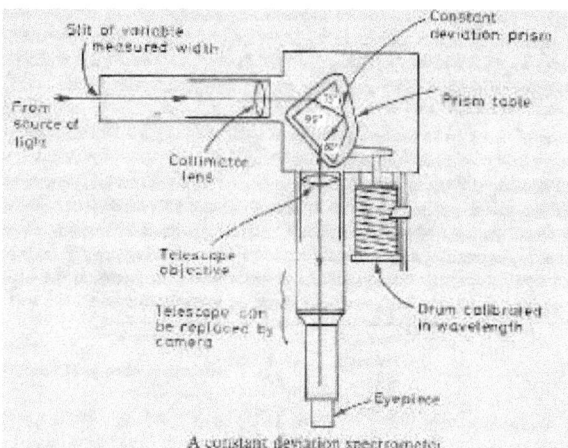

A constant deviation spectrometer.

The film may be soaked in water to swell the gelatin layer, facilitating the action of the subsequent chemical treatments.

1. The **developer** converts the latent image to macroscopic particles of metallic silver.
2. A stop bath, typically a dilute solution of acetic acid or citric, halts the action of the developer. A rinse with clean water may be substituted.
3. The **fixer** makes the image permanent and light-resistant by dissolving remaining silver halide. A common fixer is *hypo*, specifically ammonium thio-sulphate.
4. Washing in clean running water removes any remaining fixer (Residual fixer can corrode the silver image, leading to discolouration, staining and fading). Then dry the film in free air.

The spectrometer is first calibrated with a Na discharge lamp.

25.6.3.2. **Photography of Spectra** – Hartmann equation

Cd discharge lamp is used as the source. A Panchromatic photographic plate is loaded in the camera plate holder. Using the Hartman diaphragm in the spectrograph, four or five exposures of spectra may be made parallel to one another on the same photographic plate. A photograph of the line spectrum of Hg-discharge lamp is also taken parallel to that of the Cd spectrum.

The distance x of a spectral line λ from a fixed mark in the same recorded spectrum on the photographic plate may be expressed by a dispersion equation such as the Hartmann relation,

$$\lambda = \lambda_0 + \frac{A}{B + x}$$

λ_0, A and B are constants.

The Cd spectrum contains three convenient lines, at $643.8\ nm$, $\lambda_2 = 508.6\ nm$ and $\lambda_3 = 567.6\ nm$. Let the fiduciary mark be that due to the line $\lambda_1 = 643.8\ nm$, for which $x = x_1 = 0$. One gets three equations $\lambda_i = \lambda_0 + \frac{A}{B + x_i}$; solving which

$$A = \frac{(B + x_1)(x_2 - x_3)}{C - D}, \text{ where } C = \frac{(x_1 - x_2)}{\lambda_2 - \lambda_1}; \ D = \frac{(x_1 - x_3)}{\lambda_3 - \lambda_1}.$$

$$B = C (\lambda_2 - \lambda_0) - x_1.$$

Processing of Photographic plate / film

In a dark room, photographic films are 1) developed and 2) Fixed and 3) washed in flow of water.

25.6.3.3.   Emission spectroscopic Analysis

Qualitative analysis of a sample of alloy brass. i) Use Copper rods, ii) Zn rods, iii) Brass rods , iv) powder of unknown alloy held in between graphite rods.

25.6.3.4. **Zeeman Effect**

$$\Delta E = m_l \; \mu_B B$$

$$\mu_B = \text{Bohr Magneton}, \; \mu_B = 9.2740 \, x10^{-24} \, J.T^{-1}.$$

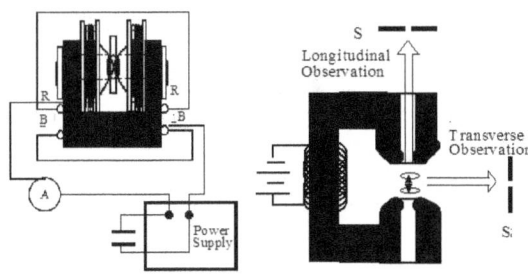

This displacement of the energy levels,in the pres3nce of a magnetic field, gives the uniformly spaced multiplet splitting of the spectral lines is known as the Zeeman effect. The experiment can be performed with a Fabry-Perot etalon of Lummer-Gehrcke plate, with a strong magnet.

For Na D- lines: $3p_{3/2} \rightarrow 3s_{3/2} (589.0 \, nm)$

$$3p_{1/2} \rightarrow 3s_{3/2} (589.6 \, nm)$$

$$\Delta\lambda = 0.6 \, nm$$

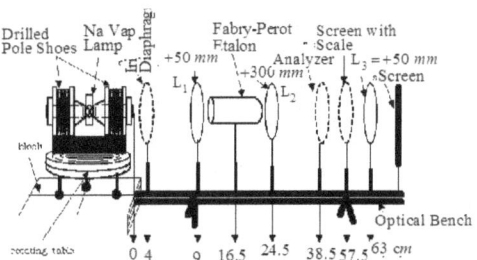

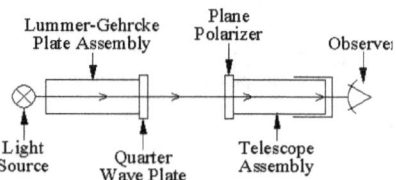

In Zeeman effect the doublet spectrum become a triplet: one π- and two σ-components on either side of π.

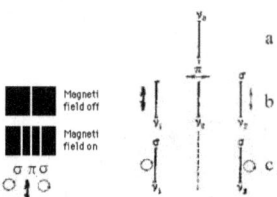

25.6.3.5. Absorption spectra.

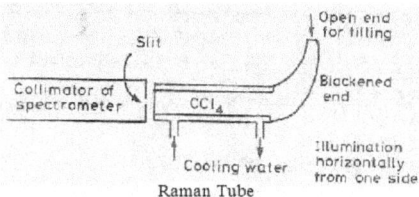

Raman Tube

Sample of Carbon tetra-chloride CCl_4 can be an ideal liquid, whose absorption spectrum can be taken. It may have as many as 40 lines in the spectrum. $435.8\ nm$ Hg source can be used. By the method described in Section 25.5.3.2., using the appropriate Hartmann dispersion relation, Raman shift can be found.

25.6.3.5. Determination of λ and $\delta\lambda$ by Michelson's Inrterferometer

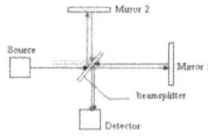

$$\lambda = \frac{2x}{N}$$, N = Number of fringes born as a result of displacement of moveable mirror position by x.

$$\delta\lambda = \frac{\lambda^2}{2d}$$, d = distance through which the moveable mirror is moved for one complete extinction of the ring system of interference.

21.7. GEOMETRICAL OPTICS (RAY OPTICS)

OPTICAL IMAGES:

If the pencil of rays originating in a single point is made to converge to or appear to diverge from some other point, the second point is called the optical image of the first. If the rays actually passing through the second point then the image is said to be Real, if not Virtual.

21.7.1 PLANE MIRROR
(i) Image of an object gets reversed.
(ii) Reflection do not change size with distance

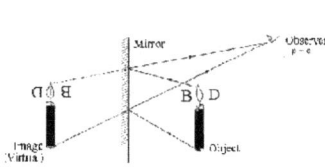 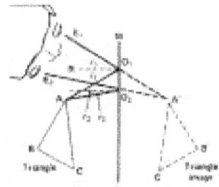

21.7.2 SPHERICAL MIRRORS:
The distance between the pole, P, and principal focus, F is the focal length, f.

Focal length of spherical mirrors, $\boxed{f = \frac{1}{2}R}$

Radius of Curvature, R

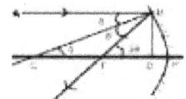

21.7.3 Mirror Formulae:

$$\boxed{\frac{1}{u} + \frac{1}{v} = \frac{1}{f}}$$

when applied with correct sign for distances for concave and convex cases.

21.7.3.1 Magnification,

$$\boxed{\text{Linear } m = \frac{\text{Image height}}{\text{Onject height}}}$$

$$\boxed{\text{Area } m = \frac{v^2}{u^2}}$$

21.7.3.2 Newton's Formula:
x & y : distance of the object & image from the principal foci.

$$u = f + x$$
$$v = f + y$$
$$f^2 = x$$

21.7.3.3. Movie system

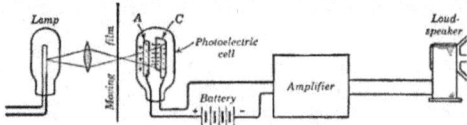

21.7.3.3 **IMAGE formed by Mirrors**:
4 lines have to be drawn to represent an image graphically for a mirror as a lens.
a)   The Principal axis,
b)   A ray from the top of the object that passes undeviated through the Centre of the lens,
c)   A ray from the top of the object, parallel to the axis that goes through the Principal axis.
d)   A ray from the top of the object through the Principal focus that emerges parallel to the
         axis.

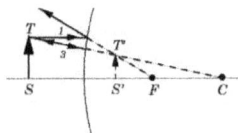

21.7.4 CONCAVE Mirror
Uses: Reflecting Telescopes, Dental mirrors, Head lamp reflectors, Shaving & makeup mirrors.
Ideal mirror has parabolic shape
Radius of curvature, R and focal length, f, of a concave mirror is

$$R = 2f$$

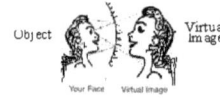
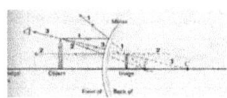

21.7.4.1 Image Formation with a Concave Mirror
Diverging mirror spreads light rays

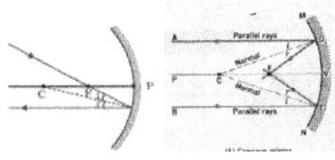

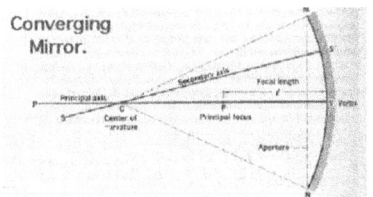

Converging Mirror.

Rules for image formation by concave mirrors.		
Position of object	Position of image	Character of image
1) At ∞	At F	Real, zero size
2) Between ∞ and C	Between F and C	Real, inverted, diminished
3) At C	At C	Real, inverted, same size
4) Between C and F	Between C and ∞	Real, inverted, magnified
5) At F	At ∞	
6) Between F and V	From -\∞ to V	Virtual, upright, magnified
7) At V	At V	Virtual, upright, same size

21.7.5 CONVEX MIRRORS
Action of Convex mirror (Diverging mirror) is as shown.

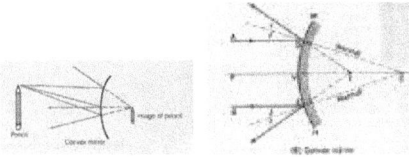

21.7.5.1 **Uses**
Safety viewers at dangerous corners of roads and on upper deck buses, anti-ship- lifting devices, Car wing mirrors

The image is between the Pole and the focus, behind the mirror, erect, diminished, and virtual for all positions of the object in front of the mirror.

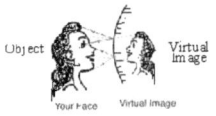

21.7.5.2 Image Formation

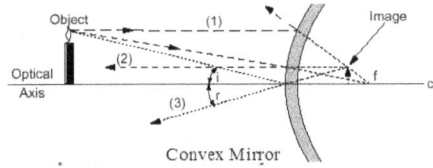

Convex Mirror

Rules for image formation by convex mirrors.		
Position of object	Position of image	Character of image
1) At ∞	At F	Virtual, zero size
2) Between ∞ and V	Between F and V	Virtual, upright, diminished
3) At V	At V	Virtual, upright, same size

21.7.5.3 Convention as to Signs:

1) All distances are measured from the Pole of the mirror. Distances measured in a direction opposite to the incident beam are considered '+ve'

2) Distances measured from the Pole in the direction in which the incident beam is traveling are considered '-ve.'

3) $f = +ve$ for concave mirrors; $f = -ve$ for convex mirrors.

21.7.6 DEFECTS of Mirrors

Defects observed in Images produced by reflection at curved surfaces are

1. Spherical aberration,
2. Curvature and distortion and
3. Astigmatism.

+*+*+*+*+*+*+

Chapter 22

OPTICS - 2 – REFRACTION,

Lens Makers Formula, Convex Lens, Image formation &
its Applications, Divergent Lens, Eye defects, Eye pieces,
Polarized light, Determination of Speed of Light

"For the wise all 'things are wiped away" - Buddha

22.1 **LENSES**

Lenses are among the most used components in optical systems.

22.1.1 Types of Lenses

A set of prisms act as converging and diverging lenses.

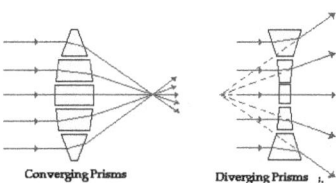

Converging Prisms   Diverging Prisms

22.1.2 Geometry of Lens Surfaces

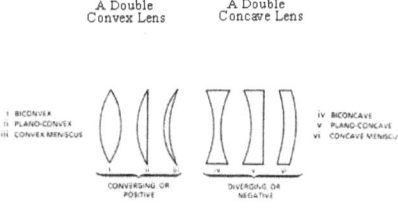

A Double Convex Lens   A Double Concave Lens

i BICONVEX
ii PLANO-CONVEX
iii CONVEX MENISCUS

iv BICONCAVE
v PLANO-CONCAVE
vi CONCAVE MENISCUS

CONVERGING OR POSITIVE   DIVERGING OR NEGATIVE

22.1.3 Lens Maker's Formula:

$$\frac{1}{f} = (n - 1)\left(\frac{1}{R_1} - \frac{1}{R_2}\right)$$

n = refractive index of glass for the D-line

R_1 and R_2 are the radii of the two surfaces of the lens.

Each R is a <u>positive</u> quantity, if it is a radius of curvature of a lens surface convex toward the incident rays, but <u>negative</u> if the surface is concave toward the incident rays. F is positive for a convex lens and negative for a concave lens.

Design of a Plano-Convex lens
The contour of the lens is a hyperbola given by

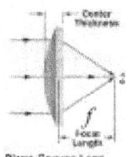

Plano-Convex Lens

$$\frac{(z-c)^2}{a^2} - \frac{x^2}{b^2} = 1 ,$$

$$a = \frac{f}{n+1} , \quad b = \sqrt{\frac{(n-1)}{(n+1)}} f , \quad c = \frac{n}{n+1} f$$

22.2.1 SIX Cardinal points of a Lens:
1. 2 Focal points: F_1 & F_2,
2. 2 Principal points: H_1 & H_2, and
3. 2 Nodal points: N_1 & N_2

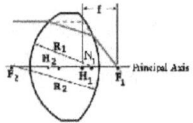

22.2.2 IMAGE formed by Lenses:
4 lines have to be drawn to represent an image graphically for a lens as in a mirror.
a) The Principal axis,

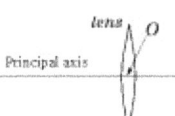

b) A ray from the top of the object that passes undeviated through the Centre of the lens,
c) A ray from the top of the object, parallel to the axis that goes through the Principal axis.
d) A ray from the top of the object through the Principal focus that emerges parallel to the axis.

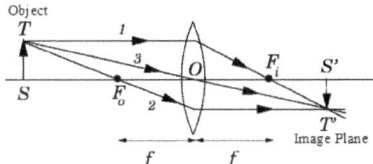

22.2.3 Conventions of Signs in Lenses

Both the convex and concave lenses have the same 'Lens formula', viz.,

$$\frac{1}{f} = \frac{1}{v} - \frac{1}{u}$$

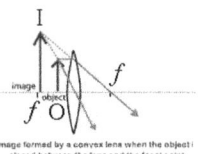

Distances used to describe image formation

a) The focal length of a <u>concave lens is negative</u>; the focal length of a convex lens is –ve.

b) Distances are measured from the center of the lens,

c) Distances that are measured opposite to that of the incident beams positive while measured in the same direction is negative,

d) Conversely, if on evaluating the equation, the measured value of v is positive, the image is virtual, and is situated on the same side of the lens as the object; and if v is negative the image is real.

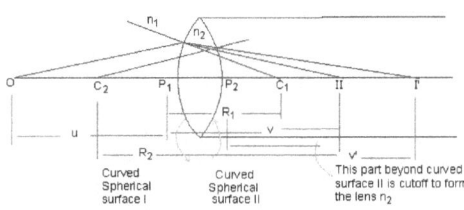

Image formed by a convex lens when the object i placed between the lens and the focal point.

22.2.4 Lens Maker's Formula

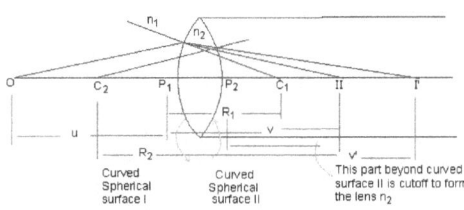

$$\boxed{\frac{1}{f} = \frac{1}{d_o} + \frac{1}{d_i}}$$

$$\boxed{M = \frac{h_i}{h_o} = -\frac{d_i}{d_o}}$$

22.2.5. Depth of Field and Depth of Focus and Aperture

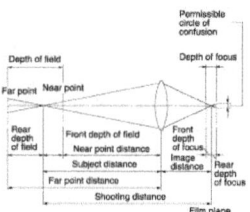

Depth of Field and Depth of Focus

Relationship Between Depth of Focus and Aperture

22.3 CONVERGING LENSES

22.3.1 Magnification (m)

$$\boxed{m = \frac{\text{Size of image, I}}{\text{Size of object, O}} = \frac{v}{u}}$$

$$\boxed{\frac{v}{f} = m + 1}$$

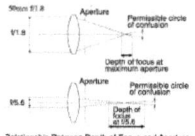

22.3.2 Converging Lens Constructions: **The Six cases**

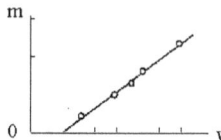

Object from Infinity

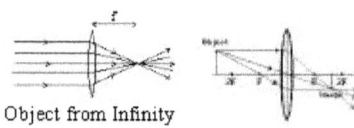

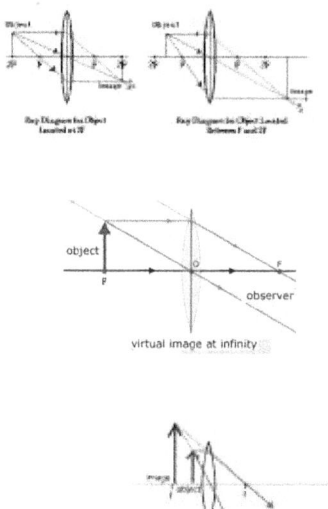

22.3.2.1. Fourier Transform

If an object is in front of a lens, then at the focus is the Fourier transform of the lens and the image is produced at the same time as the focus.

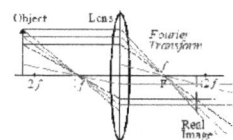

22.3.3 Flood Light

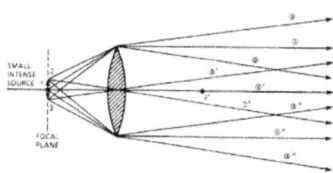

a) Lenses to correct 'long sight' and 'short sight'
b) Microscope
c) Telescope objective
d) Camera (single lens system)
e) Projector
f) Flood light

22.3.4 A Magnifier Glass (A Simple Microscope)

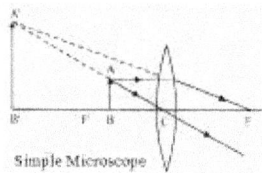

Simple Microscope

$$m = \beta / \alpha = [1 + \frac{D}{f}]$$

D =Least distance of distinct vision = 0.25m.

22.3.5.1. A Compound Microscope

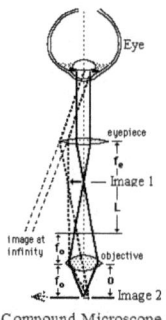

Compound Microscope

If NA = numerical aperture of the optical system,

Optical resolution, $d = \frac{\lambda}{2(NA)}$

$|D|$ = Distance between Eyepiece and virtual image, $M_O = \frac{v}{u}$

Magnification $\boxed{M = M_O . M_E} = \frac{v}{u}(1+\frac{|D|}{f_E})$.

22.3.5.2. Electron Microscopy

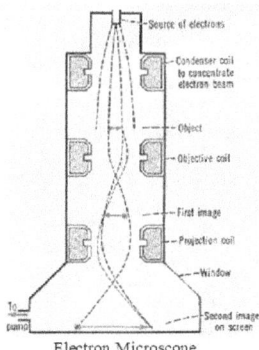

Electron Microscope

22.3.5.3. Phase Contrast Microscope

Phase contrast microscopy, first described in 1934 by Dutch physicist Frits Zernike, is a contrast-enhancing optical technique that can be utilized to produce high-contrast images of transparent specimens, such as living cells (usually in culture), microorganisms, thin tissue slices, lithographic patterns, fibers, latex dispersions, glass fragments, and sub-cellular particles (including nuclei and other organelles).

The essential element of this microscope is the 'phase plate'. Phase plate is a transparent-glass plate having a small section whose optical thickness is $\frac{\lambda}{4}$ greater than the remainder of the plate. The thicker section is located in the central part of the μv -plane, *i.e.*, in the region of low spatial frequencies. Inserting the plate gives a new image function given by the FT of $U(v)$.

22.3.6 Telescopes

Image Formation by a Terrestrial Telescope

22.3.6.1 Galilean Telescope (Opera Glass)
It consists of a positive objective O and a negative lens eyepiece E. Their focal points F_E and F_O are in coincidence.

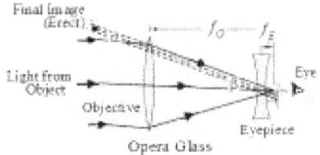

Opera Glass

The dotted lines at angle β indicate the direction along which the tip of the virtual image is seen with the Galilean telescope, the image is erect and magnified.

$$M = \frac{f_O}{f_E} = \frac{\beta}{\alpha}$$

22.3.6.2 Refracting Astronomical (or celestial) Telescope

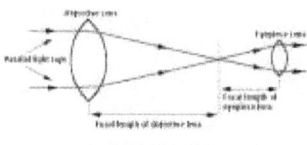

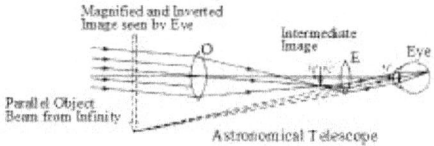

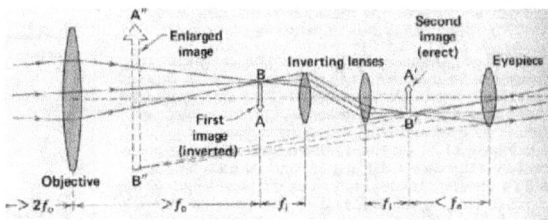

$$M = \frac{f_O}{f_E} = \frac{\beta}{\alpha} = \frac{\text{Diameter of Object}}{\text{Diameter of Eye lens}}$$

Find the size of the photograph of Moon (at distance $D = 3.84 \, x10^{10}$ cm and diameter $3.48 \, x10^8$ cm), if $f_O = 120$ cm .

Prism binoculars are short astronomical telescopes each.

22.3.6.3 Converting a Telescope into Microscope:

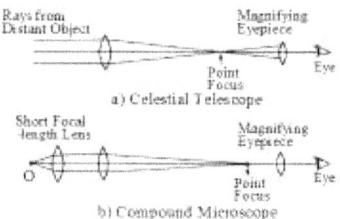

a) Celestial Telescope

b) Compound Microscope

22.3.6.4 Camera (Single lens system)

A camera has
(i) The lens that produces the image,
(ii) The shutter that controls the time for which light is incident on the film.
(iii) The diaphragm which controls the amount of light

Aperture of the lens of a camera is described by

$$\boxed{\text{f - number}} = \text{f \# (or, relative aperture)} = \frac{\text{(Focal length of lens)}}{\text{(diameter of aperture)}}.$$

A camera with $\frac{f}{8}$ lens is one with a focal length $\boxed{\frac{f}{8} \rightarrow f = 8\,x \text{ (diameter of the lens)}}$.

(ii) Film that collects the image, in a conventional camera, or a CCD device, in the digital camera.

22.3.6.5 Telephoto-lens

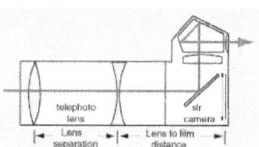

22.3.7 **DEFECTS** of Lenses

Defects observed in Images produced by refraction at curved surfaces are
1. Spherical aberration,
2. Chromatic aberration,
3. Coma,
4. Astigmatism, and
5. Curvature and distortion.

22.3.7.1 Spherical Aberration of Convex Lens

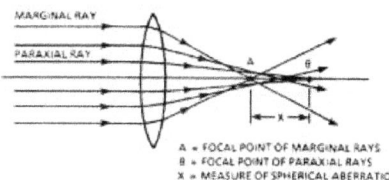

A = FOCAL POINT OF MARGINAL RAYS
B = FOCAL POINT OF PARAXIAL RAYS
X = MEASURE OF SPHERICAL ABERRATION

a) Points A and B are 'aplanatic' and

$$AC = \frac{R}{n}$$
$$BC = R\,n$$

such that image of a point object at A is formed at B and the vice-versa, and is free from spherical aberration.

b) Condition for minimum spherical aberration,

$$x = f_1 - f_2$$

c) **Remedy**:

Use a lens which has reduced lens aperture.

Use a Plano-convex lens, with its convex surface towards the incident rays or the emergent light whichever is more parallel to the axis.

d) Circle of Least Confusion

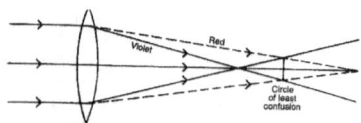

22.3.7.2 Chromatic Aberration of a Convex Lens

Blue light comes to a focus nearer to the lens than red light.

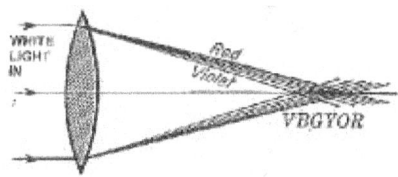

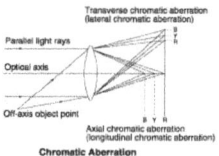

Chromatic Aberration

22.4.1 ACHROMATIC DOUBLET

The best correction for chromatic aberration is the use of a negative lens, made of a glass with a different index of refraction, in combination with a positive lens

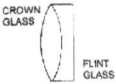

22.4.2 Light Gathering Power (LGP) of a lens

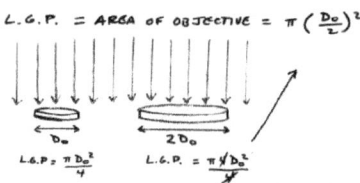

22.4.2 Human Eye (Variable focal length)

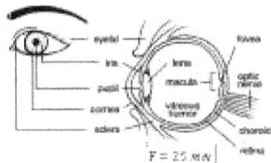

22.4.3 Human Eye Disorders
Normal eye
Eye lens has +60 D for distant vision, whereas for near vision it is +64 D

222.4.3.1 Near- (Myopia) and Far-sightedness (Hyper-metropia)

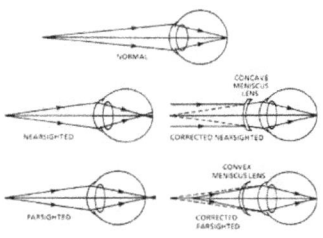

Divergence lens of -5D, say if used, relaxed eye has +55 D, and near vision point will be at $u = 0.11$ m.

22.4.3.2 Hyperomia:

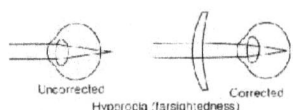

Uncorrected
Corrected
Hyporopia (farsightedness)

If convex lens of +3D is used, relaxed eye will have +63 D, and near vision eye will have +63 D.

22.4.3.3 <u>Presbyopia</u>:

The eyelens becomes harder with age, and becomes too stiff to accommodate.

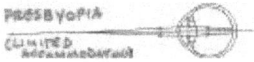

An eye that suffers from myopia as well as from hypermetropia is said to suffer from presbyopia. A person with this defect cannot see objects distinctly placed at any distance from him contact lenses with <u>bifocals or progressive lenses</u> are used.

22.4.3.4 Astigmatism:

Astigmatism simply means that the eye is not perfectly round or spherical. Glasses with special cylindrical lenses (negative or positive) are the most common ways to correct.

22.4.3.5 Coma

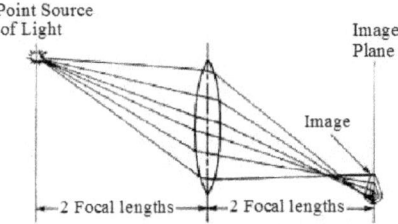

Point Source
of Light
Image
Plane
Image
—2 Focal lengths— —2 Focal lengths—

22.5. DIVERGING Lenses:

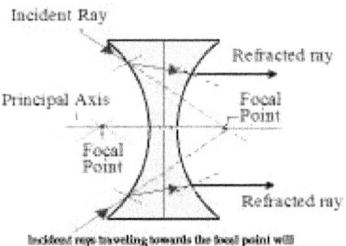

Incident Ray
Refracted ray
Principal Axis
Focal
Point
Focal
Point
Refracted ray
Incident rays traveling towards the focal point will
refract and travel parallel to the principal axis.
Refraction by Diverging Lens

22.5.1 **Applications**:
    1. Wide-angle spy hole in doors
    2. Lenses to correct for 'short sight' (Myopia).

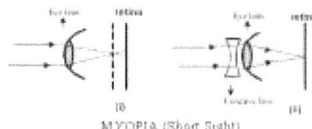

MYOPIA (Short Sight)

    3. Wide-angle lens in coach rear windows
    4. Eye lens in Galilean telescopes.

22.5.2   **POWER of a lens**:

22.5.2.1 Converging Power $\boxed{P = -\dfrac{1}{f(m)}}$ Unit   *Dioptres*

22.5.2.2 Convex lenses have +ve powers,
22.5.2.3 Concave lenses have −ve powers.

22.6 Combination of two thin lenses:
    If two thin lenses of focal lengths, f_1 and f_2 are combined by placing them in contact, they will act as a single lens of focal length F

$$\frac{1}{F} = \frac{1}{f_1} + \frac{1}{f_2}$$

22.6.1 **MAGNIFIER**
22.6.1.1 Equivalent lens: Combination of two Lenses

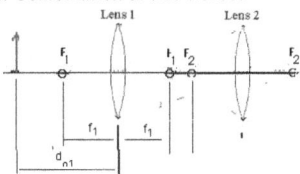

The focal length of two lenses of focal lengths f_1 and f_2 placed at a distances d apart is given by

$$\frac{1}{F} = \frac{1}{f_1} + \frac{1}{f_2} - \frac{d}{f_1 f_2}$$

and the positions of the principal planes are

$$\alpha = \frac{d\,F}{f_2}, \quad \beta = \frac{d\,F}{f_1}$$

For the combination to be <u>telescopic</u>, their separation should be

$$d = f_1 - f_2$$

22.6.1.2 **Eye Pieces (Oculars)**
    Eyepieces contain mainly a field lens and an eye lens
22.6.1.3 **Ramsden** Eyepiece:
    (i) This eyepiece is a positive eyepiece because a real image is formed.
    (ii) The eye lens produces the final image at infinity because the field lens forms an image at a distance f in front of the eye lens.

(iii) Three lens arrangement is not entirely achromatic. Hence eye lens and field lens are separately of the Kellner's eye piece type.
(iv)   Spherical aberration is minimized.
(v)   It is very commonly used since its cross wire is placed outside it.
(vi)   The first eyepiece designs, the Ramsden and Huygenian, only contain two lenses and are very poor eyepieces by modern standards. They have very narrow fields of view, short eye relief and many aberrations. Cheap telescopes often include these inexpensive eyepieces.

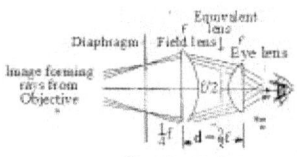

Ramsden eyepiece

The **equivalent lens** has

focal length $$F = \frac{3f}{4}$$.

The equivalent lens is located at a distance f/2 away from the field lens in the lines joining the two lenses.

The image is located at $\frac{f}{4}$ in front of the field lens.

22.6.1.4 Huygens lens:

The two lenses should have their focal lengths in the ratio 3:1 for achromatic combination.

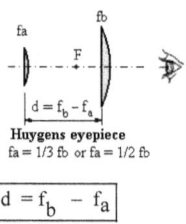

Huygens eyepiece
fa = 1/3 fb or fa = 1/2 fb

$$d = f_b - f_a$$

22.6.1.5 Kellner's Eye piece:

The Kellner is the best of the inexpensive eyepieces Both the field lens and the eyelens have the same focal length f one doublet (two lenses together) and one singlet lens
The two lenses are separated by a distance d = a fair field of view (45 degrees).

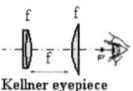

Kellner eyepiece

22.6.2.1 **Symmetrical** Eyepiece:

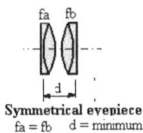

Symmetrical eyepiece
fa = fb   d = minimum

22.6.2.2   Plossl eyepiece:

A wide field of view (55 degrees), very good eye relief and are well corrected for aberrations. They cost more than Kellners, but they are worth it.

22.6.2.3   Orthoscopic eyepieces
These are generally not named for their inventors, Mittenzwey and Abbe, field of view (50 degrees).

22.6.2.4   Erfle eyepiece
Invented to provide a wide apparent field of view and they do that (65 degrees) is a combination of three doublet lenses.

22.6.2.5. Magnification

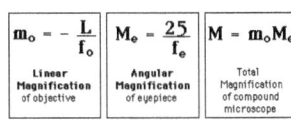

$$m_o = - \frac{L}{f_o} \qquad M_e = \frac{25}{f_e} \qquad M = m_o M_e$$

Linear Magnification of objective | Angular Magnification of eyepiece | Total Magnification of compound microscope

22.7   POLARIZATION OF LIGHT

22.7.1.1   Unpolarized Light
Sun light and other natural lights are <u>unpolarized.</u>

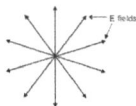

22.7.1.2   <u>Polarized Light</u>
Three Types
a)   Linear (Plane) polarization
b)   Circular polarization
c)   Elliptical polarization

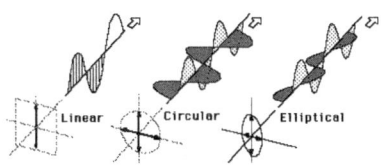

22.7.1.3   Plane of polarization:

The plane containing the E-vector and the direction of propagation is called "*the plane of polarization*"

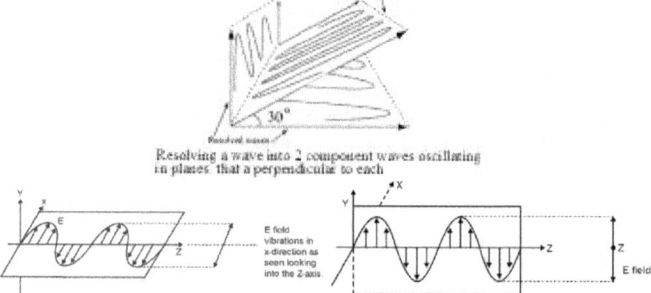

Resolving a wave into 2 component waves oscillating in planes that a perpendicular to each

E field vibrations in x-direction as seen looking into the Z-axis

E field

22.7.2 POLARIZATION by Reflection

a) <u>BREWSTER ANGLE</u>, θ_B

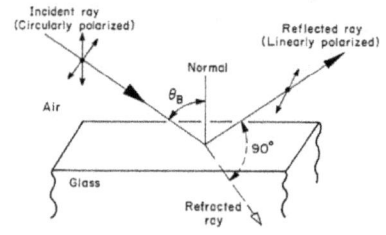

$$\boxed{\tan\theta_B = 2^{n_1}}$$

b) For a glass plate in air,

$$\boxed{\theta_B = \tan^{-1}\frac{1.52}{1.00} \approx 56.7^\circ}$$

22.7.3 POLARIZATION by Selective Absorption (DICHROISM)

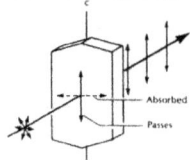

Absorbed

Passes

22.7.3.1 Malus Law

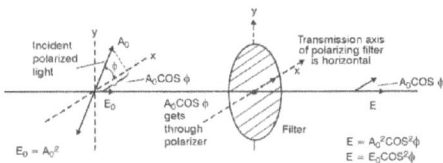

According to Malus, when completely plane polarized light is incident on the analyzer, the component of the E-field E_0 transmitted by the polarizer into the transmission axis of the analyzer is $E_0 \cos\theta$, where I_0 = intensity of the light transmitted by the polarizer, and θ = azimuthal angle, the orientation of the transmission axis of the polarizer with respect to that of the analyzer, then the analyzer transmits only light of intensity

$$I = I_0 \cos^2 \theta$$

22.7.3.2 Degree of polarization, P

$I_{\parallel}$ and $I_{\perp}$ are intensity components of the transmitted light when the analyzer $\theta = 0$ and $\pi / 2$

$$P = \frac{I_{\parallel} - I_{\perp}}{I_{\parallel} + I_{\perp}}$$

22.7.3.3 **Tourmaline crystal** and cross Polarization

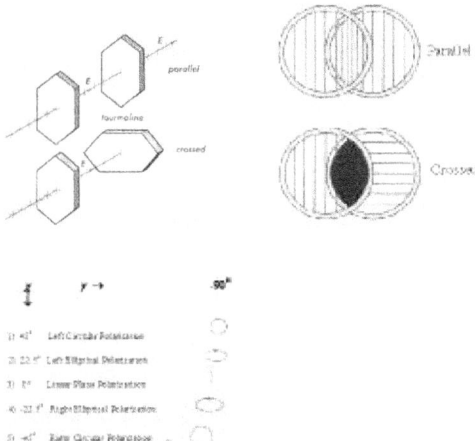

22.7.4.1 POLARIZATION BY DOUBLE REFRACTION (Birefringence)

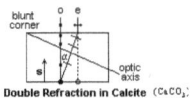

Double Refraction in Calcite $(CaCO_3)$

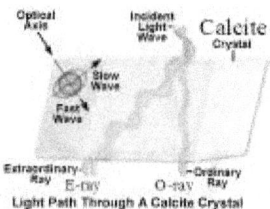

Light Path Through A Calcite Crystal

Two images are produced by a very clear cleavage rhomb of calcite ($CaCO_3$) crystal placed over an image If the calcite crystal is rotated, the image formed by the ordinary rays (ω) does not move, and the image formed by the E-rays (ε) rotate. The c-axis is labeled on each image.

22.7.4.2   Nicol Prism

It is made of two triangular wedges of calcite ($CaCO_3$).

$$RP = \lambda / \Delta\lambda$$

The angle over which incident light can enter a Nicol polarizer and successfully pass through is about 14°.

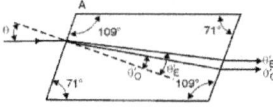

For calcite, $n_E = 1.4864; n_O = 1.6584;$

$n_{balsam} = 1.526$

22.7.4.3   Rochon Prism

In the first half of the Rochon prism, both the O-ray and the E- ray travel with the same velocity. In the second half, the O-ray continues at the same velocity. But the extraordinary ray travels more rapidly and therefore is deviated by an amount that depends on the angle of the interface.

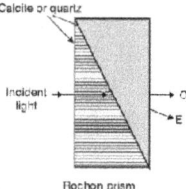

Rochon prism

22.7.4.4   Wollaston Prism

The O-ray in the first half of the prism becomes the E-ray in the second half, and *vice versa*.

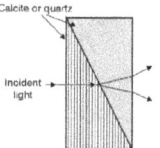

22.7.4.5 Glan-Foucalt Polarizer

22.7.4.6 Glan Polarizer

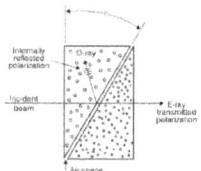

22.7.4.7. Analysis of Polarized light by the <u>Babinet Compensator</u>

   Jamon type Babinet compensator consists of two thin quartz wedges whose optic axes are mutually at right angles.

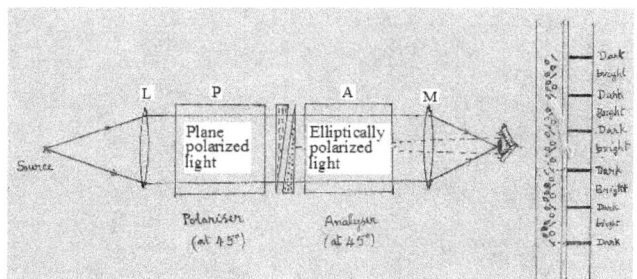

   Phase difference $\varphi = \dfrac{2\pi\delta}{\lambda}$ is introduced between the incident plane polarized light of the 1st prism and the emergent component at the 2nd prism.

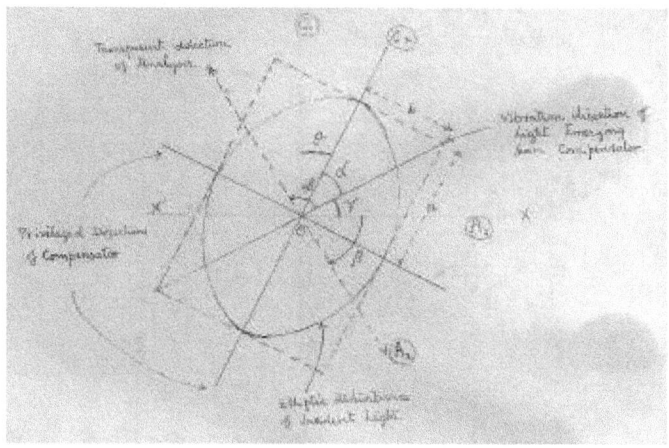

$$90° - \theta = C_2 - C_1 = 83.5°; \ \beta = 33° \sim 284° = 109°; \ \alpha = \beta - \theta = 102.5°.$$

$$\frac{a}{b} = \tan \alpha = -\text{Cot } 12.5° = 4.5107$$

22.7.5.1 Luminous Flux

It is the visible (luminous) energy emitted from a source per second.

Luminous intensity (L) = luminous flux per unit solid angle.

$$L = \frac{Lumen}{Steradian} = Candela$$

$$1 \ Candela = \frac{1}{60,000} m^2$$

of the surface of a black body at freezing point of platinum ($1169°C$).

Illumination of a surface

$$E = \frac{1}{d^2} = \frac{Luminus \ Flux}{m^2}$$

$$1 \ Lux = \frac{Lumen}{m^2}.$$

22.7.5.2 Principle of Photometry: Photometric distance Law

Illuminance E, on a surface to the intensity I_0 of an illuminating point source, θ is angle between the normal to the surface and direction of illumination.

$$E = \frac{I_0 \cos \theta}{d^2}$$

$$\frac{I_1}{I_2} = \frac{I_1^2}{I_2^2}$$

307

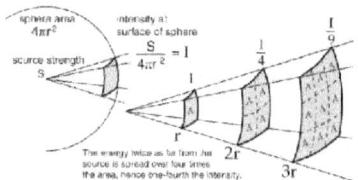

22.8 SPEED of Light: c
22.8.1 Speed of Light Determination
22.8.1.1 Roemer's (astronomical) method (1676)
Roemer noticed a 16 minutes difference in the time for an eclipse of a moon of Jupiter at half year intervals.

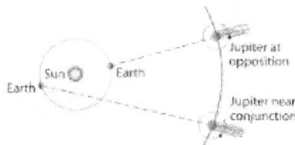

$$\boxed{T_1 \text{ or } T_2 = nT \pm \frac{2r}{v}}, \quad (v = c)$$

n = Number of revolutions made by the satellite around Jupiter, from conjunction to opposition,
r = Radius of orbit of earth,

$$\boxed{T_1 - T_2 = \frac{4r}{c}}$$

22.8.1.2 Bradley's method of aberration (1727)
Observing periodic displacements of certain fixed stars in the direction of earth's motion one gets

$$\boxed{Sin\alpha \approx \alpha = (v/c)\, Sin\beta}$$

22.8.1.3 Fizeau's method using toothed wheel (1849)

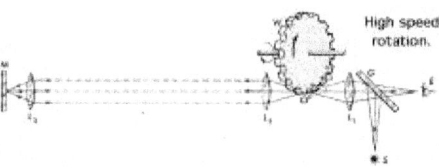

d = Distance OM = 8.6 km; m = Number of teeth; n = frequency of rotation of wheel;
n_r = frequency of the wheel when a teeth occupies the position of the rth space

$$\boxed{c = \frac{4m\, n_r\, d}{(2r - 1)}}$$

22.8.1.4 Foucault's rotating mirror method. (1850)
It is the group velocity that is measured experimentally.

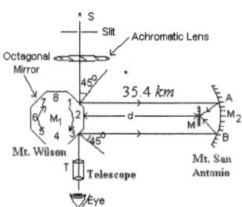

$$c = \frac{8\pi nbd^2}{x(a + d)}$$

22.8.1.5 Michelson's Null method (1926)

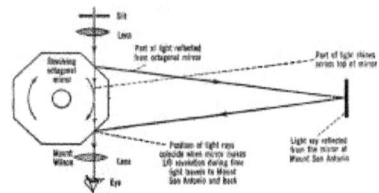

$$c = 2mnd$$

m = Number sides in the mirror; d = distance between two stations; n = frequency of rotating mirror.

22.8.1.6 Kerr cell method.

22.9. PARAXIAL OPTICS (Matrix Method of Geometrical or Ray Optics)

In geometrical optics light from an object AB passes through air for some distance, then refraction through a lens (refracting surface), further through optical systems (like in a combinations of lenses), to form an image A'B'. Any ray can be traced through the optical system by using the conditions,

a) The incident, refracted rays and normal are coplanar (lie in the same plane),

b) $$\frac{Sin\theta_1}{Sin\theta_2} = \frac{n_2}{n_1}$$

Instead of calculating step by step the position of the image for each surface, the transformation from object to image can be carried out through matrix operations. The rules of matrix multiplication have to be followed. Matrix optics is a technique for tracing paraxial rays.

Paraxial optics is a method of determining the first order properties of an optical system that assumes all ray angles to be small. A paraxial ray-trace is linear with respect to ray angles and heights since paraxial angles are defined to be the tangent of the actual angle U. Rays in the vicity of the optical axis are used and the surface sag is ignored or negligible.

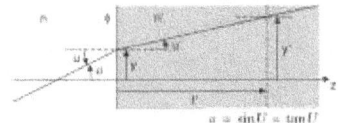

$$n_1 \sin \theta_1 = n_2 \sin \theta_2$$
$$\theta_1 = \theta_2$$

Reflection (or refraction) occurs at an interface between two optical spaces. The transfer distance t' allows the ray height y' to be determined at any plane within an optical space (including virtual segments).

Three fundamental matrices used are, the translation matrix, the refraction matrix and the reflection matrix.

a) Effect of Translation [Translation Matrix (T)]

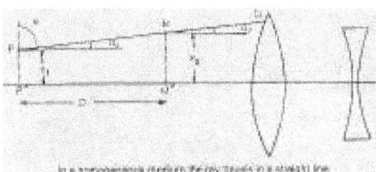

In a homogeneous medium the ray travels in a straight line

If the incident ray (2x1) matrix is $\begin{bmatrix} \lambda_i \\ x_i \end{bmatrix}$, the outgoing ray by $\begin{bmatrix} \lambda_o \\ x_o \end{bmatrix}$

$$\begin{bmatrix} \lambda_o \\ x_o \end{bmatrix} = \begin{bmatrix} 1 & 0 \\ D/n_1 & 1 \end{bmatrix} \begin{bmatrix} \lambda_i \\ x_i \end{bmatrix},$$

then the effect of translation through a distance D in a homogeneous medium of refractive index n_1 is given by the (2x2) matrix,

where $T = \begin{bmatrix} 1 & 0 \\ D/n_1 & 1 \end{bmatrix}$.

b) Effect of Refraction [Refraction Matrix]

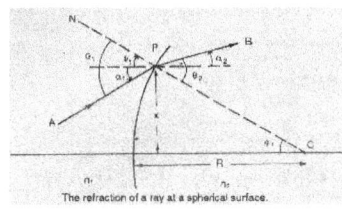

The refraction of a ray at a spherical surface.

$$\begin{bmatrix} \lambda_o \\ x_o \end{bmatrix} = \begin{bmatrix} 1 & -P \\ 0 & 1 \end{bmatrix} \begin{bmatrix} \lambda_i \\ x_i \end{bmatrix},$$

where $P = \dfrac{n_2 - n_1}{R}$ and $\varphi_1 = \dfrac{x}{R}$.

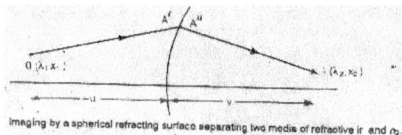

imaging by a spherical refracting surface separating two media of refractive n_1 and n_2.

c) <u>Imaging by Spherical refracting surface</u>

$$\begin{bmatrix} \lambda_o \\ x_o \end{bmatrix} = \begin{bmatrix} 1 + \dfrac{u\,P}{n_1} & -P \\ 0 & 1 - \dfrac{v\,P}{n_2} \end{bmatrix} \begin{bmatrix} \lambda_i \\ x_i \end{bmatrix}, \quad x_2 = (1 - \dfrac{v\,P}{n_2})\, x_1.$$

Magnification, $m = \dfrac{x_2}{x_1} = (1 - \dfrac{v\,P}{n_2}) = \dfrac{n_1 v}{n_2 u}$

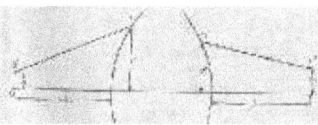

----o-0-o-0-o-0-o-0-o---

Chapter 23

OPTICS – 3: WAVE THEORY,

Interference, Young's Double-slit, Diffraction, Fraunhofer and Fresnel, Grating

"The most exciting phrase to hear in science, the one that heralds the most discoveries, is not "Eureka!" (I found it!) but "That's funny..." ~Isaac Asimov

23. WAVE THEORY of Light

23.1 INTERFERENCE

Depending on the nature of the light beams and when they meet, the two beams might enhance each other, to give a brighter beam, or they might interfere in a way that makes the total beam less bright. The former is called **constructive** interference, whereas the latter is called **destructive** interference.

23.1.1 Conditions for Interference

1) The two sources must emit continuous waves of the same λ and periodic time T,
2) The two waves must have the same phase or a constant phase difference,
3) The two sources should be very close to each other.
4) The two sources should be very narrow so as to avoid overlap, and
5) The amplitudes of the two waves should be preferably small.

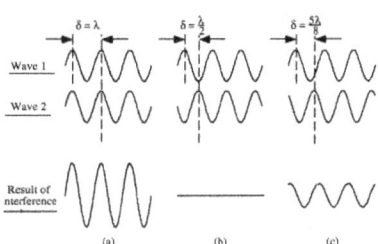

23.1.2 Types of Fringes

23.1.2.1 **Non-Localized Fringes**

If the interference of the overlap of waves from two virtual sources occur at any place where the screen is placed, then they are called Non-localized fringes at infinity. They are straight line pattern, and are fringes of equi-distant. Young's slit provides such a pattern.

23.1.2.2 **Localized Fringes**

a) The fringe pattern depends on the position of the virtual sources and the screen. The result is due to <u>wave front splitting</u>.

b) When the screen is parallel to intersection of D (virtual sources) with the Source, the fringes will be straight, and they are localized; appear to come from the region located at the coherent sources. These are called fringes of <u>equal thickness</u> or <u>Fizeau's fringes</u>.

For example, Michelson's Interferometer is due to <u>amplitude splitting</u> of the two waves.

If the Virtual sources are parallel, then the fringes are of equal inclination, and are circular in pattern. The virtual sources intersect at some point in point of view.

23.2 YOUNG'S DOUBLE SLIT

The original idea that the wavelength of light can be determined by interference effect was due to Grimaldi; Thomas Young in 1801 employed the double slit method.

23.2.1 Set up

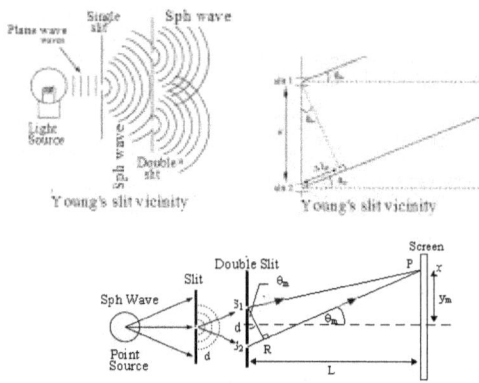

Young's slit vicinity

$$S_1P_2 = (x_n - d/2)^2 + L^2.$$
$$S_2P_2 = (x_n + d/2)^2 + L^2.$$
$$[S_2P_2 - S_1P_2] = 2 x_n \, d.$$
$$(S_2P - S_1P)(S_2P + S_1P) = 2x_n \, d.$$

Path difference $\boxed{S_2P - S_1P = \dfrac{x_{in} \, d}{L}}$

Condition for interference, $\boxed{d \, Sin\theta_m = m\lambda}$

23.2.2 **Destructive** Interference

$$\boxed{\text{Dark fringe} \Rightarrow (2m+1)\frac{\lambda}{2} = \frac{x_{in} \, d}{L}}$$

23.2.3 **Constructive** when the extra distance is an integer number (m) of wavelengths:

$$\boxed{\text{Bright fringe} \;\Rightarrow\; m\lambda = \frac{x_{in}\, d}{L}}$$

$m =$ <u>order</u> of fringe.

23.2.4 **Line spacing**: When θ_m is small, approximate

$$\boxed{\text{Sin } \theta_m \;\approx\; \tan \theta_m \approx \frac{x_m}{L}}$$

$$x_m = \frac{L}{d} m\lambda$$

23.2.5 **Fringe width, Δ**

$$\boxed{\Delta = x_2 - x_1 = \frac{L}{d}\lambda}$$

(Sharper pattern for smaller d or larger L)

23.2.6 **Measuring λ** experimentally

Measure $\Delta = x_2 - x_1$, separation d between slits, and L..

$$\boxed{\lambda = \frac{d}{L}\,\Delta(= x_2 - x_1)}$$

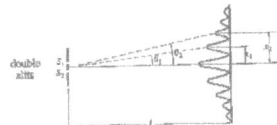

23.2.7 **Bright and Dark Bands for double slits**

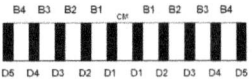

An interference pattern produced by two ideal slits.

B4 B3 B2 B1   CM   B1 B2 B3 B4

D5 D4 D3 D2 D1   D1 D2 D3 D4 D5

When the incident light is white light, the different colours create images of the slit at different locations on the screen,

since different colours have different wavelengths.

All colours interfere constructively at the central fringe m = 0, so that one is white.

Fringe m =1 for all colours looks like the full spectrum.

The **wavelength** for each colour is <u>measured.</u>

$$\boxed{\lambda_{Colour} = x_1(\lambda_{Colour}) \times \frac{d}{L}}$$

<u>Note</u>: the above patterns occur for coherent sources (*i.e.,* sources in phase).

By contrast two light bulbs are incoherent sources, and no such patterns are seen

<u>Non-Localized fringes</u>

These are straight line equi-distant fringes in Young's slit

23.3.1 **NEWTON'S RINGS** to measure λ

A plano-convex lens with its spherical surface (radius of curvature, R) in contact with a flat, horizontal plane glass plate shone in monochromatic light (λ).

Path difference, $\boxed{\Delta = 2\mu\, t + \frac{\lambda}{2} = m\lambda}$

t = distance AM

μ = Index of refraction of glass of lens.

Viewing the reflected light contains a pattern *viz.*, Newton's rings (circular interference fringes) of radius r (r << R),

$$\boxed{r = \sqrt{(m + \tfrac{1}{2})\,\lambda R}}, \qquad m = 0, 1, 2, 3, \ldots$$

$$\boxed{2R\,t = \frac{d^2}{4}}$$

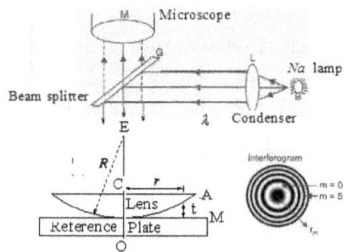

For m^{th} bright ring, $\boxed{\dfrac{d_m^2}{4R} = (2m-1)\dfrac{\lambda}{2\mu}}$

$$\boxed{\lambda = \frac{\mu\,(d_n^2 - d_m^2)}{4R(n-m)}}$$

d = diameter of interference ring,

For m^{th} dark fringe, $\boxed{2Rm\,t = \dfrac{d_m^2}{4}}$

23.3.2 WEDGE for fringe spacing

Two optical plane glass plates (ABDC & CEFG) of length L, in contact at C one end ($\approx 2 - 5°$), and separated by a separation of DE = d at the other end, gives straight line interference pattern, when $d \ll L$, and shone by light of λ

$$\boxed{t = x_1\theta}$$

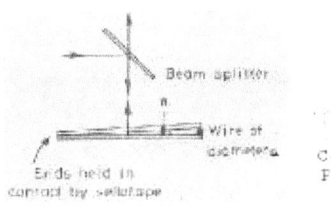
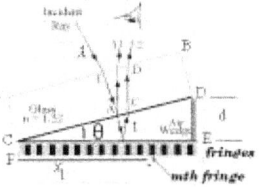

$$\boxed{\Delta\text{dark fringe} = 2\mu\, t = m\lambda}$$

Fringe spacing = $\boxed{\Delta = x_2 - x_1 = \dfrac{L}{d}\lambda}$

23,3,3 Lloyd's Mirror and Fringe width
A pattern similar to that of the Young's double-slit is produced using Lloyd's mirror.

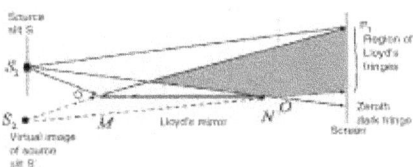

Constructive when the extra distance is an integer number of wavelengths:

$$\boxed{d\,Sin\theta_m = m\lambda} \qquad m = 0,\ 1,\ 2,\ 3,\ ....$$

Line spacing: When θ_m is small, approximate

$$\boxed{Sin\,\theta_m \approx \tan\theta_m \approx \dfrac{x_m}{L}}$$

$$x_m = \dfrac{L}{d}m\lambda$$

Fringe width $\qquad \boxed{\Delta = x_2 - x_1 = \dfrac{L}{d}\lambda}$

23.3.4 FRESNEL BIPRISM
An alternative method to the classic Young's slits experiment for measuring the wavelength of light is that due to Fresnel. It produces a double slit

Path difference, $\boxed{\Delta = \dfrac{x\,d}{D}}$

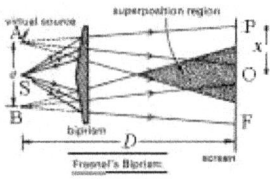

For P to lie on a dark fringe,

$$x = \frac{D}{d}(2n+1)\frac{\lambda}{2}$$

Fringe width, $\quad \beta = x_1 - x_2 = \frac{D}{d}\lambda$

$$\beta_1 - \beta_2 = \frac{D_1 - D_2}{d}\lambda.$$

$$d = \sqrt{d_1\, d_2}$$

23.3.5. Edser Butler Fringes aside Hg spectrum to determine Wavelength

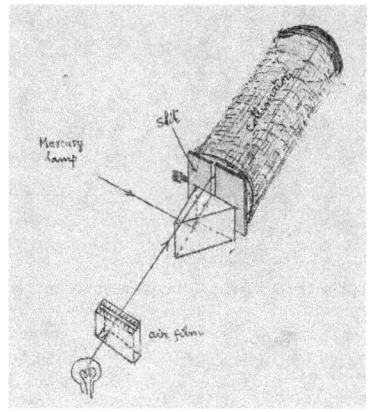

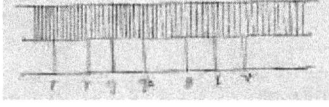

$2\mu t = M\lambda_r$

$2\mu t = (M + n)\lambda_v$

n = Number of bands consisting from the band chosen in the red.

$$2\mu t \left[\frac{1}{\lambda_v} - \frac{1}{\lambda_r}\right] = n.$$

From graph, slope $= n/[\dfrac{1}{\lambda_v} - \dfrac{1}{\lambda_r}] = = \dfrac{32}{4000} = 0.008 \; cm$, for air $\mu = 1$, and so

$t = 0.04 \; mm$.

To determine λ of Hg spectrum, plot of n versus $\dfrac{1}{\lambda_y}$ gives $2\mu t$, whence $\lambda_y = ?$

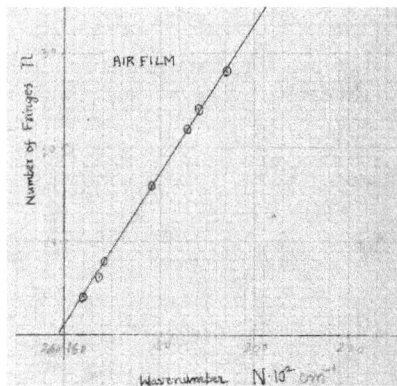

23.3.6. Moire Fringes

Moiré patterns are created whenever one semitransparent object A with a repetitive pattern is placed over another. A slight motion of one of the objects creates large-scale changes in the Moiré pattern. These patterns can be used to demonstrate wave interference. In essence, Moiré occurs when two patterns are overlaid and result in a new, third pattern.

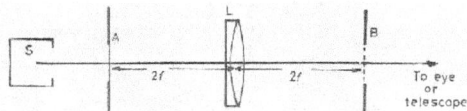

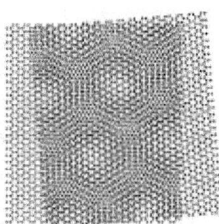

Moiré pattern occurs when a scene or an object that is being photographed contains repetitive details (such as lines, dots, *etc.*) that exceed the sensor resolution. As a result, the camera produces a strange-looking wavy pattern.

23.3.7.1. Fabry-Perot Etalon (Multiple-beam interference)

A Fabry-Perot interferometer consists of two optically flat (of the order of $\frac{1}{20}$ to $\frac{1}{100}$ λ), partially reflecting glass or quartz with their reflecting surfaces held accurately parallel. If the plate spacing can be mechanically varied the device is called an interferometer. If the plates are held fixed by spacers it is called an Etalon.

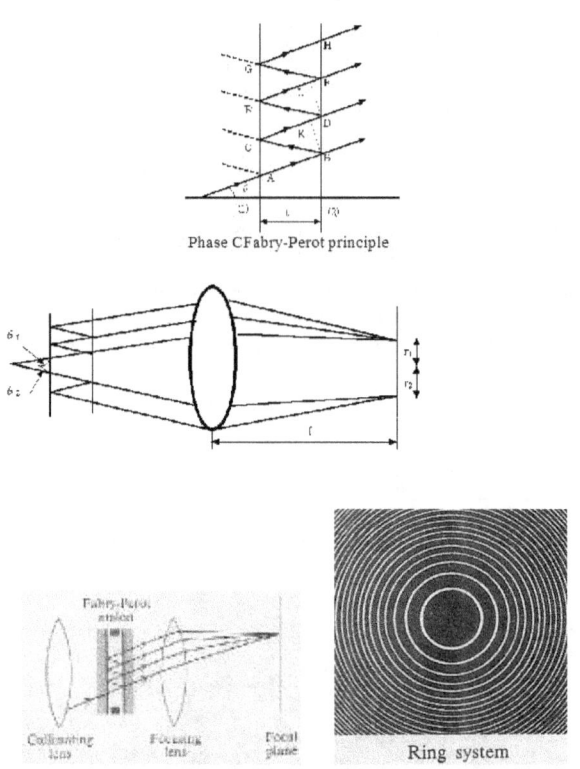

Phase CFabry-Perot principle

Ring system

To find the Resolving Power Rayleigh's criterion is used. In FP etalon, two maxima are widely separated, with minimum occurring mid-way between two successive orders of maxima. $2d = m\lambda$, d = separation of plates, m th order of bright fringe for λ.

$$\text{R.P.} = \frac{\lambda}{\Delta\lambda} = \frac{m\pi R}{(1-R^2)} = \frac{2d}{\lambda} \cdot \frac{\pi R}{(1-R^2)} \cdot$$

Finesse coefficient of the etalon, $f = \frac{4R^2}{(1-R^2)^2}$

where R = reflection coefficient of the plate of the etalon.

23.3.7.2. Lummer Gehrcke Plate

Lummer Gehrcke plate is a plane parallel glass or quartz plate of 10 to 20 cm length l and 1 to 2 cm width, and uses the total internal reflection. Prism P is in optical contact with one face of the plate. The angle of incidence is close to the critical angle. If φ = angle of emergence,

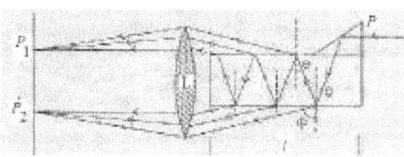

R.P. $\dfrac{\lambda}{\Delta\lambda} = \dfrac{2\,l}{3\,\lambda}.(\mu^2 - 1)$

Dispersion, $\dfrac{d\varphi}{d\lambda} = \dfrac{2\,\mu\lambda\dfrac{d\mu}{d\lambda} - 2(\mu^2 - \sin^2\varphi)}{\lambda\sin 2\varphi}$

23.4   FRESNEL & FRAUNHOFER DIFFRACTION

23.4.1   Kirchhoff Integral Theorem

Consider a monochromatic point source at $\mathbf{P_0}$ which illuminates an aperture in a screen. U is the complex amplitude of the disturbance at the surface, $k = \frac{2\pi}{\lambda}$ is the wavenumber and r is the distance from $\mathbf{P}$ to the surface.

$$U_P = -\frac{1}{4\pi}\iint\left(U\nabla_n\frac{e^{ikr}}{r} - \frac{e^{ikr}}{r}\nabla_n U \right)dA$$

23.4.1.1 Fresnel-Kirchhoff Integral Formula

$$U_P = -\frac{ikU_0\,e^{-i\,\omega t}}{4\pi}\iint\left(\frac{e^{ik(r+r')}}{rr'}[Cos(n,r) - Cos(n,r')] \right)dA$$

23.4.1.2 Fresnel-Kirchhoff Diffraction Formula

$$U_P = -\frac{i\,k}{4\pi}\iint\left(\frac{U_A\,e^{i(kr - \omega t)}}{r}[Cos(n,r) + 1] \right)dA$$

Diffraction manifests itself in the apparent bending of waves around small obstacles and the spreading out of waves past small openings. The Fraunhofer diffraction pattern (as opposed to the Fresnel diffraction pattern) is mathematically identical to the Fourier transform, at least within certain approximations

Fraunhofer Diffraction	Fresnel Diffraction
1) Nature of wave fronts Both incident and diffracted waves are effectively plane $U_P = U_A\, e^{i(kr - \omega t)}$ Usually, a lens before aperture	If incident or diffracted wave has spherical wavefront $U_P = U_A\, \dfrac{e^{i(kr - \omega t)}}{r}$
2) Observation distance *Screen* distance is infinite. . In practice, screen is at focal point of convex lens near aperture	Finite distances of the diffracting *aperture* from both the source . and screen
3) Nature of diffraction pattern Fixed in position	Move in a way that directly corresponds with any shift in the object.
4) Surface of formation Fraunhofer diffraction patterns on . spherical surfaces	Fresnel diffraction patterns . on flat surfaces
5) Appearance of Diffraction pattern Shape and intensity of a . Fraunhofe*r* pattern remain constant	Shape and *I* change as they propagate further 'downstream' of the source of diffraction

23.4.2 FRAUNHOFER (Far-field) DIFFRACTION

When the incident and diffracted light waves are effectively <u>plane waves</u> it is Fraunhofer diffraction, after its discoverer Joseph von Fraunhofer.

At a point P, $\boxed{U_P = U_A e^{i(kr - \omega t)}}$

U_A = Amplitude at the Aperture.

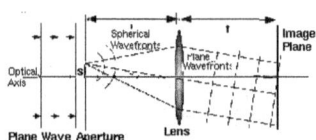

23.4.3 FRESNEL (Near field) DIFFRACTION

When either the source or the screen is very close to the diffracting aperture so that the curvature of the wave front is significant then one uses Fresnel diffraction. <u>Spherical wave</u>

$$\boxed{U_P = U_A \frac{e^{i(kr - \omega t)}}{r}}$$

23.4.3.1   Diffraction of Light by SINGLE SLIT

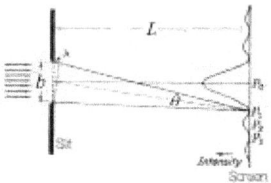

An interference pattern caused by diffraction through a single slit.

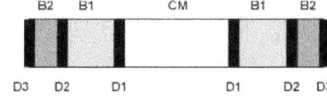

Fringe intensity

$$I = |U|^2 = Io\left(\frac{Sin\beta}{\beta}\right)^2$$

$I_0 = |C\ L\ b|^2$, which is the irradiance at $\theta = 0$;

$$\beta = \tfrac{1}{2}kb = \pm\pi, \pm 2\pi, ...etc$$

The diffraction pattern contains thus a central maximum, and on either side by dark and bright bands.

The first minimum, at $\beta = \pm\pi$, gives

$$Sin\theta = \frac{2\pi}{kb} = \frac{\lambda}{b}$$

23.4.2.2 Rectangular Aperture

$$I_{Rect\ Apert} = Io\left(\frac{Sin\alpha}{\alpha}\right)^2\left(\frac{Sin\beta}{\beta}\right)^2$$

$$\alpha = \tfrac{1}{2}ka; \quad \beta = \tfrac{1}{2}kb$$

23.4.2.3 Circular Aperture

The Diffraction pattern contains a central bright disc called <u>Airy's Disc,</u> surrounded by concentric circular bands.

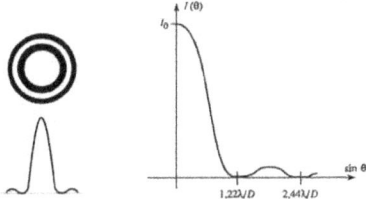

The first dark ring is

$$Sin\theta = \frac{3.832}{kR} = \frac{1.22\lambda}{D} \approx \theta$$,

$D = 2R$ is the diameter of the aperture.

23.4.2.4 **Rayleigh criterion (Optical Resolution)**

The image of a distant object (Source) formed at the focal plane of an optical telescope or a camera is actually the Fraunhofer diffraction pattern, for which the aperture is the lens opening. This is a superimposition of many Airy discs. $\frac{1.22\lambda}{D} \approx \theta$. The condition for optical resolution is called the Rayleigh criterion, which is $\frac{\lambda}{b}$ for a single-slit.

Accordingly, "two sources are just resolved, if the central maximum of the diffraction pattern of one falls or overlaps on the other".

Saddle point

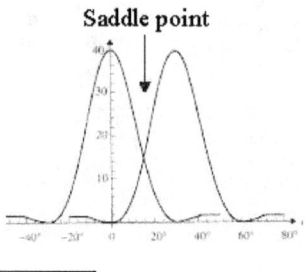

$$I_{saddle} = \frac{8}{\pi^2} = 0.81$$

23.5.1 The DOUBLE SLIT Diffraction

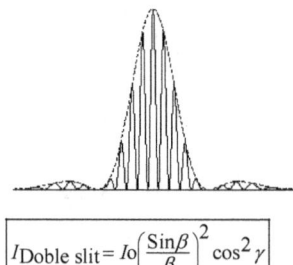

$$I_{Doble\ slit} = Io\left(\frac{Sin\beta}{\beta}\right)^2 \cos^2\gamma$$

which is interference pattern having an envelope

i.e., $\boxed{I_{Doble\ slit} \propto Cos^2\gamma}$

This is similar to the interference fringes. Bright fringes occur at

$\gamma = 0, \ \pm\pi, \ \pm\pi,...etc$

23.5.2 DIFFRACTION BY MULTIPLE (MANY) SLITS (Transmission Grating)

If slits of width b and separated by h, then for light of λ, then for a grating aperture containing N slits

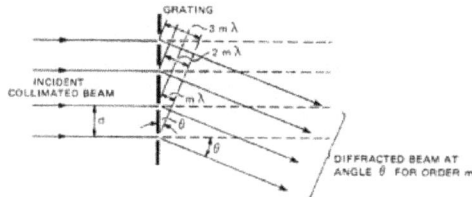

θ_i =Angle of incident rays measured with respect to the grating normal

θ_m =Angle of diffracted order m with respect to the grating normal. When θ_m is located on the opposite side of the grating normal from θ_i, then it is negative.

h = distance between successive grooves.

$$I_{\text{Grating}} = I_0 \left(\frac{Sin\beta}{\beta}\right)^2 \left(\frac{Sin(N\gamma)}{N\,Sin\gamma}\right)^2$$

$I_{\text{Grating}} = I_0$, when $\theta = 0$.

$$\gamma = m\pi, \quad m = 0,1,2,3,...$$

i.e., when $\quad m\lambda = h\,Sin\theta$

Constructive when the extra distance is an integer number of wavelengths:

Grating formula

$$(Sin\theta_i + Sin\theta_m) = m\lambda\,; \quad m = \text{the "order" } 0, 1, 2,$$

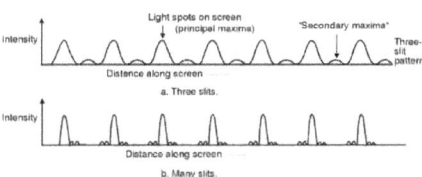

a. Transmission grating.

a. Three slits.

b. Many slits.

$$I_{\text{Grating}} = I_0 \left(\frac{Sin\beta}{\beta}\right)^2 \left(\frac{Sin(N\gamma)}{N\,Sin\gamma}\right)^2$$

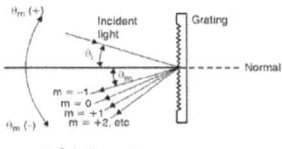

b. Reflection grating.

23.5.3 NARROW WING OF PRINCIPAL MAXIMA WITH INCREASING DIFFRACTION SLITS

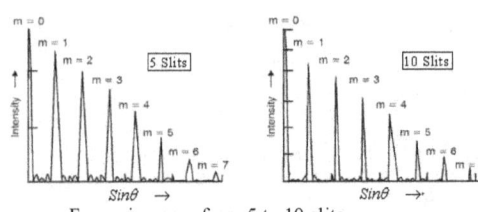

For an increase from 5 to 10 slits.

23.5.4 Resolving power (RP) of grating

Separation between the peak and adjacent maximum gets when

$$\boxed{N\gamma = \pi}\,,$$

i.e.,
$$\Delta\theta = \frac{\gamma\lambda}{Nh\,Cos\theta}\,;$$

When N = large, and from the Grating formula,

$$\Delta\theta = \frac{n\,\Delta\lambda}{h\,Cos\theta}\,,\ \text{whence}$$

$$\boxed{RP = \frac{\lambda}{\Delta\lambda} = Nm}$$

N = Number of grooves, m = order.

23.5.5. Grating Diffractometer

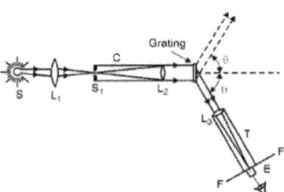

23.6. Application of Fourier Transform to Diffraction

The diffracting aperture lies in the xy plane, and the diffraction pattern in the XY plane (the focal plane of the focusing lens). L = focal length of the lens. The path difference δr between the ray from $Q(x,y)$ and the ray from $O(0,0)$,

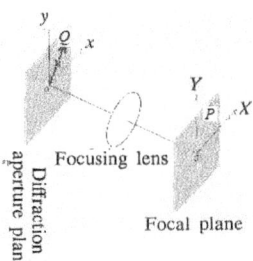

$$\delta r = x\frac{X}{L} + y\frac{Y}{L}$$

The diffraction pattern in the XY plane, for uniform aperture, is

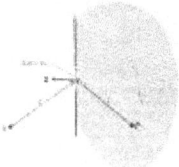

$$U(X,Y) = \iint \left(e^{ik\delta r}\ dA\right) = \iint e^{ik(xX+yY)/L}\ dxdy$$

For non-uniform aperture, then aperture function $g(x, y)$ modifies the pattern

$$U(X,Y) = \iint g(x,y).\, e^{ik(xX+yY)/L}\ dxdy\ .$$

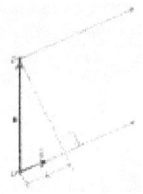

Introducing spatial frequencies, $\mu = \frac{kX}{L}$ and $\nu = \frac{kY}{L}$,

$$U(\mu,\nu) = \iint g(x,y).\, e^{i\ (\mu x+\nu y)}\ dxdy$$

$U(\mu,\nu)$ and $g(x,y)$ for a 2-D FT pair.

23.6.1 Spatial Filtering

Coherently illuminated object at the xy plane, and object function $g(x, y)$ with its diffraction pattern $U(\mu, \nu)$ in the $\mu\nu$ plane. So $U(\mu, \nu)$ and $g(x, y)$ for a 2-D FT pair. The image function $g'(x', y')$ in the $x'y'$ plane is the FT of $U(\mu, \nu)$.

Lens defects, aberrations, *etc.* of the optical system modifies (transforms) the $U(\mu, \nu)$ yielding $U'(\mu, \nu)$. This is implicit by writing

$U'(\mu, \nu) = T(\mu, \nu). U(\mu, \nu)$, or,

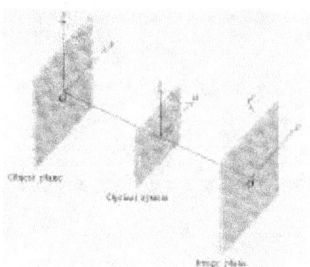

$$g'(x', y') = \iint_{\pm\infty} T(\mu, \nu). U(\mu, \nu). e^{i\,(\mu x' + \nu y')}\, d\mu d\nu$$

Aperture function for a grating and its Fourier transform.

The transfer function $T(\mu, \nu)$ can be modified by placing various screen and apertures in the $\mu\nu$ plane. This is called 'spatial filtering'.

In 1-D problem, a step function $g(y)$.

$$g'(y') = \int_{-\nu_{max}}^{+\nu_{max}} U(\nu)\, e^{-i\nu y'}\, d\nu,$$

and $T(\nu) = 1$, a step function, with $-\nu_{max} < \nu < +\nu_{max}$.

Object function
(grating)

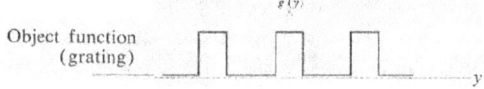

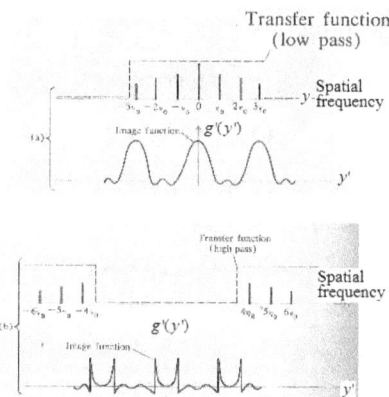

23.6.2. Phase Contrast Microscope

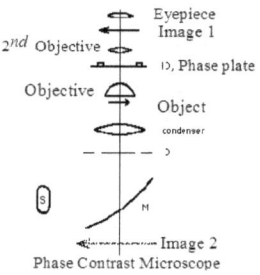

Phase Contrast Microscope

-o-0-o-0o-0o-0o-0o-0o

Chapter 24

OPTICS - 4

VECTORIAL REPRESENTATION, CRYSTAL OPTICS, LIGHT SCATTERING

"That theory is worthless. It isn't even wrong" ~Wolfgang Pauli

24.1 Vectorial representation of Light and Elements.

24.1.1 Unpolarized beam:

In optics, polarized light can be described using the **Jones calculus**, invented by R. C. Jones in 1941 The electric field of <u>any polarized beam propagating</u> along the z-axis may be written Unpolarized ray is represented by

$$\vec{E} = E_0 \, e^{i(kz - \omega t)}$$

Unpolarized
sunlight

or, in complex vector form as $\vec{E} = \hat{i} \, E_x + \hat{j} \, E_y$,

where

$$E_x = A_x \, e^{i(kz - \omega t)},$$
$$E_y = A_y \, e^{i(kz - \omega t + \varphi)}.$$

24.1.2 Jone's vector

Conveniently, this is written in matrix notation as a <u>column vector</u>,

$$\vec{E} = \begin{bmatrix} E_x \\ E_y \end{bmatrix} = \begin{bmatrix} A_x \, e^{i(kz - \omega t + \varphi_x)} \\ A_y \, e^{i(kz - \omega t + \varphi_y)} \end{bmatrix}$$

24.1.3 <u>The most general Jones vector of a polarized beam propagating along the z-axis</u>, is

$$\vec{E} = \begin{bmatrix} E_x \\ E_y \end{bmatrix} = \begin{bmatrix} A_x \, e^{i\varphi_x} \\ A_y \, e^{i\varphi_y} \end{bmatrix}$$

with intensity, $I = A_x^2 + A_y^2$

24.1.4. The Jones vector for LINEAR horizontally polarized light

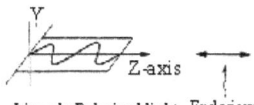

Linearly Polarized light   End view

$$\vec{E} = \begin{bmatrix} E_x \\ 0 \end{bmatrix} = \begin{bmatrix} A_x \, e^{i\varphi_x} \\ 0 \end{bmatrix} = A_x \, e^{i\varphi_x} \begin{bmatrix} 1 \\ 0 \end{bmatrix}$$

24.1.5 The Jones vector for LINEAR vertically polarized light

End view

Linearly Polarized light

$$\vec{E} = \begin{bmatrix} 0 \\ E_y \end{bmatrix} = \begin{bmatrix} 0 \\ A_y \, e^{i\varphi_y} \end{bmatrix} = A_y \, e^{i\varphi_y} \begin{bmatrix} 0 \\ 1 \end{bmatrix}$$

24.1.6   The **normalized** Jones vector for light polarized at $45°$ ($\varphi = \frac{\pi}{4}$)

Normalization means that

$E_x^2 = E_y^2 = E^2 = 1$.

$A_x = A_y = A = 1$ and

$\varphi_x = \varphi_y = \varphi = \frac{\pi}{4}$.

$$\vec{E}_{\varphi=\pi/4} = \begin{bmatrix} A e^{i\varphi} \\ A \, e^{i\varphi} \end{bmatrix} = \frac{1}{\sqrt{2}} \begin{bmatrix} 1 \\ 1 \end{bmatrix}$$

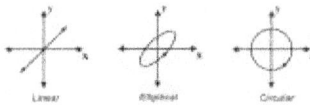

Linear   Elliptical   Circular

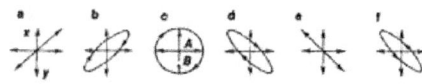

24.1.7 The **normalized** Jones vector for right-hand circularly polarized light:

$$\vec{E}_{\circlearrowright R} = \begin{bmatrix} A\,e^{i\varphi} \\ A\,e^{i(\varphi-\pi/2)} \end{bmatrix} = \frac{1}{\sqrt{2}}\begin{bmatrix} 1 \\ -i \end{bmatrix} \equiv \frac{1}{\sqrt{2}}\begin{bmatrix} i \\ 1 \end{bmatrix}$$

24.2.1 **Effect of Polarizing Element, Jone's Calculus**

If a polarized beam with field vector **E** is incident on a polarization-changing medium such as a polarizer or a wave plate, then the result is a beam in another polarization state given by **E'**

$$\vec{E}_{\varphi} = \begin{bmatrix} E_x' \\ E_y' \end{bmatrix} = \begin{bmatrix} m_{11} & m_{12} \\ m_{21} & m_{22} \end{bmatrix}\begin{bmatrix} E_x \\ E_y \end{bmatrix}$$

The (2 x 2) transformation matrix M is called the **Jones matrix**. The Table below lists the Jones matrices for common optical elements.

Optical Element	Jones Matrix
1) horizontal linear polarizer	$\begin{bmatrix} 1 & 0 \\ 0 & 0 \end{bmatrix}$
2) vertical linear polarizer	$\begin{bmatrix} 0 & 0 \\ 0 & 1 \end{bmatrix}$
3) linear polarizer at θ	$\begin{bmatrix} Cos^2\theta & Cos\theta Sin\theta \\ Cos\theta Sin\theta & Sin^2\theta \end{bmatrix}$
4) quarter wave plate (fast axis vertical)	$e^{i\pi/4}\begin{bmatrix} 1 & 0 \\ 0 & -i \end{bmatrix}$
5) quarter wave plate (fast axis horizontal)	$e^{i\pi/4}\begin{bmatrix} 1 & 0 \\ 0 & i \end{bmatrix}$

24.2.2 To find the Jones matrix for an Optical Element

An optical element rotated through an angle θ with respect to the direction given in the table above, we must multiply the above matrix by the usual matrices for rotation.

$$M(\theta) = R(\theta)\,M\,R(-\theta) \;,$$

where $\qquad R = \begin{bmatrix} Cos\theta & -Sin\theta \\ Sin\theta & Cos\theta \end{bmatrix}$

If an incident beam of light with field vector **E** passes through a sequence of four polarizing elements, M_1, followed by M_2, M_3 and M_4, then the resultant field vector **E'** is given by

$$\vec{E}_{\varphi} = [M_4\ M_3\ M_2\ M_1]\,\vec{E}_{\parallel}$$

24.2.3 **Production of Circular Polarized Light:**

Quarter Wave ($\frac{\lambda}{4}$) Plate, made of <u>doubly-refractive</u> crystal, if

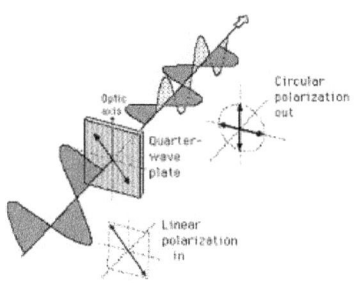

d = thickness of plate,

$$d = \frac{\lambda_0}{4(n_1 - n_2)}$$

n_1 and n_2 are indices of refraction for λ_0 along the slow and fast axis of the plate.

Linear Polarizer	Jones Matrix
1) Linear polarized in the x-direction	$\begin{bmatrix} 1 & 0 \\ 0 & 0 \end{bmatrix}$
(Typically called 'Horizontal')	
2) Linear polarized in the y-direction	$\begin{bmatrix} 0 & 0 \\ 0 & 1 \end{bmatrix}$
(Typically called 'Vertical')	
3) Linear polarized at $\varphi = \pm\frac{\pi}{4} = \pm45°$ from the x-axis	$\frac{1}{2}\begin{bmatrix} 1 & \pm1 \\ \pm1 & 1 \end{bmatrix}$
(Typically called 'Diagonal' L±45)	
4) Quarter Wave $\frac{\lambda}{4}$-Plate (Fast axis, Vertical)	$\begin{bmatrix} 1 & 0 \\ 0 & -i \end{bmatrix}$
5) Quarter Wave $\frac{\lambda}{4}$-Plate (Fast axis, Horiz)	$\begin{bmatrix} 1 & 0 \\ 0 & +i \end{bmatrix}$
6) Quarter Wave $\frac{\lambda}{4}$-Plate (Fast axis, ±45°	$\frac{1}{\sqrt{2}}\begin{bmatrix} 1 & \pm i \\ \pm i & 1 \end{bmatrix}$
7) Half wave $\frac{\lambda}{2}$-Plate (Fast axis, Vert or Horiz)	$\begin{bmatrix} 1 & 0 \\ 0 & -1 \end{bmatrix}$
8) Circular Polarizer (Right/ Left)	$\frac{1}{2}\begin{bmatrix} 1 & \pm i \\ \mp i & 1 \end{bmatrix}$
9) Phase Retarder (Isotropic)	$\begin{bmatrix} e^{i\varphi} & 0 \\ 0 & e^{i\varphi} \end{bmatrix}$

24.2.4 Addition of two circular light vectors

Two beams of equal amplitude, $\begin{bmatrix} 1 \\ -i \end{bmatrix}$ and $\begin{bmatrix} 1 \\ i \end{bmatrix}$

$$\begin{bmatrix} 1 \\ -i \end{bmatrix} + \begin{bmatrix} 1 \\ i \end{bmatrix} = \begin{bmatrix} 1+1 \\ -i+i \end{bmatrix} = 2\begin{bmatrix} 1 \\ 0 \end{bmatrix}$$ is a linearly polarized beam

24.2.5 Polarized light is represented by a *Jones vector*, and linear optical elements are represented by *Jones matrices*

Note that Jones calculus is only applicable to light that is already *fully polarized*. Light which is randomly polarized, partially polarized, or incoherent must be treated using *Mueller calculus*.

24.3 FARADAY EFFECT

Michael Faraday discovered the phenomenon, in which the presence of a magnetic field H causes an isotropic dielectric to become <u>optically active</u>. The amount of rotations θ_F of the plane of polarization of the light passing through the dielectric plate of thickness d is

$$\theta_F = V\,Hd$$

V = Verdet constant for the material, in (minutes of angle) $Oe^{-1}cm^{-1}$

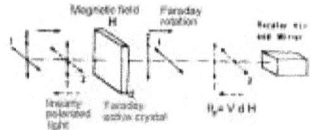

Dielectric	Verdet Constant, V (minutes of angle)$Oe^{-1}cm^{-1}$
1) Fluorite	0.0009
2) Diamond	0.012
3) Crown glass	0.015 - 0.025
4) Flint glass	0.030 - 0.050
6) Sodium Chloride	0.036

24.4. OPTICAL SOURCES

Lighting Technology Timeline

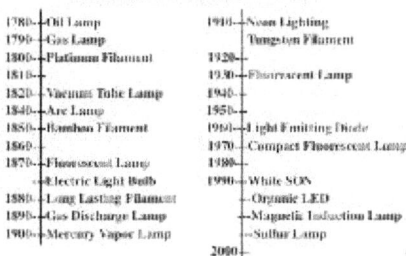

1780	Oil Lamp	1910	Neon Lighting
1790	Gas Lamp		Tungsten Filament
1800	Platinum Filament	1920	
1810		1930	Fluorescent Lamp
1820	Vacuum Tube Lamp	1940	
1840	Arc Lamp	1950	
1850	Bamboo Filament	1960	Light Emitting Diode
1860		1970	Compact Fluorescent Lamp
1870	Fluorescent Lamp	1980	
	Electric Light Bulb	1990	White SON
1880	Long Lasting Filament		Organic LED
1890	Gas Discharge Lamp		Magnetic Induction Lamp
1900	Mercury Vapor Lamp		Sulfur Lamp
		2000	

24.4.1 Incandescent Lamp

Traditionally, light is generated by heating something, whether it is a lighted candle or an incandescent lamp. In an incandescent lamp, electricity is passed through a filament, which is usually made of tungsten. This in turn heats it to a point where it glows and produces light. To prevent the filament from burning, the whole setup is in a sealed vacuum bulb.

In incandescent lamps most of the energy is converted to heat and the light produced is just a byproduct of the whole process. It is not surprising then that only about 10 to 12% of the electrical energy consumed is converted to light and the rest is just wasted as heat. In physical terms, an incandescent or halogen lamp is very poor when it comes to lighting efficiency

Sir Humphrey Davy in 1802 discovers incandescence and the carbon arc lamp for the first time. The most profound invention was by Thomas Alva Edison, after the man-made fire. This includes the traditional tungsten filament bulb and halogen lamp. Advantages are: a) Great for small area lighting, b) Good colour rendering, c) cheap to produce d) No toxic materials to dispose of e) easy to use in strobe or dimming circuits. Disadvantages: a) 90% spent for heat, only 10% for visible light Tungsten lamps are not good for illuminating large area. But halogen lamps are useful.

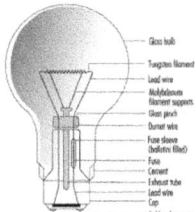

Features: Colour temperatures are around 5000K. CRI 100, 8-24 lumens per Watt, Lamp life 750 -1000 *hrs*.

Tungsten, tantalum, molybdenum, carbon are filament materials, Incandescence is thermal radiation. In 1850 Joseph Swan worked on incandescent bulb, by I Edison succeeded by using carbonized swing thread as filament.

Ductile tungsten filaments are used since 1908 to date and vacuum in the bulb.

24.4 1   FLUORESCENT LIGHT

The central element in a fluorescent lamp is a **sealed glass tube**. The tube contains a small bit of **mercury** and an inert gas, typically **argon**, kept under very low pressure. The tube also contains a **phosphor powder**, coated along the inside of the glass. The tube has two **electrodes**, one at each end, which are wired to an electrical circuit.

Electrons emerging out from the heater strike Hg atoms. Outer electron of Hg jump to a higher orbit; thereby UV radiations are emitted; on striking the phosphor coats on the glass tube stimulate emission of visible light.

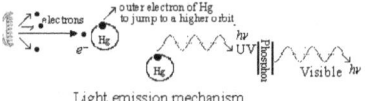

Light emission mechanism

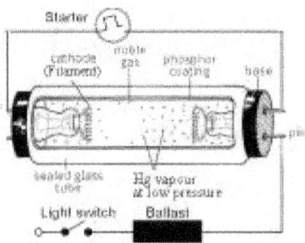

Fluorescent Lamp

Starter: A bimetallic strip with gap filled with Ar.

Choke: causes a high voltage of 1000 V AC to start the discharge.

Fluorescent starters are there to help the lamp light. When voltage is applied to the fluorescent lamp, here's what happens:

1. The starter (which is simply a timed switch) allows current to flow through the filaments at the ends of the tube.

2. The current causes the starter's contacts to heat up and open, thus interrupting the flow of current and the tube lights.

3. Since the lighted fluorescent tube has a low resistance, the ballast now serves as a current limiter.

24.4.2 Compact Fluorescent Lamp (CFL)

Fluorescent lamps and CFLs are very similar. In fact, a CFL is just a compact version of a fluorescent lamp that is smaller and easier to install. The glass tube is bent and both its ends are fixed onto a base that holds the ballast and can fit into standard incandescent bulb sockets. Therefore, there is no major difference between the working principle of a fluorescent lamp and a CFL.

24.4.2.1 Advantages of discharge lamp over incandescent lamp

    (i) A very low power is enough.

    (ii) The life period is 5 times that of a incandescent lamp.

    (iii) Gives more intense light.

(iii)   Majority of electricity convert to light

(iv)   Disturbance due to shadow is very less.

24.4.3. Arc Lamp:

Hg arc lamps are conventional sources of light used in a spectroscopy research laboratory. It works on 220 or 110 V DC, at around 1.5 A. It is used to produce monochromatic light beam, say 546.1 nm in a Hg lamp.

24.4.4 LED Bulb

The light-emitting diode (LED) is one of today's most energy-efficient and rapidly-developing lighting technologies. Quality LED light bulbs last longer, are more durable, and offer comparable or better light quality than other types of lighting. The $10W$ LED units ($42V$ supply voltage) use 6 times less energy than the $60W$ incandescent sources ($230V$ supply voltage). The efficiency is almost 50 times greater than a simple tungsten lamp. The response time of the LED is also known to be very fast in the range of 0.1 μs when compared with 100 ms for a tungsten lamp. Due to these advantages, the device wide applications as visual indicators and as dancing displays.

The electrons dissipate energy in the form of heat for silicon and germanium diodes. But in GaAsP and GaP semiconductors, the electrons dissipate energy by emitting photons. If the semiconductor is translucent, the junction becomes the source of light as it is emitted, thus becoming a light emitting diode (LED). But when the junction is reverse biased no light will be produced by the LED, and, on the contrary the device may also get damaged.

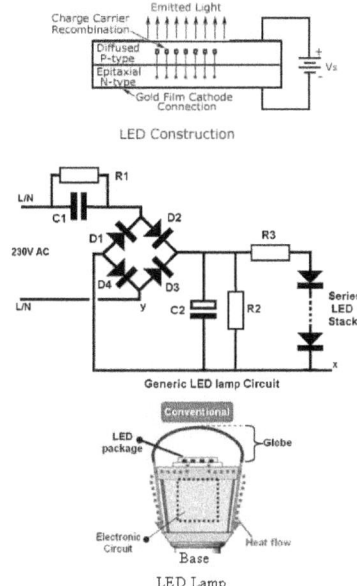

LED Construction

Generic LED lamp Circuit

LED Lamp

24.4.1. Advantages of LED's
- Very low voltage and current are enough to drive the LED.
- Voltage range – 1 to 2 V.
- Current – 5 to 20 mA.
- Total power output will be less than 150 mW.
- The response time is very less – only about 10 ns.
- The device does not need any heating and warm up time.
- Miniature in size and hence light weight.
- Have a rugged construction and hence can withstand shock and vibrations.
- An LED has a life span of more than 20 years.

24.4.2. Disadvantages
- A slight excess in voltage or current can damage the device.
- The device is known to have a much wider bandwidth compared to the laser.
- The temperature depends on the radiant output power and wavelength.

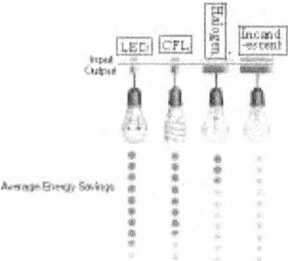

24.6 Crystal Optics

The **indicatrix** is a geometric figure, constructed so that the indices of refraction are plotted as radii that are parallel to the vibration direction of light. In isotropic minerals the indicatrix is a sphere, because the refractive index was the same in all directions.

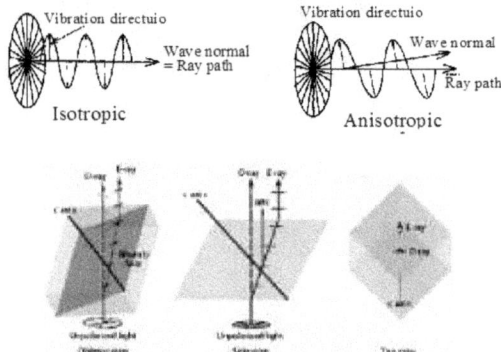

24.5.1. Uniaxial minerals

The indicatrix is an ellipsoid, the shape of which is dependent on its orientation with respect to the optic axis. In positive uniaxial minerals, the Z indicatrix axis is parallel to the c-crystallographic axis and the indicatrix is a prolate ellipsoid, *i.e.,* it is stretched out along the optic axis.

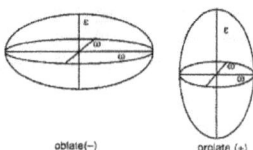

oblate(−)   prolate (+)

24.5.2. Biaxial optical indicatrix: two optic axes

In the crystals belonging to the Orthorhombic, Monoclinic and Triclinic systems, the section perpendicular to axis c (vertical) is not the same size, and the equatorial section turns into an ellipse with different axes. The optic indicatrix is an ellipsoid with three axes.

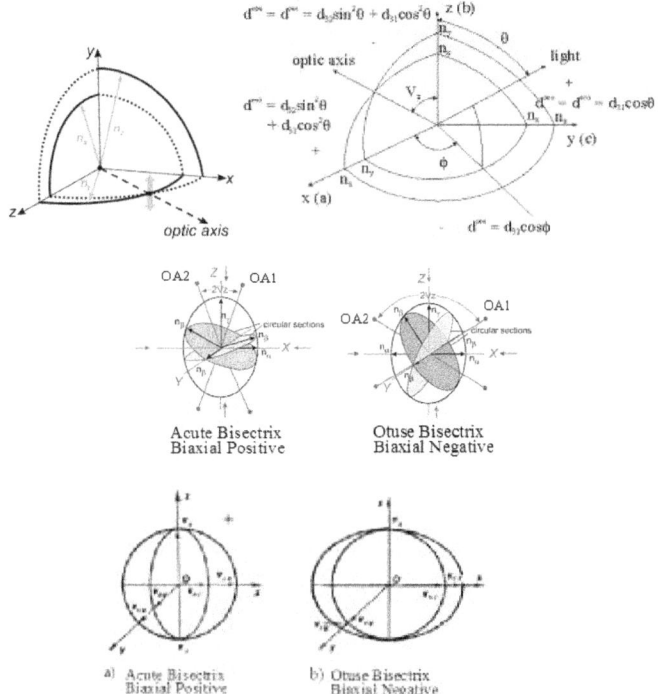

$$d^{ox} = d^{ox} = d_{2y}\sin^2\theta + d_{21}\cos^2\theta$$

optic axis

$$d^{ox} = d_{2y}\sin^2\theta + d_{31}\cos^2\theta$$

optic axis

$$d^{ox} = d_{31}\cos\phi$$

$$d^{ox} = d^{ox} = d_{21}\cos\theta$$

Acute Bisectrix
Biaxial Positive

Otuse Bisectrix
Biaxial Negative

a) Acute Bisectrix
Biaxial Positive

b) Otuse Bisectrix
Biaxial Negative

The refractive indices which coincide with the axes of the ellipsoid are known as n_α, n_β and n_γ which correspond to the three dimensions (for simplicity in the diagram, α, β and γ have been represented instead of n_α, n_β and n_γ). It is always true that the smallest refractive index is n_α, the greatest is n_γ and the intermediate one is n_β, i.e., $n_\alpha < n_\beta < n_\gamma$.

The indicatrix of these crystals presents two inclined sections which are circular, that is, isotropic.

Perpendicular to each of these sections there is an Optic Axis (OA). The angle which they form is called the optic angle (2V).

25.5.3. Interference Patterns, Isogyres

Selective absorption as a function of both wavelength and crystallographic direction gives rise to pleochroic minerals (such as mica biotite). Pleochroism may only be observed in plane polarized light. The tri-octahedral micas are strongly pleochroic and change color when rotated in plane polarized light. This is not to be confused with **interference** or **birefringence** which causes pronounced color changes in cross-polarized light.

338

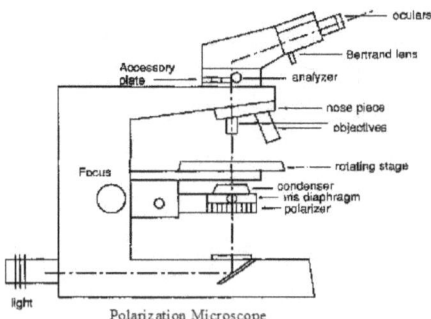

Polarization Microscope

The conoscope can be used to obtain optical information not readily available in the orthoscope. Setup the conoscope on the microscope
1. Condenser in
2. High powered objective in
3. Bertrand lens in or ocular removed
4. analyser in (*i.e.*, crossed nicols)

In the uniaxial optic axis figure, the limbs of the curves are called **isogyres**, the centre the **melatope**. Lines of equal retardation circle the centre are **isochromes**. The melatope corresponds to the optic axis of the sample. If the OA is not exactly vertical, the figure will appear off centre, and the centre will precess about the centre of the field of view as the stage is rotated.

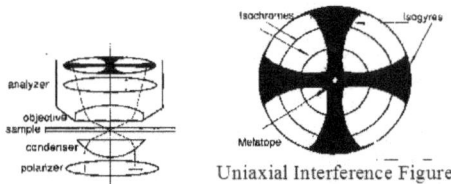

Uniaxial Interference Figure

Centred Acute bisectrix (Figure). If 2V << 60°, the two melatopes remain in the field of view for a full rotation.

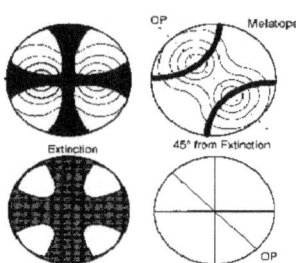

Biaxial Interference Figure

24.7 SCATTERING OF LIGHT

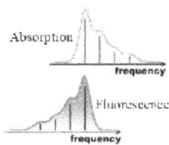

24.6.1 Rayleigh scattering (Elastic Scattering)

24.6.1.2 **Scattering of light by the atmosphere**

24.6.1.3 Observer looking at the sky, perpendicular to sunlight

Unpolarized Sun light, with electric (E) field perpendicular to ray, has equal horizontal and vertical components. It sets in motion charges in molecules. Oscillating charges emit EM wave. In the ray that comes to the observer the E field is perpendicular to the ray. Therefore the horizontal motion of the charges is the most important for this ray, while the vertical motion cannot contribute to it. Therefore the original vertical component of the E field is filtered out, and the light reaching the observer is polarized.

Hence 90° above in the sky, light is polarized in the plane perpendicular to the direction of the sun (check it out with Polaroid glasses), while at other angles it is partially polarized.

24.6.1.4 **The sky is blue,**

This is because blue light is scattered more than red light by the molecules in the atmosphere. More blue is reflected from the sky.

(Molecules in the atmosphere are much smaller than wavelengths of visible light.)

24.6.1.5 **Red Sunsets**

When the light travels a long distance through the atmosphere only long wave Red light makes it thru giving Red sunsets.

24.6.2.1 Scattering of a large wave by smaller particles

Stoke's formula

Intensity, $\boxed{I \propto \dfrac{1}{\lambda^4}}$

24.6.2.2 **Sunsets are red,**

Sunsets are red because rays in the direction of the sun pass through a thicker layer of atmosphere, and the blue light gets scattered away in other directions more than the red light. By contrast, when the sun is higher it looks orange, because of the thinner layer of atmosphere (less scattering away of blue).

24.6.2.3 **Clouds are white,**

This is because they contain water droplets that are larger than wavelengths of visible light. All wavelengths are scattered equally, hence white. Clouds become darker when water droplets grow too much and begin to absorb energy from the sun.

24.6.2.4 **Phase change on Reflection**

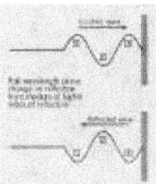

24.6.3 Tyndall scattering

When the dimensions of the particles are larger than the wavelength of the radiation, as it passes through the particles will trace a visible path through a genuine colloidal suspension, e.g. a headlight on a car shining through figure. This is known as the Tyndall effect.

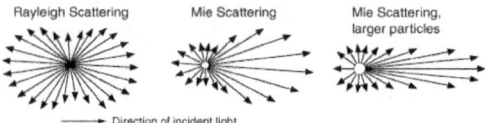

24.8 INTERACTION OF LIGHT WITH MATTER

The interaction of light with a material can be thought of in a classical picture as being due to the action of the electric field of the light wave on the charges in the material'

24.7.1 MOLECULAR SPECTROSCOPY is the study of absorption of light by molecules

24.7.2 INFRA RED ABSORPTION

Infra red (IR) spectroscopy deals with the interaction between a molecule and radiation from the IR region of the EM spectrum.

IR region = 4000 - 400 cm⁻¹.

A molecule containing N atoms has $(3N - 3)$ vibrational and rotational degrees of freedom.

The condition that any one of these molecular vibration causes a **non-zero change in dipole moment**, as and when IR radiation of frequency equal to the concerned molecular vibration or rotation then **infra-red (IR) absorption** is possible (hence, this vibration or rotation would be called an *"IR-active" mode*). An often used term in spectroscopy is the

> | *Wave number* $(\bar{v})$ *scale*   *Unit* cm^{-1} $\left(M^0 L^{-1} T^0 \right)$ |
> |---|
>
> | $\bar{v}(cm^{-1}) = \dfrac{1}{\lambda(cm)}$ |
> |---|
>
> | $\bar{v}(cm^{-1}) = \dfrac{v}{c = 2.997925 \times 10^{10} cm/s}$ |
> |---|

In general terms it is convenient to split an IR spectrum into two approximate regions:

- $4000 - 1000 \ cm^{-1}$ known as the *functional group region*, and

- $< 1000 \ cm^{-1}$ known as the *fingerprint region*

The *fingerprint* region, however, can be useful for helping to confirm a structure by direct comparison with a known spectrum

24.7.3 RAMAN SCATTERING

Sir CV Raman discovered the Raman Effect in 1928, and he received the Nobel Prize in 1930.
Incident light on a molecule induces a dipole moment which oscillates at the same frequency as the light, so that light of that frequency is re-radiated: this is *elastic* or **Rayleigh scattering.** If any one of the (3N-3))molecular vibration or rotation cause a non-zero change in the polarizabilty of the molecule then light incident on the molecule interacts with the molecular vibration leading to *Stokes* and *Anti-Stokes* components (both *inelastic*) in the scattered light and result in Raman scattering.

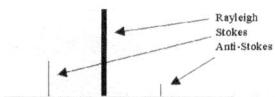

That is, due to this so-called *Raman active mode frequency* two new lines (Stokes and anti-Stokes) $(\overline{V}_0 \pm \overline{V}_i)$ appear on either side of the Rayleigh (incident) line $\overline{V}_0$ in the spectrum.

$\overline{V}_i$ is the *Raman Shift*

$$\text{Stokes} \qquad \overline{V}_S = (\overline{V}_0 - \overline{V}_i);$$

$$\text{Anti-Stokes} \qquad \overline{V}_{AS} = (\overline{V}_0 + \overline{V}_i)$$

24.7.3.1 A hypothetical system

With two vibrational modes (which are assumed to be both Raman-active and IR-active), the absorption bands appear at frequencies of light equal to the vibrational frequencies $\overline{V}_1 \,\&\, \overline{V}_2$ (at the left of the figure).

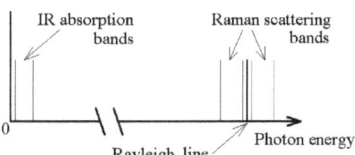

The Rayleigh scattering appears at the frequency of the incident light $\overline{V}_0$ and the Stokes and anti-Stokes. Raman bands appear near and on either side the energy $\overline{V}_0$ of the incident light (at the right of the figure).

Studies on Raman spectroscopy leads to a better understanding of the chemical composition of the sample, as in many organic compounds different chemical bonds have very characteristic vibrational frequencies.

+*+*+*&+*+*+*^+*+*^+*+*&+*+*+

Chapter 25

STATIC ELECRTRICITY

Insulators & Conductors, Electric charge, Coulomb Law,
Electric field, Gauss Law, Electric potential, Capacitance

"Science without religion is lame; religion without science is blind"   Albert Einstein

25.1   INSULATORS AND CONDUCTORS

- Thales (about 600 B.C.) discussed static electricity.
- Gray in 1729 discovered that static electricity could be discharged from an object through the human body thus giving the idea of conductors and insulators.
- Du Fay in 1733 found he could charge metallic objects using an insulated handle.
- Du Fay in 1745 discovered +ve and –ve electric charges, confirmed later by Benjamin Franklin (1752).
- Glass rubbed with *silk* becomes *positively* charged; to about 10 *nC*.
- Ebonite rubbed with fur becomes positively charged.
- Polythene rubbed with duster becomes negatively charged.
- Uncharged pieces of paper will be attracted to a charged rod because of movement of charge within the paper.
- There is no electric field within charged conductors.

25.1.1   UNIT OF ELECTRIC CHARGE

* Coulomb (C) is the SI unit.

* The electronic charge (e) is the smallest quantity of charge observed. All real charges come in multiples of e

$$q_{electron} = -e$$

$$q_{proton} = +e$$

Electric Charge, q Unit $e = 1.6021 \, x10^{-19} C$ (Scalar) $(M^0 L^0 T^0)$

- HA Lorentz postulated the concept of electron in 1895, whereas JJ Thomson experimentally found in 1897
- The ampere (A) is the basic electrical unit Hence 1 C is the electronic charges flowing in one second when 1 A current flows
- A macroscopic object possesses charge q if it has an imbalance in its proton and electron populations, N_p and N_e :

$$q = (N_p - N_e) = e$$

25.1.2 COULOMB'S LAW of Force between two Charges (1785)
(INVERSE SQUARE LAW)

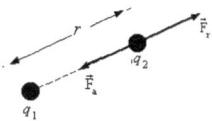

Using a torsion balance Coulomb established that the distance r and charge dependence of the electric force $\vec{F}$ between two stationary particles with charges q_1 & q_2 as an *inverse-square force*.

| $\vec{F} = k\dfrac{q_1 q_2}{r^2}$ | Unit | N | (Vector) | $(M^1 L^1 T^{-2})$ |

Coulomb's Law Constant,

$$k = 1/4\pi\varepsilon_0 = 8.9875 \times 10^9 \, Nm^2 C^{-2}$$

$$k = 1/4\pi\varepsilon_0 = 8.9875 \times 10^9 \, F^{-1} m$$

- **Permitivity of Free Space**,

| Permittivity | Unit | $\varepsilon_0 = 8.8542 \times 10^{-12} \, Fm^{-1}$ | (Scalar) | $(M^1 L^{-1} T^{-2})$ |

- *Inverse-square force graph*

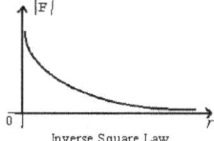

Inverse Square Law

- Most matter is electrically neutral -- no net charge
- "Like charges repel; unlike charges attract".

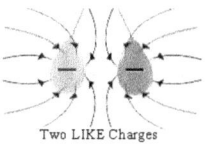

Two LIKE Charges

25.2. THE ELECTRIC FIELD, $\vec{E}$

25.2.1 A charged distribution will always produce an Electrostatic field, in the surrounding medium. The magnitude of the ES field ($\vec{E}$) of attraction / repulsion on a 'test' charge (q) placed in that region divided by the charge on the 'test' particle; and the direction of $\vec{E}$ will depend on the positive and negative charges responsible for the field.

$$\boxed{\vec{E}= \lim_{q\to 0} \frac{\vec{E}}{q}} \quad \text{Unit} \quad N\ C^{-1} \quad \text{(Vector)} \quad (M^1L^1T^{-2})$$

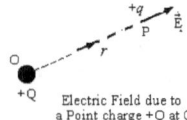

Electric Field due to
a Point charge +Q at O

25.2.2 Electric field due to a point charge, $+Q$, at the origin

The important features of $\vec{E}$:

(i) $\vec{E}$ is proportional to $+Q$

(ii) $\vec{E}$ is proportional to $\frac{1}{r^2}$

(iii) $\vec{E}$ directly points out away from a positive charge, or directly inward a negative charge.

25.2.3 To calculate the ES Field: *for a Point charge*:
Any particle having charge $+Q$

$$\boxed{\vec{E} = \frac{Q}{4\pi\varepsilon_0 r^2}} \quad \text{Unit } N\ C^{-1} = \frac{V}{m} \quad \text{(Vector)} \quad (M^1L^1T^{-2})$$

25.2.4 Lines of the Electric Field:
Spatial characteristics of an electric field of different situations are shown
25.2.4.1 Radial fields due to a positive charge Q are directed outward and that of a negative charge are directed inward.

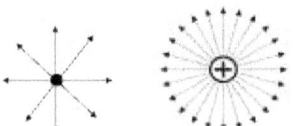

25.2.4.2 Field between two charged parallel plates

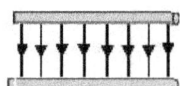

25.2.4.3 Field between two charges

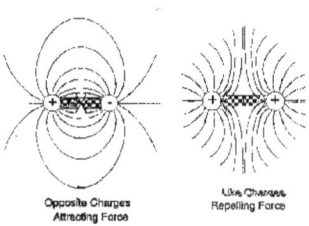

**Opposite Charges
Attracting Force**

**Like Charges
Repelling Force**

25.2.4.4 Dipole

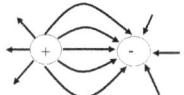

25.2.4.5 Uniformly charged Disc

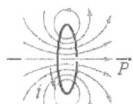

25.2.4.6 Two charged particles

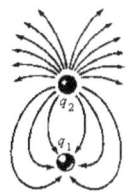

Two charged particles q_1 & q_2

25.2.5 **Note**: The electric field is

1) Uniform (*i.e.*, it does not depend on position),
2) Perpendicular to the charged plane, and
3) Oppositely directed on either side of the plane.
4) Always points away from a positively charged plane, and *vice versa*.

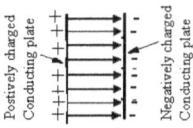

Field Lines between two charged
conducting plates

25.2.5.1 **Divergence**:

346

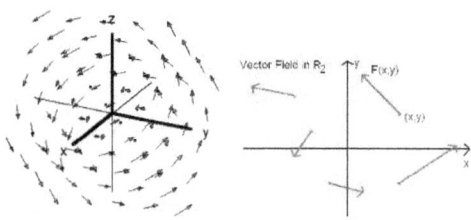

25.2.5.2 A charged particle in a Uniform field: $\vec{E}$

When a charged particle moves in a uniform field the motion described by constant acceleration motion.

$$q\,\vec{E} = m\,a$$

$$a = \frac{q\,\vec{E}}{m}$$

25.2.6 GAUSS'S LAW:

The flux Φ_E of the electric field for a closed (Gaussian) surface

$$\Phi_E = \oiint \vec{E}.d\vec{S}$$

$d\vec{S}$ is surface element.

25.2.6 Gauss's Law: The flux Φ_E of the electric field **at a point** is

$$\Phi_E = \frac{\Sigma q}{\varepsilon}$$

$$\oiint \vec{E}.d\vec{S} = \frac{\Sigma q}{\varepsilon}$$

In electrostatics, Gauss's law is equivalent to Coulomb's law.

25.2.7 Gauss's law can be used to find:

(i) near a long straight, uniform line charge far from the edges:

$$\vec{E} = \frac{\lambda}{2\pi\varepsilon R}$$

(ii) Near a planar sheet of uniform surface charge far from the edges

$$\vec{E} = \frac{\sigma}{2\varepsilon}$$

(iii) Inside and outside a sphere of uniform volume charge density
(iv) Inside and outside a spherical shell of uniform surface charge density
Total charge enclosed by the surface $q(r)$.

$$q(r) \begin{cases} Q & r \geq a \\ Q(\frac{r}{a})^3 & r < a \end{cases}$$

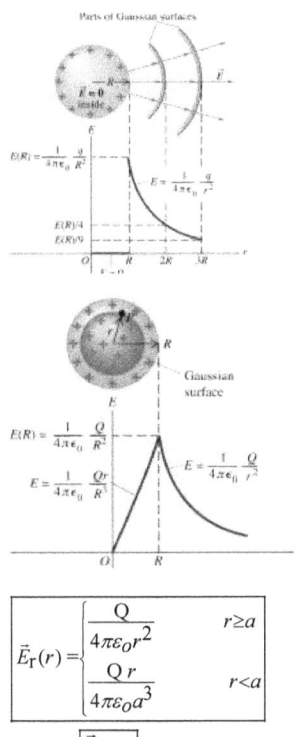

$$\vec{E}_r(r) = \begin{cases} \dfrac{Q}{4\pi\varepsilon_0 r^2} & r \geq a \\[2mm] \dfrac{Q\,r}{4\pi\varepsilon_0 a^3} & r < a \end{cases}$$

(v) Inside a conductor $\boxed{\vec{E} = 0}$

(vi) Just outside a conductor: $\boxed{\vec{E} = \dfrac{\sigma}{\varepsilon}}$.

The electric field immediately above the surface of a conductor is directed perpendicular to the surface.

25.3.1 ELECTRIC POTENTIAL (V) *due to point charges q*

$\vec{E}$ is called a conservative field force. The potential energy U of a test particle in the field of Q is

$$\boxed{U = qV} \qquad \text{Unit} \quad J \qquad \text{(Scalar)} \qquad (M^1 L^2 T^{-2})$$

$$\boxed{V = \frac{1}{4\pi\varepsilon_0} \sum_i \frac{Q_i}{r_i}}$$

$$\boxed{V = \frac{1}{4\pi\varepsilon_0} \sum_{\Delta qi} \frac{\Delta q_i}{r_i}; \text{(as } \Delta q_i \to 0) \quad V = \frac{1}{4\pi\varepsilon_0} \int \frac{dq_i}{r_i}}$$

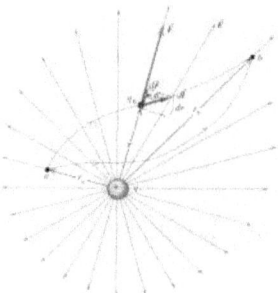

$$\boxed{V = \frac{Q}{4\pi\varepsilon_0 r}} \qquad \text{unit } Volt \ (= JC^{-1}) \qquad \text{(Scalar)} \qquad (M^1 L^2 T^{-2})$$

where for a system of charged particles

$$\frac{Q}{r} = \sum \frac{q_i}{ri}, \qquad \text{for a continuous distribution of charges.}$$

25.3.2   POTENTIAL DIFFERENCE (P.D):

In the field of a point charge, the potential difference ($V_Q = V_Q - V_P$) between two points P and Q is the work done (W), or energy expended, when a unit charge is transferred between them, electric potential energy of a charge when it is taken between two different points in an electrostatic field E from the work done in moving the charge between these two points.

$$\boxed{V_Q = -\int_P^Q \vec{E}.d\vec{\ell}} \quad \text{unit } Volt \ (= JC^{-1}) \qquad \text{(Scalar)} \qquad (M^1 L^2 T^{-2})$$

the increase in PE of the charge q.

$$\boxed{\Delta U = q \cdot \Delta V}$$

Thus, if an electron is moved, for which $\boxed{q = -1.6 \ x \ 10^{-19} C}$, through a $\boxed{I = \frac{\Delta Q}{\Delta t} = \frac{Q}{t}}$ potential difference of $(-1V)$ then one must do $1.6 \ x \ 10^{-19} J$ of work. This amount of work (or energy) is called an **electron volt** (eV); i.e.,

$$\boxed{\begin{array}{l} 1 \ JC^{-1} = 1 \ V \\ 1 \ eV = 1.6 \ x \ 10^{-19} J \end{array}}$$

The 'eV'is a convenient measure of energy in atomic physics.[The energy required breaking up a hydrogen atom into a free electron and a free proton is $13.6. \ eV$].

The definition of potential is not unique, any constant can be added to it. It is usual to consider V to go to zero at an infini9te distance.

Electrical energy

$$\boxed{E = VQ \qquad \text{unit } VC^{-1} \qquad \text{(Scalar)} \ (M^1 L^2 T^{-2})}$$

25.3.3 RELATION BETWEEN POTENTIAL DIFFERENEC AND ELECTRIC *FIELD*

For a uniform electric field (E) (the work W which we perform in moving the charge) the potential difference between plates A and B a distance d apart is

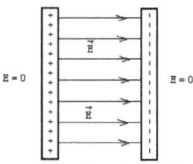

Uniform electric field

$$\text{Work } W = \frac{(\text{Force } \vec{F})\cdot(\text{distance } d)}{\text{Charge } q}$$

$$W = -q \int_A^B \vec{E}\cdot d\vec{r}$$

$$V_Q = -\int_P^Q \vec{E}.d\vec{\ell}$$

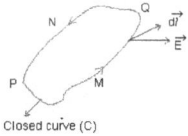

Closed curve (C)

25.3.4 The electric flux through any closed surface
is equal to the total charge enclosed by the surface divided by distance.

$$\boxed{\vec{E} = -grad V}\quad \text{since } curl(grad V) = 0$$

$$\vec{E} = -grad V = -\vec{\nabla}V = \frac{dV}{dx} = \frac{V}{d}$$

$$\vec{\nabla}V = \hat{i}\frac{dV}{dx} + \hat{j}\frac{dV}{dy}\,\hat{k}\frac{dV}{dz}$$

25.3.5 <u>Free charge exists</u>

$$div(-\vec{\nabla}V) = -\vec{\nabla}^2 V = \frac{\rho}{\varepsilon_0}$$

$$\vec{\nabla}^2 V = \frac{\partial^2 V}{\partial x^2} + \frac{\partial^2 V}{\partial y^2} + \frac{\partial^2 V}{\partial z^2} = -\frac{\rho}{\varepsilon_0}$$

25.3.6 Poisson's equation.

$$\boxed{div\vec{E} = \vec{\nabla}\cdot\vec{E} = \frac{\rho}{\varepsilon_0}}$$

Or, if the free charge is zero one gets

25.3.7 Laplace's equation:

$$\boxed{\vec{\nabla}^2 V = 0}.$$

Solution of Laplace's equation for two co-axial conductors

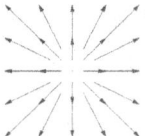

25.3.8 EQUIPOTENTIAL SURFACE

The equipotential surfaces are spheres centred on the charge and having constant V.

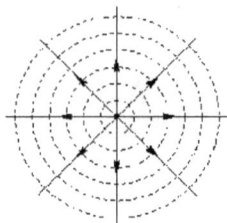

Under static conditions the potential V = uniform inside a conductor; so the surface of a conductor is an equi-potential surface..

25.3.9 ELECTRIC DIPOLE

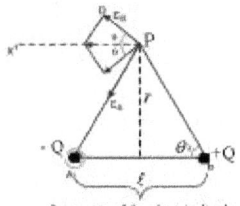

Geometry of the electric dipole

Dipole moment, P

$$\boxed{\vec{p} = Q\vec{\ell}}$$

At a distance $r \gg \ell$ (the charge separation), their fields cancel to first order, leaving a second order field,

$$\boxed{V = \frac{\vec{p} \cdot \vec{r}}{4\pi\varepsilon_0 r^3}},$$

$$\boxed{\vec{E}_r = \frac{2p\cos\theta}{4\pi\varepsilon_0 r^3}},$$

$$\boxed{\vec{E}_\theta = \frac{p\sin\theta}{4\pi\varepsilon_0 r^3}},$$

$$\vec{E}_\phi = 0$$

25.3.10 ELECTRIC DIPOLE POTENTIAL
CONTOURS OF V AND $\vec{E}$ FOR A DIP*OLE*

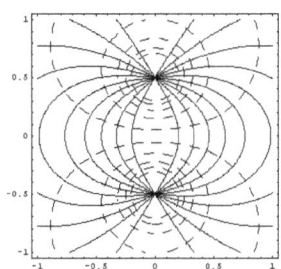

25.4 CAPACITANCE, C

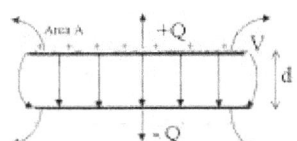

Two concentric spheres,

Capacitance $\boxed{C = \dfrac{Q}{V}}$ unit F (CV^{-1}) (Scalar) $(M^{-1}L^{-2}T^2)$

$$\boxed{V = E_\perp d = \frac{\sigma d}{\varepsilon_0} = \frac{Q d}{\varepsilon_0 A}}$$

$$\boxed{C = \frac{\varepsilon_0 A}{d}; \quad C = \frac{\varepsilon A}{d}; \quad \varepsilon = \kappa \varepsilon_0}$$

Note :
(i)   The capacitance is proportional to the area of the plates'
(ii)   Inversely proportional to their perpendicular spacing.
It follows that a good parallel plate capacitor possesses closely spaced plates of large surface area.

25.4.1 THE STORAGE OF CHARGE
25.4.1.1 Gauss' law:
One of the most useful results in electrostatics is named after the celebrated German mathematician Karl Friedrich Gauss (1777-1855)
The electric flux through any closed surface is equal to the total charge enclosed by the surface divided by distance.

$$\boxed{\Phi_E = \Sigma E_\perp \Delta A = \Sigma \frac{q}{\varepsilon_0}}$$

25.4.1.2 ENERGY STORED IN A CAPACITOR

$$\boxed{W = \frac{1}{2}\frac{Q^2}{C} = \frac{CV^2}{2} = \frac{\varepsilon_0 E^2 A d}{2}}$$

25.4.1.3 CAPACITANCES IN PARALLEL

$$C_{Eq} = \frac{Q}{V} = \frac{(Q_1+Q_2)}{V} = \frac{Q_1}{V}+\frac{Q_2}{V} = C_1 + C_2$$

The equivalent capacitance of two capacitors connected in parallel is the sum of the individual capacitances

$$C_{Eq} = \sum_{i=1}^{N} C_i$$

25.4.1.4 CAPACITORS IN SERIES:

$$\frac{1}{C_{Eq}} = \frac{V}{Q} = \frac{V}{Q_1}+\frac{V}{Q_2} = \frac{1}{C_1}+\frac{1}{C_2}$$

The reciprocal of the equivalent capacitance of two capacitors connected in series is the sum of the reciprocals of the individual capacitances

$$\frac{1}{C_{Eq}} = \sum_{i=1}^{N} \frac{1}{C_i}$$

25.4.1.5 **CAPACITY of an Isolated Charged Sphere**:
$$C = k\ r$$

25.4.1.6 **Capacity of two Concentric spheres, *a* & *b*:**
$$C = k\frac{ab}{(b-a)}$$

25.4.1.7 **Capacity of a Parallel plate Condenser**
$$C = k\frac{A}{4\pi d}$$

$$C = \frac{\varepsilon_0 A}{d - \ell + (\ell/k)}$$

25.4.1.8 Capacity of a Cylindrical Condenser of length ℓ

$$C = k \frac{\log \ell}{2\log(b/a)}$$

25.4.2 APPLICATIONS OF CAPACITORS:

(i)   Capacitors do not play an important role in dc circuits because it is impossible for a steady current to flow across a capacitor.

(ii)   If an uncharged capacitor C is connected across the terminals of a battery of voltage V then a *transient* current flows as the capacitor plates charge up. However, the current stops flowing as soon as the charge Q on the positive plate reaches the value $Q = CV$. Thus, if a capacitor is placed in a dc circuit then, as soon as its plates have charged up, the capacitor effectively behaves like a *break* in the circuit.

25.5. VOLTAGE SOURCES

Anode: The electrode in an electrochemical/galvanic cell that experiences oxidation, or gives up electrons.

Cathode: The electrode in an electrochemical/galvanic cell, at which a reduction reaction occurs, or the electrode that receives electrons from an external circuit.

Electrolyte: A solution through which an electric current may be carried through the motion of ions.

25.5.1. The Lead Acid battery

It is made up of plates, lead, Pb and lead oxide, PbO_2 (various other elements are used to change density, hardness, porosity, *etc.*) with a 35% sulfuric acid (H_2SO_4) and 65% water (H_2O) solution. This solution is called electrolyte, which causes a chemical reaction that produce electrons.

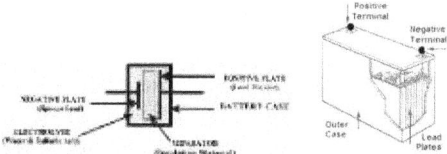

25.5.2. Leclanche Battery System (Dry Cell- A cell with an immobilized electrolyte)

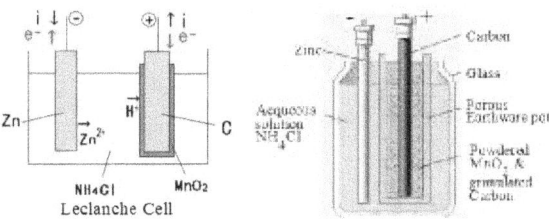

Leclanche Cell

The Leclanche or zinc-carbon dry cell battery has existed for over 100 years and has been the most widely used of all dry cell batteries because of a combination of low cost, ready availability, and relatively strong performance.

25.5.3.1. Daniel Cell

One of the many early cells used for a voltage standard was the Daniel Cell (1836). It consists of a zinc plate immersed in a zinc sulfate solution, and a copper plate immersed in a copper sulfate solution, and produced about 1.08 V.

25.5.3.2. Weston Standard Cell (Cadmium cell)

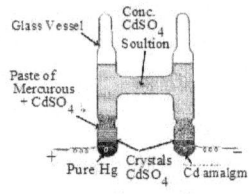

Weston Cell

In the Weston cell the $CdSO_4$ solution acts as the electrolyte, Hg as positive electrode, Cd amalgm as negative electrode and paste of $(Hg,Cd)SO_4$ acts as de-polarizer. The voltage is 1.0183 V.

25.5.4. The Conventional Battery

The Zinc Alkaline Manganese Dioxide Battery System

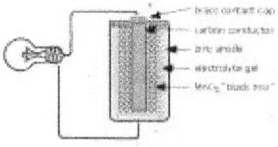

Cathode/Anode/Electrolyte: Manganese Dioxide/Zinc/Aqueous Potassium Hydroxide.

Chemistry:
$$(c + H_2O + e^- \rightarrow MnOOH + OH^-) \times 3 \quad \text{cathode}$$
$$3MnOOH + e^- \rightarrow Mn_3O_4 + OH^- + H_2O$$
$$Zn + 4OH^- \rightarrow Zn(OH)_4 + 2e^- \qquad \text{anode - early reaction}$$
$$Zn + 2OH^- \rightarrow Zn(OH)_2 + 2e^- \qquad \text{anode - late reaction}$$
$$(Zn(OH)_2 \rightarrow ZnO + H_2O) \times 2$$
$$3MnO_2 + 2Zn \rightarrow Mn_3O_4 + 2ZnO \qquad \text{(discharge reaction)}$$

25.5.5. Primary Battery

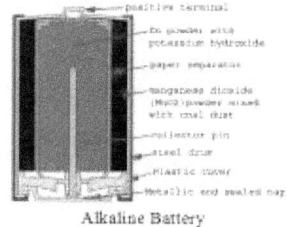

Alkaline Battery

Primary Battery	Anode	Cathode	Reaction Mechanism	Voltage%	
Leclanche	Zn	MnO_2	$Zn + 2MnO_2 \rightarrow ZnO + Mn_2O_3$	1.6	
Magnesium	Mg	MnO_2	$Mg + 2MnO_2 + H_2O \rightarrow Mn_2O_3 + Mg(OH)_2$	2.8	
Alkaline	MnO_2	Zn	$MnO_2\ Zn + 2MnO_2 \rightarrow ZnO + Mn_2O_3$	1.5	
Mercury	Zn	HgO	$Zn + HgO \rightarrow ZnO + Hg$	1.34	
Mercad	Cd	HgO	$Cd + HgO + H_2O \rightarrow Cd(OH)_2 + Hg$		0.91
Silver Oxide	Zn	Ag_2O	$Zn + Ag_2O + H_2O \rightarrow Zn(OH)2 + 2Ag$	1.6	
Zinc/O_2	Zn	O_2	$Zn + 0.5O_2 \rightarrow ZnO$	1.65	
Zinc/Air		Zn	Air	$Zn + 0.5O_2 \rightarrow ZnO$	1.65
Li/$SOCl_2$	Li	$SOCl_2$	$4Li + 2SOCl_2 \rightarrow 4LiCl + S + SO_2$	3.65	
Li/SO_2	Li	SO_2	$2Li + 2SO2 \rightarrow Li2S2O4$	3.1	
$LiMnO_2$	Li	MnO_2	$Li + MnIVO_2 \rightarrow MnIVO_2\,(Li+)$	3.5	
Li/FeS_2		Li	FeS_2	$4Li + FeS_2 \rightarrow Li_2S + Fe$	1.8
Li/$(CF)_n$	Li	$(CF)_n$	$nLi + (CF)_n \rightarrow nLiF + nC$	3.1	
Li/I_2 (3)	Li	$I_2\,(P_2 VP)$	$Li + 0.5I_2 \rightarrow LiI$	2.8	

25.5.6. Secondary Battery

Secondary Battery	Anode	Cathode	Reaction Mechanism	Voltage%
Lead-Acid	Pb	PbO_2	$Pb + PbO_2 + 2H_2SO_4 \rightarrow 2PbSO_4 + 2H_2O$	2.1
Edison	Fe	NiO_2	$Fe + 2NiOOH + 2H_2O \rightarrow 2Ni(OH)_2 + Fe(OH)_2$	1.4
Nickel-Cadmium	Cd	NiO_2	$Cd + 2NiOOH + 2H_2O \rightarrow 2Ni(OH)_2 + Cd(OH)_2$	1.35
Nickel-Zinc	Zn	NiO_2	$Zn + 2NiOOH + 2H_2O \rightarrow 2Ni(OH)_2 + Zn(OH)_2$	1.73
Nickel-Hydrogen	H_2	NiO_2	$H_2 + 2NiOOH \rightarrow 2Ni(OH)_2$	1.5
Nickel Metal Hydride	MH	NiO_2	$MH + NiOOH \rightarrow M + Ni(OH)_2$	1.35
Silver-Zinc	Zn	AgO	$Zn + AgO + H2O \rightarrow Zn(OH)2 + Ag$	1.85
Silver-Cadmium	Cd	AgO	$Cd + AgO + H_2O \rightarrow Cd(OH)_2 + Ag$	1.4
Zinc/Chlorine	Zn	Cl_2	$Zn + Cl_2 \rightarrow ZnCl_2$	2.12
Zinc/Bromine	Zn	Br_2	$Zn + Br_2 \rightarrow Zn\,Br_2$	1.85
Lithium Ion	Li_xC_6	$Li_{1-x}CoO_2$	$Li_xC_6 + Li_{1-x}CoO_2 \rightarrow LiCoO_2 + C_6$	4.1
Lithium/MnO_2	Li	MnO_2	$Li + MnIVO_2 \rightarrow MnIVO_2\,(Li+)$	3.5
Lithium/FeS_2	Li(Al)	FeS_2	$2Li(Al) + FeS_2 \rightarrow Li_2 FeS_2 + 2Al$	1.73
Lithium/FeSv	Li(Al)	FeS	$2Li(Al) + FeS \rightarrow Li_2S + Fe + 2Al$	1.33
Sodium/Sulfer	Na	S	$2Na + 3S \rightarrow Na_2S_3$	2.1
Sodium/$NiCl_2$	Na	$NiCl_2$	$2Na + NiCl_2 \rightarrow 2NaCl + Ni$	2058

25.5.7. Button (Coin) cell

These highly specialized cells are known for their very long battery life and a low self-discharge rate. They are categorized as silver oxide, alkaline, lithium, and zinc air. They have shapes typically 5 to 25 mm in diameter and 1 to 6 mm thickness.

    a) Silver: capacity 200 mAh to an end-point of 0.9 V, internal resistance 5–15 Ω, weight 2.3 g,

    b) Alkaline (resp., manganese dioxide): 150 mAh (0.9), 3–9 Ω, 2.4 g,

    c) Mercury: 200 mAh, 2.6 g,

    d) Zinc-air: 620 mAh, 1.9 g.

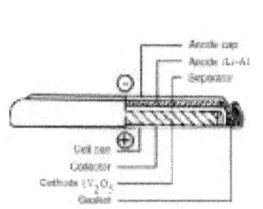 

Code	Name	+ve electrode	Electrolyte	− ve electrode	Normal (V)	End-point (V)
L	Alkaline	MnO_2	Alkali	Zinc	1.5	1.0
S	Silver	Silver oxide	Alkali	Zinc	1.55	1.2
P	Zinc-air	Oxygen	Alkali	Zinc	1.4	1.2
C	Lithium	MnO_2	Organic	Lithium	3	2.0
B		Carbon monofluoride		Organic	Lithium 3	2.0
G		Copper oxide		Organic	Lithium 1.5	1.2
Z	Nickel oxyhydroxide battery	MnO_2, nickel oxyhydroxide	Alkali	Zinc	1.5	?
M, N (withdrawn)	Mercury	Mercuric oxide	Alkali	Zinc	1.35/1.40	1.1

+*+*+*+*+*+*+*+*+*+_ *+

Chapter 26

DIRECT CURRENT ELECTRICITY

Current, Resistance, Charging circuit, Kirchhoff's rules,
Biot-Savart Law, Solenoid, Magnetic field due to flow of current

"A fact is a simple statement that everyone believes. It is innocent, unless found guilty.

A hypothesis is a novel suggestion that no one wants to believe.

It is guilty, until found effective. ~Edward Teller

26.1 ELETRIC CURRENT (I) AND RESISTANCE (R)

26.1.1 Electric Current I is the flow of a charge:

If a Potential Difference (PD) is applied between the two ends of a conductor, electrons move, *i.e.*, a transient current will flow (a current lasting for a short time) from the lower to the higher potential.

The direction of the current flow will be opposite in direction to that of the electrons. The sense of the current corresponds to the direction of the <u>drift velocity</u> v_D of positive charge carriers in the material.

$$\bar{q}v_D \text{ or } I$$

$$\ominus \longrightarrow$$

$$\longleftarrow$$
$${}^{+}I$$

A steady state current (I) will result if the p.d is maintained constantly.

$$I = \frac{\Delta Q}{\Delta t} = \frac{Q}{t} \qquad \text{Unit } 1A = 1\ Cs^{-1} \quad \text{(Scalar)} \qquad (M^0L^0T^{-1})$$

26.1.2 <u>Transport Equation</u>

$$I = \frac{dQ}{dt} = n\ q\ v_D A$$

A = Area of cross section of the current carrying conductor,

n = carrier density (number of charge carriers in the material), that free to move m^{-3}.

q = charge carried per particle

26.1.3 DRIFT VELOCITY: v_D

Free electrons in a conductor collide with vibrating ions in the lattice and so overall effect of the stop start motion on the electrons under an electric field is being a drift velocity being superimposed on the random thermal motion of the electrons.

Drift velocities in Materials		
Material	n (m^{-3})	v_D (ms^{-1})
1) Metal	10^{29} electrons	10^{-6}
2) Semi-conductor	10^{18} Charge carriers	10

Thermal speed at random $\approx 10^{6}\, ms^{-1}$

26.1.4 RESISTANCE (R)

Electric field when a voltage V is applied across a length d is

$$\vec{E} = \frac{V}{d}$$

Average speed of charges carriers, $v_{av} \propto \vec{E}$

$$I = n\, q\, v_{av} A$$

$$\frac{V}{I} = \frac{d}{A} \frac{1}{nq\text{(constant)}} = \frac{d}{A}\rho = R$$

26.1.4.1 OHM'S LAW

Good conductors (like metals) obey OHM'S LAW;
They are LINEAR or Ohmic. *i.e.*, $V = I\,R$

Semi-conductors and Thermistors violates OHM'S LAW;
They are NON-LINEAR and non- Ohmic.

Ohm's law is

$$V = IR$$

$$R = \frac{V}{I} \qquad \text{Unit } 1\Omega = 1 \; VA^{-1} \qquad \text{(Scalar)} \qquad (M^{1}L^{2}T^{-3})$$

26.1.4.1.1 Magic Triangle

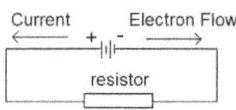

26.1.4.2 Resistance (R) and Resistivity (ρ) of Wire:

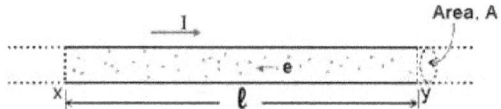

$$R = \frac{V}{I} = \frac{\rho \ell}{A}$$

$$q(t) = VC \, [1 - e^{-t/\chi}]$$

$$\rho = \frac{2m}{q^2 N \Delta \ell} \qquad \text{Unit } \Omega m \qquad \text{(Scalar)} (M^1 L^1 T^3)$$

$$\rho = \frac{1}{4} \pi d^2$$

The Table below shows the resistivity (ρ) of some common metals at $0°C$.

Material	$\rho(\Omega m)$
1) Silver	1.5×10^{-8}
2) Copper	1.7×10^{-8}
3) Aluminium	2.6×10^{-8}
4) Iron	8.85×10^{-8}

26.1.4.3 Conductance G

$$G = \frac{1}{R} = \frac{\sigma A}{\ell}$$

$$G = \frac{I}{V} \qquad \text{Unit } 1\Omega^{-1} \qquad \text{(Scalar)} \qquad (M^{-1} L^{-2} T^3)$$

26.1.4.4 Conductivity, σ

$$\sigma = \frac{1}{\rho}$$

26.1.4.5 Current Density, J

What if the current doesn't flow uniformly in a conductor? Then we should be interested in the **current density** or current per unit area:

$$J = \frac{I}{A}.$$

$$J = \sigma E = \frac{E}{\rho}$$

26.1.5 Temperature ($\theta°C$) dependence of Resistance, R_θ

$$R_\theta = R_0[1 + \alpha \, \theta]$$

$$\rho_\theta = \rho_0[1 + \alpha(\theta - \theta_0)]$$

where α is temp coefficient of resistance.

26.1.6 Drude Model of a Metal (conductor) yields Ohm's Law,

$$\sigma = \frac{n e^2 \tau}{m}$$

where $\sigma = \frac{1}{\rho}$, and τ mean time between collisions.

26.1.7 Resistors in Series

The total resistance between A and B is

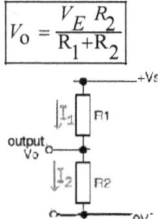

$$R_{eq} = R_1 + R_2$$

For N resistors connected in series, generalizes fairly obviously

$$R_{eq} = \sum_{i=1}^{N} R_i$$

26.1.8 Resistances in parallel

The total resistance between A and B is

$$\frac{1}{R_{eq}} = \frac{I}{V} = \frac{I_1 + I_2}{V} = \frac{I_1}{V} + \frac{I_2}{V} = \frac{1}{R_1} + \frac{1}{R_2}$$

$$\frac{1}{R_{eq}} = \sum_{i=1}^{n} \frac{1}{R_i}$$

26.1.9 Potential Divider (or Voltage Divider)

To obtain a range of potential differences from one Supply, form the circuit shown, by making R_2 a variable resistor. The output

$$V_O = \frac{V_E R_2}{R_1 + R_2}$$

26.1.10 USES OF RESISTORS:

(i)   A resistor limits magnitude of the flow of current, in an active circuit, the larger the R the smaller the I.

(ii)   A resistor in an active circuit has a p.d. $V = IR$ across it, and so it can be placed in a circuit to provide a desired p.d.

(iii)   A resistor can convert electrical energy to heat or other forms of energy.

26.1.11 Electromotive force (EMF) and Resistance

The EMF (ε) is the energy a cell or source imparts in each unit of charge passing through it.

The EMF is equivalent to the pd between the terminals of a cell or 'active source' when in 'open circuit'.

For a complete electrical circuit with a cell (battery) having internal resistance *r the law of conservation of energy* holds good.

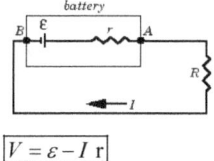

$$V = \varepsilon - I\, r$$

It follows that if one short circuits a battery ε and $R = 0$ the current drawn from the battery is limited by its internal resistance (r), I = the maximum possible current I_0

26.2 ELECTRICAL ENERGY (E) and CURRENT in DC CIRCUITS

$$E = VQ = VIt = I^2 Rt = \frac{V^2}{R}t$$

26.2.1 Power: (P)

$$P = \frac{\text{Energy, } E}{\text{Time,} t} = VI = I^2 R = \frac{V^2}{R}$$

26.2.2. KIRCHHOFF'S RULES:

1) <u>POINT Rule</u>: *The sum of all the currents entering any junction point is equal to the sum of all the currents leaving that junction point,*

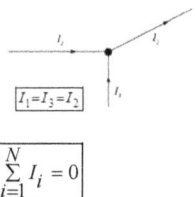

$$\sum_{i=1}^{N} I_i = 0$$

2) <u>LOOP Rule</u>: *The algebraic sum of the changes in electric potential encountered in a complete traversal of any closed circuit (loop) is equal to zero.*

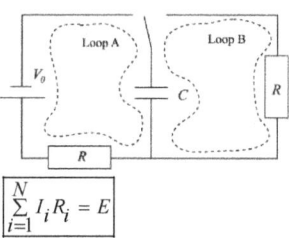

$$\sum_{i=1}^{N} I_i R_i = E$$

26.2.2.1 Uses of Kirchhoff's rules:

They are used in the semiconductor industry to analyze the incredibly complicated circuits, etched onto the surface of silicon wafers, which are used to construct computer CPUs.

26.3.1 **RC Circuits**:
26.3.1.1 Charging RC circuit:

The time dependence of charge q(t)

$$q(t) = EC \, [1 - e^{-t/\tau}]$$

$$q(t = 0) = Q_0 \, [1 - e^{-t/RC}]$$

time constant, $\tau = RC$

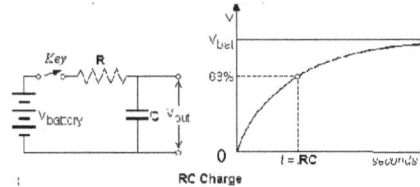

RC Charge

26.3.1.2 Discharging RC Circuit:

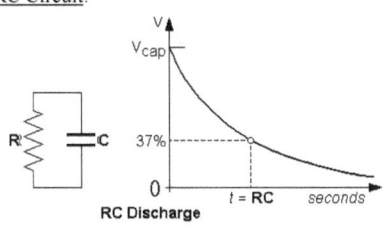

RC Discharge

$$q(t) = Q_0 \, e^{-t/RC}$$

26.4 METHODS OF MEASUREMENT OF RESISTANCE

(1) Ammeter – Voltmeter method
(2) Ohmmeter: Substitution method
(3) D.C. Potentiometers
(4) The Wheatstone bridge
(5) The Metre bridge
(6) Kelvin's Double Bridge

26.4.1. D.C. Potentiometers

Potentiometer is a device used to measure the internal resistance of a cell, to compare the e.m.f. of two cells and potential difference across a resistor. It consists of a long wire of uniform cross sectional area and of 10 m in length. The material of wire should have a high resistivity and low temperature coefficient. The wires are stretched parallel to each other on a wooden board. The wires are joined in series by using thick copper strips. A metre scale is also attached on the wooden board.

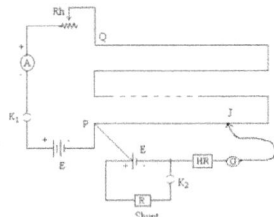

26.4.1.1. Determination of Resistance of a Potentiometer (10 m length)

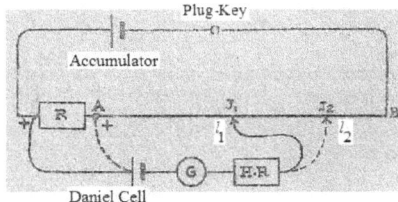

Resistance of potentiometer wire $= \dfrac{R}{l_2 - l_1} 1000 \ \Omega$.

26.4.1.2. Calibration of potentiometer

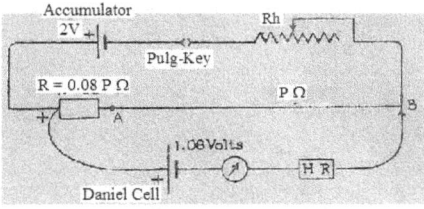

26.4.1.3. Calibration of Low Range Voltmeter

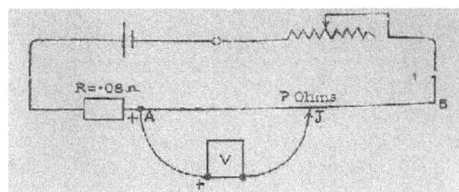

26.4.1.4. Given below in Fig. show the circuit arrangements in using a slide-wire potentiometer to
   a) compare the e.m.f. of two cells,
   b)   compare two resistances,
   c)   calibrate a voltmeter of f.s.d. greater than that of the supply voltage across the potentiometer,

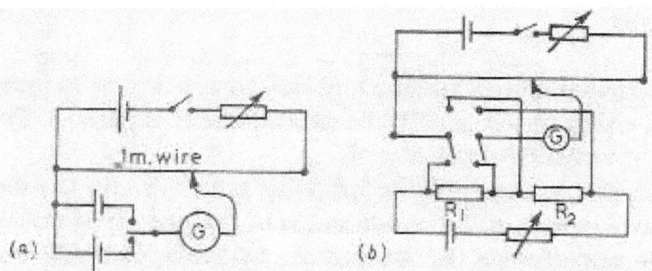

   d)   calibrate an ammeter,
   e)   measure the internal resistance of a cell, and
   f)   measure a very small e.m.f..

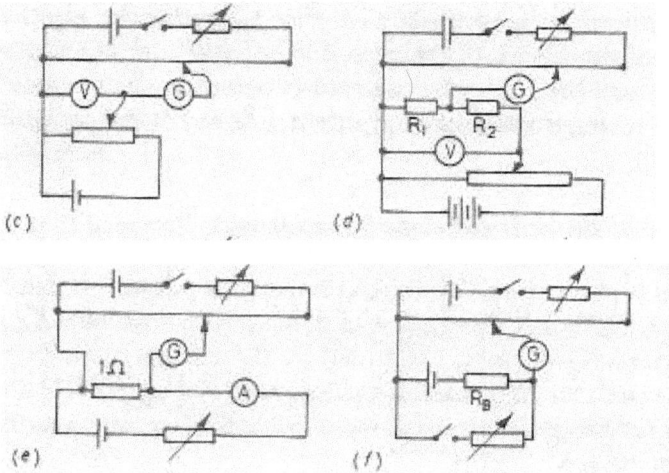

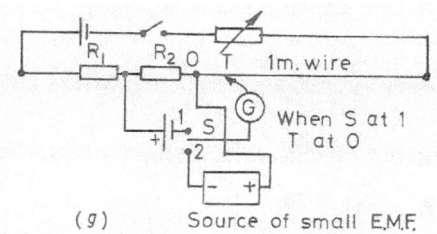

(g)　　　　Source of small E.M.F.

26.4.2　WHEATSTONE BRIDGE

The bridge circuit consists of four resistors, R_1, R_2, R_a R_x, a battery and a galvanometer, G. Wheatstone called the circuit a "Differential Resistance Measurer". Decade box (A resistance standard variable in discrete steps) or Slide wire resistance find as one of the 4 resistors.

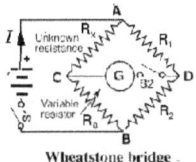

Wheatstone bridge

Bridge circuit is balanced when there is null current through G. Then

$$\boxed{\dfrac{R_a}{R_x} = \dfrac{R_1}{R_2}}$$

The unknown resistance value R_x can thus be determined.

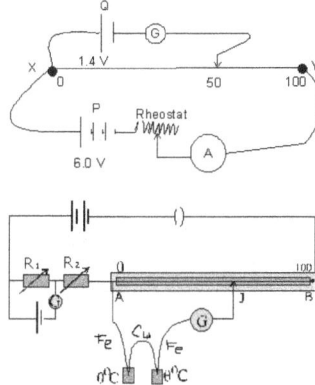

Today, Wheatstone bridge circuits are not usually used to measure resistance values, but they *are* used in designing sensor circuits.

A **variometer** is an instrument used in gliders to detect changes in air pressure due to sudden changes in altitude. One type of variometer uses thermistors to monitor pressure changes: A heating element in the flow passage heats air which arrives at different temperatures at a thermistor sensor upstream and downstream of the heating element depending on the rate of air flow.

28.4.2.1 Maxwell-Wein Bridge

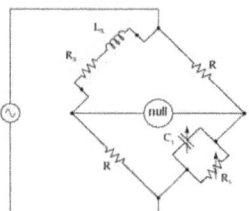

The Maxwell-Wien bridge

26.4.2.2. Temperature coefficient of Resistance by <u>Carey Foster's Bridge</u>

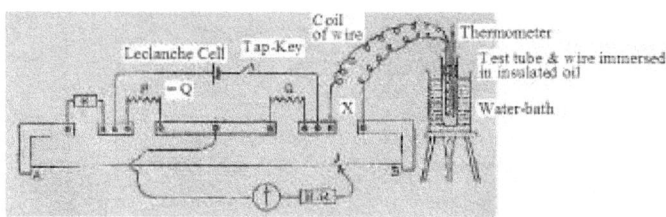

$$X = R\frac{l_2}{l_1} \ \Omega$$

Temperature coeff. of resistance, $\alpha = \dfrac{X_2 - X_1}{X_1\theta_2 - X_2\theta_1} \ \Omega/°C$

26.4.3 METRE BRIDGE

The Metre Bridge is an instrument based on the Wheatstone bridge circuit. It is generally used to determine the value of an unknown resistance.

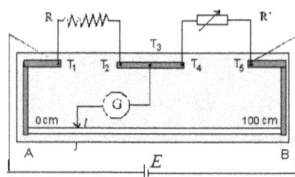

R is the unknown resistance

R' is a resistance box
J is a jockey key
G is a galvanometer.
AB is a one metre wire fixed on a wooden ruler of length 1.0 m. Each end of the wire AB is soldered to an 'L' shaped strip of brass at the other end of which is the resistance R (or R')

$$\frac{R}{R'} = \frac{(\rho\ell/A)}{\rho(100 - \ell)/A}$$

$$\boxed{\frac{R}{R'} = \frac{\ell}{(100 - \ell)}}$$

The advantage of this method over the Ohm's Law method of determination of unknown resistance lies in the fact that it is far more accurate.
Since the smallest measurable value on a metre ruler is 1 mm, the %-age error would be of the order of $\boxed{\frac{1\ mm}{333\ mm} 100\% \approx 0.3\%}$.

26.4.4. Kelvin's Double Bridge

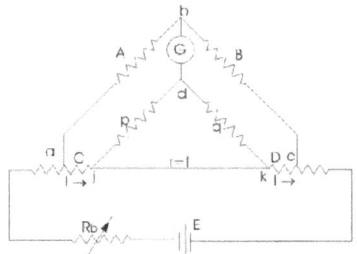

$$C = D\frac{A}{B}$$

+*+*+*+*+*+*+*

Chapter 27

MAGNETISM

Biot-Savart Law, Straight wire current, Solenoid,
Force between two conductors, Gauss's Law, Lenz's Law,
DC & AC Generators, Galvanometers

The important thing in science is not so much to obtain new facts as to discover

new ways of thinking about them. ~William Lawrence Bragg

27.1 SOURCES OF MAGNETIC FIELD $\vec{B}$ ((MAGNETIC EFFECTS of a Current):

27.1.1 Electric Current – Its Role.

(!) | A stationary electric charge produces an electric field $\vec{E}$ in the region around it. |

(2) | A Moving Charge produces a Magnetic Field H in that vicinity! |

(3) | Thus any current I will have an associated magnetic field $\vec{H}$ | .

27.1.2 Laplaces Law

$$\delta\vec{F} = \frac{i \, d\ell \, \text{Sin} \, \theta}{x^2}$$

27.1.3 Biot-Savart Law
* Electric current or moving charges are source of magnetic field.

* A Small current carrying conductor of length $d\ell$ (length element) carrying current i is a elementary source of magnetic field .The force on another similar conductor can be expressed conveniently in terms of magnetic field dB due to the first.

* The dependence of magnetic field dB on current i, on size and orientation of the length element $d\ell$ and on distance r was first guessed by Biot and Savart.

* The magnitude of the magnetic field dB at a distance r from a current element $d\ell$ carrying current i is found to be proportional to i, to the length $d\ell$ and inversely proportional to the square of the distance $|\vec{r}|$.

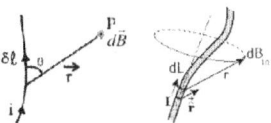

* The direction of the magnetic Field is perpendicular to the line element $d\ell$ as well as radius a.

* Mathematically, Field dB is written as

$$\delta\vec{B} = \left(\frac{\mu_0}{4\pi}\right)\left[\frac{i\ d\ell\ \sin\theta}{r^2}\right]$$

$$\mu_0 = 4\pi \times 10^{-7}\ Hm^{-1}$$

$$\vec{H} = \left(\frac{I}{4\pi}\right)\oint\left[\frac{d\ell\ \hat{r}}{r^2}\right]$$

27.1.4   RH Cork Screw Rule (Finger Current Theorem) for magnetic field used to find the direction of the $\vec{H}$.

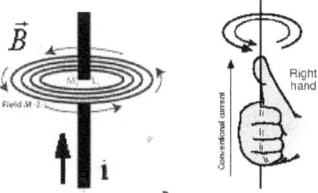

27.1.5   The intensity of the $\vec{H}$ due to an infinitely long straight conductor with current i is

$$\vec{H} = \frac{2i}{a}\ Oersted$$

$$i\ emu\ current = 10I\ Amp$$

$$\vec{B} = \frac{2I}{10\ a}\ Oersted$$

$$\vec{B}_{Earth} = 5 \times 10^{-5}\ Tesla .$$

Infinite wire	$\vec{B} = \frac{\mu_0 i}{2\pi a}$
Semi-infinite wire	$\vec{B} = \frac{\mu_0 i}{4\pi a}$
Wire segment	$\vec{B} = \frac{\mu_0 i}{2\pi a}(\sin \vartheta_i + \sin \delta)$
Arc wire	$\vec{B} = \frac{\mu_0 i}{4\pi a}\varphi$
Full wire circle	$\vec{B} = \frac{\mu_0 i}{2a}$

27.2. Magnetic field of Current loop

Intensity of $\vec{H}$ field along the axis of a circular coil carrying current

27.2.1   At point P distant x

$$\vec{F} = \frac{2\pi n a^2 i}{(a^2 + x^2)^{3/2}} = \frac{2\pi n a^2 I}{10(a^2 + x^2)^{3/2}} \quad Oersted$$

n = # of turns in the coil of radius a.

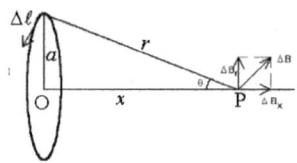

27.2.2   At the centre of the coil

$$\vec{F} = \frac{2\pi n\, i}{a} = \frac{2\pi n I}{10\, a} \quad Oe$$

$$I = \frac{10a}{2\pi n} F \quad Amp$$

27.3.1   AMPERE'S FORCE LAW

The magnetic field $\vec{B}$ due to a group of current carrying conductors is given by the law:

$$N\, i = \oint \vec{H}.d\ell \quad , \text{ or } \oint \vec{B}.d\ell = \mu_0 I$$

where N is the number of conductors carrying a current i and ℓ is a line vector. The integration must form a closed line around the current.

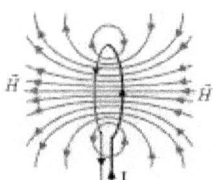

The field is circular and concentric with the current so B can be integrated around the current at a distance r to give in situations with enough symmetry.

Ampere's law alone can be used to find the magnitude of **B**. The flux of **B** through any closed surface is zero.

$$\int_A \vec{B}.\hat{n} \, dA = 0$$

One *Ampere* is the magnitude of the current which, when flowing in each of two long parallel wires one meter apart, results in a force between the wires of exactly $2x10^{-7}$ N per metre of length

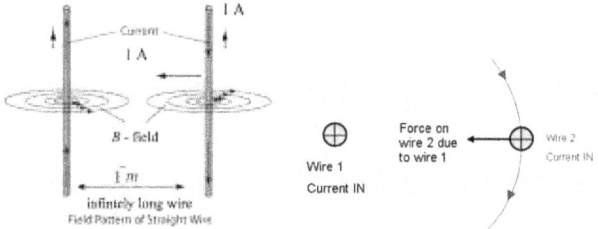

27.3.2 Intensity of $\vec{H}$ at a point on the axis of a Solenoid

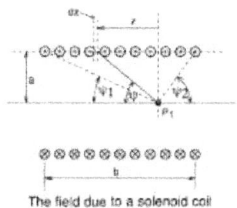

The field due to a solenoid coil

a) At a point on the axis

$$\vec{F} = 4\pi n \; i = \frac{4\pi n \, I}{10} \; Oe$$

b) At the end of the Solenoid

$$\vec{F} = 2\pi n \; i = \frac{2\pi n \, I}{10} \; Oe$$

27.3.3 Force on a conductor carrying a current, when placed in a Magnetic Field $\vec{H}$.

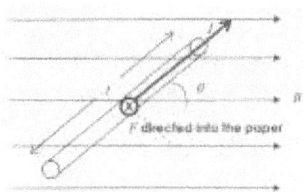

$$F = BI\ell \sin\theta$$

$$\boxed{\vec{F} = q\ \vec{v}\Lambda\vec{B}},$$

$$\boxed{i\ell = q\ \vec{v}}$$

27.4. MAGNETOSTATICS

27.4.1 Magnetic dipole moment $\vec{m}$ of a coil with current

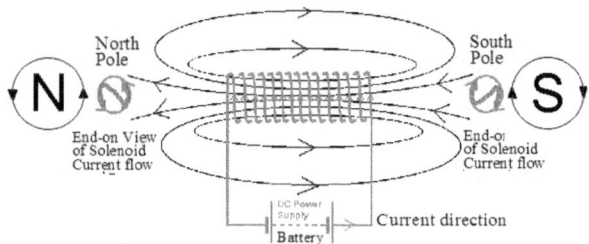

Magnetic Dipole moment of a Coil carrying current

$$\boxed{\vec{m}\ =\ IA\hat{n} = \frac{1}{2}\int_V \vec{r}\Lambda\ \vec{j}(r)\ dV}$$

27.4.2 Force on a dipole

is $$\boxed{\vec{F}\ =\ \vec{\nabla}(\vec{m}.\vec{B})}$$

$$\boxed{F = p\ H}, \quad \boxed{i = \frac{dq}{dt}}$$

27.4.3 For a bar magnet,

A bar magnet having pole strength, $\vec{p}$, and length, d

$$\boxed{\vec{m} = \vec{p}.\vec{d}}$$

$$\boxed{W = \vec{m}.\vec{H}}$$

$$\boxed{\vec{F} = \frac{\mu p_1 p_2}{4\pi r^2}}\ dynes$$

$$\boxed{\vec{F} = \vec{H}\tan\theta}$$

27.4.4 <u>Torque on a magnetic dipole in a uniform B field is</u>

$$\vec{\tau} = \vec{m} \wedge \vec{B}$$

27.4.5 **Force on a current carrying straight conductor**

$$\vec{F} = \int i d\vec{\ell} \wedge \vec{B}$$

$$\boxed{\vec{F} = i\vec{\ell}\wedge\vec{B}}$$

27.4.5.1. **Force between two straight conductors carrying current.**

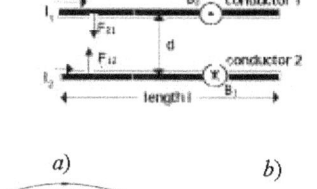

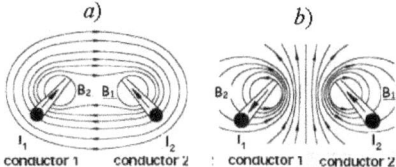

a)   b)

conductor 1   conductor 2   conductor 1   conductor 2

Consider

(a) wire 1 will produce a field B_1 at all nearby points .The magnitude of B_1 due to current I_1 at a distance d, *i.e.* on wire b is

$$\vec{B}_1 = \frac{\mu_o I_1}{2\pi d}$$

c) According to the RH rule the direction of $\vec{B}_1$ is in downward as in fig (a)

d) Consider length ℓ of wire 2 and the force experienced by it will be ($I_2\ell\,\vec{B}_1$) whose magnitude i

$$\vec{F}_2 = I_2 \ell\,\vec{B} = \frac{\mu_o \ell I_1 I_2}{2\pi d}$$

e) Direction of $\vec{F}_2$ can be determined using vector rule. $\vec{F}_2$ lies in the plane of the wires and points to the left,

f) From fig (b) we see that direction of force is towards A if I_2 is in same direction as I_1 fig(a) and is away from 1 if I_2 is flowing opposite to I_1 (fig b).

g) Conclusion is that the conductors attract each other if the currents are in the same direction and repel each other if currents are in opposite direction.

27.4.5.2 Gradient coils

The requirement of the gradient coils is twofold. First they are required to produce a linear variation in field along one direction, and secondly to have high efficiency, low inductance and low resistance, in order to minimize the current requirements and heat deposition.
Maxwell coils used to produce a linear field gradient in B_z along the z-axis.

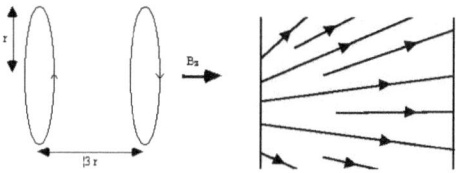

27.4.5.3 Potential Energy of a magnetic dipole in an external B field

$$\boxed{U_{mech} = - \vec{m}.\vec{B}}$$

27.5 MOTION OF CHARGES in EM Fields

The combined force on q by $\vec{E}$ and $\vec{B}$ is

$$\vec{F} = q(\vec{E} + \vec{v} \wedge \vec{B})$$

The speed of the particle with mass m and charge q and the radius of the circular path perpendicular to $\vec{B}$ is

$$\boxed{q\vec{v}\,\vec{B} = \frac{m\,v^2}{r}}$$

$$\boxed{r = \frac{m\,v}{qB}}$$

In a crossed $\vec{E}$ and $\vec{B}$ velocity selector set up, charged particle enter the region with v perpendicular to both $\vec{E}$ & $\vec{B}$ and pass through undeviated, .if

$$\boxed{\vec{v} = \frac{\vec{E}}{\vec{B}}}$$

The magnitude of the strength of $\vec{B}$ at any point will depend on (i) the value of I, (ii) or rate of flow of charge, (iii) the distance of the point from the conductor and (iv) the medium.

27.5.2 Materials for Permanent Magnets:

Certain ferromagnetic materials like, steel or alloys of iron, nickel and cobalt, such as alnico. The circulation and spinning moments of the electrons in the atoms are responsible for this magnetism.

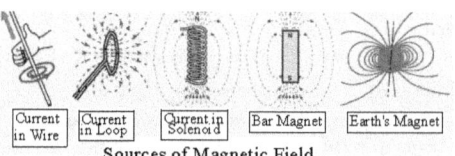

Sources of Magnetic Field

27.5.3 Right Hand Corkscrew Rule (Fleming's RH rule)
The direction of B is given by Fleming's RH rule, illustrated below.

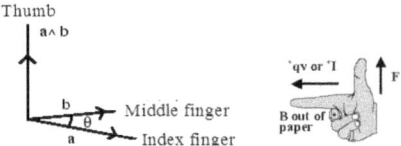

The magnitude of the strength of $\vec{B}$ at any point will depend on
- (i)   the value of I,
- (ii) or rate of flow of charge,
- (iii) the distance of the point from the conductor and
- (iv) the medium.

27.6   GAUSS'S LAW for Magnetic field:

$$\oiint \vec{B} \cdot d\vec{S} = 0$$

The magnetic flux Φ for any closed surface is zero, corresponding to the absence of magnetic free charges (monopoles). A line representing $\vec{B}$ closes on itself.

Magnetic flux density has the unit Tesla (T),

Magnetic Flux	Unit	$1\ Tesla = 1\ Wbm^{-2} = 1\ NA^{-1}m^{-1}$

The displacement current and Ampere's law:
The modified form of Ampere's law is

$$\oint \vec{B} \cdot d\vec{s} = \mu_0 I_{cnc} + \mu_0 \varepsilon_0 \frac{d\Phi_M}{dt}$$

$$\oiint \vec{B} \cdot d\vec{S} = \mu_0 \left(\Sigma i + \varepsilon_0 \frac{d\Phi_B}{dt} \right)$$

where $i_d = \varepsilon_0 \dfrac{d\Phi_M}{dt}$ is the displacement current.

27.7   **FARADAY'S LAW** *OF ELECTROMAGNETIC INDUCTION*
(FORCE ON A CONDUCTOR IN A MAGNETIC FIELD OF A MAGNET)
The EMF ε induced in a circuit is proportional to the time rate of change of the magnetic flux Φ_B linking that circuit

$$\varepsilon = \frac{\Delta\Phi_B}{\Delta t}$$

27.7.1   LENZ'S LAW

The EMF ε induced in an electric circuit always acts in such a direction that the current it drives around the circuit opposes the change in magnetic flux Φ_B which produces the EMF This result is known as *Lenz's law*.

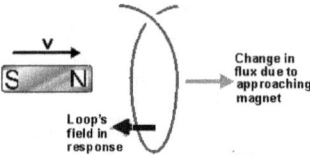

The induced E is caused by a changing magnetic field according to Faraday's Law:

$$emf = \oint \vec{E} \cdot d\vec{\ell} = -\frac{d}{dt} \oint \vec{B} \cdot d\vec{A}$$

Faraday's law can be expressed in terms of this EMF:

$$emf = \oint \vec{E} \cdot d\vec{\ell} = -\frac{d}{dt} \oiint \vec{B} \cdot d\vec{S} = -\frac{d\Phi_B}{dt}$$

27.7.2   MOTIONAL EMF ε

An EMF is generated around a *fixed* circuit placed in a time varying magnetic field, $\vec{B}$. But, according to Faraday's law, an EMF is also generated around a *moving* circuit placed in a magnetic field which does not vary in time. This electric field is the ultimate origin of motional EMF, A motional EMF is induced if part or all of a circuit moves through a region of B field. In a sliding wire circuit the EMF ε is

$$\varepsilon = -\vec{B}\ell\frac{dA}{dt} = \frac{B}{v}$$

$$\varepsilon = \frac{\Delta\Phi_B}{\Delta t} = \vec{B}\ell v$$

27.7.3   GENERATORS & ALTERNATORS:

A Generator is a device which converts mechanical energy into electrical energy.

$$\Phi = \vec{B}A$$

$$\Phi = \int \vec{B} \cdot dA$$

where Φ is the flux (Wb), B is the field strength (T) and A is the area (m²)

From Faraday's law the EMF induced in a rotating coil generator is

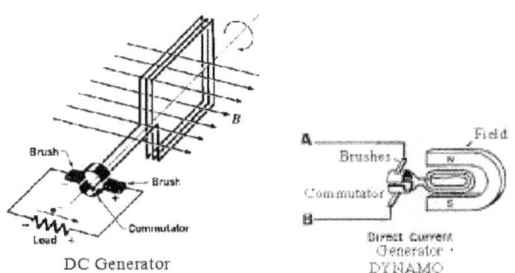

$$\Phi_B = \vec{B}A\ Cos\theta = \vec{B}A\ Cos\omega t$$

$$emf, \varepsilon = -N\frac{d\Phi_B}{dt} = N\vec{B}A\omega\ Sin\omega t$$

$$\varepsilon_{Max} = NAB\omega$$

$$\varepsilon_{Max} = 2\pi BAf$$

where f is the frequency of revolution

Magnetic flux Φ density has the unit Tesla (T),

$$\text{Unit} \quad 1\ Tesla = 1\ Wbm^{-2} = 1\ NA^{-1}m^{-1}$$.

$$\varepsilon_{Max} = 2NvB\ell$$

where v is the velocity of rotation

27.7.4   DC GENERATORS (DYNAMOS)

A DC generator is essentially a DC motor in which the coil is rotated by an external force, causing electricity to be produced. The commutator reverses the connections to the coil every half cycle, producing a pulsating DC output.

DC Generator

27.7.5 AC GENERATORS (ALTERNATORS)

A.C Generator consists of armature: slip rings: and Carbon brushes:

An AC generator has a rotating coil in a magnetic field, or a rotating magnetic field positioned inside a coil. Instead of a commutator, slip rings are used to keep contact with the brushes and the direction of the induced EMF changes every half cycle - an AC output is produced.

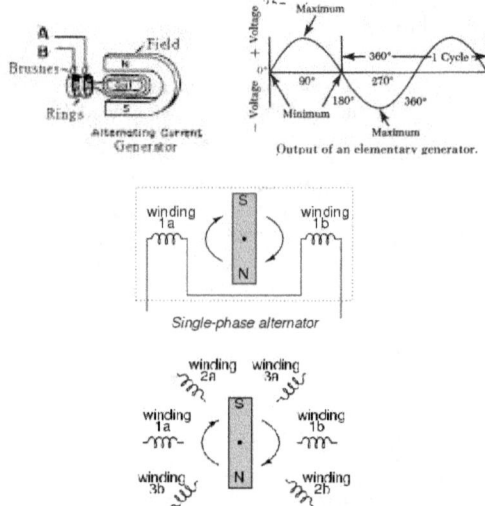

Output of an elementary generator.

Single-phase alternator

Three phase alternator

The output varies *sinusoidally* and has a maximum value of

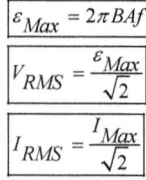

$$\varepsilon_{Max} = 2\pi BAf$$

$$V_{RMS} = \frac{\varepsilon_{Max}}{\sqrt{2}}$$

$$I_{RMS} = \frac{I_{Max}}{\sqrt{2}}$$

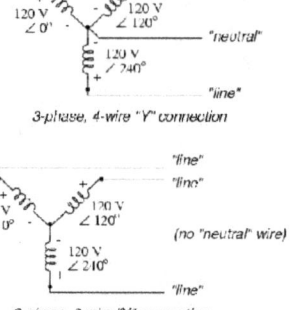

3-phase, 4-wire "Y" connection

3-phase, 3-wire "Y" connection

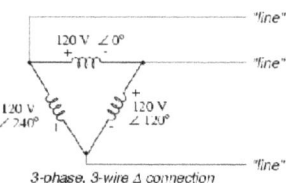

3-phase, 3-wire Δ connection

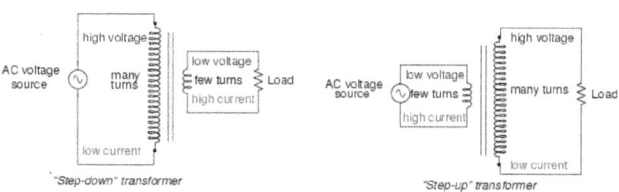

"Step-down" transformer   "Step-up" transformer

27.7.6 <u>TORQUE</u> on a current carrying wire loop in B field

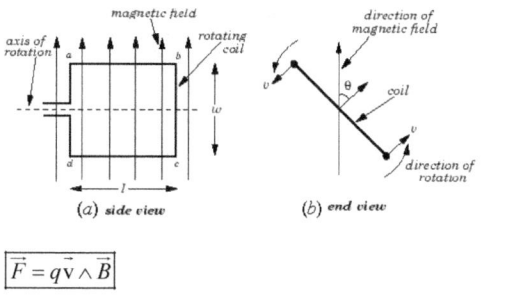

(a) side view   (b) end view

$$\vec{F} = q\vec{v} \wedge \vec{B}$$

27.8. **GALVANOMETERS (G)**

An analog measuring device, denoted by G, that measures current flow I using a rotation of needle deflection caused by a $\vec{B}$ (magnetic) field force $\vec{F}$ acting on either side of a current-carrying wire. The torque produced causes the coil to rotate. This turning effect resisted by the attached spring attached to the spindle and the pointer gets deflected on a scale.

27.8.1 <u>Moving Coil Galvanometer</u> (Jacques D'Arsonval-Marcel Deprez, 1880s)

I = main current,

k = constant of the galvanometer, is the rotating torque per unit angkle per unit twist of the strip

θ =deflection of the mirror

R_G =Resistance of the galvanometer

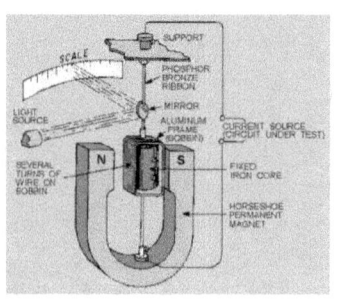
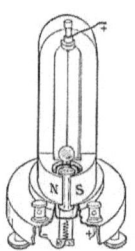

Restoring torque $T_r = k\theta$

Deflecting torque $T_d = BINACos\theta$

$T_r = T_d$ gives $I = \dfrac{k}{BNA}\theta$

This means $I \propto \theta$

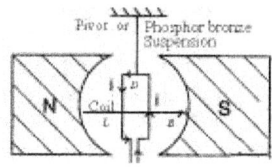

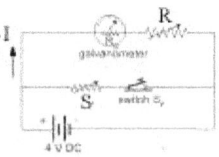

$$i = \left(\frac{c}{nAH}\right)\frac{\theta}{Cos\theta}\, emu$$

$$i = I\left(\frac{S}{(R_G + R + S)}\right) = k\theta$$

As $S \ll R_G$ and R, $\boxed{\dfrac{IS}{k\theta} = R_G + R}$.

27.8.1.1. <u>Comparison of High Resistances</u> by G

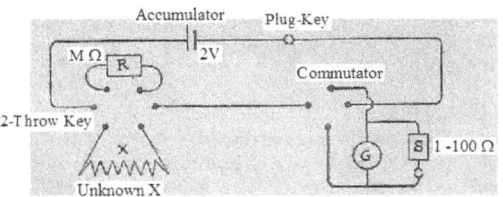

d_1 = defelection G when X is not included,

d_2 =deflection in G when X is included.

$$X = R\frac{d_2}{d_1} \; \Omega .$$

27.8.1.2. <u>Comparison of Low Resistances</u> by G

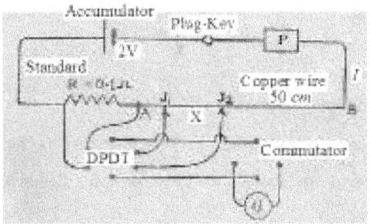

$$IR \propto d_1 \; ; IX \propto d_2 ,$$

$$X = R\frac{d_2}{d_1} .$$

Specific resistance of the wire of radius r,

$$\rho = \frac{X\pi \, r^2}{\text{liength of wire}} \; \Omega/cm^3 .$$

27.8.2. <u>Tangent Galvanometer</u> (T.G.)

Large diameter vertically mounted multi-tapped coil surrounding a horizontally mounted magnetic compass. The amount of current in the coil is proportional to the tangent of the angle of deflection of the compass. $I \propto \tan\theta$.

27.8.3 Ballistic Galvanometer (B.G.)

A normal galvanometer measures current. But a B.G measures charge due to impulse in the coil (sudden flow of charges for a short interval of time). A ballistic galvanometer is used to measure flux and current and is a type of mirror galvanometer. On this type of devise the moving part has a large moment of inertia (compared to current-measuring galvanometers) and, as such, has a long oscillation period. The coil is wound on a **bamboo** core. The suspension material is quartz or phosphor bronze.

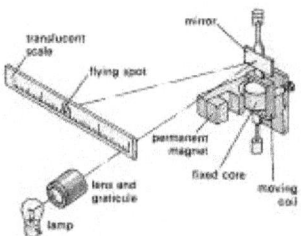

27.8.4.1 Figure of merit of B.G.

It is the amount of current required to produce a deflection of 1 division. This is current sensitivity of the G.

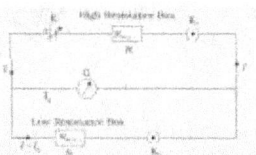

Figure of merit $K = \dfrac{E}{(R+G)\,\theta\, I} \mu C / \text{scale division}$.

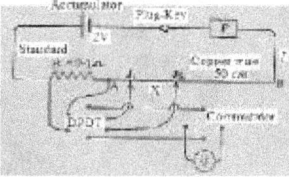

$$K = \frac{EP}{(P+Q)} \frac{1}{(R+G)} \frac{1}{d} \cdot 10^6 \ \mu A/cm$$

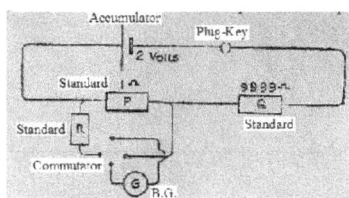

$$K = I \frac{EP}{(P+Q)} \frac{1}{(R+G)} \frac{1}{\theta} .10^6 \quad \mu Coul / \text{scale division}.$$

where if damping correction to θ is added, $\theta = \theta_1 (\frac{\theta_1}{\theta_2})^{1/4}$.

27.8.5. Dead Beat Galvanometer

When current is passed through a galvanometer, the coil oscillates about its mean position before comes to rest. For the coil to come to rest immediately, it is wound on a *metallic* frame. When the coil oscillates, eddy currents are set up (being in a metallic frame), which opposes further oscillations of the coil, thereby enabling the coil to attain its equilibrium position almost instantly. Since the oscillations of the coil die out instantaneously, the galvanometer is called <u>dead beat</u> galvanometer.

$$q = \frac{T}{2\pi} \left(\frac{c}{nAH} \right) \theta \ emu$$

$$q = \frac{T}{2\pi} \left(\frac{i}{\alpha} \right) \theta \ emu$$

α = steady deflection when a steady current passes.

Correction for damping

Decrement, $\dfrac{\theta_1}{\theta_2} = \dfrac{\theta_2}{\theta_3} = \dfrac{\theta_3}{\theta_4} = ... = d ; \quad \lambda = \ln d ; \quad \dfrac{\theta_s}{\theta_n} = e^{(n-s)\lambda}$

From 1st and 11th throw, $\dfrac{\theta_1}{\theta_{21}} = e^{20\lambda} ;$

Logarithmic decrement, $\lambda = \dfrac{1}{20} \ln(\dfrac{\theta_1}{\theta_{21}})$

$$q = \frac{T}{2\pi} \left(\frac{c}{nAH} \right) \theta (1 + \tfrac{\lambda}{2}) \ emu$$

$\theta^c = \dfrac{\theta}{x}$, x = distance between mirror and scale.

27.8.6.1. Constant of a B.G.

a) Hibbert's magnetic standard

It is a permanent magnet of hard steel in the shape as shown in Fig. The two Poles are separated by a narrow cylindrical annular air-space. A coil of insulated copper wire of 10 or 20m turns wound over a thin brass tube can feely drop through the air-space between the two Poles. Thus an induce e.m.f. is produced.

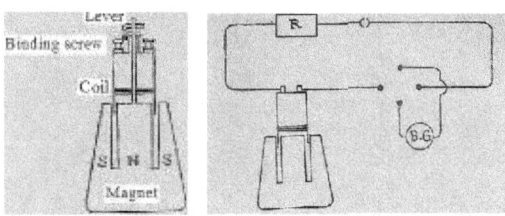

Induced charge, $=\dfrac{Nn}{(R+G)10^8} Coul$,

n= # turns in the coil, N= magnetic flux of the Hibbert's standard.

Constant of the B.G. , $K=\dfrac{Nn}{(R'-R)10^8}[\dfrac{1}{\theta'}-\dfrac{1}{\theta}]$.

b) Solenoidal Inductor Method

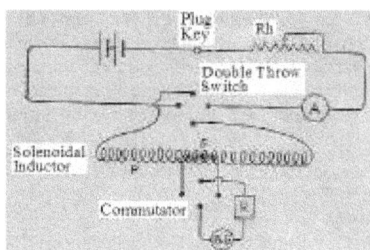

$$K=\dfrac{8\pi n\,p^n S\,AI}{(R'-R)10^9}[\dfrac{1}{\theta'}-\dfrac{1}{\theta'}]\ Coul/cm.$$

27.8.7.1. Galvanometer with shunt as an Ammeter

At the heart of most analog meters is a galvanometer, an instrument that measures current flow using the movement, or deflection, of a needle. The needle deflection is produced by a magnetic force acting on a current-carrying wire.

Since Galvanometer is a very sensitive instrument it can't measure heavy currents. To convert it into an Ammeter, a very low resistance known as "shunt" resistance R_S is connected in parallel to Galvanometer. Value of shunt is so adjusted that most of the current passes through the shunt R_S. In this way a Galvanometer is converted into Ammeter and can measure heavy currents without fully deflected.

$$\boxed{R_S=\dfrac{I_G}{I-I_G}R_G}$$

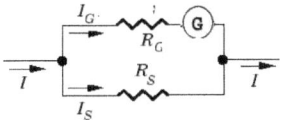

27.8.7.2.   Galvanometer with a multiplier in Series as a Voltmeter

Since Galvanometer is a very sensitive instrument,it can not measure high potential   difference. To convert it into voltmeter, a very high resistance R_X known as "series   resistance" is connected in series with the galvanometer.

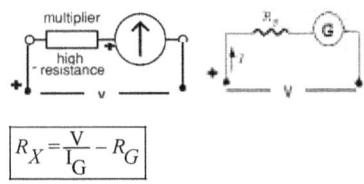

$$R_X = \frac{V}{I_G} - R_G$$

I_G = current flow for full scale deflection

$$I_G = \frac{V}{R_X + R_G}$$

$$R_X = \frac{I_G}{I - I_G} R_G$$

27.8.8.1. SENSITIVITY OF GALVANOMETERS:

(i) Current sensitivity: (θ / I)

Torque (C) due to B-field on a coil of N turns and cross section A when I flows

$$\vec{C} = \vec{B}ANI$$

The opposing torque on the suspension of torsion constant (k) giving a twist θ is $k\theta$:

$$(\theta / I) = BAN / k$$

(ii) Voltage sensitivity (θ / V)

$$\theta / V = BAN / kR_G$$

27.8.8.2. Measurement of the Resistance (R) of a G

Wheatstone bridge method.

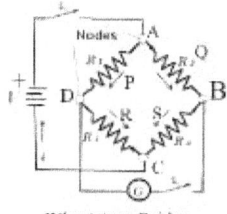

Wheatstone Bridge

It is a four arm bridge circuit, consisting of electrical resistors P, Q, R and S. P and Q have known fixed values in the ratio arms. S is a variable resistor. The galvanometer shows null when the bridge is balanced.

Adjust resistance S until the defection of G is unaltered by making ON and OFF in K.

Then $\boxed{R = \dfrac{P}{S}Q}$.

27.8.9. Electrodynamometer:

An instrument that measures large amount of electric current by indicating the strength of repulsion or attraction between the magnetic fields of two sets of coils, one fixed and one movable, whereas Galvanometer is an instrument for detecting and measuring small electric currents. It is an instrument used for measuring the electric power.

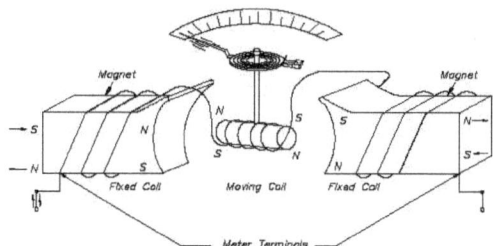

It has the same basic operating principle as the D'Arsonval meter movement, except that the permanent magnet is replaced by fixed coils (See Fig.). It can be used in both AC and DC systems to measure current. Some voltmeters and ammeters use the electrodynamometer. However, its most important use is in the Wattmeter.

27.9.1 Faraday's Law of Electrolysis

Michael Faraday (1833) enunciated this law. Mathematically, it can be expressed as

$$\boxed{Q \propto \dfrac{z\,m}{M}}$$

$$\boxed{Q = It = zFn}$$

where Q = electrical charges passed, t = time for which I current passed,

z = Electro-chemical equivalent (change in oxidation state), m and M are mass and Molar mass of oxidized and reduced species, $\boxed{F = 96485\ C\ mol^{-1}}$ = Faraday constant, the charge of 1 mole of electrons. n = amount of substance oxidized / reduced.

$$\boxed{m = z\,I\,t}$$

27.9.2. ELECTROLYSIS

The process of using electricity to split water into Hydrogen and Oxygen.

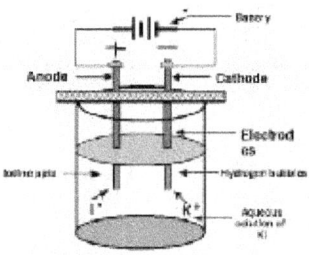

$$2H_2O(l) \rightarrow 2H_2(g) + O_2(g)$$

Note the 2/1 ratio of Hydrogen and Oxygen water.
Electrolysis is a way of <u>removing individual hairs</u> from face or body, though painful.

If one reverses the flow of electricity within a voltaic cell by exceeding a required voltage,
it results in an electrolytic cell)

Voltaic	Electrolytic
Oxidation: $X \rightarrow X^+ + e^-$ (Negative Anode)	$Y \rightarrow Y^+ + e^-$ (Positive Anode)
Reduction: $Y^+ + e^- \rightarrow Y$ (Positive Cathode)	$X^+ + e^- \rightarrow X$ (Negative Cathode)
Overall: $X + Y^+ \rightarrow X^+ + Y$ (G < 0)	$X^+ + Y \rightarrow X + Y^+$ (G > 0)
This reaction is spontaneous and will release energy	This reaction is non-spontaneous and will absorb energy

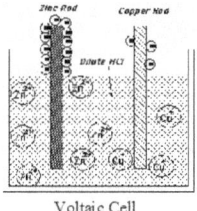

Voltaic Cell

27.9.3. ELECTROPLATING

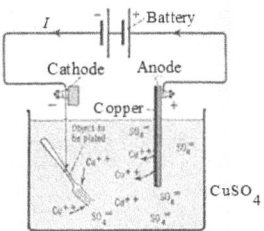

Electroplating is the process of applying one or more layers of a metal,.say gold, to a object by passing a positively charged electrical current through a solution containing dissolved metal ions (anode) and a negatively charged electrical current through object to be plated (cathode).

The electrolyte is copper sulphate, $CuSO_4$.

Cadmium, Copper, Gold, Hard Chrome, Nickel, Silver, Tin, Tin-Lead, Zinc, Zinc-Iron, Black Nickel, Black Chrome, Rhodium are types of electro-plating.

Features of electroplating are Corrosion resistance, Wear Resistance, Appearance, Lubricity, Solderability.

+&+&+&+&+&+&+

Chapter 28

ALTERNATING CURRENT

Inductance, DC RL circuit, Transformer,
Dia-, Para-, Ferro-magnetism, Resonant Circuits

Science is facts; just as houses are made of stones, so is science made of facts;

but a pile of stones is not a house and a collection of facts is not necessarily science. ~Henri Poincaré

28.1 INDUCTANCE, L

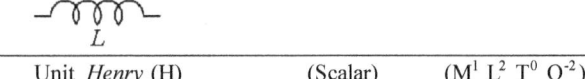

L	Unit *Henry* (H)	(Scalar)	$(M^1 L^2 T^0 Q^{-2})$

Self-induced EMF ε and Self inductance L

If the current changes in a **circuit element** such as a coil, then a self induced emf exists:

28.1.1 Faraday's Law,

It states that a voltage is induced in a conductor when that conductor is moved through a magnetic field, $\vec{H}$ or when the magnetic field moves past the conductor. When the EMF is induced in Wire B, a current will flow whose magnetic field opposes the change in the magnetic field that produced it

For this reason, an induced EMF is sometimes called *counter EMF (or CEMF).*

28.1.2 Lenz's Law,

It states that the induced EMF opposes the EMF that caused it three requirements. Induced EMF in Coils inducing an EMF are

1. A conductor,

2. A magnetic field, $\vec{H}$ and

3. Relative motion between the two faster the conductor moves, or the faster the agnetic field collapses or expands, the greater the induced EMF

The induction can also be increased by coiling the wire in either Circuit A or Circuit B, or both, as shown.

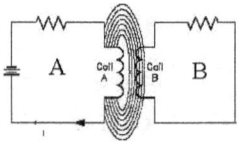

28.1.2 Self induced EMF: ε

A coil of wire is called an inductor (L). As current flows through the circuit, a large magnetic field $\vec{H}$ is set up around the coil. Since the current is not changing, there is no EMF produced. If we open the switch, the field around the inductor collapses. This collapsing magnetic field produces a voltage in the coil. This is called self-induced EMF ε

An inductor tends to oppose a change in current flow.

$$\varepsilon = -L\frac{di}{dt}$$

$$\varepsilon_{ind} = -L\frac{d(N\Phi_B)}{dt}$$

The self-inductance L depends on the geometry of the element. For a coil, of N turns

$$N\Phi_B = Li$$

$$L = -\frac{\varepsilon_{ind}}{di/dt}$$

where Φ_B is the flux linking each of the N turns of the coil.

28.1.3 L of a long tightly wound solenoid (Helical coil)

A = cross section, N = Number of turns, ℓ = length of solenoid

$$\Phi_B = N\vec{B}A = N\frac{\mu_0 NI}{\ell}A = \mu_0 AN^2\frac{I}{\ell}$$

$$L_{solenoid} = -\frac{d\Phi_B}{dt} = -\frac{\mu_0 AN^2}{\ell}$$

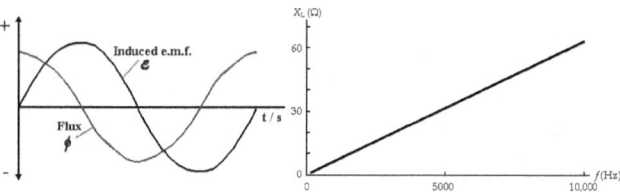

28.1.5 For a Toroid

R – radius of revolution,

$$L_{Toroid} = -\frac{d\Phi_B}{dt} = -\frac{\mu_0 AN^2}{2\pi R}$$

28.2.1 **DC LR CIRCUIT**

In an LR circuit the time scale for appreciable changes in the current is set by the inductive time constant, τ

$$\tau = \frac{L}{R}$$

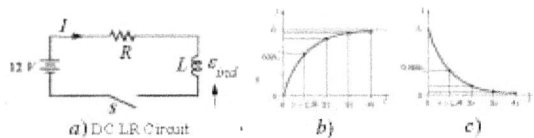

a) DC LR Circuit   b)   c)

$$i(t) = \frac{\mathcal{E}}{R}(1 - e^{-t/\tau}) = \frac{\mathcal{E}}{R}(1 - e^{-R\,t/L})$$

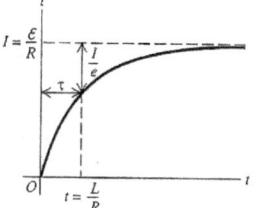

For a PASSIVE LR circuit, the current decreases according to

$$i(t) = I_0 \frac{\mathcal{E}}{R} e^{-R\,t/L}$$

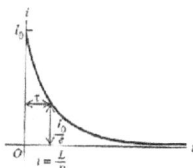

28.2.2   Current in an LR circuit

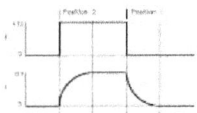

(Vide Section 26.3 for RC circuit)

Comparison of Equations			
Exponential responses of Capacitors and Inductors			
	Discharging	Charging	Time Constant, τ
Capacitor	$v_C(t) = V_0 e^{\{-t\,/RC\}}$	$v_C(t) = v_0(1 - e^{\{-t\,/RC\}})$	RC,
Inductor	$i_L(t) = I_0 e^{\{-R\,/\,L\}t\}}$	$i_L(t) = I_0(1 - e^{\{-R\,/L\}t})$	L/R

28.2.3 ENERGY TRANSFERS IN LR CIRCUITS

The energy E_L stored in an inductor with current I is

$$E_L = \tfrac{1}{2} L\, I^2$$

The energy u_B stored in the magnetic field $\vec{B}$ due to the current I. The energy density of the magnetic field energy is

$$u_B = \frac{1}{2\mu_0} B^2$$

28.2.4 MUTUAL INDUCTANCE (M)

$$M \qquad \text{Unit } \textit{Henry} \text{ (H)} \qquad\qquad \text{(Scalar)} (M^1\ L^2\ T^0\ Q^{-2})$$

Consider two long thin solenoids, one wound on top of the other. The length of each solenoid is ℓ, and the common radius is r. Suppose that the bottom coil has N_1 turns per unit length, and carries a current I_1. The magnetic flux Φ passing through each turn of the top coil is

$$\Phi = \mu_0 N_1 I_1 \pi r^2 ,$$

and the total flux Φ_2 linking the top coil is

$$\Phi_2 = N_2 \ell \mu_0 N_1 I_1 \pi r^2 ,$$

where N_2 is the number of turns per unit length in the top coil. The mutual inductance M of the two coils

$$\Phi_2 = \left(\mu_0 N_1 N_2 \ell \pi r^2 \right) I_1 = M I_1$$

$$L_M \equiv M = \sqrt{L_1 L_2}$$

The changing currents in two nearby coils mutually induce emf sin each other

$$u_{12} = M \frac{di_1}{dt}$$

$$u_{21} = M \frac{di_2}{dt}$$

where M is the mutual inductance of the pair.

28.2.5   TRANSFORMERS:

In a transformer the voltages and currents in the primary and secondary coils depend on the number of winding turns.

$$P_{input} = P_{output}$$

$$V_P I_P = V_S I_S$$

$$\frac{V_P}{V_S} = \frac{N_P}{N_S} = \frac{I_S}{I_P}$$

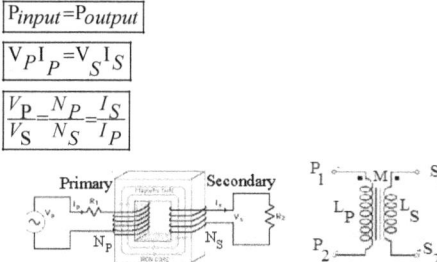

28.3.1   **MAGNETIC FIELD IN MATTER**
Atomic currents:

The orbiting electron e in an atom has a magnetic moment μ_L

$$\mu_L = \frac{e}{2m_e}\hat{L}$$

and a similar contribution due to the spin angular momentum $\vec{\mu}_s$

$$\vec{\mu}_s = \frac{e}{2m_e}\hat{S}$$

Magnetic moment in the material M is the magnetic moment per unit volume

$$M = \frac{\sum \mu}{\Delta V}$$

$$M = \oint \vec{H} \cdot dS$$

28.3.2   Magnetic Substances

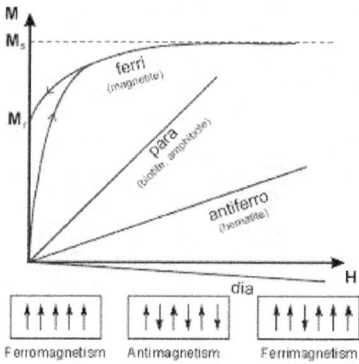

1) Diamagnetic,
2) Paramqgnetic,
3) Ferromagnetic,
4) Antiferromagnetic, and
5) Ferrimagnetic.

28.3.3 DIAMAGNETISM

Diamagnetism is a fundamental property of all matter, although it is usually very weak

The magnetic dipole moments are induced in molecules by the magnetic field; and $\vec{M}$ and $\vec{B}$ have opposite directions.

$$M = \chi \vec{H}$$

χ is the magnetic susceptibility $\chi < 0$, $eg.$ Au, Cu

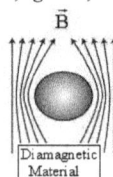

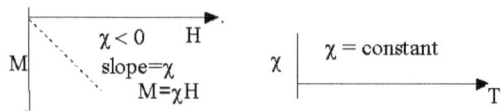

28.3.4   PARAMAGNETISM

Paramagnetism exists in a substance only in the presence of a magnetic field $\vec{H}$. $eg.$, Al, Sb, etc.

This weak induced magnetic field in in the same direction as the applied $\vec{H}$

The five different types of magnetic materials are:

Diamagnetic atoms have only paired electrons,
  whereas paramagnetic atoms, which can be made magnetic, have at least one unpaired electron

Diamagnetic atoms repel magnetic fields.
The unpaired electrons of paramagnetic atoms realign in response to external magnetic fields and are therefore attracted.
Paramagnets do not retain magnetization in the absence of a magnetic field.
The density of field lines is proportional to the field intensity, arrows indicate the direction of force acting on diamagnetic substances. The field around a spherical permanent magnet diminishes with increasing distance, resulting in a strong gradient when the sphere is small (**A**). Diamagnetics would be repelled from the sphere. Magnetic field in the vicinity of a ferromagnetic sphere magnetized by a (uniform) external magnetic field (**B**). Diamagnetics would be repelled from the sphere in "polar" regions and attracted to the sphere in "equatorial" area.

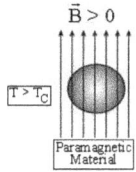

28.3.3.1. Quinke's Method

$$B = \mu_o (H + M)$$

B = Magnetic flux density, H = Magnetic field strength ($A\ m^{-1}$),

M = Magnetization ($A\ m^{-1}$), μ_o = Magnetic permeability of free space,

χ = Magnetic susceptibility of the sample, $\chi = \dfrac{M}{H}$.

$$mg = (0.5\pi r^2)(\chi - \chi_o)H^2$$

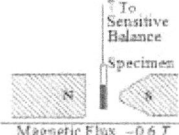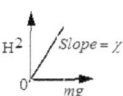

Samples: Paramagnetic or diamagnetics likeTransition metals like Cr, Mn, Ni, Co, Fe.
Metal oxides like CuO, VO, (a cylindrical rod shaped of length at least 16 *cm*) or a test tube filled with a liquid sample, say manganese sulphate solution.

The permanent magnetic dipole moment of an unpaired electron in a paramagnetic tends to align with the $\vec{B}$. The vectors M and $\vec{B}$ are parallel.

28.3.5 CURIE'S LAW:

$$\boxed{\chi = \frac{C}{T}}$$

$$\boxed{\chi = \frac{C}{T - \theta}}$$

Paramagnetic susceptibility is proportional to the total iron content
Many iron bearing minerals are paramagnetic at room temperature. Some examples, in units of $10^{-8}\ m^3 kg^{-1}$, include:

Montmorillonite (clay)	13
Nontronite (Fe-rich clay)	65
Biotite (silicate)	79
Siderite(carbonate)	100
Pyrite (sulfide)	30

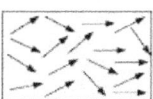

Randomly oriented domains

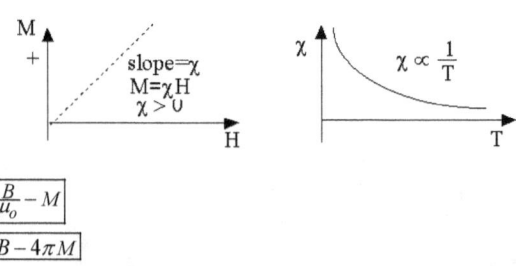

$$H = \frac{B}{\mu_o} - M$$

$$H = B - 4\pi M$$

valid below a critical temperature. $\chi \ll 1$.
Eg., β-Sn , Pt, Mn.

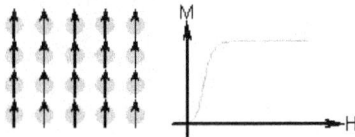

28.3.5 FERROMAGNETISM

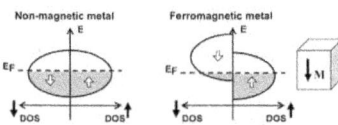

Such materials are characterized by a possible permanent magnetization, and generally have a profound effect on magnetic fields (*i.e.*, $\mu \gg \mu_o$). Unfortunately, ferromagnetic materials *do not* generally exhibit a linear dependence between $\vec{M}$ and $\vec{H}$, or $\vec{B}$ and $\vec{H}$, with constant values of χ_m and μ .

28.3.5.1 Magnetization $\vec{M}$ Curve

Consider an unmagnetized sample of ferromagnetic material. If the magnetic intensity, which is initially zero, is increased *monotonically*, then the $\vec{B}$ - $\vec{H}$ relationship traces out a curve such as that shown This is called a *magnetization curve*.

The maximum permeability occurs at the "knee" of the curve. In some materials, this maximum permeability is as large as $10^5 \mu_o$ because $\boxed{\vec{B} = \mu_o(\vec{H} + \vec{M})}$.

The maximum value of $\vec{M}$ is called the *saturation magnetization* of the material.

If the magnetic intensity $\vec{H}$ is decreased, the $\vec{B}$ - $\vec{H}$ relation does not follow back down the curve. The *hysteresis loop* of the material.

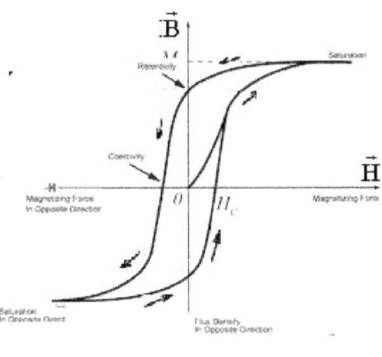

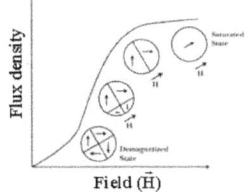

Ferromagnetic materials are used either to channel magnetic flux (*e.g.*, around transformer circuits) or as sources of magnetic field (*e.g.*, permanent magnets).

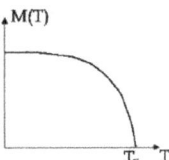

Molecular magnetic dipoles in a magnetic domain tend to bre aligned in this. Such materials are used in permanent magnets.

$$\chi > 0, \quad \boxed{\chi = \frac{C}{T-\theta}}$$

Magnetic intensity:

$$\boxed{B = \mu_o(H + M)}$$

$$\boxed{H = B - 4\pi M}$$

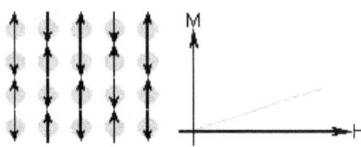

A PERIODIC TABLE showing the type of magnetic behaviour of each element at room temperature

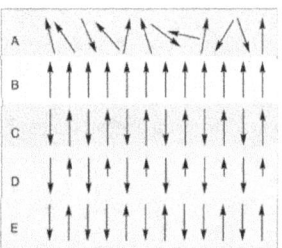

28.3.6 ANTIFERROMAGNETICS:
Eg., Cr.

28.3.7 **FERRIMAGNETICS**: *Eg.*, Ba ferrite

Magnetic character of materials is typically analyzed relative to its magnetic susceptibility (χ). Magnetic susceptibility is the ratio of magnetization (M) to magnetic field (H). The type of magnetic behavior of a compound can be defined by its value of χ **Table** for a comparison of magnetic behavior versus χ.

Magnetic Behavior	Value of χ
1) Diamagnetic	small and negative
2) Paramagnetic	small and positive
3) Ferromagnetic	large and positive
4) Antiferromagnetic	small and positive

28.3.8. Magnetic Hysteresis by Experiment

Magnetizing field, $H = \dfrac{4\pi N_m I_m}{10 L_R}$

Magnetic Induction, $B = \dfrac{4\pi n_p NA}{10 N_s A_R} \cdot \dfrac{I}{\theta}\theta_m$

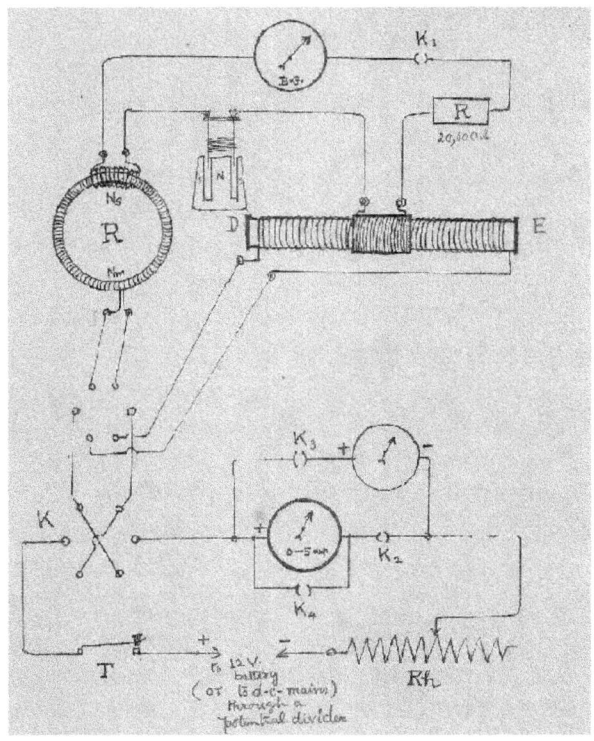

28.4.1. Heydweiller's M-C Bridge

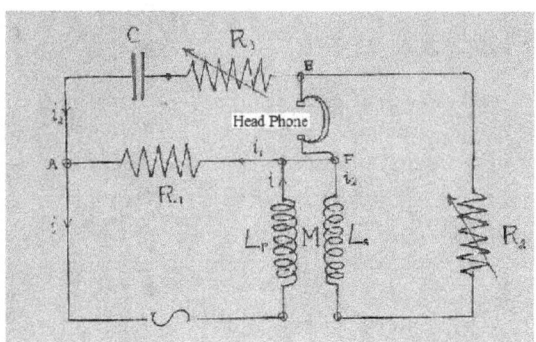

$$\frac{L_S}{M} = \frac{R_1 + R_3}{R_1}$$

$$M = CR_1(R_2 + R_{LS})$$

28.4.2. Kirchoff's method for Mutual Inductance

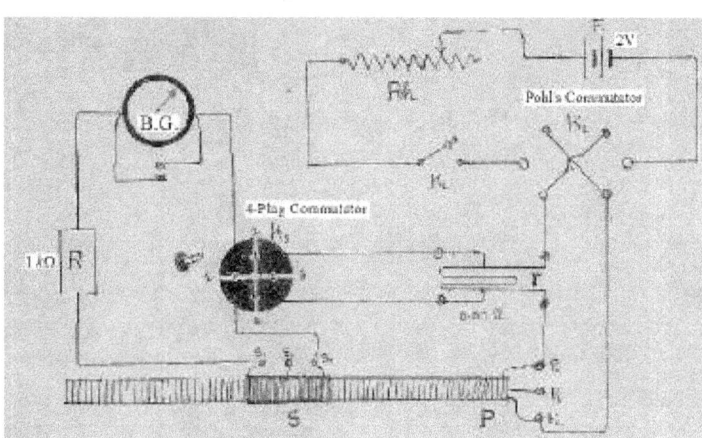

a) First throw in B.G.= θ is noted,
b) Steady deflection in B.G. = α,
c) The period of B.G. in open circuit, T is noted,
d) Logarithmic decrement, $\lambda = \frac{1}{2(n-s)} \log_e (\frac{x_s}{x_n})$; extreme excursions in B.G., x_n.

$$M = \frac{T}{4\pi}.r.\frac{\theta}{\alpha}(1 + \frac{\lambda}{2}).$$

If A =cross-sectional area of secondary,

Also $\quad M = 4\pi n_p N_s A.10^{-9} H$

28.4.3. Kelvin's Double Bridge for Low R

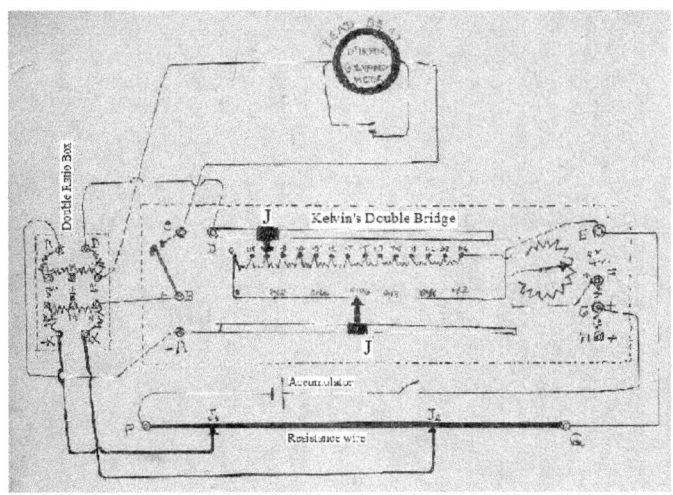

28.4.4. Loss of charge Method for very high R

$$X = \frac{t}{C \log_e(\frac{\theta_o}{\theta_t})}$$

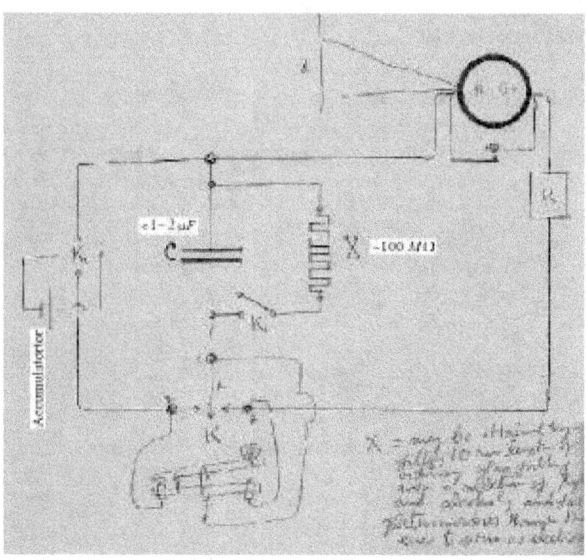

28.4.5. Mathissen's Method for Low R

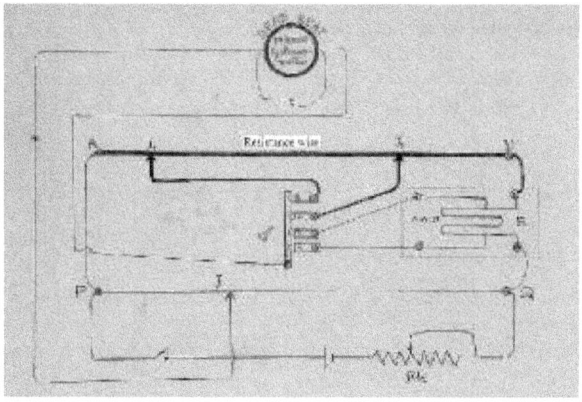

$$r = \frac{L_1 - L_2}{(l_1 - l_2)} \frac{R}{L}$$

28.4.6. Rayliegh's Bridge for L
   a) First throw in B.G.= θ is noted,
   b) Steady deflection in B.G. = α,
   c) The period of B.G. in open circuit, T is noted,

d) Logarithmic decrement, $\lambda = \dfrac{1}{2(n-s)} \log_e(\dfrac{x_s}{x_n})$; extreme excursions in B.G., x_n.

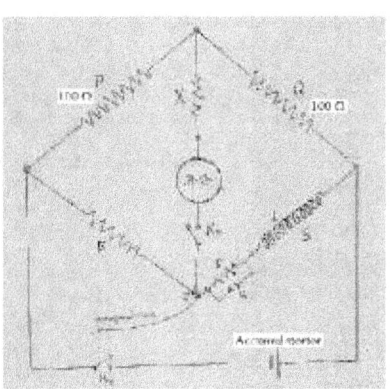

$$\lambda = \dfrac{1}{2(n-s)} \log_e(\dfrac{x_s}{x_n})$$
$$L = \dfrac{T}{4\pi}.r.\dfrac{\theta}{\alpha}(1+\dfrac{\lambda}{2})$$

28.4.7. Remington's LC Bridge

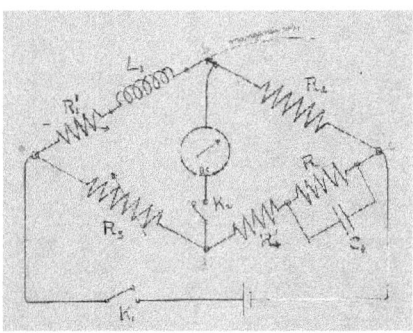

$$L_1 = \dfrac{C_4 R_1 R^2}{(R+R_4')}$$

28.4.8. Stroude and Oate's Method of L
$$L = C[R_2 R_3 + R(R_3 + R_4)]$$

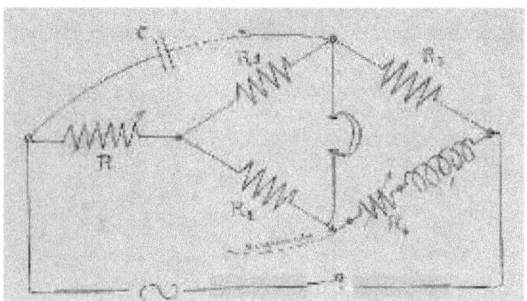

28.4.9. Substitution Method for Very High R

If α and β are steady deflections of the galvanometer, $\beta \approx \frac{\alpha}{2}$

Resistance of Galvanometer, $G = Q\dfrac{\beta}{(\alpha-\beta)}$.

a) K_1, K_3 both closed, with X, one gets θ_1 and $Q(=Q_1)$

b) With K_3, K_2, K_1 keys closed, substituting R for X, adjust $Q(=Q_1)$ & S to get deflection θ_1 .

c) The current is reversed and find θ_2 and Q_2 .

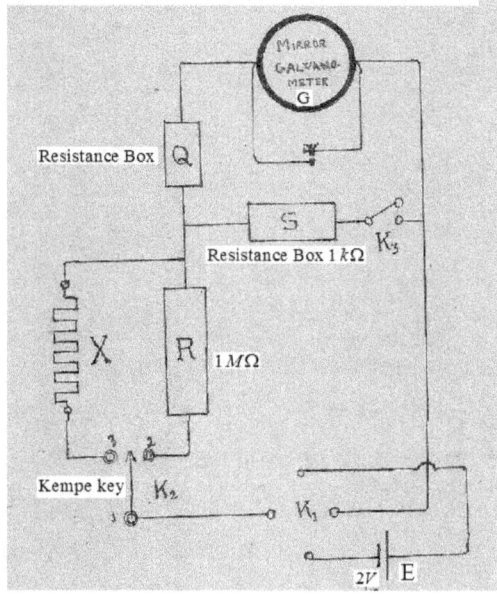

28.4.9.1. Unknown High resistance,

$$X = \frac{\theta_2}{\theta_1} \{R(1+\frac{Q_2+G}{S}) + Q_2 + G\} - (Q_1 + G)$$

28.4.10. TERRESTRIAL MAGNETISM:
Outside the surface the earth's magnetic field is approximately a dipole field.

$$B_{earth} \sim 10^{-4}$$

28.5. ALTERNATING CIRCUITS:
SYMBOL OF AC SOURCE

 V

28.5.1 PHYSICAL QUANTITIES FOR AC.
The voltage across the source

$$V = V_m \sin \omega t$$

Current sustained in the source be

$$I = I_m \sin(\omega t + \varphi)$$

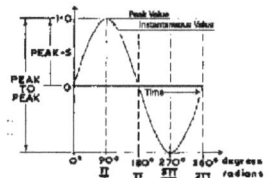

The <u>RMS values of the current and voltage in a resistance</u> are

$$V_{rms} = \frac{V_m}{\sqrt{2}} \text{ or } \frac{V_p}{\sqrt{2}}$$

$$I_{rms} = \frac{I_m}{\sqrt{2}} \text{ or } \frac{I_p}{\sqrt{2}}$$

Average power

$$P_{av} = I_{rms}^2 R$$

$$P = V_{rms} I_{rms} \cos\varphi$$

28.5.2.1 Ohm's Law in a Nutshell

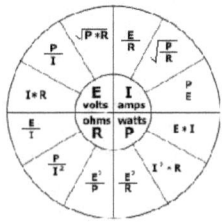

28.5.2.2. Advantages of 120 V, 60 Hz AC Power

1) With an RMS value of 120V, AC Power = 120 V DC power in heating effect.
2) When V> 120 VAC, there would be more danger of fatal electric shock.
3) V<120 VAC would be less efficient in supplying power.
4) A higher voltages has the advantage of less I_{rms}^2 R loss, since the same power can be produced with less I.

I_{rms}^2 R loss increases with I_{rms}^2. Therefore, for industrial use, where large amounts of power are used, the main line is often 208-240 V, 3-phase.

5) 3-phase power is more efficient for the operation of large motors.
6) AC power is more efficient than DC in distribution from the generating station.60 Hz is convenient for commercial AC power, as lower frequencies transformers have to be large in size.
7) Too low a frequency will cause lamps light to flicker.
8) Too high frequencies would cause eddy current losses and hysteresis in transformers.

28.5.3 LC oscillations:

In an LC circuit the charge on the C and current in the circuit oscillate sinusoidally with the same

$$f_o = \frac{1}{2\pi}\frac{1}{\sqrt{LC}}$$

SUM of electric energy in C + the magnetic energy in L = a constant

28.5.4 THE VECTORS OF AC:

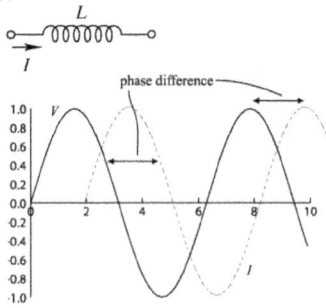

28.5.5 PURELY RESISTIVE CIRCUIT,

$$\boxed{\varphi = 0 \text{ or } i\pi}$$

$$\boxed{i = i_m \ \sin(\omega t + \varphi)}$$

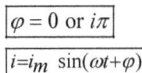

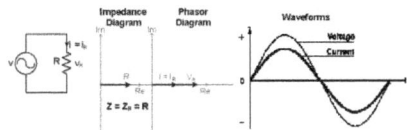

28.5.6   PURELY INDUCTIVE CIRCUIT,

$$\boxed{\varphi = -\frac{\pi}{2}}$$

$$\boxed{i = i_m \ \sin(\omega t + \varphi)}$$

$$\boxed{X_L(\Omega) = 2\pi f(Hz)L(H)}$$

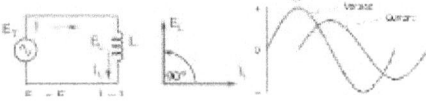

28.5.7   PURELY CAPACITIVE CIRCUIT,

$$\boxed{\varphi = \frac{\pi}{2}},$$

$$\boxed{i = i_m \ \sin(\omega t + \varphi)}$$

$$i = i_m \qquad \boxed{X_C = \frac{1}{2\pi fC}}$$

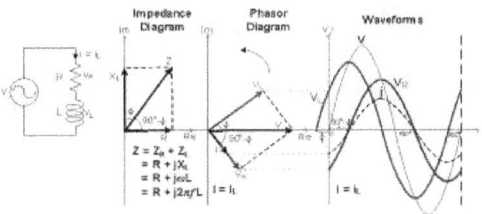

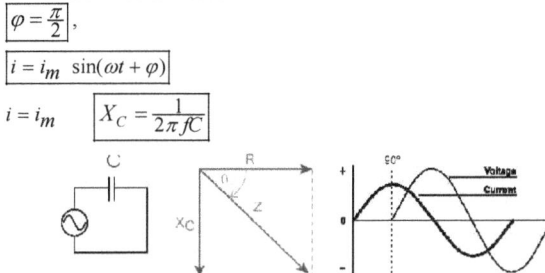

28.5.8   **PHASE RELATIONSHIPS in AC**

> In a Resistive circuit, the Voltage and Current are IN PHASE
> In a Capacitive circuit, the Current LEADS the Voltage by 90°
> In an Inductor circuit, the the Current LAGS Voltage by 90°

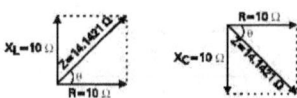

28.5.9 DIFFERENCES BETWEEN DC AND AC

Comparison Chart on AC & DC		
	Alternating Current	Direct Current
1) Quantity of Energy that can be carried	Safe to transfer over longer city distances and can provide more power.	Voltage of DC cannot travel very far until it begins to lose energy .
2) State of Flow of Electrons	Electrons keep switching directions - forward and backward.	Electrons move steadily in one '. direction or 'forward'
3) Cause of the Direction of flow of electrons	Rotating magnet along the wire.	Steady magnetism along the wire.
4) Frequency of I	The frequency of ac current is $50Hz$. or $60Hz$ depending upon the country	The frequency of direct current is zero.
5) Direction I Flow	Reverses its direction while flowing in a circuit..	Flow is one direction in the circuit
6) Magnitude of I	I of magnitude varying with time	The current of constant magnitude.
7) Source of Power	AC Generator and Mains, Hydro-electric Thermal Plant, NuclearReactors	Cell or Battery.
8) Passive Parameters	Impedance, Z (R, L, & C).	Resistance(R) only
9) Power Factor	Lies between 0 & 1.	it is always 1.
10) Signal Type	Sinusoidal, Trapezoidal, Triangular, Square.	Pure and pulsating.

28.510 DAMPING IN SERIES RLC CIRCUIT:

When $\boxed{R < \sqrt{\frac{4L}{C}}}$, then the charge on the C and the current in circuit oscillate, and the circuit is called lightly (under) damped.

$\boxed{R > \sqrt{\frac{4L}{C}}}$, the circuit is heavily (over) damped.

$\boxed{R = \sqrt{\frac{4L}{C}}}$, it is critically damped.

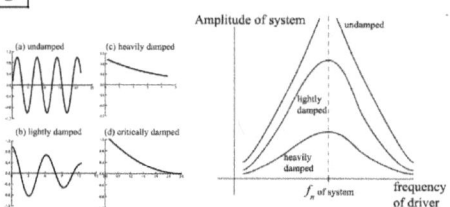

28.5.11 Reactance of the circuit $X = (X_L - X_C)$

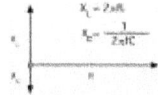

28.5.12 The impedance Z of the circuit,

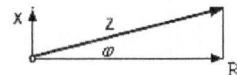

Impedance

$$Z = \sqrt{R^2 + (X_L - X_C)^2} = \sqrt{R^2 + (\omega L - \frac{1}{\omega C})^2}$$

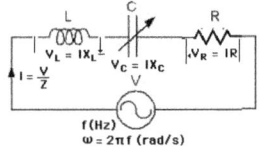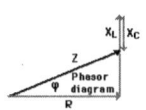

Phase $\quad \boxed{\varphi = \tan^{-1} \dfrac{(X_L - X_C)}{R}}$

28.5.13. Power for an RLC circuit driven by an ac source:
AC Power

$$\boxed{P_{av} = I_{rms}^2 \ R}$$

$$\boxed{P = V_{rms} I_{rms} Cos\varphi}$$

Power Factor

$$\boxed{\text{Power Factor} = \frac{\text{True Power}}{\text{Apparent Power}}}$$

28.5.14. RESONANT CIRCUITS:

X_L Stops High frequency waves,
    Stops Low frequency signals

X_C Prevents LowHigh frequency waves,
    Allows passage of High frequency signals

28.5.15 Series Resonant Circuit

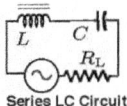

Series LC Circuit

At $(X_L - X_C)$, Passes the Resltant frequency waves,
    Prevents All other frequency signals

   a) Example: Tuning to a radio receiver ot TV

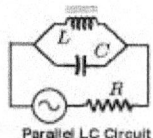

Parallel LC Circuit

| Higher frequencies: Pass through the Capacitor, C, |
| Lower frequncies: Pass through the INDUCTOR, L |

b) Example: Filtering out a noise frequency or Blocking an Undesirable channel
c) The resonant frequency

$$f_o = \frac{1}{2\pi}\frac{1}{\sqrt{LC}} \ (Hz)$$

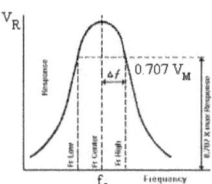

d) Quality Factor

The frequency response of the circuits current magnitude above, relates to the "sharpness" of the resonance in a series resonance circuit. The sharpness of the peak is measured quantitatively and is called the **Quality factor, Q** of the circuit

$$Q = \frac{X}{R}$$

$$Q = \frac{X_L}{R} = \frac{R}{X_C} = \frac{1}{R}\sqrt{\frac{L}{C}}$$

e) Bandwidth

Bandwidth, (BW) is the range of frequencies over which at least half of the maximum power and current is provided as shown

$$BW = \frac{f_o}{Q} = \frac{R}{L}(rad) = \frac{R}{2\pi L}(Hz)$$

$$f_L = f_o - \frac{1}{2}BW$$
$$f_H = f_o + \frac{1}{2}BW$$

Series Resonant Circuits are used in designing and building of Bandpass Filters and indeed, resonance is used in 3-element mains filter design to pass all frequencies within the "passband" range while rejecting all others.

28.5.16   Parallel Resonance Circuit

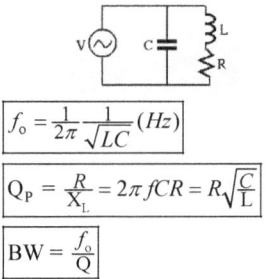

$$f_o = \frac{1}{2\pi} \frac{1}{\sqrt{LC}} \, (Hz)$$

$$Q_P = \frac{R}{X_L} = 2\pi fCR = R\sqrt{\frac{C}{L}}$$

$$BW = \frac{f_o}{Q}$$

28.5.17. Impedance Bridge

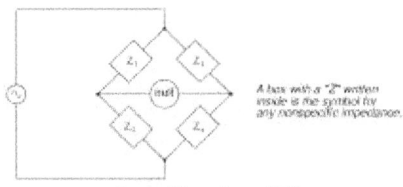

A box with a "Z" written inside is the symbol for any nonspecific impedance.

General Impedance Bridge

28.6. Heat Sources using Electricity

28.6.1. Electric Heaters / Ovens
   Use Ohmic (Resistive) heating for all cook wares, bath room heaters,

28.6.2 Electric Iron
   Use tungsten tape wound over mica sheet.

28.6.3 Microwave Oven - How does a microwave oven heat without heat?
   (Percy Spencer invented in USA, 1950s)
   How does a microwave turn electricity into heat? Like this!

1. Inside the strong metal box there is a MW generator called a magnetron. While cooking starts, the magnetron uses its input power and converts it into high-powered, operate at MW frequencies of about $2.45 \, GHz$, i.e., $\lambda = 12.23 \, cm$.

2. The magnetron blasts these waves into the food compartment through a wave guide. This cuboid chamber has metallic walls (acting as a Faraday cage). The front door, made of glass, and the light bulb cavity are both covered by metal grids. The holes in the grids are small compared with the λ of the MW, hence the grids act just like metal plates.

3. The food placed on a rotating turntable, spinning slowly round so that MWs cook it evenly.

4. The MWs bounce back and forth off the reflective metal walls of the food compartment and penetrate inside the food, causing the molecules inside it to rotate more energetically. This induces polar molecules (eg., water) in the food to rotate and produce thermal energy in a process known as dielectric heating, characterized by the penetration depth which drops the MW power by a factor $1/e$., which is absorption of 63% MW power. . Microwave heating is more efficient on liquid water than on ice, where the movement of molecules is more restricted.

5. Vibrating molecules with heat energy make the food hotter. Thus the MW pass their energy onto the food molecules, rapidly heating it up. There is a clear maximum in the dielectric loss for water at a frequency of approx. $20 \, GHZ$. The $2.45 \, GHz$ operating frequency of domestic ovens is selected to be some way from the maximum in order to limit the efficiency of the absorption

Frequently asked questions:

Does the rate of heating depend on the type of food? The main factor determining the rate of heating in a food is its water content. However, the salt added to many foods may allow an additional heating mechanism. In solution, there are charged Na+ and Cl− ions, and so it is possible to heat salty food more quickly than pure water

Can ice be melted in a microwave oven? All microwave ovens come with a defrost facility. However, it takes much longer to get ice to raise its temperature by 1 °C than it does for a similar amount of liquid water. In ice the molecules of H2O are held in a crystal lattice in fixed positions. Therefore, there is less heating effect.

Are microwave ovens only used for heating food? MW ovens are used in many other industrial fields, *e.g.*, pasteurization of vegetables, drying of paper or textiles, thermal treatment of pharmaceutical products and vulcanization of rubber and elastomers.

Precautions are only specially manufactured container of synthetic, glass, or paper which do not absorb or reflect MWs are suitable as cooking containers

Heat is generated as a result of MW waves penetrate the food and water molecules vibrate rigorously increasing internal hating. Boiling of water occurs when the VP within the bubbles is greater than the pressure from the liquid plus the pressure above the water within the cooker. Thus increasing the pressure above the water and elevating the B.P.

28.6.3.1. Travelling Wave Tube

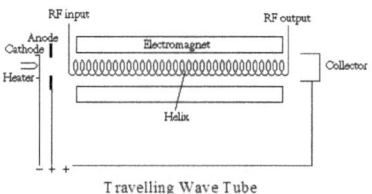

Travelling Wave Tube

28.6.4. Induction Heating

Induction (Electro-magnetic) heating takes place in an electrically conducting object (not necessarily magnetic steel) when the object is placed in a varying magnetic field. Induction heating is due to the hysteresis and eddy-current losses.

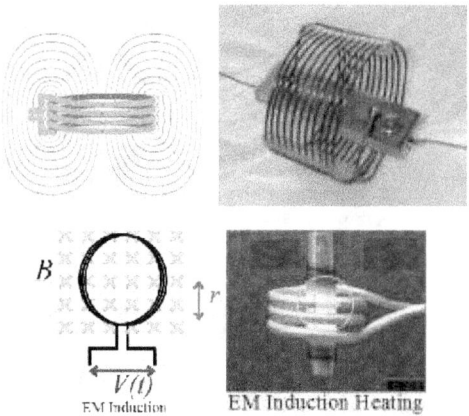

EM Induction   EM Induction Heating

Eddy-current losses occur in any conducting material in a varying magnetic field. Eddy-current losses are much more important than hysteresis losses in induction heating. Note that induction heating is applied to nonmagnetic materials, where no hysteresis losses occur.

Two basic things for induction heating to occur:
a. A changing magnetic field and
b. An electrically conductive material placed into the magnetic field

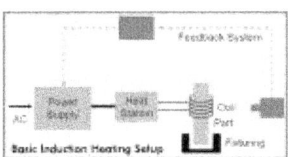

Simply stated, induction heating is fast, is the most clean, efficient, cost-effective, precise, and repeatable method of material heating available to the industry, today.

28.7. NETWORK THEORY

Network Theory is the study of how to solve electric circuit problems. The connecting of elements in series and parallel is the most basic type of network.

28.7.1\1 Mesh Analysis by Loop current Method (Maxwell's method)

Form this form mesh or loop equations using Kirchoff's voltage law. Solve for the currents.

28.7.1.2. Nodal Analysis

Here Kirchoff's current law is made use of to form nodal equations.

Phasor Algebra for AC circuit analysis.

i) Schematic

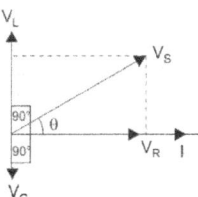

Series LC with
resistance in series

ii) Rectangular form: $Z = R + j(X_L - X_C)$

iii) Phasor diagram

iv) Polar form

$$Z = |Z| \ \Omega \ \angle\varphi, \ \boxed{\varphi = \tan^{-1}\frac{(X_L - X_C)}{R}}$$

28.7.2. Network Theorems

28.7.2.1. Superposition Theorem

The response of a linear network containing several independent sources is found by considering each generator separately and then adding each individual responses. While evaluating the response due to one source, one replaces each of the other independent generators by its internal impedance.

Network

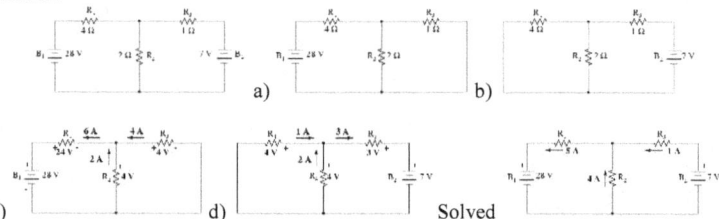

a)   b)

c)   d)   Solved

28.7.2.2. Compensation Theorem

Any impedance of a network, composed of impedances and generators, can be replaced by a generator of zero internal impedance and of e.m.f equal to the instantaneous p.d. across the replaced impedance.

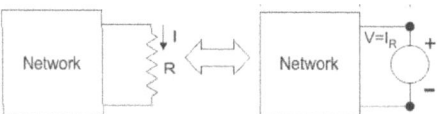

28.7.2.1. Thevenin's theorem

It states that any linear circuit can be simplified to an equivalent circuit with just a single voltage source V_{oc} (open circuit voltage $\equiv V_{Th}$) and a series resistance Z_0 (when all sources of voltage are short circuited) connected to a load Z_L.

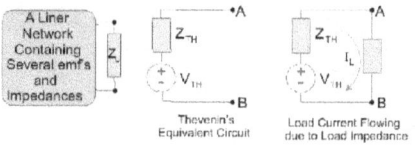

Thevenin's Equivalent Circuit   Load Current Flowing due to Load Impedance

$$I_L = \frac{V_{Th}}{R_{Th} + R_L}, \ \text{or} \ I_L = \frac{V_{oc}}{Z_0 + Z_L}$$

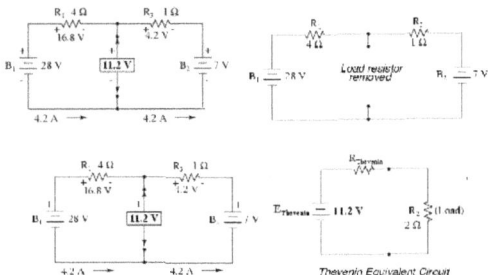

28.7.1.2. Norton's Theorem

It is possible to simplify any linear circuit, no matter how complex, by an equivalent circuit with just a single current source I_N (shot-circuited current source, $I_{sc} \equiv I_N$) and parallel resistance $R_N (= Z_0)$ connected to a load R_L.

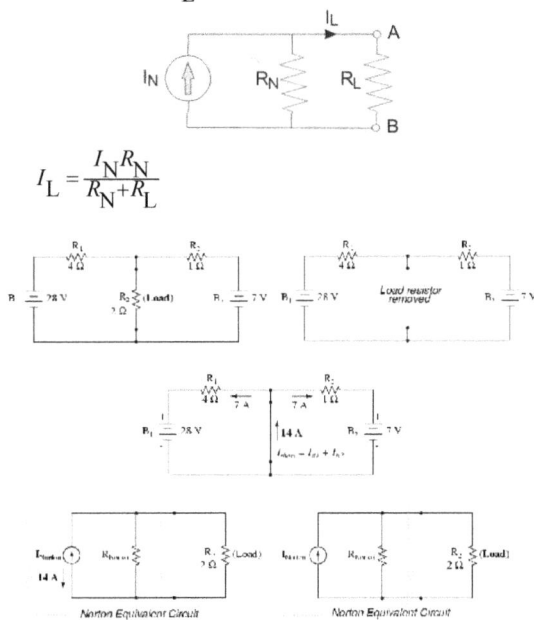

$$I_L = \frac{I_N R_N}{R_N + R_L}$$

28.7.1.3. Maximum Power Transfer theorem

Simply stated, the maximum amount of power will be absorbed by one network from another, joined at two terminals, when the impedances looking into the networks at the said port are conjugate to each other. *i.e.*, from a given generator of internal impedance $|Z_G| \; \Omega \angle \varphi$ the maximum power will be obtained if the value of load impedance connected across it is $|Z_G|\ \Omega\ \angle -\varphi$. This is the "matching condition".

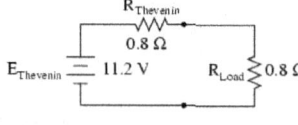

$$P_Q = \frac{\mu^2 v_i^2 R_L}{(r_p + R_L)^2}$$

$$P_{Q\,max} = \frac{\mu^2}{4\,r_p}v_i^2 \quad \text{when } R_L = r_p$$

Jacobi's Maximum-Power-Transfer Theorem

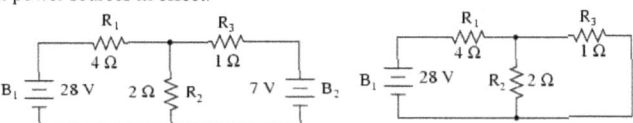

	$R_{Thevenin}$	R_{Load}	Total	
E	5.6	5.6	11.2	Volts
I	7	7	7	Amps
R	0.8	0.8	1.6	Ohms
P	39.2	39.2	78.4	Watts

	$R_{Thevenin}$	R_{Load}	Total	
E	6.892	4.308	11.2	Volts
I	8.615	8.615	8.615	Amps
R	0.8	0.5	1.3	Ohms
P	59.38	37.11	96.49	Watts

	$R_{Thevenin}$	R_{Load}	Total	
E	4.716	6.484	11.2	Volts
I	5.895	5.895	5.895	Amps
R	0.8	1.1	1.9	Ohms
P	27.80	38.22	66.02	Watts

28.7.1.4.   Superposition theorem

The theorem states that a circuit can be analyzed with one source of power at a time, the corresponding component voltages and currents algebraically added to find out what they will do with all power sources in effect.

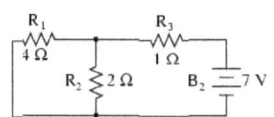

	R_1	R_2	R_3	$R_2//R_3$	$R_1 + R_2//R_3$	Total
E	24	4	4	4	28	Volts
I	6	2	4	6	6	Amps
R	4	2	1	0.667	4.667	Ohms

28.7.1.5. Reciprocity Theorem

If an e.m.f. is located at any point of a network composed of linear, bilateral elements, produces a current at any other point of the circuit, then the same e.m.f. acting at the second point will produce the same current at the first point. This theorem implies interchangeability of source and its effect.

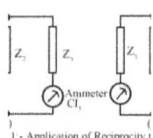

1:- Application of Reciprocity t

28.7.2. Four-Terminal Networks (Coupled Circuits) (A black box)

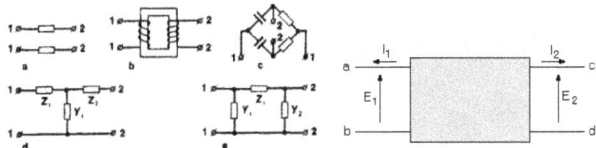

Any complex network (black box) can be reduced to an equivalent basic type of network (attenuator) say, T or π , by the equations involving open-circuited and short-circuited impedances.

28.7.2.1. T-Network (Star)

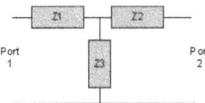

28.7.2.2. π-network (Delta)

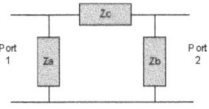

The impedances of the π network Z_A , Z_B and Z_C from the T network from the following equations:

$$Z_A = (Z_1.Z_2 + Z_2.Z_3 + Z_3.Z_1)/Z_2$$
$$Z_B = (Z_1.Z_2 + Z_2.Z_3 + Z_3.Z_1)/Z_3$$
$$Z_C = (Z_1.Z_2 + Z_2.Z_3 + Z_3.Z_1)/Z_1$$

Conversion equations for π to T

$$Z_1 = (Z_A.Z_B)/(Z_A.+Z_B+Z_C).$$
$$Z_2 = (Z_B.Z_C)/(Z_A.+Z_B+Z_C)$$
$$Z_3 = (Z_C.Z_A)/(Z_A.+Z_B+Z_C).$$

28.7.2.3. Network Terminations
Image Impedance

A network is said to be operated under image impedance conditions, when the internal impedance Z_S of the sourc3e of power acting at the input terminals and the load impedance Z_L at the output terminals , are so related to the network that Z_{12} with Z_L connected equals Z_S , and similarly, Z_{34} with Z_S connected equals Z_L .

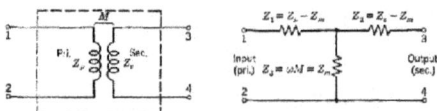

Image transfer constant θ

$$\frac{E_{34}}{E_{12}} = \frac{I_2}{I_1} = \sqrt{\frac{Z_{I1}}{Z_{I2}}}\, e^{-\theta}$$

Characteristic Impedance Z_O

For symmetric T and π networks $Z_1 = Z_2$ and $Z_A = Z_C$,

$$Z_{12} = Z_L = Z_{k1}, \; Z_{34} = Z_S = Z_{k3}$$
$$Z_O = Z_I,$$

28.7.3. Ladder Networks
28.7.3.1. Characteristic Impedance of Basic T section of the fundamental Ladder

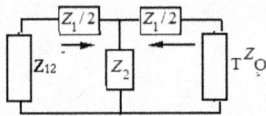

$$_T Z_O = \sqrt{\frac{Z_1^2}{4} + Z_1 Z_2}$$
$$Z_{OC} = (Z_1/2)\,\|\,Z_2;$$
$$_T Z_O = \sqrt{Z_{OC} \cdot Z_{SC}}$$

28.7.3.2. Characteristic Impedance of Basic π section of the fundamental Ladder.

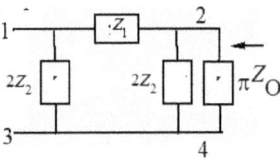

$$Z_{OC} = 2Z_2\,\|\,[Z_1 + 2Z_2]; \; Z_{SC} = Z_1\,\|\,2Z_2$$
$$_\pi Z_O = \sqrt{Z_{OC} \cdot Z_{SC}}$$

28.7.3.3. Two-port network

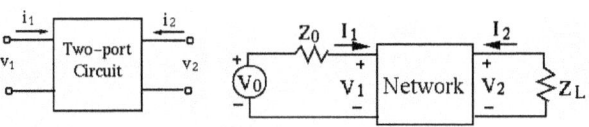

28.7.4. Wave Filters

An ideal filter would produce no attenuation in the band desired to 'transmit' till the cut off frequency f_C and would give an attenuation $= \infty$ at all other frequencies, $f > f_C$, the first one being called the 'Pass Band', and the other 'attenuation band'(or Stop Band).

$$Z_O = \sqrt{Z_{OC} \cdot Z_{SC}}$$

28.7.4.1. LP Filter

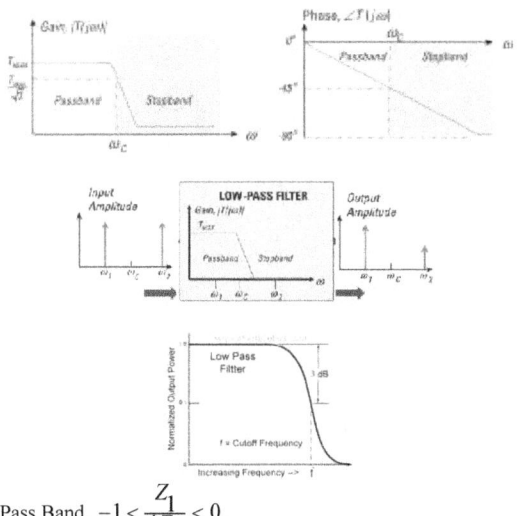

For a Pass Band, $-1 < \dfrac{Z_1}{4Z_2} < 0$.

28.7.4.2. HP Filter

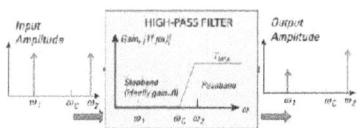

28.7.4.3. BP Filter

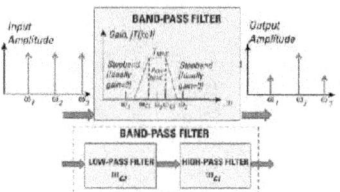

28.7.4.4. Band Stop Filter

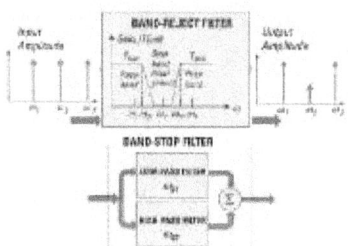

28.7.5. Constant-k Ladder Filters (LC Filters)

If $Z_1 . Z_2 = R_o^2 = k$, where R_o = purely resistive, called Nominal impedance of the Section.

For T-Section, $_T Z_O = R_o \sqrt{(1 - \dfrac{f^2}{f_c^2})}$

For π Section, $_\pi Z_O = \dfrac{R_o}{\sqrt{(1 - \dfrac{f^2}{f_c^2})}}$

If $Z_O = \sqrt{\dfrac{L_1}{C_2}} = R_o$ and $f_C = \dfrac{1}{\pi \sqrt{L_1 C_2}}$, one gets $L_1 = \dfrac{R_o}{\pi f_c}$ and $C_2 = \dfrac{1}{\pi R_o f_c}$

28.7.6. Typical Filters

Twin-T Null Circuit (Parallel TRC network) RC Band Stop Filter

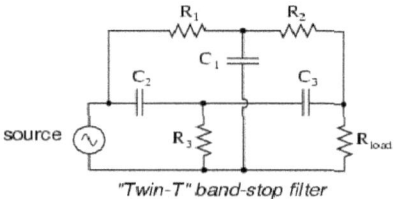

"Twin-T" band-stop filter

$f_o = \dfrac{1}{2\pi RC}$

28.7.6.1. Parallel Resonant Filters

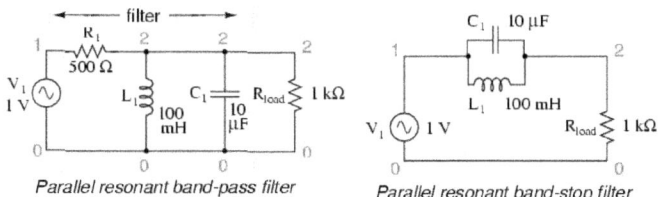

Parallel resonant band-pass filter   Parallel resonant band-stop filter

28.7.6.2. Wien Bridge

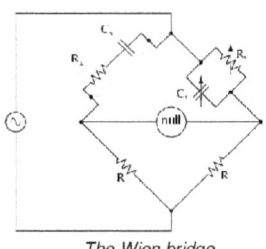

The Wien bridge

28.7.6.3 Maxwell-Wien Bridge

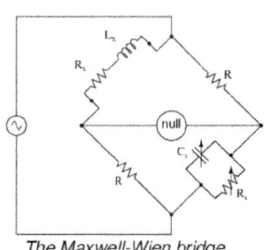

The Maxwell-Wien bridge

-o-0-o-0-o-0-o-0-o-0-o-0-o-

Chapter 29

DEVELOPMENTS OF ATOMIC
THEORY & RELATIVITY

Quantum Physics came from the Vedas???? (https://shar.es/1BwzIb)
"The principles of physics, as far as I can see, do not speak against the possibility of maneuvering
thins atom by atom"   Richard Feynman

29.1   **CATHODE RAYS**

29.1.1 ELECTRIC DISCHARGE OF GAS
        Dalton, in 1808 proposed that matter is made of atoms. All substances were
either made of single atoms or combinations of atoms (molecules); thought that atoms
were indivisible.
In the 19th century, experiments showed that atoms were divisible. As a result, new
particles and forces were found.

29.1.2 CATHODE RAYS
Geissler in the 19th century, Invented a new vacuum pump.
He produced discharges of electricity in evacuated tubes of varying shape. He also
produced difference colors of discharge by placing different gases in the tubes.

29.1.2 GAS DISCHARGE TUBE: A tube that allows an electric current to pass through a gas at
low pressure.

29.1.2.1 ELECTRODES:
Metal plates sealed in the ends of a gas discharge tube. (+ is the anode and - is the
cathode.)
When air is pumped out of the tube,
the discharge across an induction coil stops, and the electrodes in the tube are connected
by one or more violet streamers.

29.1.2.2 Appearance of the gas in a Discharge tube under gradually diminishing pressure:

$$1\ atm = 760\ mm\ \text{Hg} \simeq 760\ torr = 101.30\ kPa = 1013\ mb$$
$$Pa = Nm^{-2}, \qquad 1\ bar = 10^5\ Pa$$

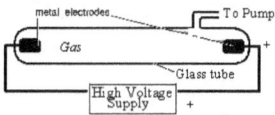

1) Above 10 *mm* pressure of Hg column:
    No discharge, but sparking occurs with cracking sound.

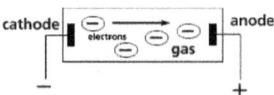

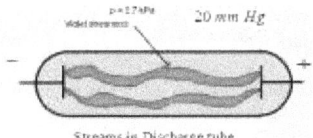

Streams in Discharge tube

2) At 10 *mm* pressure:
   A thin streak of light, violet red in colour passes from one electrode to the other with a cracking noise.

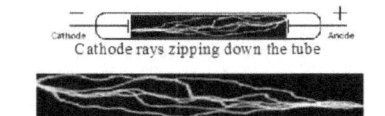

Cathode rays zipping down the tube

Without gas in the tube, the rays can travel great distances

3) At 4 *mm*:
   Positive column fills the whole tube.

For	Air:	Orange red (Mauve).
	H_2:	Bluish red
	N_2:	Red
	He:	Pink
	CO_2:	Bluish white.
	Cl:	Green.

29.1.2.3. At low enough pressure (< 5 *mm*),
   A pink glow fills the entire tube. Continued decreases in pressure cause the pink glow to concentrate around the anode and a blue glow around the cathode. The space between the glows is dark (called **Faraday's dark space**).

Faraday dark space

Streams Faraday Dark space

At 1.65 *mm*:
   Cathode glow, bright blue.
   Faraday's dark space + Positive column.

20.1.2.4. Continued reduction in pressure causes the dark space to expand, and the color at the electrodes to fade until the tube is dark, except for a faint green or violet glow around the anode. The sides of the tube fluoresce (usually green). The dark region is now called **Crookes' dark space**.
   At 0.8 *mm*:
   Cathode glow + Crooke's dark space + negative glow + Faraday's dark space + positive column.
   At 0.37 *mm*:

The **positive column** breaks up into striations which are pink coloured discs of light separated by dark region.

At 0.1 *mm*:

Positive column disappears, leaving the others; **Faraday's dark space** fills this luminous streaks leaves the cathode, the walls of the tube begins to fluoresce.

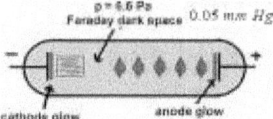

At 0.02 *mm*:

Negative glow with Faraday's dark space disappears; **Crooke's dark space** fills the tube. At first the electrodes and later the whole tube begins to shine.- stage for the production of X-rays.

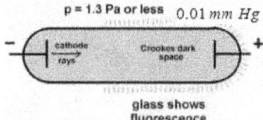

At 10^{-4} *mm*:

No discharge stage, being no ions to carry the charge.

29.1.2.5 Investigations centered on what was happening in the dark space and decided that the glow in the gas originated at the cathode. For this reason, the discharge was called CATHODE RAYS.

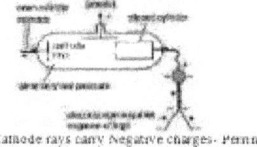

Cathode rays carry Negative charges- Perrin

29.1.3 MALTESE CROSS TUBE:

Plucker made an anode into a Maltese cross, and this produced a shadow in the glow at the end of the tube to show that the cathode rays traveled in straight lines.

29.1.4 PADDLE WHEEL DISCHARGE TUBE:

Crookes reported that a paddle wheel placed in the path of the cathode rays turned, proving that they carried energy and that the rays (particles) moved from the cathode to the anode.

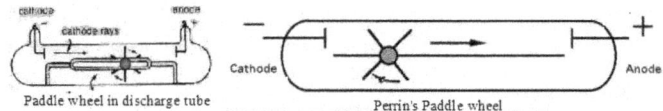

Paddle wheel in discharge tube   Perrin's Paddle wheel

Paddle Wheel Tube: Jean Perrin showed that the cathode rays have mass and momentum.

29.1.5 CATHODE RAYS IN A MAGNETIC FIELD:

Crookes showed that the rays were deflected by a magnetic field, and experience a force, *i.e.*, cathode rays behaved like negatively charged particles.

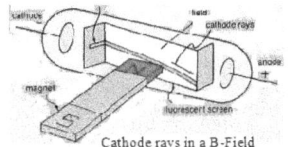

Cathode rays in a B-Field

29.1.6 CATHODE RAYS IN ELECTRIC FIELDS:
Arthur Schuster noticed that the particles were repelled from a negative plate and attracted to a positive plate, proving that cathode rays are negatively charged particles.

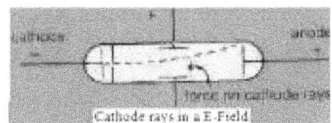

Cathode rays in a E-Field

29.1.7 CATHODE RAYS CARRY A NEGATIVE CHARGE:
Jean Perrin constructed an apparatus that had an anode made of a hollow aluminum cylinder that was open at both ends. At the end opposite to the cathode, was a closed cylinder (connected to an electroscope) which collected the cathode rays. The electroscope showed that the cathode rays were negatively charged.

29.2.1 THOMSON'S TUBE: Determination of ($\frac{q}{m_e}$) by Parabola Method:

The deflecting plates deflected the particles in one direction.
Magnetic coils deflected the particles in the other direction.

$$\boxed{\frac{q}{m_e} = 1.7588x10^{11} Ckg^{-1}}.$$

($\frac{q}{m_e}$) was the same regardless of the potential difference used to accelerate particles.

$\frac{q}{m_e}$ was the same for different cathode materials.

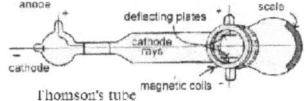

Thomson's tube

Motion of of +ve rays along y-axis, applying electric field (X)

$$z = \left(\frac{Xe}{M}\right)\left(\frac{\ell}{v}\right)\left(\frac{L}{v}\right) z'$$

Applying a magnetic field along z-axis,

$$y = \left(\frac{Hev}{M}\right)\left(\frac{\ell}{v}\right)\left(\frac{L}{v}\right) y'$$

$$\boxed{\frac{e}{M} = \left(\frac{XA}{H^2 B^2}\right)\left(\frac{y^2}{z}\right)}$$

29.2.2 Similar experiments with hydrogen ions, showed that

$\frac{q}{M}$ = 1836 times smaller than for cathode rays.

Thomson called the cathode ray particle, the ELECTRON.

29.2.3 PROPERTIES OF CATHODE RAYS:

They are produced by the negative electrode, or cathode, in an evacuated tube, travel towards the anode, travel in straight lines, cast sharp shadows, have energy and can do work; are deflected by electric and magnetic fields and have a negative charge. They are the electrons.

29.2.4 **QUANTIZATION of charge: (1910,** Robert Millikan)

Millikan's OIL DROP experiment **(Robert A. Millikan, Nobel Prize in 1923)**

An Electric field (E) is applied between the plates, to overcome the gravitational field by viscous force.

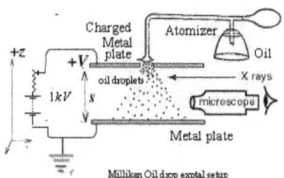

Using non-volatile oil, viscosity (η), density (ρ), and droplets of size radius (a) using an atomizer, moving down with *terminal velocity*, v_1

$$(v_1 = \frac{\ell_1}{t_1})$$

With both metal plates (optically plane) earthed,

Viscous drag = Up thrust

$$\boxed{6\pi\eta a\, \bar{v}_1 = \tfrac{4}{3}\pi a^3 (\rho - \sigma)g}\ ,$$

σ = density of air, ρ = density of oil,

When a potential difference V is applied, $E = \frac{V}{s}$, s = separation between the plates,

$$6\pi\eta a v_2 = Ee - mg$$

$$E\, e_n = m\, g\, \frac{(v_2 + v_1)}{v_1}$$

After ionization $e_n' = mg\dfrac{(v_2' + v_1)}{E v_1}$

$$e = (e_n - e_n') = mg \frac{(v_2 - v_2')}{E v_1}$$

29.3.   QUANTIZATION OF ENERGY OF EM RADIATION,

29.3.1 Planck's fabulous insight (1900)

Planck postulated that matter can absorb and emit EM radiation only in energy bundles called quanta, whose size is

| Quanta | $\propto$ (Frequency of radiation) .

BLACK BODY: an ideal emitter and absorber of radiation Cavity radiation approximates blackbody thermal radiation.

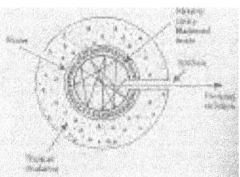

An empirical mathematical formula that would accurately reproduce the shape of the curve of Emission versus wavelength of radiation from the cavity for different temperatures are:

29.3.2 The Wien Law (In 1894, Wilhelm Wien)

$$u(\lambda, T) = \lambda^{-5} f(\lambda, T)$$

$$\lambda_{Max} = \frac{b}{T}$$

$$b = \lambda_{Max} T = 0.29 \; cmK \;, \quad b = 2.90 \; x \; 10^{-3} m \; K$$

29.3.3 The R-J Law:

$$u(v, T) = \rho(v) \, \overline{E}(v) = 8\pi \frac{v^2}{c^3} k_B T$$

29.3.4 The ULTRA-VIOLET CATASTROPHE:

While R-J Law failed at the UV end of this spectrum of thermal distribution, the Wien' Law failed at the red end of the spectrum as shown by the Otto Lummer-& E Pringsheim experiment.(1900).

This disagreement in the UV region is referred to as the *ultraviolet catastrophe* in thermal radiation, a serious paradox for classical mechanics.

29.3.5 PLANCK'S RADIATION DISTRIBUTION LAW:

29.3.5.1 Max Planck (1900) found a formula for an ingenious interpretation between the high frequency Wien law and the low frequency R-J law. Statistical method for waves by analogy with Maxwell's gas particles using equipartition of energy was applied.. By using the MAXWELL BOLTZMANN distribution, *viz.*,

$$N = N_0 e^{-\varepsilon/k_B T}$$

Boltzmann constant $= \boxed{k_B = 1.3805x10^{-23} \, JK^{-1}}$

29.3.5.2 Quantum hypothesis,

viz., $\boxed{\varepsilon = n \, \hbar \, \omega \equiv n \, h \, v}$, where $n = 0, 1, 2, \ldots$

$\boxed{\text{Planck constant } \hbar = (h/2\pi) = 1.054 \times 10^{-34} \, J - s \quad \text{(Scalar)} \quad (M^1 \, L^2 \, T^{-1})}$

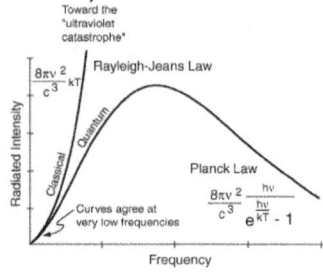

$\boxed{\hbar = (h/2\pi) = 1.054x10^{-34} \, J - s}$,

29.3.5.3 The *angular frequency* of the oscillation

$\boxed{\omega = 2\pi v}$,

29.3.5.4 Planck's distribution law

is the expression for the distribution of the maximum intensity of radiation in the spectrum of the blackbody.

$$\boxed{du(v,T) = \left[\frac{8\pi \, h \, v^3}{(e^{hv/k_B T} -1)} \right] \frac{1}{c^3} dv}$$

$$S_v = \left[\frac{8\pi \, h \, v^3}{(e^{hv/k_B T} -1)} \right] \frac{1}{c^3}$$

$$S_\lambda = \left[\frac{8\pi \, h \, c}{(e^{h \, c/\lambda \, k_B T} -1)} \right] \frac{1}{\lambda^5}$$

Planck's hypothesis was a revolutionary break from classical electromagnetic theory based on Maxwell's equations. The Wien's law and the R-J law were found to be the short wave and long wave limits, respectively, of the Planck law. Accordingly, the energy density u_λ.

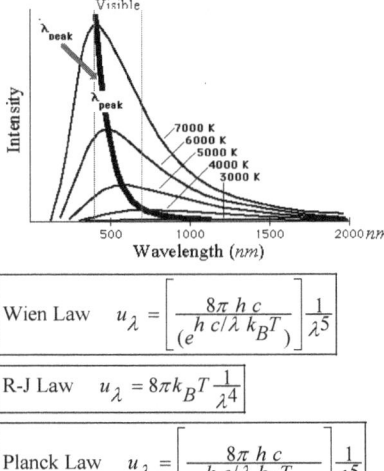

Wien Law	$u_\lambda = \left[\dfrac{8\pi\,h\,c}{(e^{h\,c/\lambda\,k_B T})} \right]\dfrac{1}{\lambda^5}$
R-J Law	$u_\lambda = 8\pi k_B T \dfrac{1}{\lambda^4}$
Planck Law	$u_\lambda = \left[\dfrac{8\pi\,h\,c}{(e^{h\,c/\lambda\,k_B T}-1)} \right]\dfrac{1}{\lambda^5}$

This fits experimental data perfectly!

29.4.1 Discovery of the PHOTON (Photon hypothesis) / Photoelectric Effect (Albert Einstein, 1905: The Nobel Prize in Physics in1921).

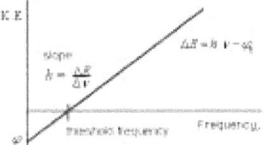

29.4.1.1 THRESHOLD Equation:
Kinetic energy of an electron = photon energy – work function

$$\tfrac{1}{2}m\,v^2 = h\,v - \varphi_0 = e\,V_0$$

V_0 is the *STOPPING Potential*.

$$\boxed{eV_0 = h(v - v_0)}$$

Features:
(i) Maximum kE of electrons (= $e\,V_0$), is independent of intensity of light.
(ii) Maximum kE of electrons is depends linearly of v of light.

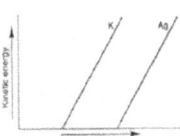

(iii) There is a threshold frequency, v_o, such that for $v < v_o$, no emission of electrons.

(iv) No time delay for emission of electrons.

29.5.1. PHOTONS as PARTICLES (X-RAYS and the COMPTON Effect): (A.H. Compton, in 1923)
29.5.1.1 Predictions:

   (i)   Intensity, $I \propto (1 + \cos^2 \vartheta)$

   (ii) I does not depend on the incident ($\lambda_{inc} \equiv \lambda$ and $\lambda_{sc} \equiv \lambda'$), i.e. $\lambda_{sc} = \lambda_{inc}$.

i.e. COHERENT scattering.
29.5.1.2 Compton Wavelength

$$E = c \sqrt{(m_0^2 c^2 + p^2)}$$
$$\Delta\lambda = \lambda_C \ (1 - \cos\vartheta) = 2\lambda_C \ \mathrm{Sin}^2 \tfrac{\vartheta}{2}$$
$$\Delta\lambda = \lambda_C (1 - \cos\theta)$$
$$\Delta\lambda = 2\lambda_C \sin^2(\theta/2)$$
$$\lambda_C = h / m_0 c = 0.002426 \ nm, \text{ approx.}$$

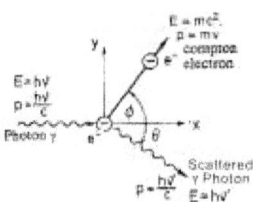

λ_C is called the *COMPTON WAVE LENGTH* of the electron.

29.6.1 BACKGROUND on an atomic view:
Isaac Newton was the first to resolve white light into separate colours by dispersion with a glass prism.
In 1752 Th.Melvill showed emission lines from light emitted by incandescent gases.
29.6.2 The existence of a spectrum:
Balmer in 1885 represented all these lines of the series by a simple formula bearing his name, viz., frequency, ν

$$\boxed{\nu = cR(\tfrac{1}{2^2} - \tfrac{1}{n^2})};\qquad n = 3, 4, 5, \dots}$$

R is called the RYDBERG CONSTANT, in 1906 Lyman, and later Paschen.
29.6.3 J J Thomson (in 1897) discovered the electrons.

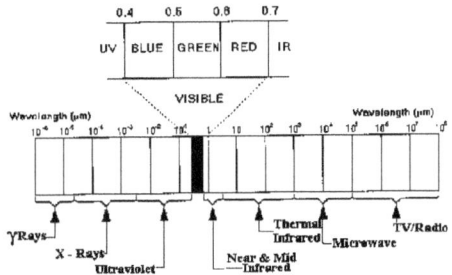

29.7. SPECIAL THEORY OF RELATIVITY:

Relativity is a widely used term. It is generally used to describe everything from the comical version of $E = mc^2$ to concepts about time travel. Here, we are referring to the theory called the Special Theory of Relativity, first asserted by Albert Einstein..

29.7.1.1 Galilean Transformation

A transformation connects the observations in different reference frames.

Viewers in the same reference frame, the space-time coordinates of an event are identical:

Coordinates:
$$x' = x - u\,t$$
$$y' = y$$
$$y' = z$$
$$t' = t$$

Volocities:
$$v_{Ax}' = v_{Ax} - u$$
$$v_{Ay}' = v_{Ay}$$
$$v_{Az}' = v_{Az}$$

29.7.1.2 The Michelson-Morley Experiment (1887)

for showing the invariance of the speed of light in vacuum.

29.7.2 The Lorentz Transformations:

29.7.2.1 In Einstein's Special Theory of Relativity, he laid down two postulates:

29.7.2.2 (i) The laws of physics have the same mathematical form in all inertial reference frames.

(ii) The speed of light through a vacuum ($c = 2.997925 x 10^8\, ms^{-1}$ or 186,000 $miles/s$) is constant as observed by any observer, moving or stationary:

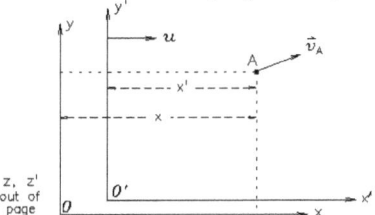

29.7.2.3 For two observers

Using two coordinate systems with origins separated by a fixed distance, a, the transformations are:

Coordinates:

$$x' = \gamma(x - u\ t)$$
$$y' = y$$
$$y' = z$$

$$t' = \gamma(t - \tfrac{x\beta}{c})$$

Velocities:
$$V_A = \left(\frac{V_A{}' + u}{1 + \frac{V_A{}'\beta}{c}} \right)$$

where $\boxed{\beta = \tfrac{u}{c}}$

and $\boxed{\gamma = \dfrac{1}{\sqrt{1 - \beta^2}}}$

$$x' = \frac{x - v\ t}{\sqrt{1 - \frac{u^2}{c^2}}}$$

$$y' = y$$

$$z' = z$$

$$\boxed{t' = \frac{t - v\,x/c^2}{\sqrt{1 - v^2/c^2}}}$$

A new view of Space and Time:

29.7.2.3 The Time Dilation equation for Relativity is:

$$\boxed{t' = \frac{t_o}{\sqrt{1 - v^2/c^2}}}$$

Thus moving clocks run slow.

29.7.2.4 **The length contraction**

$$\boxed{L = L_o\sqrt{1 - v^2/c^2}}$$

$$\boxed{\gamma = \sqrt{1 - \beta^2}}$$

$$\boxed{\beta = \tfrac{u}{c}}$$

The length of a moving object is shortened.

29.7.2.5 *Equivalence of mass and energy relation*

:

$$\boxed{E_O = m_o c^2}$$

$$\boxed{E = m\ c^2}$$

29.7.2.6 Rest mass energy

$$\boxed{m_O = m\sqrt{1 - v^2/c^2}}$$

$$\boxed{E = E_o\,\frac{1}{\sqrt{1 - v^2/c^2}}}$$

$$\boxed{m = m_o\,\frac{1}{\sqrt{1 - v^2/c^2}}}$$

29.7.2.7 Energy-momentum relationship

$$E^2 = p^2 + m^2$$

$$E = c\sqrt{m_o^2 c^2 + p^2} \, ,$$

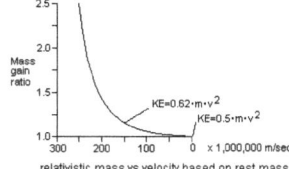

relativistic mass vs velocity based on rest mass

+&+&+&+&+&+&+&+

Chapter 30

SOLID STATE PHYSICS - 1
CRYSTAL LATTICES, POINT GROUPS,
CRYSTAL SYSTEM, CRYSTAL DIFFRACTION
AND CRYSTAL GROWTH

*"The true purpose of education is to train the mind to think,
for that reason it is priceless"* - Albert Einstein

30.1 **Crystals and Lattices**:
30.1.1 Crystal
A **crystalline state** is a solid composed of atoms arranged in an orderly repetitive array.
A class of solids showing neither reticular nor granular structure is termed as non-crystalline or amorphous.
Lattices demonstrate *Discrete Translation Symmetry.*
Different arrangements also include varying degrees of rotational and parity inversion symmetry.
A mineral is a **'naturally occurring homogeneous solid'** with a **definite (but not generally fixed) chemical composition** and a **'highly ordered atomic arrangement'**, usually formed by an **inorganic process.** One of the consequences of this ordered internal arrangement of atoms is that all crystals of the same mineral look similar.
Nicolas Steno (1669) states *Law of constancy of interfacial angles* - angles between corresponding crystal faces of the same mineral have the same angle, even if the crystals are distorted as illustrated by the cross-sections through 3 quartz crystals

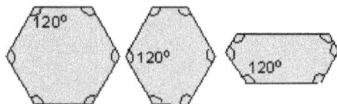

30.1.1.1. Motif (Basis)
The **motif** is a list of the atoms associated with each lattice point, along with their fractional coordinates relative to the lattice point. Since each lattice point is, by definition, identical, if the motif is added to each lattice point, one will generate the entire structure.

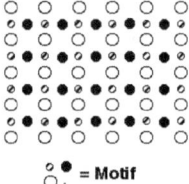

30.1.2 Periodicity in Crystals

30.1.2.1 <u>Space Lattice</u>

A lattice is defined by three fundamental translation vectors, $\hat{a}_1$, $\hat{a}_2$ and $\hat{a}_3$ such that the atomic arrangement look the same in every respect when viewed from a point r, as when viewed from another point r'

$$r' = r + u_1\hat{a}_1 + u_2\hat{a}_2 + u_3\hat{a}_3$$

where u_1, u_2, u_3 are integers. The set of points defined by equation above for all values of u_1, u_2, u_3 define a lattice.

Lattice + Motif = Crystal structure

30.1.2.2 Lattice Translation Vector, $\vec{T}$

$$\vec{T} = u\,\hat{a}_1 + v\,\hat{a}_2 + w\,\hat{a}_3$$

In shorthand, lattice vectors are written in the form:

T = [uvw]

Negative values are not prefixed with a minus sign. Instead a bar is placed above the number to denote that the value is negative

$$\vec{T} = -u\,\hat{a} + v\,\hat{b} - w\,\hat{c}$$

This lattice vector would be written in the form

T = [$\bar{u}$v$\bar{w}$]

Lattice directions are written the same way as lattice vectors, in the form [UVW].

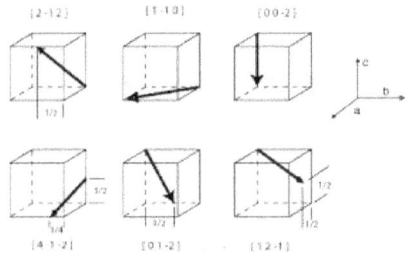

Many crystal systems have elements of symmetry. In these systems, certain sets of directions are symmetrically equivalent to each other. The <u>set of directions</u> that are symmetrically related to the direction [uvw] are written <uvw>.

30.1.2.3 **Primitive Lattice Cell**

A parallelepiped defined by primitive lattice vectors, $\hat{a}_1$, $\hat{a}_2$ and $\hat{a}_3$, is called a primitive cell.

Ideally, the most stable arrangement of polyhedra in a crystal will be that which will MINIMIZE the ENERGY per unit volume

i.e., i) preserves electrical neutrality,
ii) satisfies the directionality and discreteness of all covalent bond,
iii) minimizes strong ion-ion repulsion, and
iv) packs the atoms as CLOSELY as possible, consistent with i), ii) and iii) above.

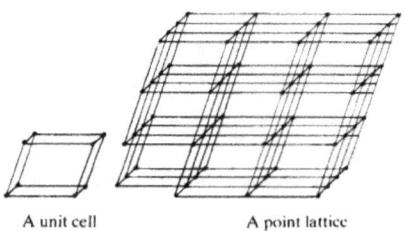

A unit cell   A point lattice

30.1.3 Fundamental Lattice Types, Bravais lattices
A distinct type of lattice is called a Bravais lattice.
a) In 2-D space there are five types of Bravais lattices.

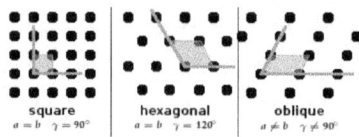

square   hexagonal   oblique
$a = b$ $\gamma = 90°$ $a = b$ $\gamma = 120°$ $a \neq b$ $\gamma \neq 90°$

b) In 3-D, there are <u>fourteen types</u> of Bravais lattices.

20.1.2.1 The Primitive Unit Cell
The most common types of unit cell is the **primitive (P) unit cell** with one lattice point per unit cell;

30.1.2.2 The **face centred** (F) unit cell
An additional lattice points **at the centre of each face** and four lattice points per unit cell; and

30.1.2.3 The **body centred** (I) unit cell
It has a lattice point in the middle of the unit cell and two lattice points per unit cell.

30.1.2.4 Other cell types are the C face centred unit cell and the rhombohedral unit cell.

30.2.1 Lattice Geometry
To define the geometry of the unit cell in 3-Ddimensions, choose a right-handed set of crystallographic axes, x, y, and z, which point along the edges of the unit cell. The origin of our coordinate system is at one of the lattice points

30.2.1.1 Unit Cell
If you know the motif, an easy way to find the number of atoms per unit cell is to multiply the number of atoms in the motif by the number of lattice points in the unit cell

30.2.2 Lattice parameters (Unit cell parameters)
The length of the unit cell along the *x, y,* and *z* direction are defined as *a, b,* and *c,* alternatively, the sides of the unit cell in terms of vectors *a, b,* and *c.*

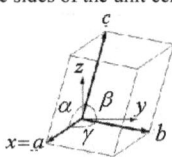

Volume,

$$V = |\vec{a} \cdot \vec{b} \wedge \vec{c}|$$

$a, b, c, \alpha, \beta, \gamma$ are collectively known as the **lattice parameters** (often also called 'unit cell parameters', or just 'cell parameters'.

Sl. No	Bravais Lattice type	Lattice cell Parameters	Crystal System / Characteristic Symmetry
1)	Primitive Cubic (P)	$a = b = c$ $\alpha = \beta = \gamma = 90°$	Cubic
2)	Face Centered Cubic (F)	$a = b = c$ $\alpha = \beta = \gamma = 90°$	Four 3-fold axes along $a+b+c$, $-a+b+c$
3)	Body Centered Cubic (I)	$a = b = c$ $\alpha = \beta = \gamma = 90°$	Four 3-fold axes along $a+b+c$, $-a+b+c$
4)	Primitive Orthorhombic(P)	$a \neq b \neq c$ $\alpha = \beta = \gamma = 90°$	Orthorhombic 3 mutually
5)	Face Centered Orthorhombic(C)	$a \neq b \neq c$ $\alpha = \beta = \gamma = 90°$	perpendicular 2-fold rotation or
6)	Face Centered Orthorhombic(F)	$a \neq b \neq c$ $\alpha = \beta = \gamma = 90°$	perpendicular roto-inversion axes
7)	Body Centered Orthorhombic(I)	$a \neq b \neq c$ $\alpha = \beta = \gamma = 90°$	along a, b and c
8)	Primitive Tetragonal(C)	$a = b \neq c$ $\alpha = \beta = \gamma = 90°$	Tetragonal A single 4-fold
9)	Body Centered Tetragonal (I)	$a = b \neq c$ $\alpha = \beta = \gamma\ 90°$	rotation or $roto$ $-$ inversion
10)	Simple Monoclinic (P)	$a \neq b \neq c$ $\alpha = \gamma = 90°$, $\beta \neq 90°$	axis along c Monoclinic A single 2-fold
11)	B-Face Centred Monoclinic (C)	$a \neq b \neq c$ $\alpha = \gamma = 90°$, $\beta \neq 90°$	rotation or roto-inv along b
12)	Hexagoal (P)	$a \neq b \neq c$ $\alpha = \beta = 90°$, $\gamma = 120°$	A Hexagonal, a 6-fold rota or a roto-inversion, c
13)	Triclinic (P)	$a \neq b \neq c$ $\alpha \neq \beta \neq \gamma \neq 90°$	Triclinic axis in any direction
14)	Primitive Rhombohedral (P)	$a = b = c$ $\alpha = \beta = \gamma \neq 90°$	Trigonal 3 – fold axis along c

30.2.2.1 <u>Wigner-Seitz primitive</u> cell

A primitive unit cell may also be constructed as follows
(i)   Start with an array of points in the (direct) lattice,
(ii)   Connect any one lattice point to all the neighbouring lattice points with lines.

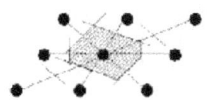

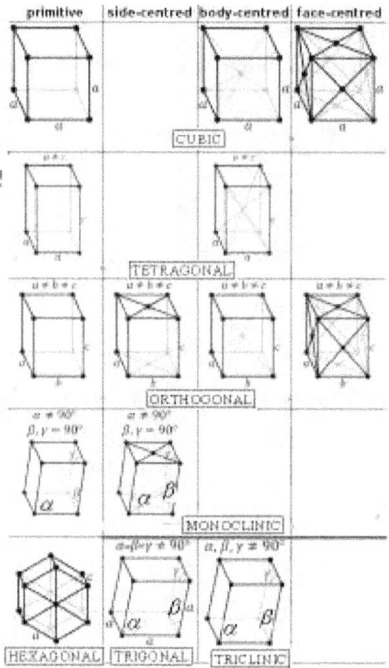

(iii)   At the mid-point of these lines we draw normals (if we started out with a two dimensional lattice) or normal-planes (if started out with a 3-D lattice). The smallest area (or volume) enclosed in this way is called the Wigner-Seitz primitive cell of the direct lattice. All space may be filled up without leaving any gap by joining these Wigner Seitz primitive cells.

30.3   Some common Simple Crystal Structures
30.3.1   Simple Cubic (SC)

This structure is relatively rare amongst the metallic elements and only ^{209}Po appears to crystallize in the SC structure at room temperature and pressure. This is likely because the packing efficiency and coordination number for this structure are low at 0.52 and 6, respectively. Crystallographic data shows the length of a side of the unit cell to be 3.34 $\overset{\circ}{A}$.

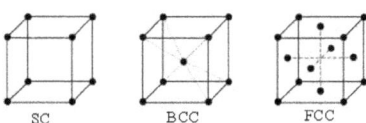

30.3.2   The Body-Centred Cubic (BCC) lattice
is another common crystal structure adopted by metallic solids such as Fe, Cr, Mo and W.

30.3.3 The Face-Centred Cubic (FCC) lattice

This structure is very common amongst metallic elements because it maximizes nearest neighbor interactions (coordination number of 12). The unit cell has a packing efficiency of 0.74. The FCC structure is also known as the cubic close-packed (CCP) structure

30.3.4 The Sodium Chloride (NaCl) Structure

This cell can be described as a simple FCC lattice with a two atom (Na, Cl) basis or two interpenetrating FCC lattices, one of Na and one of Cl, displaced from each other by 0.5 of the body diagonal. This is a common structure for ionic compounds including LiH, KCl, PbS, AgBr, MgO and MnO.

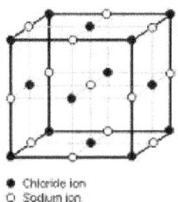
● Chloride ion
○ Sodium ion

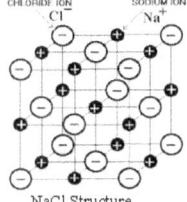

NaCl Structure

30.3.5 The Diamond Structure

Adopted by C (diamond), Si, Ge and grey Sn, which have a strong covalent bonding tendency. The structure is composed of C atoms with tetrahedral bonds. The tetrahedron has the geometric shape of a pyramid with four triangular faces forming isosceles triangles (has the same side length). The tetrahedral (T_d) molecule has four atoms. All the bond angles from the center atom are $109.5°$ The structure is a FCC lattice with two atoms associated with each lattice point, one atom at $(0,0,0)$ and another at $(\frac{1}{4},\frac{1}{4},\frac{1}{4})$.

These two atoms form a basis of diamond structure, and two atoms are the same. Many semiconductors, such as silicon (Si), Germanium (Ge), and Gallium Arsenide (GaAs), diamond structure.

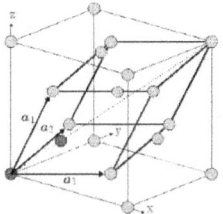
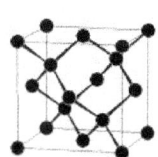

30.3.5 The conventional CaF_2 (fluorite) unit cell.

The fluoride ions form a FCC lattice while the Ca ions are placed in a simple cubic arrangement in the tetrahedral holes. The anti-fluorite structure has the atomic positions reversed.

30.3.6 The cubic ZnS unit cell (Zinc blende)

This is closely related to the diamond unit cell, and has two interpenetrating FCC lattices, one lattice is composed of Zn atoms and the other of S atoms.

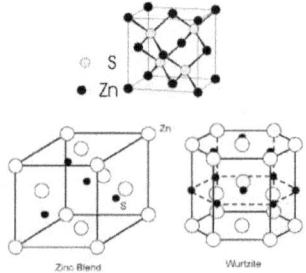

○ S
● Zn

Zinc Blend   Wurtzite

30.3.6 Hexagonal Close Packed (HCP) Structure

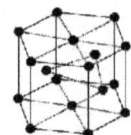

Number of atoms per unit cell of a cubic lattice,

$$n = \frac{a^3 \rho N_A}{M}$$

Simple Crystal structure Data						
Type	SC	BCC	FCC	HCP	NaCl	Diamond
Coordination # N	6	8	12	12		
NN distance 2r	a	$\frac{a\sqrt{3}}{2}$	$\frac{a\sqrt{2}}{2}$			$2\frac{a\sqrt{3}}{8}$
Lattice constant, a	2r	$\frac{4r}{\sqrt{3}}$	$\frac{4r}{\sqrt{2}}$	2r	a	
# of atoms / cell, n	$(\frac{1}{8}x8)$	$\{(\frac{1}{8}x8)+1\}$	4	6		8
# of lattice points, Z	1	2	4			
V of all atoms in cell,$v(\frac{4}{3}\pi r^3)x1$		$(\frac{4}{3}\pi r^3)x2$	$(\frac{4}{3}\pi r^3)x4$	$\left[\pi a^3\right]$	$(\frac{8}{3}\pi r^3)$	
Volume of unit cell, V	$a^3=(2r)^3$	$\frac{64r^3}{3\sqrt{3}}$	$\frac{64r^3}{2\sqrt{2}}$	$\frac{3\sqrt{3}a^2c}{2}$	$(\frac{4r}{\sqrt{23}})^3$	
APF	$\frac{v}{V}$	$[\frac{\pi}{6}=52\%]$	$(\frac{\sqrt{3}}{8\pi}=68\%)$	$\{\frac{\pi}{3\sqrt{2}}=74\%\}$	$(\frac{\pi}{3\sqrt{2}}=74\%)$	$\{\frac{\sqrt{3}}{8\pi}=68\%\}$	$(\frac{\pi\sqrt{3}}{16}=34\%)$
Example		Po	{Cu,Al,Pb,Ag}	Mg	-	(LiH, KCl, PbS, MgO,MnO)	[Ge, Si, GaAs]

30.4 SYMMETRY OPERATIONS

Each operation is performed relative to a point, line, or plane - called a symmetry element.

30.4.1.1 What is a symmetry element?

A *symmetry element* is a geometrical entity (a line, plane or point) with respect to which one or more symmetry operations may be carried out

30.4.1.2 The set of 5 kinds of symmetry operations associated with the symmetry of a molecule. They are: rotation, reflection, and inversion.

	Symmetry Elements and Operations:		
Notation	Symmetry Element	Symmetry Operation	Description
1) E	Identity	Zero	Nothing changes.
2) i	Center of symmetry or inverted center	Inversion	Projects the object through the center (inverts about the center)
3) C_n	n-fold proper axis of rotation	Rotation	Rotates $(360/n)°$ in the clock-wise or anticlockwise direction about the axis
4) $\sigma_h, \sigma_v, \sigma_d$	Mirror plane	Reflection	Reflects across a plane $\perp$, $\parallel$ and iagonal to principal axis
5) S_n	n-fold improper axis of rotation with a plane of reflection	Rotation followed by a reflection	Rotates $(360/n)°$ in the clockwise or anticlockwise direction about the axis followed by a reflection across a plane perpendicular to the rotation axis

$\hat{\sigma}_h$ (*horizontal* plane); in a plane $\perp$ principal axis ,

$\hat{\sigma}_d$ (*dihedral* plane); in a plane containing and $\parallel$ principal axis and bisecting lower order axes, *viz.* dihedral plane of symmetry

$$e.g., \quad \hat{\sigma}_{xy}(x, y, z) \rightarrow (x, y, -z)$$

Note: $\hat{\sigma}^{2n} = \hat{E}$, n = integer

$$\hat{i}\ (x, y, z) \rightarrow (-x, -y, -z)$$

A tetrahedral Structure has the following 24 symmetry operations: 1 E, 3 C_2, 8 C_3 (= 4 $C_3 + 4 C_2^{-1}$), 6σ, and 6 S_4 (= 3 $S_4 + 3 S_4^{-1}$).

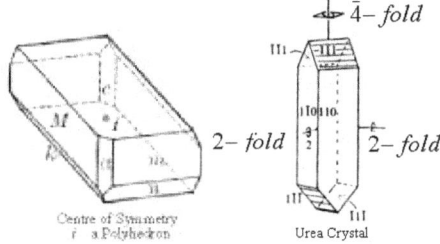

Urea Crystal

Improper Rotations: $\hat{S}_n^{\ k}$

Rotation about an *n*-fold axis followed by reflection through a plane perpendicular to

Improper n-fold Rotation, $\hat{S}_n^{\ k}$, $k = 1, \dots, n$.

When $k = 1$, $n = 1$ $\hat{S}_n^{\ k} = \hat{\sigma}$, Reflection Operation

When $k = 1$, $n = 2$ $\hat{S}_n^{\ k} = \hat{i}$, Inversion Operation

20.4.1.3 What is a symmetry operation?

Each of these **Symmetry Operations** is associated with a Symmetry Element which is a point, a line, or a plane about which the operation is performed such that the crystal orientation and position before and after the operation are indistinguishable

It defines the movement which results in a lattice indistinguishable from the original.

30.4.1.4 Combinations of Symmetry Operations

What is a Point Group?

All symmetry elements of a finite object, passing by a point, define the total symmetry of the object, which is known as the **point group symmetry** of the object.

A Point Group consists of a set of all the elements of symmetry possessed by a lattice and which intersect at a common point. *i.e.,* no translations, everything operates about Centre of Mass of the lattice

There are many **symmetry point groups**, but in crystals they must be consistent with the crystalline periodicity (repetition by translation). On the other hand, for instance, the symmetry axes of order **5** (5-fold axes) are not possible in crystals and therefore only **32 point groups** are allowed in the crystalline state of matter. These **32 point groups** are also known in Crystallography as the **32 crystal classes**.

Point Group • Crystal Translational Periodicity = 32 Crystal Classes

32 Crystal Classes • CentreofSymmetry = 11 Laue Groups

Crystal Translational Periodicity • 32 Crystal Classes = 14 BRAVAIS LATTICES

32 Crystal Classes • 14 Bravais Lattices = 230 Space Groups

32 CRYSTAL CLASSES			SYSTEM
G	GUG	HU(G?H)l	System
1	$\bar{1}$	-	Triclinic
2	$\frac{2}{m}$	m	Monoclinic
3	$\bar{3}$	-	Hexagonal
4	$\frac{4}{m}$	$\bar{4}$	Tetragonal
6	$\frac{6}{m}$	$\bar{6}$	Hexagonal
222	$\frac{2}{m}\frac{2}{m}\frac{2}{m}$	2mm	Orthorhombic
322	$\bar{3}\frac{2}{m}\frac{2}{m}$	3mm	Hexagonal
422	$\frac{4}{m}\frac{2}{m}\frac{2}{m}$	4mm and $\bar{4}$2m	Tetragonal
622	$\frac{6}{m}\frac{2}{m}\frac{2}{m}$	6mm and $\bar{6}$2m	Heaxagonal
332	$33\frac{2}{m}$	-	Cubic
432	$\frac{4}{m}\bar{3}\frac{2}{m}$	$\bar{4}$3m	Cubic

30.4.1.5. Crystal Systems

The rotational symmetry of a crystal places constraints on the shape of the conventional unit cell we choose to describe the structure. On this basis we divide all structures into one of 7 crystal systems. For example, for crystals with 4 fold symmetry it will always be possible to choose a unit cell that has a square base with $a = b$ and $\gamma = 90°$.

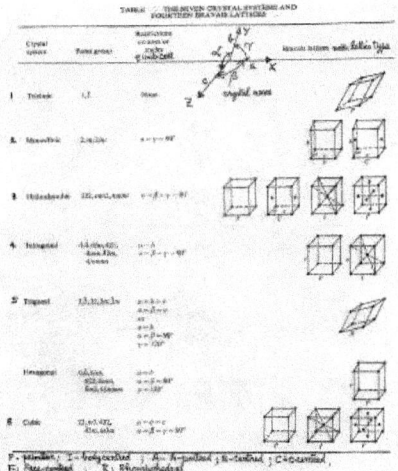

Seven Crystal Systems

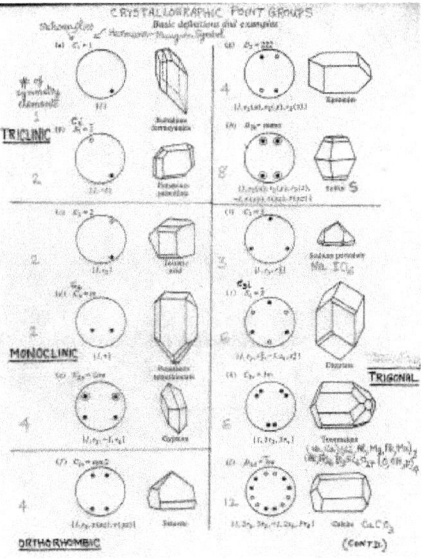

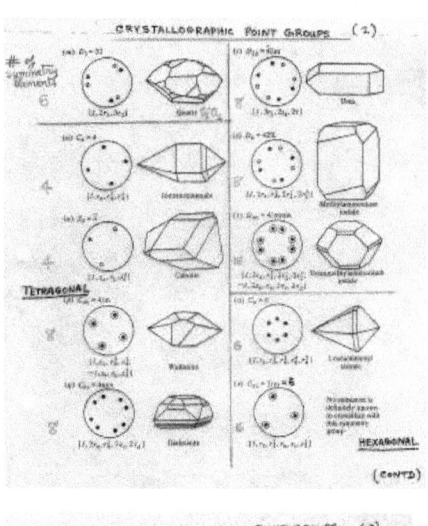

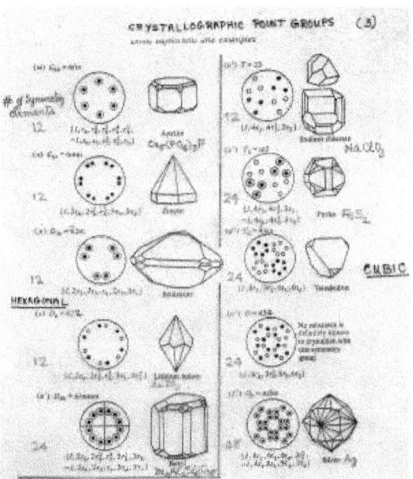

30.4.1.6 Miller Indices

Equation of a plane in space

$$\frac{x}{a}+\frac{y}{b}+\frac{z}{c}=1$$

Intercepts a, b and c

$$h=\frac{1}{a}, k=\frac{1}{b}, and\ l=\frac{1}{c}$$

1. Extend the plane to make it cut the crystal axis system at points (a_1, b_1, c_1)

2. The reciprocals of the intercepts, *i.e.*, $\left(\frac{1}{a_1},\frac{1}{b_1},\frac{1}{c_1}\right)$

3. Multiply or divide by the highest common factor

4. Replace negative integers, say $-h$ by $\bar{h}$

5. If the plane is parallel to an axis, it cuts it ∞, and $\frac{1}{\infty}=0$.

6. Use ordinary braces for single plane, and by double braces to denote a family of planes.

If a plane has intercepts $\infty, 1, \infty$, this means the plane is represented by $(\frac{1}{\infty}\frac{1}{1}\frac{1}{\infty}) = (010)$

Generally a plane in a crystal lattice is denoted by $(h\ k\ l)$

And a family of planes by $\{h\ k\ l\}$.

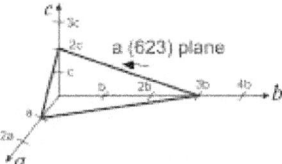

a (623) plane

Angle between two planes, (h_1, k_1, ℓ_1) and (h_2, k_2, ℓ_2)

$$Cos\varphi = \frac{(h_1 h_2 + k_1 k_2 + \ell_1 \ell_2)}{(h_1^2 + k_1^2 + \ell_1^2)(h_2^2 + k_2^2 + \ell_2^2)}$$

30.4.2 Reciprocal Lattice (RL) :

The inverse scaling between real and reciprocal space is based on Fourier transforms. Mathematical representation of reciprocal lattice.

Reciprocal lattice vector is the inverse in magnitude of the real vector and is normal to the planes separating the original vector.

If $\vec{a}, \vec{b}, \vec{c}$ are linearly independent triad set

$$\vec{r} = \lambda\,\vec{a} + \mu\,\vec{b} + \nu\,\vec{c} \neq 0$$
$$\lambda = \mu = \nu \neq 0$$

which is linearly independent set $V_3(F)$, orthogonal and are non-coplanar. then there exists a reciprocal triad, $\vec{a}^*, \vec{b}^*, \vec{c}^*$ defined by

$$\vec{a}^* = [\vec{b} \wedge \vec{c}] / [\vec{a}\ \vec{b}\ \vec{c}]$$
$$\vec{b}^* = [\vec{c} \wedge \vec{a}] / [\vec{a}\ \vec{b}\ \vec{c}]$$
$$\vec{c}^* = [\vec{a} \wedge \vec{b}] / [\vec{a}\ \vec{b}\ \vec{c}]$$

where $\boxed{[\vec{a}\ \vec{b}\ \vec{c}] = \vec{a} \cdot (\vec{b} \wedge \vec{c}) = \vec{b} \cdot (\vec{c} \wedge \vec{a}) = \vec{c} \cdot (\vec{a} \wedge \vec{b})}$

$$[\vec{a}\ \vec{b}\ \vec{c}] = (\hat{i}\ a_x + \hat{j}\ a_y + \hat{k}a_z)\begin{vmatrix} \hat{i} & \hat{j} & \hat{k} \\ b_x & b_y & b_z \\ c_x & c_y & c_z \end{vmatrix}$$

$$|\vec{a}^*| = \frac{1}{d_{100}} = \frac{1}{|a|\ \cos(\gamma - \pi/2)}$$

Reciprocal Lattices of Direct CUBIC lattices					
Direct	Lattice constant	Volume	RL	Lattice constant	Volume
1) sc	a	a^3	sc	$\frac{2\pi}{a}$	$(\frac{2\pi}{a})^3$
2) bcc	$\frac{a}{2}$	$\frac{a^3}{2}$	fcc	$\frac{2\pi}{a}(\pm i \pm j)$, $\frac{2\pi}{a}(\pm j \pm k)$ $\frac{q}{a}$ $\frac{2\pi}{a}(\pm i \pm k)$	$2(\frac{2\pi}{a})^3$
3) fcc	$\frac{a}{2}$	$\frac{a^3}{4}$	bcc	$\frac{2\pi}{a}(+i-j+k)$	$4(\frac{2\pi}{a})^3$

Ewald construction is used to simplify problems in X-ray diffraction, in which RL centred at O as a sphere of radius $\frac{1}{\lambda}$. This gives $Sin\theta_{hkl} = n\, \frac{\lambda}{2d_{hkl}}$.

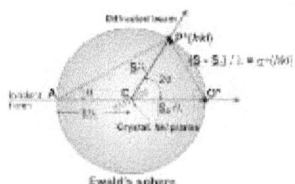

Ewald's sphere

30.4.4 X-ray Diffraction from scattering lattices:

RL structure can be determined by diffraction techniques, Lattice has many 'Bragg Planes' (*hkl*) of atoms in various directions, Each lattice plane will constructively reflect radiation of the proper wavelength λ when incident at the proper angle θ, The interatomic distances of solids are perfectly matched for X-ray wavelengths, The minimum wavelength of X-rays is

$$\lambda_{min} = \frac{1.24\ x10^{-6}}{V}Vm = \frac{1.24\ x10^{-6}}{V(=5\ x10^{-4})} = 0.25A$$

The spacing or distance between parallel planes of atoms in a cubic crystal, d_{hkl} is

$$d_{hkl} = \frac{a}{\sqrt{h^2 + k^2 + \ell^2}}$$

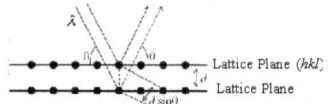

Lattice Plane (*hkl*)
Lattice Plane

Wavelengths of Characteristic X-rays of Targets			
Type	$K\alpha$	$K\beta$	Filter
1) Copper	1.542 A	1.392 A	Ni
2) Chromium	2.291 A	2.085 A	V
3) Iron	1.937 A	1.757 A	Mn
4) Cobalt	1.790 A	1.621 A	Fe
5) Molybdenum	0.711 A	0.632 A	Nb or Zr

William Bragg diffraction condition

$$\boxed{2d_{hkl}\ Sin\theta = n\lambda}$$

For the various **orders of reflection** from a given plane, θ_{hkl}, can be readily found from

$$\boxed{Sin\theta_{hkl} = n\ \frac{\lambda}{2d_{hkl}}}$$

Useful adhesives:

Quartz wax (MP = 150°C)
Sticky Wax (MP = 75°C)
Bayberry Wax (MP = 60°C)
Cheese Wax
Water soluble cement   Rochelle salt

Two common methods for material analysis and characterization are:
1) Powder X-ray Diffraction (PXRD) Method
            In Debye-Scherrer powder method a rectangular strip of X-ray film is mounted in a cylindrical camera. The sample is a finely ground powder, *eg.*, filled in a thin walled capillary tube, and set exactly on the incident X-ray beam, gives the powder pattern exposed by a beam of monochromatic X-rays (W or Cu target X-ray source) collimated by a pinhole system. There are three types of film mounting. In the Straumanis type, the two ends of the film meet midway between the entry and exit of the rays. The diffraction obtained is between $\theta = 0^{\circ}$ to $\theta = 90^{\circ}$

$$\boxed{\theta = \left(\frac{180}{\pi}\frac{1}{4R}\right)S^{\circ}}$$

The camera is designed to have diameter

$$\boxed{2R = \left(\frac{180}{\pi}\right) = 57.3\ mm}$$

So $\theta = \frac{S^{\circ}}{2}$, where S = distance between the arcs of particular to θ. For this camera, S = 1 *mm* corresponds to $\theta = 1^{\circ}$.
            Missing Reflections in the Cubic System

		Missing X-ray Reflections (hkℓ)		
		Type of Cubic Structure		
	P	BCC(I)	FCC	Diamond
Example	CsCl	Vanadium (V)	Coppeer(Cu)	Diamond, Si
Condition	All h, k, ℓ allowed	Only (h+k+ℓ) = even or all odd	h,k and ℓ all must be odd	like in FCC

Information obtained can be
    i)   The average size of crystals or crystallites,
    ii) The degree of cold working (in metals),
    iii) Phase relations in inorganic solid solutions,
    iv) Structural imperfections.

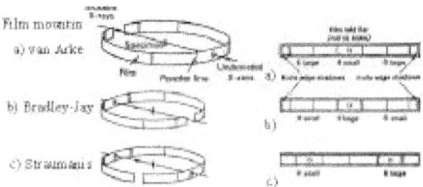

2) The Laue Back reflection Diffraction method.

 A plane X-ray film mounted in a Laue camera gives the Laue diffraction spots of a <u>single crystal</u> on diffraction, using X-rays of heterogeneous wavelengths.

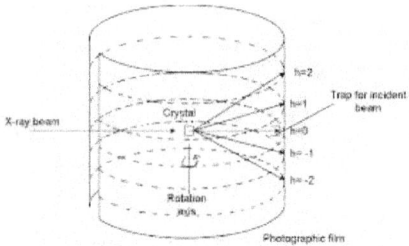

Laue pattern of single crystalline KCl.

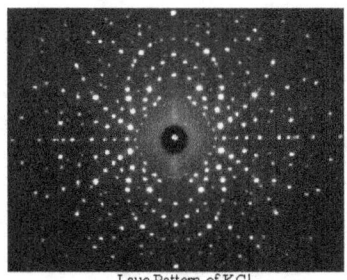

Laue Pattern of KCl

$g(x, y)$ = Diffracting aperture details in XY plane,

$U(X, Y)$ = Diffraction pattern in XY plane of Screen.

$$U(\mu, v) = \iint g(x, y) \cdot e^{i(\mu x + vy)} dx dy$$

$U(\mu, v)$ and $g(x, y)$ are functions forming Fourier Transform pair.

Diffraction Pattern in RL space = FT (Crystal structure)

Inverse FT (Diffraction Pattern in RL space) = Crystal structure

$\mu = kX / L$ and $v = kY / L$ are spatial frequencies,

Greninger Net

The angle between the direct beam (centre) and any spot can be obtained with the Greninger net (EA. Wood, "Crystal Orientation Manual", NY, 1963).

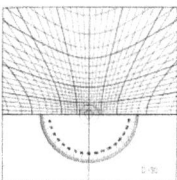

Wulff's net

Wulff's nets and pole figures are two types of stereographic projections. A stereographic projection gives information pertaining to the 3d structure and angles relating structural features in a two dimensional graphic. For crystallography we are interested mainly in the orientation of crystallographic planes through their normals, and the angles between planes. In most stereographic projections a sphere is considered that surrounds a crystal, for instance. Normals to the crystallographic planes intersect the sphere at a pole, P, as shown below to the left for a pole figure (top) and a Wulff's net (bottom).

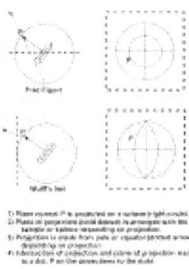

3) Rotation Method

In this method the crystal is rotated at uniform speed (or rotator oscillation between $15°$ and $15°$) about a vertical axis (usually one of the unit cell axes) through its centre (Fig.).

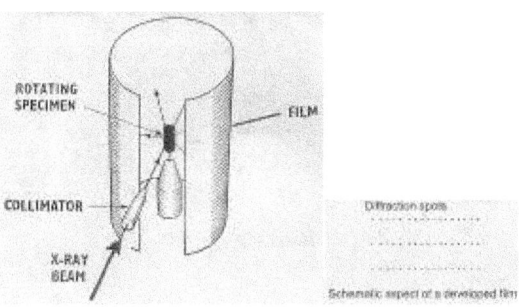

A beam of monochromatic X-rays is collimated falls on the crystal, and pattern recorded on the film inside surface a cylindrical cassette. The diffraction pattern will be a series of spots on straight lines.

The length of the crystal unit cell axis, say a, from

$$a = \frac{h\,\lambda}{Sin\ tan^{-1}(y/r)}$$

Where r = radius of the camera, λ = wavelength of X-rays, y = distance on the film of diffraction line for $h = 0$ to h for spot corresponding to straight-line at y.

4) Weissenberg method - For crystal structure analysis

The Weissenberg camera and its working principle is shown below schematically.

The Weissenberg photograph is extremely easy to interpret.

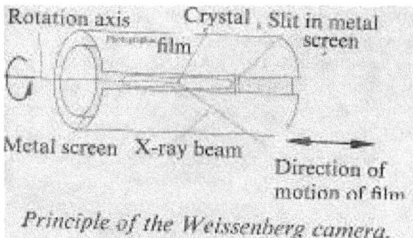

Principle of the Weissenberg camera.

Generally, the axes - the lines containing spots such as h00 and 0k0 - are readily recognized, and from standard charts (or even sometimes without them) the indices of all the other spots can be read off. Information that might take weeks to acquire by the oscillation method could take only minutes with the Weissenberg method, and it would be more certain. The intensities of the spots can be measured more accurately and weaker spots can be detected. Weissenberg's goniometer had everything to commend it.

30.5 **Brillouin Zone construction in 2-D**

The *reciprocal lattice* basis vectors span a vector space that is commonly referred to as reciprocal space, or often, **k** space.

Step # 1 Use the real space lattice vectors to find the reciprocal lattice vectors and construct the reciprocal lattice. When constructing Brillouin zones, they are always centred on a reciprocal lattice point.

Step # 2 Draw a line connecting this origin point to one of its nearest neighbours. This line is a reciprocal lattice vector as it connects two points in the reciprocal lattice.

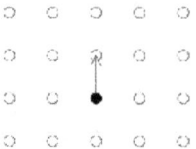

Step # 3 Draw on a perpendicular bisector to the first line. This perpendicular bisector is a Bragg plane

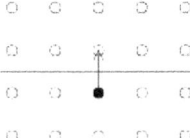

Step # 4 Add the Bragg Planes corresponding to the other nearest neighbours.

The locus of points in reciprocal space that have no Bragg Planes between them and the origin defines the first Brillouin Zone. It is equivalent to the Wigner-Seitz unit cell of the reciprocal lattice. In the picture below the first Zone is shaded.

Step # 5 Draw on the Bragg Planes corresponding to the next nearest neighbours.

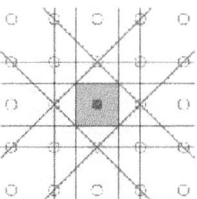

The second Brillouin Zone is the region of reciprocal space in which a point has one Bragg Plane between it and the origin. This area is shaded yellow in the picture below. Note that the areas of the first and second Brillouin Zones are the same.

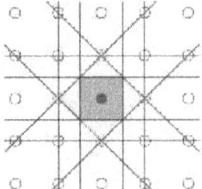

30.6 CRYSTAL GROWTH

Advancement of solid state science depends on the availability of perfect, defect-free single crystals. Crystal grown at elevated temperatures has certain inherent difficulties. Crystalline imperfections are apt to be present due to thermal vibrations, lattice imperfections, etc. During crystal growth, orderly arrangement of atoms takes place followed by evolution of heat. Entropy decreases. The heat is liberated by the system as a result of crystallization. When in dynamic equilibrium between the crystal and in its present phase energy is at the minimum, and no more growth of crystal will occur. Factors like temperature, pressure, chemical potential or strain in the system enable further growth process.

Single crystals can be obtained through one of the following techniques:

(1) Solid-solid phase transition,

(2) Vapour growth by vapour-solid phase transition,

(3) Melt growth,

(4) Solutions growth, from liquid-solid phase transition, and

(5) Gel growth..

30.6.1 Solution Technique

Flux growth: This is a high temperature solution growth of crystals. Here a given high temp a supersaturated solution of the compound is slowly lowered in temp so that saturation of solution and crystal growth takes place. Desired component materials are dissolved in a solvent in a heated crucible forming so-called flux.

30.6.2 Hydrothermal growth

At ambient temp, insoluble compound is made to form solution by increasing temp and simultaneously pressure for crystal formation on lowering temp. Diamond, Calcite, (Quartz) II-VI compounds, etc are crystallized so.

30.6.3 Aqueous solution growth

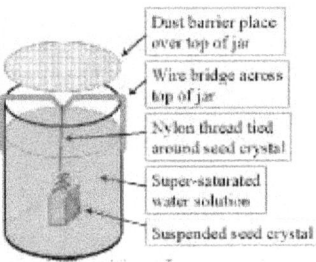

Dust barrier place over top of jar

Wire bridge across top of jar

Nylon thread tied around seed crystal

Super-saturated water solution

Suspended seed crystal

30.6.3 Melt Technique

Bridgeman-Stockbarger used two-zone furnace, Flame-fusion method by Vernuil, crystal pulling method, by Czochralskii, by spontaneous cooling or crystallization on a seed crystal as in Kyrapoulos method.

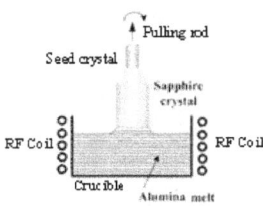

Fig. Czochralski method
Crystal Pulling

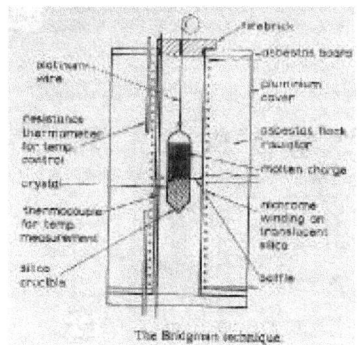

The Bridgman technique.

30.6.4 Bridgeman method (Fig above)
30.6.5 Epitaxial Method
        Liquid phase epitaxy requires a vacuum coating unit and a substrate in which thin crystal films are produce.
30.6.6 Electro-crystallization
30.6.7 Crystal growth under micro-gravity conditions.
30.6.8 **Gel Growth**

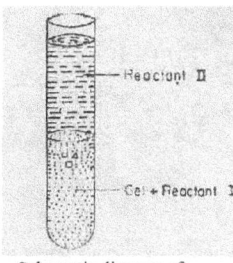

Schematic diagram of
test tube apparatus for crystal
growth by reaction

Chemical gels like gelatin gel (gelatin powder + water at $50°C$ (add formaldehyde. Tetramethoxysilane (TMS) gel, agar-agar gel *etc*, also are used, Single test tube method or a "U-tube" method can be chosen.

In the 'test tube' method, reactant I is incorporated inside the gel and reactant II is diffused into the gel to interact with reactant I and causing nucleation of crystals and growth.

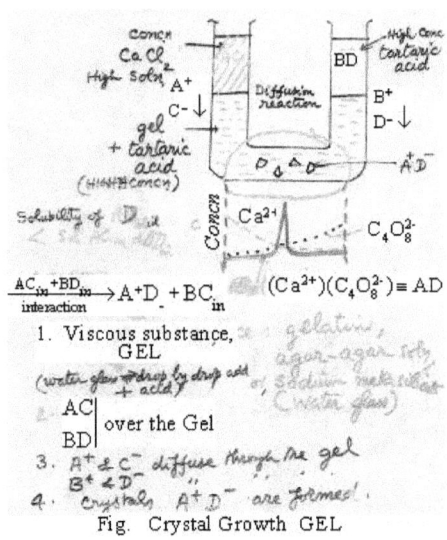

$$\underset{interaction}{AC_{in} + BD_{in} \longrightarrow} A^+D^- + BC_{in} \qquad (Ca^{2+})(C_4O_8^{2-}) \equiv AD$$

1. Viscous substance, GEL
   (water glass + drop by drop add + acid)
   $\left.\begin{array}{c} AC \\ BD \end{array}\right|$ over the Gel
3. $A^+ \& C^-$ diffuse through the gel
   $B^+ \& D^-$ " "
4. crystals $A^+ D^-$ are formed.

Fig. Crystal Growth GEL

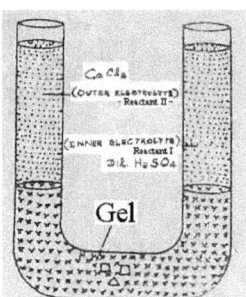

Fig. Schematic diagram of U-tube apparatus for crystal growth by reaction

In the 'U-tube' method, have the silica gel (mix in 10 *cc* at pH = 5, titrating with glacial acetic acid, which is to be mixed with 5 *cc* of 95% Ethanol and allowed to colloidal gel, in 90 *hrs.*), then add to one arm enough of the reactants, plus ethanol (BD)

carefully over the gel top, and to the other arm the nutrient (AC) over the gel. Diffusion takes place A^+ and C^- ions and B^+ and D^- ions. Slowly crystals of AD are formed and grow into perfection.

30.6.9 pH

The pH of an acid solution formed when 0.005 mole of HNO_3 is added to 0.5 l of water.

H-ion concentration = 0.01 mole/l
$$pH = -\log(0.01) = 2$$

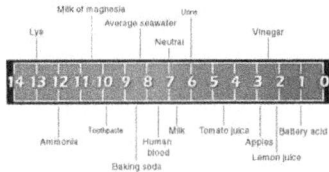

30.7. NON-CRYSTALLINE MATTER
30.7.1. ALLOYS

A material made up of a metal and one other element forms an alloy.

$$\text{Alloy} \equiv \frac{\text{Base Metal}}{\text{Large amount}} + \frac{\text{Metal or non-metal}}{\text{Small amount}}$$

Panchaloha (or Panchadatu)

Panchaloha is a traditional five metal alloy of Gold (Au), Silver (Ag), Copper (Cu), Iron (Fe), Tin (Sn). It has been used for idol making, and other ornaments.

FERROUS ALLOYS
1) ALNICO   Fe (50%) + Al (20%) + Ni (20%) +Co (10%) Powerful Magnets
2) Carbon STEEL Fe (99%) + C (1%)   Strucural beam, Food container, Car body
3) Alloy STEEL   Fe + Ni + Cr + Mo   Bicycle gear, aircraft landing gear
4) Stainless STEEL   Fe (73%) + Cr (18%) + Ni (8 %) + C(1%) Kitchen utensils,
                                                    Pans, hospital equipments
5) Tool STEEL   Fe(79 - 85%) + W (10%) + C (1%)   Machine tools
6) Mang STEEL   Fe(83 - 84%) Mn (15%) + C (1 to 1.5%)   Safe, Rock cutter,
                                                    Armour
Non_FERROUS ALLOYS (Base metal - Aluminium)
   DURAL   Al (95%) + Cu (4%) + Mn (0.5%) + Mg (0.5%) Vehicle and aircrafts,
                                           Beverage can, bicycle rim
            Al + Zn + Mg + Li   "   "
ORNAMENTAL ALLOYS
1) Yellow GOLD   Au + Cu + Ag
2) GERMAN SILVER   Cu (50 - 60%) + Zn (25-35%) + Ni (15-35%)+ Pb (1%)
Cupro-Nickel ALLOYS   Cu +Ni   Coinage
Aluminium-Bronze ALLOYS   Cu (90%)+ Al (10%)
BRONZE   Cu (88%) + Zn (10%) +Sn (10%)
BRASS   Cu (60 - 80%) + Zn (40 - 20 %)
BELL Metal   Cu (80%) + Sn (20%)   Soun producing Bells
MONEL Metal   Cu (28%) + Ni (67%) + Fe (5%)   Ductile, maleeable,
                    anti-crrosion, Door, WindowsScrews, boilers, Sinks
SOLDER   Sn + Pb
WOOD'S Metal   XCu (88%) + Sn (10%) + Zn (1%)
INVAR   Fe (63%) + Ni (36%) + C (1%)
ALNICO + Sm   Magnetic properties 100% more than ALNICO
Phosphor-Bronze   P + Cu + Sn
POTIN   Cu + Zn + Pb + Sn   Coinage by Satavahanas in Hindu Era

30.7.2.   Unusual States of Matter
30.7.2.1. GLASS (Super cooled Liquids)
     crystalline solids: molecules are ordered in a regular lattice
     fluids: molecules are disordered and are not rigidly bound.
     glasses: molecules are disordered but are rigidly bound.

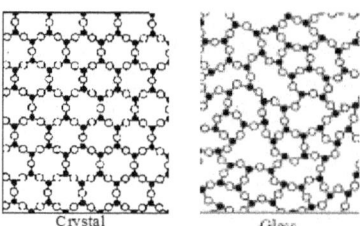
Crystal   Glass

Usually when a liquid is cooled to below its melting point T_m, crystals form and it solidifies; but sometimes it can become *supercooled* and remain liquid below its melting point because there are no nucleation sites to initiate the crystallization. If the

viscosity rises enough as it is cooled further, it may never crystallize. The viscosity rises rapidly and continuously, forming thick syrup and eventually an amorphous solid. The molecules then have a disordered arrangement, but sufficient cohesion to maintain some rigidity. In this state it is often called an <u>amorphous solid or glass</u>.
The glass transition is NOT the same as melting.

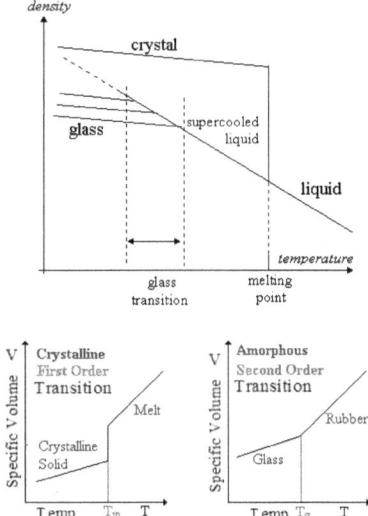

Glass Transition
* Property of the amorphous region.
* Below T_g : Disordered amorphous solid with immobile molecules.
* Above T_g: Disordered amorphous solid in which portions of molecules can wiggle around.
* A second order transition.

When an amorphous polymer is heated, the temperature at which it changes from a glass to the rubbery form is called the <u>glass transition temperature,</u> T_g. *A given polymer sample does not have a unique value of* T_g because the glass phase is not at equilibrium.

1) Soda Lime Glass (Soft glass, Silica Glass)

$$\text{Soda Glass} \equiv \frac{SiO_2}{\text{Silica}} + \frac{Na_2O_3}{\text{Washing Soda}} + \frac{CaCO_3}{\text{Lime stone}}$$

2) Boro-Silicate Glass (Heat resistant glass)

$$\text{Boro-silicate Glass} \equiv \frac{SiO_2}{\text{Silica}} + \frac{Na_2O_3}{\text{Washing Soda}} + \frac{CaCO_3}{\text{Lime stone}} + \frac{B_2O_3}{\text{Boric Oxide}} + \frac{Al_2O_3}{\text{Alumina}}$$

3) Flint Glass (Optical glass)

$$\text{Flint Glass} \equiv \frac{SiO_2}{\text{Silica}} + \frac{KCO_3}{\text{Lime stone}} + \frac{PbO}{\text{Lead oxide}}$$

4) Safety Glass (Laminated Glass)
Thin layer of plastic between two glass plates.

5) Plexi Glass (Glass substitute) (Vinyl Polymer)(Lucite, Perspex, Crystallite)

Plexiglass, is Poly-methyl Methacrylate, a name trademarked by the chemical

company Rohm and Hass, is made up of methyl methacrylate, $\begin{array}{c} CH_3 \\ | \\ CH_2 = C \ \ COOCH_3 \end{array}$.

The methyl methacrylate is processed into this hard vinyl polymer through free radical vinyl polymerization.

30.7.2.2. COLLOIDS

Thomas Graham, in 1861, discovered. Substances, in solution, such as salt, sugar, copper sulphate diffuse through parchment membranes are called 'Crystalloids'; others such as glue, gelatin, gum Arabic do not diffuse through parchment are 'Colloids'.

8 types of Colloids

TABLE Colloids		
Dispersion medium	Dispersion Phase	Colloid
1) Solid (*eg*., Glass) + Solid (*eg*., metal)	→	Slod Sol (*eg*., Coloured glass)
2) Solid (*eg*., gelatin) + Liquid (*eg*., water)	→	Gel (*eg*., Jelly)
3) Solid (*eg*., Carbon) + Gas (*eg*., Air)	→	Solid Foam (*eg*., Charcoal)
4) Liquid (*eg*., Water) + Solid (*eg*., Blood protein)	→	Sol (*eg*., Blood Plasma)
5) Liquid (*eg*., Water) + Liquid (*eg*.,Liquid fat)	→	Emulsion (*eg*., Milk)
6) Liquid (*eg*., Soap solution) + Gas (*eg*., Air	→	Foam (*eg*., Shaving cream)
7) Gas (*eg*., Air) + Solid (*eg*., Carbon)	→	Smoke (*eg*., Wood smoke)
8) Gas (*eg*., Air) + Liquid (*eg*., water)	→	Mist (*eg*., Fog)

30.7.2.3. Liquid Crystals
30.7.2.4. Plasmas
30.7.2.5. Super-conductors
30.7.2.6. Super-Fluids
30.7.2.7. Metallic Glass
30.7.2.8. Bose-Einstein Condensate

+^+&+&+&+&+&+&+&+&+^+

Chapter 31

NUCLEAR PHYSICS - I

Structure of Matter, Nuclear Decay, Binding Energy,
Segre Chart, Radioactive Decay, Dynamics of Nuclear Decay,
Decay Series, Radioactive Equilibrium,
Detectors, Rutherford Scattering

"Tell me what has politics to do with truth, goodness and beauty"   -Robert Oppenheimer
"For the present I believe that the war will be over long before the first atom bomb is built."
     – *W.Heisenberg,1939*

**"Atomic power can cure as well as kill. It can fertilize and enrich a region as well as devastate it.
It can widen man's horizons as well as force him back into the cave. "-** Alvin M. Weinberg, 1944

"Father of Indian Nuclear Programme."   Dr. Homi Jahangir BHABHA

31.1 STRUCTURE OF MATTER

'The alchemists' efforts during the middle ages resulted in the science of chemistry, which in turn led to the idea of the '**elements**', *i.e.*, the atoms. Although it was thought at first that atoms are indivisible, it is now known that they have definite structure.

31.1.1 Size of Atoms and Nuclei

It was Lord Ernest Rutherford in 1911 who established the fact that the atom consists of a nucleus surrounded by electrons. Their physical dimensions are

Diameter of an atom is $\sim 10^{-8}\, cm = 0.1\, nm$

Diameter of a nucleus is $\sim 10^{-12}\, cm = 10\, fm$

Diameter (classical)) of an electron is $\sim 5\, fm$

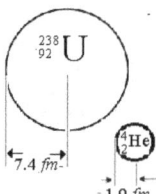

Radii of typical nuclei

31.1.2 TURNING POINTS IN THE DEVELOPMENT OF NUCLEAR PHYSICS

The idea that matter consists of an assembly of a-toms was derived from Democritus (430 BC). Chemists had no answer to the question, say, why one gram of oxygen combines always with eight grams of hydrogen, not in any other proportion, as the resulting *water molecule* contained a fixed number of **atoms** of each kind. John Dalton (1803 - 8) concluded from his work that 2 atoms of hydrogen and one of oxygen combined to create a molecule H_2O. Michael Faraday derived in 1833 the laws of electrolysis, each atom or molecular fragment carried a **fixed electric charge**. A chronological listing of discoveries that shaped the ideas on atoms and nuclei is provided below:

1. Dimitri Ivanovich Mendeleyev (1872) was responsible for finding Periodicity of valence & developing the Periodic Table of Elements.
2. Antoine-**Henri Becquerel** (1896) discovered natural radioactivity, in Uranium salts, revealing that 'atoms are no more indivisible',
3. Joseph John Thomson (1897) discovered electrons
4. **Marie Curie** (1898) chemically separated radioactive Radium from the ores.
5. Lord Ernest Rutherford (1899) identified alpha rays and beta rays
6. Paul Ulrich Villard (1900) discovered gamma rays
7. **Albert Einstein** (1905) announced the equivalence of mass and energy equation: $E = m\ c^2$ is a cornerstone in the development of nuclear energy.
8. Frederick Soddy (1910) discovered statistical decay law in nuclear physics..
9. Victor Hess (1910) discovered cosmic rays
10. Ernest **Rutherford** (1911) formulated the Nuclear model of the atom.
11. J. J. Thomson's (1912) was responsible for the discovery of isotopes.
12. Robert A. Millikan (1913) measured experimentally the charge on an electron
13. H.G.J. Moseley (1913) introduced the concept of atomic number, from X-ray spectra.
14. Niels Hendrik David **Bohr** (1913) who gave the first successful physical model for the Hydrogen atom.
15. E. Rutherford (1919) discovered the phenomenon of nuclear transmutation by the disintegration of Nitrogen (Induced nuclear transmutation).
16. Otto Hahn & Lise Meitner (1921) discovered nuclear isomers.
17. Paul A. M. Dirac (1928) predicted the positron.
18. G. Gamow, R. Gurney & E. Condon (1928) who successfully explained Alpha-decay by using Quantum Mechanics.
19. **Ernest Orlando Lawrence** (1930) invented cyclotron
20. **Wolfgang Pauli** (1930) predicted the neutrino
21. Sir James E. Chadwick (1932) discovered the neutron.
22. **Werner Karl Heisenberg** (1932) gave *n-p* hypothesis.
23. ,Carl D. Anderson (1932) discovered the positron, the first antiparticle.
24. **Enrico Fermi** (1934) gave the theory of Beta-decay.
25. Otto Hahn, Lise Meitner, Fritz Strassmann (1934) split the atom for the first time; the epochal experiment
26. Hidekei Yuckawa (1935) gave theory of Meson as exchange particle between nuclear constituents.
27. Carl D. Anderson & S.H. Neddermeyer (1936) discovered mu-mesons.
28. Lise Meitner and Otto R. Frisch (1938) are responsible for Uranium Fission.
29. Hans A. Bethe (1938) showed that nuclear fusion is responsible for power in the Sun.
30. Leo Szilard and Walter Zinn (1939) demonstrated that fission reactions to be self-sustaining due to nuclear chain reactions.
31. Enrico Fermi (1942) designed the Atomic pile (first nuclear reactor), a sustained controllable nuclear chain reaction.
32. **Manhattan Project** (1942) was set up in the USA under the command of Brigadier General Leslie Groves. Scientists recruited to produce an atom bomb included J. Robert Oppenheimer (USA), David Bohm (USA), *etc.* Use of uranium and

plutonium. The first three completed bombs were successfully tested at Alamogordo, New Mexico on 16th July, 1945.

33. Julius Robert Oppenheimer (1943), associated with about 200 of the best scientists, designed "**Little Boy**" $^{238}_{92}$U bomb (dropped over Hiroshima) and "**Fat Man**" plutonium bomb (dropped over Nagasaki).

34. Bombing Hiroshima (August 6, 1945) and Nagasaki (August 9 1945) in Japan.

35. Cecil Frank Powell (1946) discovered the Pi-meson.

36. Maria Goeppert-Mayer (1946) developed her "nuclear shell model".

37. Edward Teller (1952) leads a team to build the first Hydrogen bomb.

38. Clyde L. Cowan, Jr. and Frederick Reines (1955) observed the mysterious particle neutrino.

39. Tsung-Dao Lee & Chen NingYang (1956) observed the violation of conservation of parity in beta-decay.

40. Glenn Seaborg (1944 -1958) discovered 8 elements related to uranium, *viz.*,: americium, curium, berkelium, californium, einsteinium, fermium, mendelevium, and nobelium. When element 106 was discovered, it was named after him, seaborgium.

41. **Murray Gell-Mann** (1963) discovered the Quarks.

42. Murray Gell-Mann, George Zweig, Oscar Greenberg, Yoichiro Nambu and Yuval Ne'eman (1977) developed quantum chromo-dynamics (QCD) theory of strong interactions.

43. (1964) the first three quarks (up, down, and strange) are hypothesized.

44. **Steven Weinberg**, *et al.* (1970s) gave the Standard Model of nucleus as a fundamental and well-tested physics theory to describe the building blocks of the Universe.

45. (1974) Evidence for a fourth quark was found in November of 1974. Two experiments simultaneously announced the discovery of a meson with a mass of about 3.1 GeV/c^2, called the J meson by one group and the ψ meson by the second. It was later determined to be a combination of charm and anti-charm quarks. Since neither group had priority on the discovery, the meson is now called J/ψ. Like many particles discovered in the 20th Century, it is given the name "charmonium".

46. (1977) The discovery of the bottom quark

47. (1995) Mass of the top quark was determined. The top is so massive and short lived that it does not live long enough to combine with other quarks to form a hadron. In fact the top quark is more massive than many atoms.

48. Maxim Polyakov, Dmitri Diakonov, and Victor Petrov (1997) predicted the existence of a **pentaquark** with a mass about 50 % heavier than that of a hydrogen atom. Atoms are formed of two types of elementary particles, *viz.*, electron and quarks. The discovery of quarks – particles combination makes up the protons and neutrons present in the nuclei of atoms.

49. (2003) Strong evidence for the existence of the pentaquark.

50. **Higg's boson** (nick name The GOD particle, a quantum excitation of the Higg's field) (Nobel Prize in Physics, 2013, for Peter Higgs and Francois Englert) is so central to the state of physics today, so crucial to our final understanding of the structure of matter, yet it is elusive. It has mass $\cong 125\ GeV$. It is the primary reason for building the SSC (Superconducting Super Collider) at the CERN, near Geneva, Switzerland.

31.1.3   NUCLEUS AND NEUTRON-PROTON HYPOTHESIS
31.1.3.1 Definition of a Nucleon
Nucleon is a generic name for both proton and neutron. The various nuclei are different combinations of neutrons and protons. A nucleus with atomic number Z contains A nucleons, *i.e.*, Z number of protons and $(A - Z)$ number of neutrons.
Neutron-Proton Hypothesis for the nucleus was based on this
31.1.3.2 What is a Nuclide (or Nuclear species)?
Any specific combination of protons and neutrons is called a nuclear species
31.1.4 Classification Systems and Nomenclature
31.1.4.1 A review of some of the commonly used terms
The system for classifying nuclei, based on convenience and tradition is given below..

1.	Proton number (Atomic number)	Z
2.	Neutron number	N
3.	Mass number	$A = (N + Z)$
4.	Stable nuclei	$Z = \dfrac{A}{[1.98 + 0.0155\ A^{2/3}]}$
5.	Nuclide of an element E(A, Z)	$^{A}_{Z}\text{El}$
6.	Isotopes (*iso* $\rightarrow$ *equal; topes* $\rightarrow$ *place*)	Nuclides with identical Z but different N
7.	Isobars	Nuclides with the same A; *eg.*, $^{202}_{80}\text{Hg}$ & $^{202}_{82}\text{Pb}$
8.	Isotones	Nuclides with constant N, but different Z.; *eg.*, $^{13}_{6}\text{C}$ & $^{14}_{7}\text{N}$
9.	Isomers species)	Two nuclides (Nuclei of the same in different excited states of which at least one is '*metastable*'. $^{A}_{Z}\text{El}$ & $^{Am}_{Z}\text{El}$
10.	Light nuclei	Nuclei in which $N=Z$; *i.e.*, up to $^{40}_{20}\text{Ca}$
11.	Heavy nuclei	Nuclei having $N > Z$
12.	Isodiapheres	Nuclei having the same excess of neutrons over protons, $(A - 2Z)$; $^{37}_{17}\text{Cl}$ & $^{39}_{18}\text{Ar}$
13.	Atomic mass, M	Exact value of mass of a neutral atom In relation to that of a neutral $^{12}_{6}\text{C}$
14	Atomic mass unit	$1\ amu \equiv 1u = \frac{1}{12}\ (\text{mass of } ^{12}_{6}\text{C})$

31.1.4.2. Other physical Data
Atomic weight $= N\ M_n + Z\ (M_p + m_e) - (\text{binding energy})$

$1\ u = 931.5\ MeV/c^2 = 1.66 \times 10^{-27} kg$

$m_e = 0.00054858\ u = 0.511\ MeV = 9.1094\ x10^{-31}\ kg$

$$M_p = 1.007276 \, u = 938.27 \, MeV = 1.67262 \times 10^{-27} \, kg$$

$$M_n = 1.007825 \, u = 938.78 \, MeV = 1.67353 \times 10^{-27} \, kg$$

$$_1^1H = 1.008665 \, u = 939.57 \, MeV = 1.67493 \times 10^{-27} \, kg$$

Radius of a nucleus is $R = r_0 \, A^{1/3}$ with $r_0 = 1.2 \times 10^{-15} \, m = 1.2 \, fm$

For example,

$$R_{He} = (1.2 \, fm)(4)^{1/3} = 1.9 \, fm;$$
$$R_{Cu} = (1.2 \, fm)(64)^{1/3} = 4.8 \, fm;$$
$$R_U = (1.2 \, fm)(238)^{1/3} = 7.4 \, fm.$$
$$R_{Am} = (1.2 \, fm)(243)^{1/3} = 7.5 \, fm.$$

31.2   NUCLEAR DECAY
31.2.1   Nuclear Species

Of the 6,000 species of nuclei that can exist in the universe, about 2,700 are known, but only 270 of these are *stable*. The rest are *radioactive*, that is, they spontaneously decay. The driving force behind all **radioactive decay** is the ability to produce products of greater stability than one had initially. In other words, radioactive decay releases energy and because of the high energy density of nuclei, that energy release is substantial. The phenomenon of radioactivity has played a significant role in the development of both atomic and nuclear physics

Nuclei can undergo a variety of processes resulting in the emission of radiation of EM (X-rays and gamma rays) or corpuscular type ($\alpha-$, $\beta-$, and positrons, internal conversion electrons, Auge electrons, neutrons, protons, fission fragments, among others).

31.2.2   Nuclear Disintegration

Antoine Henri Becquerel (1896) discovered natural radioactivity in uranium. In 1898 Marie Sklodowska and Pierre Curie discover polonium and another new radioactive element, which they name "radium" The three distinct types of accelerated particles from radioactive decay are named after the first three letters of the Greek alphabet: $\alpha-$ (alpha), $\beta-$ (beta), and $\gamma-$ (gamma) separated by a magnetic field.

31.2.3   THE CHART OF THE NUCLIDES

The Periodic Table of elements is of limited use to the nuclear physicist, as it gives only limited information about the nuclear properties of an element. The Chart of the Nuclides is a plot of nuclear N *versus* Z.

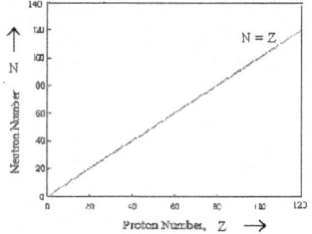

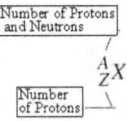

Any nuclide may be specified by the letter X or El with specific N and Z;

$$i.e., \text{ by } {}^{3}_{1}H, {}^{4}_{2}He, {}^{141}_{56}Ba, {}^{235}_{92}U, \text{ etc,}$$

Isotopes; (Nuclides with the same Z) *eg.*, ${}^{206}_{82}Pb$ and ${}^{214}_{82}Pb$.

Isobars (Nuclides with the same A); *eg.*, ${}^{234}_{91}Pa$ and ${}^{234}_{92}U$.

Isotones (Nuclides with the same N), *eg.*, ${}^{31}_{15}P$ and ${}^{32}_{16}S$

Nuclear charge: + Ze

    Size: Fermis

31.2.4 Radius *versus* Atomic Mass

Radius $R \sim A^{1/3}$	unit $Fm\ (=10^{-15}m)$	(scalar)	$(M^0\ L^1\ T^0)$

31.2.5 The Strong Nuclear Force

Protons which would otherwise strongly repel at close distances are held in place by an extremely strong, but extremely short range force called the *strong* force. The nucleon-nucleon potential is as shown:

31.2.6. Alpha-particle Potential Energy in and near Nucleus.

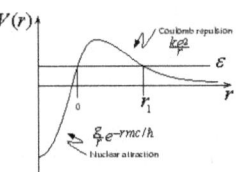

31.2.7. Einstein's mass energy relation

$$E = mc^2$$

Here the law of conservation of mass <u>does not hold good</u>; but the law of conservation of energy is valid.

31.3.1. **The BINDING ENERGY OF A NUCLEUS**

Nuclear mass (M) and Binding energy (B):

$$B = (Z\,M_p + N\,M_n)\,c^2.$$
$$= (Z\,M_H + N\,M_n - Z\,m_e)\,c^2.$$

The total mass of a nuclide is not equal to the total mass of the constituent neutrons and protons. The larger the *binding energy* of a nucleus, the more stable it is.

31.3.2. Atomic Mass Unit (*u*)

1 *amu*	unit $1u = 1.6605\,x10^{-27}\,kg$	(scalar)	$(M^1\ L^0\ T^0)$

$$1\ u = 931.5\ MeV\ /\ c^2$$

31.3.3. Binding Energy per Nucleon (B/A)
The basis for comparing binding energies is B/A
Nuclei with the largest binding energy *per nucleon* are the **most stable**.
Binding energy per nucleon curve:

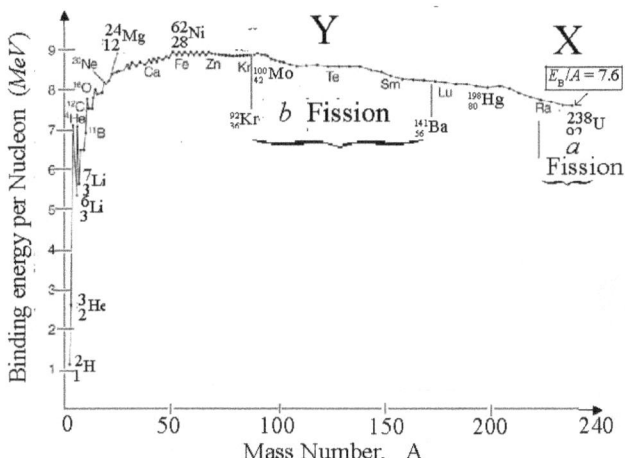

For example, in the case of $^{27}_{13}Al$ nucleus,

$$B\ /\ A = 225.0 MeV\ /\ 27 nucleons$$
$$= 8.332 MeV\ /\ nucleon$$

31.3.4. MASS DEFECT
The difference in mass between the mass of the nuclide and that of its constituent neutrons and protons is called the mass defect.

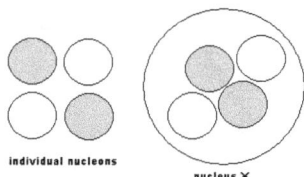

individual nucleons

nucleus X

Mass of the sum of individual nucleus > mass of nucleus X.

ΔM = mass of N neutrons + mass of Z protons – mass of nucleus, $^{Z+N=A}_{\quad Z}X$

By mass-energy relation, binding energy = $\Delta M\ c^2$.

31.3.5. PACKING FRACTION

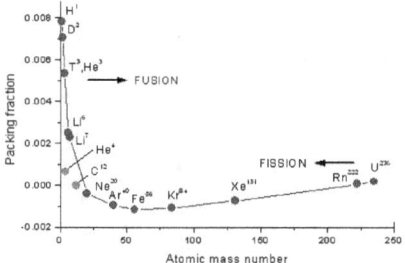

31.4.1. Neutron Number *versus* Proton Number (*N-Z* Plot)
Nuclear stability: Chart of the nuclides; **Segre Chart**:

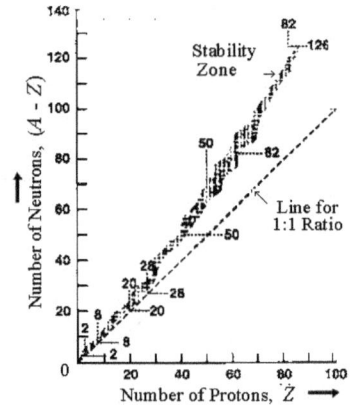

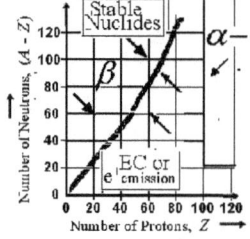

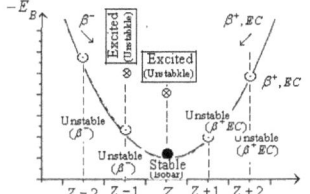

Isobar cut across the valley of stability showing the different kinds of nuclei

31.5. **RADIOACTIVE DECAY**

31.5.1 Discovered radioactivity (Henri Becquerel, 1896)

They are spontaneous - not affected by temperature chemical reactions or by any external influences.

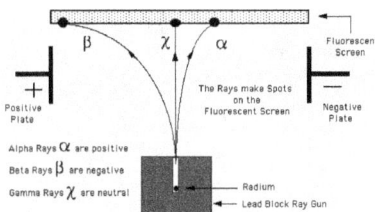

31.5.2. EXPONENTIAL DECAY OF A RADIOACTIVE SAMPLE

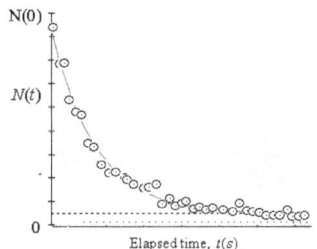

31.5.3. **Activity**

The disintegration of an unstable nucleus, at any moment, is a random physical phenomenon. $N(t)$ = The number of unstable nuclei obeys an exponential law:

$$\frac{dN_t}{dt} = \lim_{\Delta t \to 0} \frac{\Delta N}{\Delta t} = -\lambda N_t$$

31.5.4. Radioactive decay Law:

$$\int_{N_0}^{N_t} \frac{dN_t}{dN} = \int_0^t -\lambda \, dt$$

$$N_t = N_0 \, e^{-\lambda t}$$

Radioactive Decay called Spontaneous
Radioactive Decay Law called statistical

31.5.5. **Radioactivity Units**

A = number of disintegrations per second, *activity*

Marie Sklodowska Curie discovered polonium, which she named after her native country.

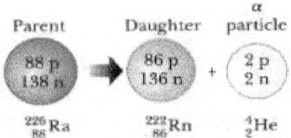

Parent   Daughter α particle

$^{226}_{88}Ra$ $^{222}_{86}Rn$ $^{4}_{2}He$

Historically, another unit was used. One curie (or $1Ci$) is the activity of 1 g of pure

Radium *i.e.*,

$$1 \; Ci = \text{The activity of 1 } g \text{ of Radium } ^{226}_{88}Ra$$
$$= 3.7 \times 10^{10} \; Bq \text{ (disintegrations per second)}$$
$$1mCi = 37MBq$$
$$500 \; kBq = 13.5\mu Ci$$

31.5.7. Mean Life (τ)

$$\tau = \left. \frac{e^{-\lambda t}}{\lambda} \right|_{0}^{\infty} = \frac{1}{\lambda}$$

31.5.8. Half-life ($\tau_{1/2}$)

An important characteristic of each unstable nuclear species is the "half-life" of the radio-nuclide. The half-life ($\tau_{1/2}$) and the decay or disintegration constant (λ) are connected by

$$\tau_{1/2} = \frac{\ln 2}{\lambda} = \frac{0.693}{\lambda} = \tau \ln 2$$

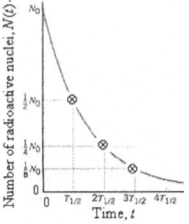

$$\tau = \tau_{1/2} / \ln 2 = 1.44 \; \tau_{1/2}$$

31.5.9. Activity calculation gives half-life.

Activity (Rate of disintegration)

$$R = - \frac{dN_t}{dt} = (\lambda \; N_0) e^{-\lambda t}$$

where $A(0) = (\lambda \ N_0)$ is the activity at the moment zero (initial activity).

Some Half-lives for Radioactive Decay

Isotope		Half-life	Decay Mode
Polonium	$^{214}_{84}Po$	1.64×10^{-4} s	α, γ
Krypton	$^{89}_{36}Kr$	3.16 min	β^-, γ
Radon	$^{222}_{86}Rn$	3.83 ca	α, γ
Strontium	$^{90}_{38}Sr$	28.5 yr	β
Radium	$^{226}_{88}Ra$	1.6×10^{3} yr	α, γ
Carbon	$^{14}_{6}C$	5.73×10^{3} yr	β^-
Uranium	$^{238}_{92}U$	4.47×10^{9} yr	α, γ
Iridium	$^{115}_{49}In$	4.41×10^{14} yr	β

31.5.10. Nuclear Level Diagram

Nuclear Energy Level diagrams provide a compact and convenient way of representing changes that takes place during a nuclear transformation.

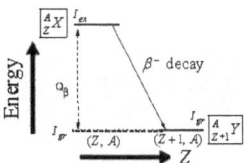

31.6. NUCLEAR DYNAMICS, ALPHA-, BETA-, GAMMA Rays

31.6.1. Radioactivity Alpha (α-) Decay

$$(Z, A) \rightarrow (Z-2, A-4) + \alpha$$

Example: $^{238}_{92}U \rightarrow ^{234}_{90}Th + ^{4}_{2}He$

$\Delta m = 238.0508$ u $- 234.0436$ u $- 4.0026$ u $= 0.0046$ u

$E = 0.0046$ u x 931.5 MeV $= 4.3$ MeV

Decay energy for an α emission:

$$[M_{Parent} - M_{daughter} - M_{He}] c^2$$

Penetrability

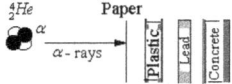

31.6.2. Radioactivity - Beta (β-) Decay

$$^{234}_{90}Th \rightarrow ^{234}_{91}Pa + ^{0}_{1}e$$

a)

$\bar{v}_e$ Anti-neutrino

$\left(^9_0 n\right)$ Neutron $\rightarrow$ $\left(P^+\right)$ Proton

e^- Electron β^-

b) $^1_0 n \rightarrow ^1_{+1}p + ^0_{-1}e + \bar{v}_e$

31.6.3. Penetrability

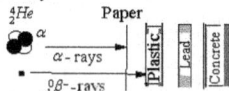

31.6.4. β^--decay.

An example of Negative Beta (β^-) radioactivity is the disintegration of $^{32}_{15}P$:

$$^{32}_{15}P \quad \rightarrow \quad ^{32}_{16}S \quad + \quad \beta^- + \quad \bar{v}_e + \quad Q_{\beta^-}$$

a)

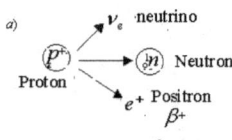

b) $_{+1}^{1}p \rightarrow {}_{0}^{1}n + {}_{+1}^{0}e + v_e$

31.6.5. Electron Capture (EC)

$$^{125}_{53}I \quad + \quad e^- \text{(K-shell)} \quad \rightarrow \quad ^{125}_{52}Te \quad + \quad v_e + \quad Q_{EC}$$

31.6.6. Radioactivity -Gamma ($\gamma-$) rays

$$^{0}_{-1}e + \quad ^{0}_{+1}e \quad \rightarrow \quad 2\gamma \quad \text{(Annihilation radiation)}$$

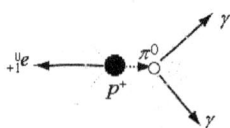

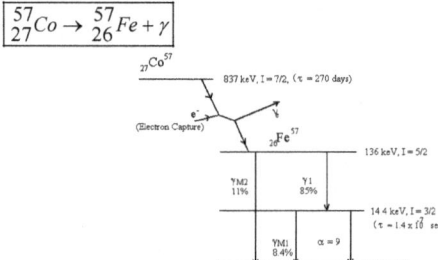

31.6.6.1 **Wavelength (λ) of a Gamma Ray:**
The 1234 Rule:

$$\lambda = \frac{1234 \; eV}{E \; (eV)} \; nm$$

31.6.6.2 Penetrability

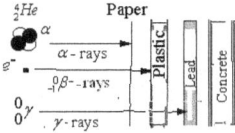

31.6.7 Internal conversion (IC)

IC is an alternative mechanism by which the excited nucleus of a gamma-emitting isotope may rid itself of its excitation energy, resulting in an electron emitted

31.6.8 Neutron Emission

$$^{13}_{4}Be \rightarrow {}^{12}_{4}Be + {}^{1}_{0}n$$

$$^{4}_{2}He\,(\alpha) + {}^{9}_{4}Be \rightarrow {}^{12}_{6}C + {}^{1}_{0}n + Q_n$$

31.6.9 PROPERTIES OF NEUTRONS

31.6.9.1 Mass of neutron:

$$M_n = 1.00866490\ u = 1.67493\ x\ 10^{-27}\ kg\ = 939.5656\ MeV\ c^{-2}$$

31.6.9.2 Charge of neutron:

Neutron is electrically neutral, and so its charge is zero.

31.6.9.3. Intrinsic spin of neutron:

Neutron spins about an axis through its own centre of mass, and has intrinsic spin angular momentum $s = \frac{1}{2}\hbar$.

31.6.9.4 The Decay of Neutron

Enrico Fermi (1934) elaborated the theory of β-decay by visualizing it as the break-up of a neutron into three components; viz., proton, electron and neutrino. This is spontaneous decay of neutron. J.M Robson (1950) determined the half-life of neutron.

$$\tau_{1/2} = 12.8\ \pm\ 2.5\ minutes$$

31.6.9.5 Wave nature of Neutron

The de Broglie wavelength $\lambda = \frac{h}{mv}$

Slow and thermal (cold) neutrons have a Maxwellian velocity distribution with average value corresponding to $0.025\ eV$, the Equi-partition energy of cold neutron

$$\boxed{\lambda\ (cold\ \text{neutron}) \approx 2\ \overset{o}{A}\ (or\ 0.2\ nm)}\ .$$

$$\boxed{\lambda\ (fast\ \text{neutron}) \approx 0.0003\ \overset{o}{A}}\ .$$

31.6.9.7. Electric Dipole Moment of Neutron

Certain theories of nucleus suggest that weak electric dipole moment should exist for a neutron. Experimentally not yet measured, it can be expressed as

Dipole moment $= (q_n) \cdot (x < 2\ x\ 10^{-24}\ m)$

The charge on a neutron is deduced to be

$$\boxed{q_n < \tfrac{1}{700}\ e}\ .$$

31.6.10 The Displacement Laws

Frederick Soddy and Kazimiers Fajans in 1913described the physical nature of the $\alpha-$ and $\beta-$ particles, by means of the Displacement Laws (also known as Fajans &

Soddy Laws).. These enabled explanation for the production of various chemical elements from radioactive processes.

Law 1: When a radioactive parent $^{A}_{Z}El$ loses an α – particle the product element $^{A-4}_{Z-2}El$ is found to be an element displaced two places to the left in the Periodic Table. and lowers the mass by four units

$$^{A}_{Z}El \xrightarrow{\ \alpha\ } ^{A-4}_{Z-2}El$$

Example: $^{226}_{88}Ra \xrightarrow{\ \alpha\ } ^{222}_{86}Rn$

Law 2: An element $^{A}_{Z}X$ is displaced by one unit to the right $^{A}_{Z+1}Y$ in the Periodic Table as a result of the loss of a β – particle, with the atomic mass remaining the same.

$$^{A}_{Z}X \xrightarrow{\ \beta\ } ^{A}_{Z+1}Y$$

Examples: $^{14}_{6}C \xrightarrow{\ \beta\ } ^{14}_{7}N$

$^{239}_{93}Np \xrightarrow{\ \beta\ } ^{239}_{94}Pu$

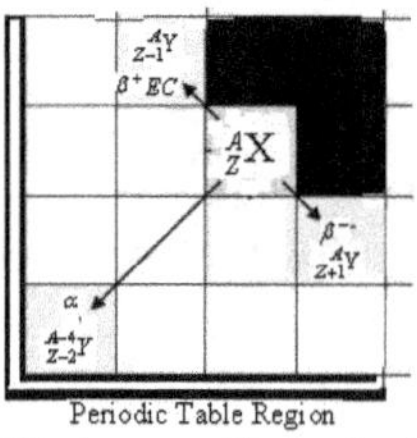

Periodic Table Region

Fig 2.11 Displacement Laws depicted

31.7.   **RADIOACTIVE DECAY SERIES** (Genealogy of Nuclides)

4 different Series of radioactive elements, clearly by a different m value in

$$A = 4n + m$$

where n-value is different for the ancestor in each Series.

31.7.1 The Thorium-232 Series (4 n – Series, $m = 0$)

$$A = 4n$$

The longest lived ancestor of this Series is $^{232}_{90}Th$ having $\tau_{1/2} = 13.9 \times 10^{9}$ $years$ (greater than 5 times the age of Earth). The end product the Series is $^{208}_{82}Pb$, which is stable.

31.7.2 Neptunium-237 Series (4 n + 1 series)

The members of the 4 n + 1 series have mass numbers specified by

$$A = 4n + 1$$

31.7.3 Uranium-238 Series

The members of this series are given by

$$A = 4n + 2$$

31.7.4 Actinium (Uranium-235) Series

$$A = 4n + 3$$

The longest lived ancestor of this Series is $^{235}_{92}U$. The sequences of the α – and β – rays that lead from parent nuclide to stable end product in this series are shown in Fig. 2.12.

The salient features of the four series are listed in Table 2.2.

The 4 Radioactive Series

Mass Number, A	Series	Ancestor	Half life, $\tau_{1/2}$, yrs	End product
1. $A = 4n$	Thorium	$^{232}_{90}Th$	1.39×10^{10}	$^{208}_{83}Pb$
2. $A = 4n+1$	Neptunium	$^{237}_{93}Np$	2.25×10^{6}	$^{209}_{83}Bi$
3. $A = 4n+2$	Uranium	$^{238}_{92}U$	4.51×10^{9}	$^{206}_{82}Pb$
4. $A = 4n+3$	Actinium	$^{235}_{92}U$	7.07×10^{8}	$^{207}_{82}Pb$

31.8 RADIOACTIVE EQUILIBRIUM

Consider the frequently occurring case where the parent A decays to product B which in turn is radioactive, and disintegrates to C.

$$A \xrightarrow{\lambda_A} B \xrightarrow{\lambda_B} C$$

where C is stable.

31.8.1 Ideal Equilibrium

$$\boxed{\lambda_B N_B(t_{Max}) = \lambda_A N_A(t_{Max})},$$

$$\boxed{t_{Max} = \frac{Ln\frac{\lambda_B}{\lambda_A}}{\lambda_B - \lambda_A}}$$

At any other time, the ratio of the daughter to its immediate parent in any three or longer chain is

$$\boxed{\frac{\lambda_B N_B}{\lambda_A N_A} = \frac{\lambda_B}{\lambda_A N_A}\left[1 - (e^{-(\lambda_B - \lambda_A)t})\right]}.$$

31.8.2 Transient Equilibrium for A and B

Transient radioactive equilibrium occurs when the parent nuclide and the daughter nuclide decay at essentially the same rate

$$\boxed{N_A \lambda_A = N_B(\lambda_B - \lambda_A)}$$

$$^{140}_{56}Ba \xrightarrow{300\ hr} {}^{140}_{57}La \xrightarrow{40\ hr} {}^{140}_{58}Ce\ (Stable)$$

31.8.3 Secular Equilibrium

When $\boxed{\tau_{1/2A} \gg \tau_{1/2B}}$,

the decay product generates radiation more quickly.

31.8.4 Permanent Equilibrium

$\boxed{\tau_{1/2A} \gg \tau_{1/2B}}$, and

$$\boxed{N_A \lambda_A = N_B \lambda_B}$$

NEUTRON (n): Detected by J. Chadwick (1932), life-time about 12 minutes.

PROTON, (p): Stablest particle- life time $\boxed{\tau = 10^{40}\ s}$.

31.9   DETECTION OF RADIOACTIVITY

    1)   Ionization chamber
    2)   Proportional counter
    3)   Geiger-Muller tube
    4)   Cloud Chamber (a) Wilson cloud chamber b) Bubble Chamber)
    5)   Photographic emulsion
    6)   Solid state detector
    7)   Scintillation counter

31.9.1. GAS IONIZATION CURVE

A characteristic curve showing the relationship between the number of ion pairs collected per event and the detector voltage for gas filled detectors is shown below.

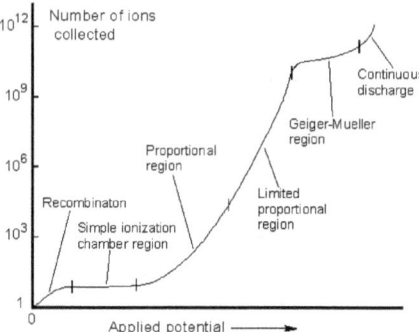

31.9.2 GAS FILLED DETECTORS

Gas filled detectors can be designed to detect any type of ionizing radiation (alpha, beta, gamma, or neutron). They may be filled with air and open to the atmosphere or they may be filled with a specific gas (like Boron Trifluoride (BF_3) gas for neutron detection) and sealed. These consist of a gas filled metal chamber (typically a cylinder) with a wire passing through the center of the chamber

31.9.2.1. Simple Ionization Chamber

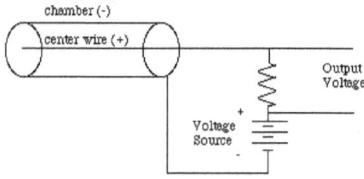

31.9.2.2   Proportional Counter

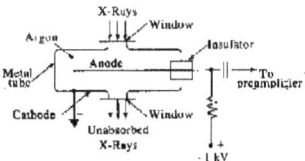

31.9.2.3   Geiger-Muller Counter

31.9.3   **VISUAL DETECTORS**
(i)   Wilson Cloud Chamber
(ii)   Diffusion Chamber
(iii)   Nuclear Emulsion method

31.9.4   **SCINTILLATION DETECTOR**
Scintillator is material which will emit photons when struck by high energy charged particles or high energy photons. Photon strikes metal plate, ejecting electrons which are pulled toward $100\ V$.
anode.

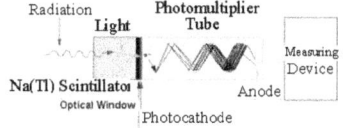

31.9.5   **SEMI-CONDUCTOR (SOLID STATE) DETECTORS**

31.9.6   **HIGH ENERGY DECTECTORS**
(I)   Bubble Chamber
(II)   Spark Chamber
(III)   Cerenkov Detector

31.10   **USES OF RADIOACTIVITY**:
(i)   Radioactive dating Very accurate measurements of the amount of ^{14}C remaining, either by observing the beta decay of ^{14}C or by accelerator mass spectroscopy (using a particle accelerator to separate ^{12}C from ^{14}C and counting the amount of each) allows one to date the death of the once-living things.
(ii)   Tracer techniques
(iii)   Radio therapy
(iv)   Industrial purposes
(v)   Nuclear power

(vi)   the automobile industry–
1.   to test steel quality in the manufacture of cars and to obtain the proper thickness of tin and aluminum
2.   the aircraft industry–to check for flaws in jet engines
3.   construction–to gauge the density of road surfaces and subsurface
4.   pipeline companies–to test the strength of welds
5.   oil, gas, and mining companies–to map the contours of test wells and mine bores, and
6.   Cable manufacturers–to check ski lift cables for cracks.

The isotope ^{241}Am, the isotope ^{252}Cf (a neutron emitter) ate popular for nowadays.

Neutron activation analysis is extremely useful in identifying the chemical elements present in coins, pottery, and other artifacts from the past. A tiny unnoticeable fleck of paint from an art treasure or a microscopic grain of pottery suffices to reveal its chemical makeup. Thus the works of famous painters can be "fingerprinted" so as to detect the work of persons behind forgery.

Magnetic field to separate charged particles with various momenta:

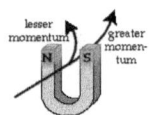

31.11. NUCLEAR MODELS

31.11.1   THE LIQUID DROP MODEL:
The liquid drop model is used to develop an expression for the Binding energy, B.

$$B = c_1 A - c_2 A^{2/3} - c_3 Z(Z-1) A^{-1/3}$$

The first two terms describe the effects of nuclear forces and the third term describes that of the electric forces. The expression gives a qualitative understanding of the B/A *versus* A curve. A more complete expression for B contains more corrective terms in B. A dynamical expression of the liquid drop model provided an early description of nuclear fission.

31.11.2 NUCLEAR SHELL MODEL
The shell model treats nucleons as independent particles; it uses quantum mechanics and the Pauli's Exclusion Principle to explain why N = Z and magic number nuclei are uncommonly stable.

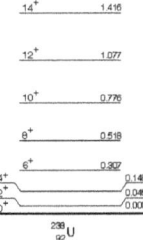

Neutrino spin.

31.12.3   Nuclear energy levels

31.12. MEASUREMENT OF NUCLEAR CHARGE
(Rutherford Scattering of Alpha Particles)

An experimentalist of extraordinary ability, Ernest Rutherford turned his attention to the Plum-pudding model of atom by J.J. Thomson (Devanarayanan, 2005, 2010).

Point mass particle are under impact.

a.   Projectiles, *viz.*, α -particles, are point mass particles
b.   Target nuclei are point mass particles
c. α -particles have positive point charges
d.   Target nuclei are positive point charges.
e.   Coulomb inverse square law force (*i.e.*, electro-static repulsive and central) is the only interaction between these colliding particles, as the distances are small.

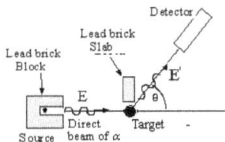

$$Cot\frac{\theta}{2} \;=\; \frac{4\,\pi\,\varepsilon_0\,r^2\,m_\alpha v_\alpha^2}{(Z_\alpha\,Z\,e^2)}\,x$$

Letting, collision radius, $\quad b \;=\; \dfrac{|\,Z_\alpha\,Z\,|\,e^2}{4\,\pi\,\varepsilon_0\,m_\alpha v_\alpha^2}$,

$$x \;=\; b\;Cot\frac{\theta}{2}$$

b is the value of x for which $\theta \;=\; 90°$.

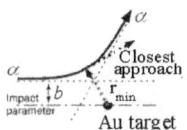

Au target

$$\sigma(\geq\theta) \;=\; \pi\,x^2 \;=\; \pi\,b^2\,cot^2\frac{\theta}{2}$$

+%&%&%&%&%&%+

Chapter 32

NUCLEAR PHYSICS - II

Nuclear Reactions, Fission, Nuclear Reactors, Breeder Reactor,
Reactors in India, Fusion, Particle Accelerators,
Particle Physics, Intrinsic Particle Property

*"We must have the capability. We should first prove ourselves and then talk of Gandhi,
non-violence and a world without nuclear weapons."-*   H. J. Bhabha

32.1. NUCLEAR REACTIONS

A nuclear reaction may be initiated by bombarding a target nuclide with a high energy
particle. Nuclear fission is a particularly distinctive reaction.

32.1.1.   Rutherford's (1918) *ARTIFICIAL TRANSMUTATION OF AN ELEMENT*:

A nuclear reaction is designated by

$$m_i + M_i \rightarrow M_f + m_f$$

or by $M_i(m_i, m_f)M_f$,

where m_i, is the bombarding / projectile, M_i is target nucleus, M_f is residual (or final or

product) nucleus and m_f the emergent particle.

A typical nuclear reaction to remember is

$$\,^4_2He + \,^{14}_7C \rightarrow \,^{17}_8O + \,^2_2H$$

32.1.2   The Q OF A NUCLEAR REACTION:

$$Q = [(M_i + m_i) - (m_f + M_f)]\, c^2$$, if masses are in *grams.*

$$Q_F = [(M_i + m_i) - (m_f + M_f)]\, (931.5 \; MeV)$$, if the masses are in *amu.*

32.1.3   NUCLEAR FISSION: (n, f) reaction

O. Hahn & F. Strassmann (1939) and L. Meitner & O.R. Frisch (1939) discovered neutron induced fission.

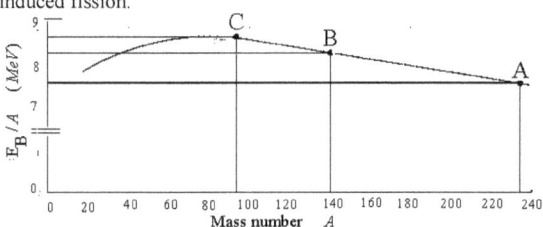

32.1.4 Two important aspects of Fission technologically are:
1.   The fission reaction is **exothermic** about 299 MeV per fission event,
2.   The incident particle $_0^1n$ which initiates the fission of the fissile nucleus $_{92}^{235}U$ is contained among the fission products. to enable Chain Reaction of fission, if the fissile material is above the CRITICAL MASS.
3.   <u>Bohr & Wheeler criterion</u> for spontaneous fission is

$$\boxed{\frac{Z^2}{A} > 47.8}$$

32.1.5 Example of a Nuclear Fission

$$\boxed{_0^1n + _{92}^{235}U \ \rightarrow \ _{55}^{140}Cs + _{37}^{93}Rb + 3\,(_0^1n)}$$

Binding energies calculated from E_a / A Curve

Nuclide	B.E per nucleon E_a / A	Mass number, A	Binding energy $(E_a / A)A$
$_{37}^{93}Rb$	8.7 MeV	93	809 MeV
$_{55}^{140}Cs$	8.4 MeV	140	1176 MeV
$_{92}^{235}U$	7.6 MeV	235	1786 MeV

32.1.6 Distortions of a nucleus undergoing Fission

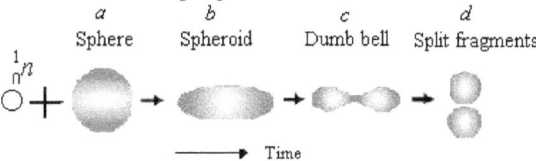

32.1.7 Fission Energy

Mass of the Reactants		Mass of the Products	
$_{92}^{235}U$	235.043924 u	$_{37}^{93}Rb$	92.91699 u
$_0^1n$	1.008665 u	$_{55}^{140}Cs$	139.90910 u
		$3\,(_0^1n)$	3.02599 u
	236.052589 u		235.85208 u

Mass difference $= 236.052589\ u - 235.85208\ u = 0.200509\ u$

$$E_{inst} = (0.200509\ \text{u}) \left(\frac{931.5\ MeV}{\text{u}} \right) = 186.8\ MeV$$

32.2.1. Fission Chain Reaction

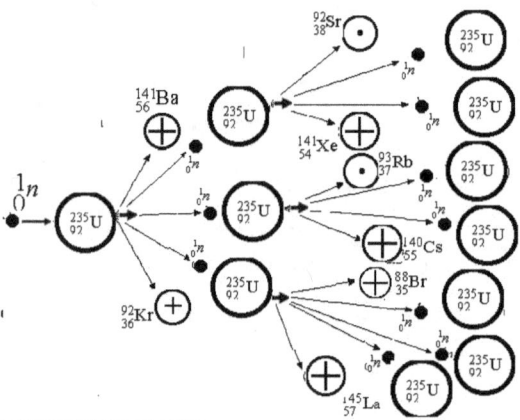

32.2.2 SPONTANEOUS FISSION (SF)

When the fissile nucleus has $\left(Z^2 / A \right) > 47.8$, $Q_F = (E_S + E_C)$ decreases by distortion,

and **spontaneous fission** (SF) occurs, within the characteristic nuclear time $t \approx 10^{-22}\ s$.

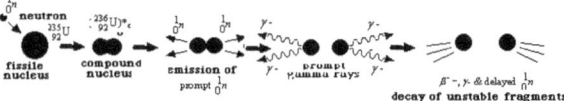

32.2.3 INDUCED FISSION

When a *fissile* nucleus has $\boxed{\dfrac{Z^2}{A} > 47.8}$, it resists fission as $E_S > E_C$.

This resistance, however, has the character of a barrier only. This barrier is the nature of a 'hump' at C, indicated in the PE versus distortion

32.2.4 TERNARY FISSION

R. Present (1941) put the possibility for tri-partition fission. This is thought to take place only 4.3 events for every 10^6 binary fissions.

.32.2.5 Other Products of Fission (fall out or β^- - and γ- radio-activity)

There are more than 30 different modes of fission, in each of which a different pair of fragment nuclei is formed. Ba, La, Br, Mo, Rb, Sb, Te, Kr, I, Xe, Cs are some of them. Fragment nuclei have their Z range from $30 \leq Z \leq 63$; $70 \leq A \leq 160$.

32.2.6 A fission chain is a series of product nuclei with the same mass number A. An example is

$$_{35}^{90}Br \xrightarrow[\tau_{1/2}=1.6\ s]{\beta^-} {}_{36}^{90}Kr \xrightarrow[23\ s]{\beta^-} {}_{37}^{90}Rb \xrightarrow[2.9\ m]{\beta^-} {}_{38}^{90}Sr \xrightarrow[28\ y]{\beta^-} {}_{39}^{90}Y \xrightarrow[64\ hr]{\beta^-} {}_{40}^{90}Zr\ \text{(Stable)}$$

$$^{90}Kr \xrightarrow[33s]{\beta^-} {}^{90}Rb \xrightarrow[2,7min]{\beta^-} {}^{90}Sr \xrightarrow[28year]{\beta^-} {}^{90}Y \xrightarrow[64h]{\beta^-} {}^{90}Zr(stable)$$

$$^{143}Ba \xrightarrow[0,5min]{\beta^-} {}^{143}La \xrightarrow[12min]{\beta^-} {}^{143}Ce \xrightarrow[33h]{\beta^-} {}^{143}Pr \xrightarrow[13,7d]{\beta^-} {}^{143}Nd(stable)$$

32.2.7   Fission of nucleus & PERCENTAGE OF FRAGMENTS produced

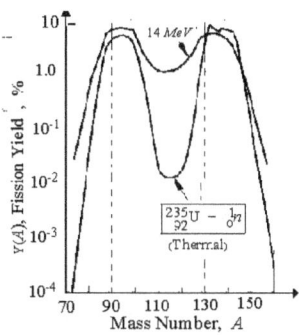

32.2.8   Prompt and Delayed Neutrons by Fission Fragments

A nucleus which has an excess of ${}_{0}^{1}n$ may decay by β^- - or by neutron emission. Neutron emission would most likely occur if, in the process of β^- - decay, the product nucleus is left in an excited state with energy in excess of the binding energy of a ${}_{0}^{1}n$, in that nucleus. The emission is a 'prompt' ${}_{0}^{1}n$, if the emission is within t $\approx 10^{-14}$ s The neutron emitter may have the same $\tau_{1/2}$ for ${}_{0}^{1}n$ as the β^- - decay of the parent nuclide, i.e., several seconds. Such neutrons are known as the 'delayed' ${}_{0}^{1}n$.

32.2.9   Fission Life time: $10^{-22}s$

Neutron fission cross section: for ${}_{92}^{235}U$ 550 barns

Crosssection, σ	Unit $1b = 10^{-28}m^2$	(Scalar)	($M^0 \ L^2 \ T^0$)

32.2.10   Fissionable Materials:

1.   Uranium-235: ${}_{92}^{235}U$ (Number of ${}_{0}^{1}n$ released =2.47).:

2.   Uranium-233: ${}_{92}^{233}U$ (Number of ${}_{0}^{1}n$ released =2.33).

3.   Natural Uranium: ${}_{92}^{234}U(0.0057) + {}_{92}^{235}U(0.7204) + {}_{92}^{238}U(99.2739)$

4.   Plutonium ${}_{94}^{239}Pu$ (Number of ${}_{0}^{1}n$ released =2.89)..

32.2.11   Transuranic Elements

$$^{238}_{92}U + ^1_0n \rightarrow [^{239}_{92}U^*] \rightarrow ^{239}_{92}U + \gamma$$

$$^{239}_{92}U \xrightarrow[23.5\,m]{\beta^-} ^{239}_{93}Np + \beta^-$$

Edwin M. McMillan and Philip H. Abelson discovered the first transuranic element neptunium,

32.3   **NUCLEAR REACTOR**
The basic unit of a nuclear reactor is the chain reaction of U-235. This is started by a slow moving   thermal neutron. To <u>sustain the chain reaction,</u> a minimum quantity of U-235 is needed, which is called the <u>critical mass</u>. A nuclear reactor produces immense amount of heat energy; we have seen   earlier that there is a mass difference between the U-235 and its fission products which leads to release of energy. This energy is converted into electricity in a nuclear reactor.

32.3.1   Atomic Pile

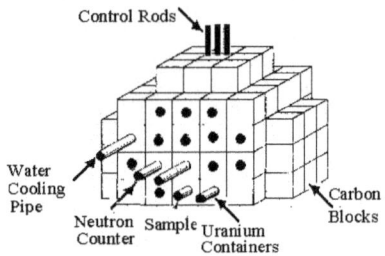

1.   The <u>Moderator is</u> named for graphite blocks that slow down neutrons to the best reaction rate. heavy water (D_2O : deuterium water) is another moderator material.

2.   The <u>Fuel</u> is a fissionable isotope like $^{235}_{92}U$.

3.   The <u>Control Rods,</u> Cd, or boron (B) absorbs neutrons to regulate the rate of sustained reaction.

4.   The <u>Coolant</u>, water, (Carbon dioxide gas, water, heavy water, liquid sodium, etc. are used as coolants) removes the heat of reaction to make steam for turbine generators.

5.   The <u>Shielding</u>, concrete, keeps the radiations inside the reactor. These strong layers prevent the escape of emitted harmful, energetic α – particles, β – particles and γ - rays.

32.3.2   Critical Mass
To maintain a fission chain reaction a certain minimum amount of fissile material must be present so that too many ^1_0n do not escape from the fuel. This amount of material is called the *critical mass.*

32.3.2.1.   World's Safest Nuclear Reactor **ATBR** (2005)

32.3.3. BREEDER REACTORS:

A reactor that produces more fissionable materials than it burns is called a breeder reactor.

(i) Consider the fuel is a mixture of $^{235}_{92}U$ and $^{238}_{92}U$. Cold (n,f) leads to

$$^{235}_{92}U + ^{1}_{0}n \rightarrow (^{236}_{92}U^*) \rightarrow X + Y + (2.5)^{1}_{0}n.$$

The emergent neutrons are absorbed by to give

$$^{238}_{92}U + ^{1}_{0}n \rightarrow (^{239}_{92}U^*)$$

$$^{239}_{93}Np + ^{0}_{1}e + v \rightarrow ^{239}_{94}Pu + ^{0}_{1}e + v$$

The end product $^{239}_{94}Pu$ is a fissionable material; i.e. the non-fissionable $^{238}_{92}U$ is converted to $^{239}_{94}Pu$.

(ii) A second breeding cycle is possible with $^{232}_{90}Th$ to get $^{233}_{92}U$.

32.3.4.1 Advantages

No moderator is required; can use $^{232}_{90}Th$ to get $^{233}_{92}U$ or a mixture of $^{235}_{92}U$ and $^{238}_{92}U$. Breeder reactor is efficient because the key factor in its fuel that gives the largest possible number of $^{1}_{0}n$ released per neutron absorbed. Such a reactor is being built use a mixture of $PuO_2 + UO_2$ as the fuel and fast $^{1}_{0}n$ (~1 MeV) activate fission. Fast neutrons carry energy of at least several keV and therefore travel more than 10^4 times faster than thermal (~1 keV) neutrons. $^{239}_{94}Pu$ in the fuel assembly on absorbing one such fast neutrons undergoes fission with the release of 3 neutrons. Through $^{1}_{0}n$ capture process $^{238}_{92}U$ in the fuel then produce additional $^{239}_{94}Pu$.

Provide limitless supply of fuel for nuclear reactors. This is known as the breeding cycle.

32.3.4.2 Disadvantages:

(i) more expensive to build than the other types of reactors.

(ii) They are also useless without a subsidiary industry to collect the fuel, process it, and transport the $^{239}_{94}Pu$ to new reactors.

(iii) It is the reprocessing of $^{239}_{94}Pu$ that concerns most of the scientists of breeder reactors.

(iv) $^{239}_{94}Pu$ is so dangerous because carcinogen that the nuclear industry places a limit on exposure to this material that assumes those workers inhale no more than 0.2 μg of Pu over their lifetimes.

(v) The $^{239}_{94}Pu$ produced by these reactors might be stolen and assembled into bombs by terrorist organizations.

But fission reactors depend on the supply of uranium, which is quite expensive and gets depleted at a rate which will deplete the supply in approximately 50 yrs.

32.3.5. Breeding Cycles

There are two breeding cycles.

32.3.5.1 Uranium Breeding Cycle

Some $^{238}_{92}U$ (Non-fissionable from *nat* U; 99.3% abundant, but FERTILE) give through the following reaction, the fissionable $^{239}_{94}Pu$:

(Fertile) $^{238}_{92}U + ^{1}_{0}n \rightarrow [^{239}_{92}U^*] \xrightarrow[24\ m]{\beta^-} ^{239}_{93}Np^* \xrightarrow[23\ d]{\beta^-} ^{239}_{94}Pu$ (Fissionable)

The $^{239}_{94}Pu$ produced then undergoes the fission reaction

(Fissionable) $^{239}_{94}Pu + ^{1}_{0}n \rightarrow ^{147}_{56}Ba + ^{90}_{38}Sr + x\ ^{1}_{0}n$ (fast); $(x > 2)$

The neutrons produced thus are then used to make more $^{239}_{94}Pu$ from $^{238}_{92}U$. This is palatable but Pu is the most toxic material known to humanity. It is widely known that **one atom of Pu can kill a man if it gets into his lungs**!

32.3.5.2. Thorium Breeding Cycle

(Fertile) $^{232}_{90}Th + ^{1}_{0}n \rightarrow [^{233}_{90}Th^*] \xrightarrow[22\ m]{\beta^-} ^{233}_{91}Pa^* \xrightarrow[27\ d]{\beta^-} ^{233}_{92}U$ (Fissionable)

$(^{233}_{92}U;\ \ \tau_{1/2} = 0.162\ MYrs)$.

(Fissionable) $^{233}_{92}U + ^{1}_{0}n \rightarrow$ Pair of Fragments $+ y\ ^{1}_{0}n$ (slow); $(y > 2)$.

32.4 REACTORS IN INDIA

32.4.1 Research Reactors

Research Reactors of INDIA

Name of Reactor	Type	Power	Modertor	Fuel	Location	Supplier	Date of Criticality
1 APSARA	PWR	1 MW t	Water enriched U		BARC	UK	Aug 1957
2 CIRUS	PHWR(Candu)	40 MWt	D_2O	Nat UO$_2$	BARC	Canada	Jul 1960
3 ZERLINA		0.1 kW					1961
4 DHRUVA	PHWR	100 MWt	D_2O	Nat UO$_2$		BARC	Aug 1985
5 PURNIMA			Na cooled	$^{239}_{94}$Pu		DAE	May 1972
6 FBTR		40 MWt	Na cooled	U-Pu carbide		DAE	Oct 1985
7 KAMINI		30 kW	Na cooled	$^{233}_{92}$U	Kalpakkom	DAE	Oct 1996
8 Prtotype FBR			Na Cooled	$^{239}_{94}$Pu	Kalpakkam	DAE	(2009)

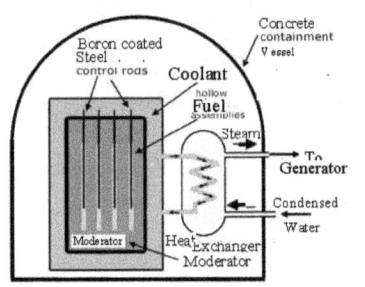

32.4.2 Power Reactors:

OPERATING POWER REACTORS OF INDIA

	Reactor	Type	MWe net	Year
1	Tarapur 1 (TAPS) (MH)	BWR	160	Oct 1969.
2	Tarapur 2 (TAPS) (MH)	BWR	160	Oct 1969
3	Tarapur 3 TAPS) (MH)	PHWR	540	Sep 2005
4	Tarapur 4 (TAPS) (MH)	PHWR	540	Aug 2006
5	Kaiga 1 (KA)	PHWR	220	Mar 2000
6	Kaiga 2 (KA)	PHWR	220	Nov 2000
7	Kaiga 3 (KA)	PHWR	220	May 2007
8	Kaiga 4 (KA)	PHWR	220	(Dec 2010)
9	Kakrapar 1 (GUJARAT)	PHWR	220	May 1993
10	Kakrapar 2 (GUJARAT)	PHWR	220	Sep 1995
11	Kakrapar 3 (GUJARAT)	PHWR	700	(Jun 2015)
12	Kakrapar 4 (GUJARAT)	PHWR	700	(Dec 2015)
13	Kalpakkam 1 (MAPS)	PHWR	170	Jan 1984
14	Kalpakkam 2 (MAPS)	PHWR	220	Mar 1986
15	Narora 1 (UP)	PHWR	220	Jan 1991
16	Narora 2 (UP)	PHWR	220	Jul 1992
17	Rawatbhata 1 (Rajasthan)	PHWR	90	Dec 1973
18	Rawatbhata 2 (RAUJASTHAN)	PHWR	187	Apr 1981
19	Rawatbhata 3 (RAJASTHAN)	PHWR	202	Jun 2000
20	Rawatbhata 4 (Rajasthan)	PHWR	202	Dec 2000
21	Rawatbhata 5 (RAJASTAN)	PHWR	202	Dec 2009
22	Rawatbhata 6 (RAJASTHAN)	PHWR	202	Mar 2010
23	Rawatbhata 7 (Rajasthan)	PHWR	700	(Jun 2016)
24	Rawatbhata 8 (Rajasthan)	PHWR	700	(Dec 2016)
25	Kudankulam 1 (TN)	PHWR	1000	2013
26	Kudankulam 2 (TN)	PHWR	1000	2013
27	Kalpakkam (MAPS)	PFBR	470	2013 ·
28	Kudankulam 3 (TN)	PHWR	1000	(TBD)

32.4.3. TYPICAL NUCLEAR POWER PLANT

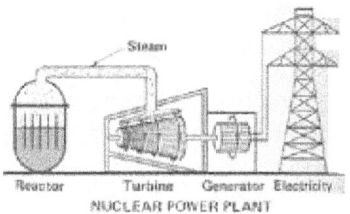

Steam

Reactor   Turbine   Generator Electricity
NUCLEAR POWER PLANT

32.4.4 ATOM BOMB:

An uncontrolled fission chain reaction is called an atom bomb. To sustain a chain reaction, a minimum quantity of fissionable material is needed (the critical mass). To contain the neutrons produced and to increase the probability of emitted neutrons to start next fission, some minimum amount of fissionable material is required.

32.4.5. Chronology of Developments and Fabricated Devices

1) On June 27, 1954, the World's first nuclear power plant generated electricity, at Obnisk near Moscow (Russia). The capacity of the generator was only 5 MWe.

2) In 1954: the World's first nuclear powered submarine, the USS *Nautilus*, was launched. It was the first vessel to complete a submerged transit to the North Pole on 3 August 1958.

3) In 1955, Arco (Idaho, USA) was the first town to be lit entirely by nuclear power. The BORAX II reactor (BWR) prototype, was used.

4) Launched on December 5, 1957, the First Nuclear-Powered Surface Ship, _Lenin_ by the USSR during the Cold War. The ice breaker never fired a shot as it had no guns, depth charges or weapons of any kind. I was built in the Admiralty Shipyards in the then Leningrad, .

5) 18 Dec 1957, '_Shippingport_', the first U.S. Nuclear Power Plant, at Beaver, near Pittsburgh, PA (USA), started production of electricity. It was a light water moderator thermal Breeder Reactor..

6) By late 1950s, AECL, Canada had developed the first nuclear reactor (CANDU, _i.e._, CA Nada Deuterium Uranium) environmentally-sensitive reactors consistently lead the world in productivity, safety, and ease of use. In Ontario, it was a pressurized heavy water (D_2O) (Deuterium oxide) coolant and moderator with fuel, $U\,O_2$ + nat U (0.7% $^{235}_{92}U$).

7) <u>A-Bomb</u>: On July 16, 1945, the first $^{235}_{92}U$ -fueled _atomic bomb_ was detonated at Alamogordo, New Mexico, and on August 6, 1945, the USA dropped an A- bomb (**'Little Boy'**)on <u>Hiroshima</u> (Japan), killing more than 100,000 people, and the $^{239}_{94}Pu$ - fueled bomb (**'Fat Man'**) dropped on <u>Nagasaki</u> (Japan).

8) I Nov 1952: The first artificial fusion reaction occurred when the hydrogen bomb was tested, at Eniwetok, Marshall Islands (USA). It is the Teller-Ulam design for the H-Bomb.

9) 21 Jul 1959, the First U.S. Nuclear-Powered Cargo passenger (commercial) vessel, _viz._, the Nuclear Ship 'Savannah', was substantial

10) India joined the nuclear club. On May 18, 1974, Indian Army detonates a 12-_kiloton_ nuclear explosive (a peaceful Bomb) in the Pokhran, Rajasthan desert. It was built using Plutonium from a research reactor.

11) India's first 40 _MWt_ Fast Breeder Test Reactor (FBTR) at IGCAR, Kalpakkam, attained criticality on 18th October 1985. It uses the Thorium Reactor design lasting for 100 years (compared to the 40 _yrs_ life of all the other reactor designs)). The fuel used is Pu-U Mono-Carbide . India has abundant supply of Thorium. Thorium provides 60% of the reactor power. It uses $^{233}_{92}U$ only for neutron radiography. India becomes the sixth nation having the BOT (build and operate technology) a FBTR besides USA, UK, France, Japan and the then USSR.

12) **Fission Disasters**: a) The famous Chernobyl disaster, in Ukraine. On April 26, 1986, the carbon control rods in the Chernobyl fission reactor near Chernobyl (Ukraine) caught on fire and caused an explosion in the reactor. A radioactive cloud spread across northern Europe and even parts of England. The incident pointed out to the world the dangers of fission power plants.

b) Following a major earthquake, a 15-_m_ tsunami disabled the power supply and cooling of three Fukushima Daiichi reactors (Japan), causing a nuclear accident on 11 March 2011.

13) In 2003, India announced that it plans to build a prototype Advanced Heavy Water Reactor (AWHR) providing 300 _MWe_, to be completed in 2016. It is located at Tarapur (Maharashtra). It is a third stage fuel cycle design, and this unique reactor will be fueled by a Th - U mix and will yield more Uranium than it consumes India has an estimated 10.7 million tons of monazite sands (containing 8.4 lakh tonnes of thorium metal (in the monazite mineral).

14) The Kudankulam (Tamil Nadu) Nuclear Power Station (KKNPP): Two 1 *GW* reactors of the VVER-1000 model are being constructed by the Nuclear Power Corporation of India (NPCIL) and Atomstroyexport, Russia. Started in Sep 2001, it is operational in 2013.

32.4.10 NUCLEAR DETONATIONS:
1) Pokhran, Rajasthan, May 11, 1974; <15 *kTonne* nuclear Fission device.
2) Pokhran, May 11 & 13, 1998 (one was 45 *kTonne* Thermo-nuclear device).

32.5.1. **NUCLEAR FUSION** (THERMO-NUCLEAR REACTION)
This is the coalescence of two lighter nuclei into heavier product nucleus. Kinetic energy required to overcome the Coulombic potential barrier of

$$E_P = \frac{Z_1 Z_2 e^2}{4 \pi \varepsilon_0 r}$$

$$r = \sim 10^{-14} \, cm$$

$$E_P = 0.15 Z_1 Z_2 \, MeV \sim 8.6 \times 10^{-5} \, TeV$$

$$1 \, k_B T = 8.6 \times 10^{-5} \, T \, eV$$

which is equivalent to a temperature of $\boxed{T \sim 10^9 \, K}$.

32.4.2 E_B / A for $A < 20$

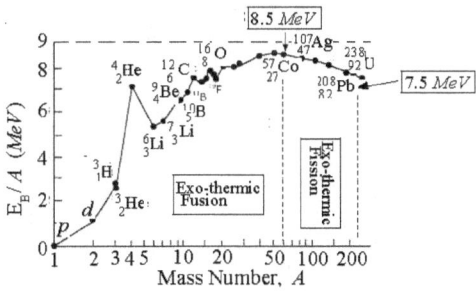

32.4.3 Source of Solar Energy, or P=P Cycle or Critchfield cycle
The source of solar energy is a series of fusion reactions. One such fusion series is the proton-proton cycle (Critchfield cycle):

$$\boxed{1 \quad p \, (p \, , e^+ + \nu_e) \, d; \ Q = + 0.42 \, MeV}$$

$$\boxed{2 \quad {}^2H \, (p \, , \gamma) \, {}^3H; \ Q = + 5.49 \, MeV}$$

the d-d reaction .

But $\boxed{d + d \rightarrow {}^3_2He + {}^1_0n + 3.2 \, MeV}$ is *exo-ergic*.

Mass of Reactants	Mass of Products
$M_d = 2.014102\,u$ $M_d = 2.014102\,u$	$= 4.028204\,u$
$M_n = 1.008665\,u$ $M(^3_2He) = 3.016029\,u$	$= -4.024694\,u$
Mass difference	$+ 0.003510\,u$

$$Q = (+\ 0.003510\ u)\,(931.5\ MeV/\ u) = +\ 3.270\ MeV$$

3	$^3_2He\,(^3He, p\ +\ p)\,^4_2He;\ \ Q = +\ 12.86\ MeV$

Each of these reactions is exothermic and there involves 5 reactions and the P-P cycle is summarized as

$$6^1H \rightarrow {}^4He + 2\,{}^1H + 2\beta^+ + 2\nu + 2\gamma + 25 MeV$$

There are PP I Chain, PP II Chain, and PP III Chain.

32.4.4. CN Cycle (Bethe cycle or Carbon cycle, or CNO cycle)

Two α – particles can combine to form a ^8_4Be nuclide, which stays together for $\sim 10^{-15} s$. In a sufficiently high density of helium, this results in an adequately high concentration of ^8_4Be, so that another most important fusion process takes place. This is the **Bethe or Carbon cycle**, which is equivalent to the fusion $4\,p \Rightarrow 1\,\alpha$.

$^8_4Be + {}^4_2He \rightarrow {}^{12}_6C + \gamma$

32.4.6. Stellar Energy

Self-sustaining fusion reactions can occur only under extreme temperature and pressure. The PP cycle is known to the belief of most astrophysicists the predominant process of energy generation in the '*Main Sequence Stars*', like the Sun (interior at $\sim 2\ x\ 10^6$ K) and cooler stars, whereas the CN cycle is responsible for the energy output of hotter hydrogen burning stars. The **neutrinos** in PP and CN chains carry away 2 – 6 % of the energy released in the reactions.

In the case of the Sun,

$$4\,p \Rightarrow 1\,\alpha + 2e^+ + 2\nu_e + 26.7\ MeV\ (\sim +6.6\,x10^{11}\ J - g^{-1}),$$

is at the rate of $5.64\ x\ 10^{11}\ kg$ - s^{-1} of hydrogen fusing in helium, with a release of $3.7\,x\ 10^{25}$ W. Of this $\sim 1.8\,x\ 10^{14}$ W only falls on the Earth in the form of photons.

Major attributes of the Sun

Mass	$M_\odot = 1.99\ x\ 10^{30}$ kg
Radius	$R_\odot = 6.96\ x\ 10^8\ m$
Luminosity	$L_\odot = 3.86\ x\ 10^{26}\ W$
#ofHatoms	$\sim 10^{56}$

32.4.7 ORIGIN OF ELEMENTS (NUCLEO-SYNTHESIS)

Step # 1	Hydrogen Burning till Main Sequence Stars.

Step # 2	Helium Burning till Red Gaint phase of Stars.

Step # 3	Carbon Burning.

Step # 4 Stars which become supernovae release so much energy that they initiate *endothermic* fusion reactions: those whose products are actually less stable than the reactants. These reactions are responsible for all the elements heavier than $^{56}_{26}$Fe , which has the highest BE /A of any possible nucleus .

Step # 5 Silicon burning (Gravitational Confinement).

Then further gravitational collapse raises temperatures to several 10^9 K .

$$^{56}_{26}\text{Fe} \rightarrow 26\ \alpha + 4\ ^1_0 n.$$

The iron core suddenly becomes a helium core with a large density $\sim 10^{24}\ ^1_0 n / cc$ present. Then slow *n*-capture, known as "**s-process**" occurs up to the synthesis of $^{209}_{83}$Bi . This s-process build-up of heavy elements terminates when

$$^{209}_{83}\text{Bi} + \ ^1_0 n \rightarrow \alpha + \ ^{206}_{81}\text{Tl}$$

However, the trans-bismuth nuclei may be built-up through the "**r-process**", till A = 254, beyond which **nuclear fusion** occurs. The reaction proceeds in an alarmingly rapid rate so as to refer it as "**supernova explosion**". Without supernovae, there would be no Ni, Cu, Zn, Ag, Au, I, Pt, Pb, Hg, U or Pu, to name some of the most familiar elements.

32.4.8 CONTROLLED NUCLEAR FUSION:
1.   Hydrogen Bomb
2.   Fusion reactors using D-D reaction or D-T reaction.
3.   The Mirror Machines (θ -Pinch apparatus. The DCX machine, Pyrotron, etc)
4.   Stellarator Machine
5.   Laser induced fusion.

32.4.9. DIFFERENCES BETWEEN FISSION & FUSION
         Both, nuclear fusion and nuclear fission reactions lead to the production of new nuclei, there are some basic differences between the two

Nuclear Fusion	Nuclear Fission
1) Two light nuclei combine to form a heavy nuclei.	A heavy nucleus breaks up to form light nuclei.
2) Never be spontaneous.	Fission reactions can be spontaneous.
3) No chain reaction is present.	Chain reaction can sustain the reaction, once started.
4) Can be started by increasing the the temperature of the nuclei . to be fused. The temperature required is very high.	Can be started by bombarding one nuclei with high energy on the . other nuclei
5) Has not been sustained in the laboratory conditions.	Can be sustained and controlled in practical situations.
6) As yet has been used for making hydrogen bombs only.	Has been used for bombs as well as power generation.

32.6. **PARTICLE ACCELERATORS**
\32.6.1 Cockroft Watson Accelerator (Electro-static Generator)
         by John D. Cockcroft & E.T.S. Walton (1932).
32.6.2 Van de Graaff Generator
         by Robert Jemison Van de Graaff in 1932.
32.6.3 Stanford Linear Accelerator (LINAC, 3.2 *km* long)

Length of sth Drift tube is

$$L_S = v_S / 2f$$

$$L_S = \lambda \sqrt{[neV_0 + c]/2mc^2}$$

32.6.4 Cyclotron (E.O. Lawrence & M.S. Livingston, 1932)

$$\boxed{q\,\overline{v} \wedge \overline{B} = \frac{m\,v^2}{\rho}}$$

$$t = \frac{\text{distance}}{\text{velocity}} = \frac{\pi\,\rho}{v}$$

Angular frequency of circulation of ions (mass m and charge q) in the magnetic field B

$$\boxed{\omega = \frac{q\,B}{m} = \text{constant}}$$

32.6.6 SYNCHRO – CYCLOTRON

In $q\,\overline{v} \wedge \overline{B} = \dfrac{m\,v^2}{\rho}$,

$$\rho = \frac{m_0\,v}{q\,B\,\sqrt{1 - v^2/c^2}}$$

and $\omega = \dfrac{q\,B\,\sqrt{1 - v^2/c^2}}{m_0}$

This was achieved by using the principle of *phase stability*, by Edwin M. McMillan and by V. Veksler (1945)

32.6.6   Circular Accelerator

32.6.7. Continuous electron beam accelerator

32.6.8 BETATRON (1940, Donald Kerst)

1) Electrons gain additional energy, due to acceleration by the sinusoidal induced emf, (V) production of an emf in its orbit by the changing magnetic flux Φ, and

2) Electrons are maintained in circular motion due to the radial force effected by the magnetic field.

The magnetic flux Φ through the electron orbit has to be chosen such that the motion of electrons will be in stable orbit of radius ρ.

Faraday's Law induced emf V

$$V = \frac{d\Phi}{dt}$$

$$\overline{F} = \int_0^{2\pi\rho} \frac{dW}{ds} = \left(\frac{e}{2\pi\rho}\right) \frac{d\Phi}{dt}$$

$$d(m\,v) = \left(\frac{e}{2\pi\rho}\right) d\Phi$$

$$q\,\overline{v} \wedge \overline{B} = \frac{m\,v^2}{\rho}$$

$$m\,v = B\,e\,\rho$$

$$\boxed{\Phi_{\text{average}} = 2\,(\pi\,\rho^2\,B_{\text{Orbit}}) = \Phi_{\text{Orbit}}}$$

32.5.9. SYNCHROTRON
In 1945 Edwin M. McMillan and independently V. Veksler.
Varying the magnetic field B while the frequency may or may not be varied.

$$\omega = \Omega = \frac{q\,B}{m_0\,c} = \text{constant}$$

32.5.9.1 Betatron-Synchrotron
32.5.9.2 Proton Synchrotron (Bevatron, or Cosmotron)

Proton Synchrotrons

Machine	Beam energy (GeV)
1. KEK, Tokyo	12
2. PS, CERN, Geneva	28
3. AGS, Brookhaven	32
4. Serpukhov	76
5. SPS, CERN	450
6. Tevatron-II, Fermilab	1000

32.5.9.3 Alternating-Gradient Synchrotrons

32.5.10 COLLIDING BEAM ACCELERATORS

COLLIDERS

Machine	Accelerated particles	
1. CESR, Cornell, NY	$e^+(6\ GeV)$	$+ e^-(6\ GeV)$
2. PEP, Stanford	$e^+(15\ GeV)$	$+ e^-(15\ GeV)$
3. TRISTAN, Tokyo	$e^+(32\ GeV)$	$+ e^-(32\ GeV)$
4. SLC, Stanford	$e^+(50\ GeV)$	$+ e^-(50\ GeV)$
5. LEP, CERN, Geneva	$e^+(60\ GeV)$	$+ e^-(60\ GeV)$
6. SppS, CERN, Geneva	$p\,(450\ GeV)$	$+ \bar{p}(450\ GeV)$
7. Tevatron 1, Fermilab, Bativia	$p\,(1000\ GeV)$	$+ \bar{p}(1000\ GeV)$
8. HERA, Hamburg	$e^-(26\ GeV)$	$+ p\,(820\ GeV)$
9. UNK, Serpukhov	$p\,(3000\ GeV)$	$+ \bar{p}(3000\ GeV)$
10. LHC, CERN	$e^+(50\ GeV)$	$+ p\,(8000\ GeV)$
	$p\,(8000\ GeV)$	$+ \bar{p}(8000\ GeV)$
11. LEP-II, CERN	$e^+(100\ GeV)$	$+ e^-(100\ GeV)$
12. SSC, Texas	$p\,(20000\ GeV)$	$+ \bar{p}\,(20000\ GeV)$

32.5.11 Large Hadron Collider (LHC) Particle Accelerator
[The CERN (European Organization for Nuclear Research) built this during 1998 – 2008.]
It lies in a tunnel 27 *kms* (17 *miles*) in circumference, below the surface of earth as 175 *m* (574 *ft*) near Geneva, Switzerland. The aim is to prove / disprove theories of nuclear physics, and especially to detect the theoretical Higgs boson, and to advancing our understanding of the laws of physics as we know to date. The principle used in the device is to prepare the collision of two beams of protons of kinetic energy at least 7 *TeV*.

32.6.12. **ACCELERATORS IN INDIA**:
1) VEC: Variable energy cyclotron at Kolkatta, (in the Salt Lake township area by BARC), in 1977.
2) 14-UD tandem Pelletron Accelerator (a Van de Graff Accelerator): at TIFR, Mumbai (1977?)
3) 15-UD tandem Pelletron Accelerator: at Nuclear Science Centre, New Delhi, 1991.
4) INDUS-1 & INDUS-II (Synchrotrons), at CAT, Indore, (1997?)

32.7. FUNDAMENTAL PARTICLES
32.7.1 CHRONOLOGY

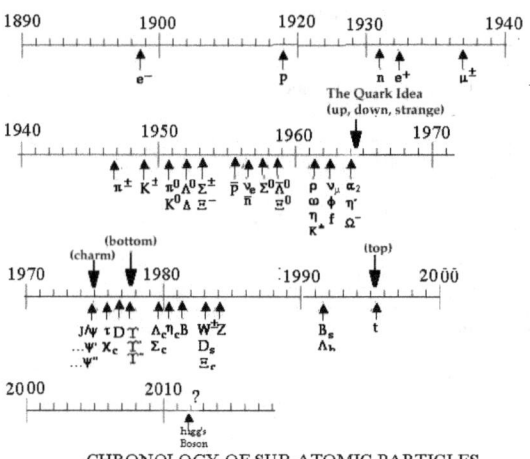

CHRONOLOGY OF SUB-ATOMIC PARTICLES

32.7.2 INTERACTIONS in the Universe

	Strength and Range of Fundamental Forces		
Force	Relative Strength	Range	Exchanged particle
1 Strong (Nuclear)	1	$10^{-15}m$	Pions (π, K)
2 Electromgnetic	10^{-2}	Infinite	Photon, γ
3 Weak (Nuclear)	10^{-13}	$10^{-18}m$	Bosons (W^-, W^+, W^0)
4 Gravity	10^{-38}	Infinite	Graviton

32.7.3 Families of Sub-atomic Particle
    1) LEPTONS
    2) BARYONS
    3) MESONS
    4) QUARKS

Family: name	leptons - *fundamental* nickname force interactions	
electron	e	E&M, gravity, weak
muon	μ	"
tauon	τ	"
electron neutrino	ν_e	weak, gravity?
muon neutrino	ν_μ	"
tauon neutrino	ν_τ	"

Family: name	hadrons - *not* fundamental nickname force interactions	
mesons	$\pi, K, \eta, ...$	(2 quarks)
baryons		(3 quarks)
proton	p	E&M, gravity, strong, weak
neutron	n	gravity, strong, weak
		... and many others, $\Lambda, \Sigma, \Xi, \Omega, ...$

quarks - *fundamental* building blocks of hadrons		
name	nickname	force interactions
up	u	E&M, gravity, strong, weak
down	d	"
strange	s	"
charm	c	"
bottom	b	"
top	t	"

32.7.4 INTRINSIC PARTICLE PROPERTIES AND CONSERVATION LAWS

Without *invariance principles*, there would be no Laws of Physics! An Invariance Principle reflects a *basic symmetry*, and is always intimately related to a Conservation Law

The mass and spin properties of particles are related to energy, momentum and angular momentum. These quantities are determined by applying these Conservation Laws. The laws are universal and apply to all, *i.e.*, EM, Strong and Weak interaction processes.

Charge conservation law is universal and applies to all interactions.

32.7.4.1. Lepton number L

It is known **that lepton number is conserved in all weak decay.**

Table 18.5 Lepton number and Family of Leptons

L	e^-	ν_e	μ	ν_μ	τ	ν_τ
L_e	+1	+1	0	0	0	0
L_μ	0	0	+1	+1	0	0
L_τ	0	0	0	0	+1	+1
	e^+	$\bar{\nu}_e$	$\bar{\mu}$	$\bar{\nu}_\mu$	$\bar{\tau}$	$\bar{\nu}_\tau$
L_e	-1	-1	0	0	0	0
L_μ	0	0	-1	-1	0	0
L_τ	0	0	0	0	-1	-1

32.7.4.2   Baryon number, B

Baryon number appears to be conserved in all reactions, thus explaining why protons, the lightest baryons, cannot decay, whereas neutrons can. Neutrons decay *via*

$$_0^1 n \rightarrow p^+ + e^- + \nu_e$$

Conservation of baryon number is not seen violated, except GUT suggests that the proton might decay in a manner that should violate this Law.

32.7.4.3 Isospin, I

The hadrons possess a **non-zero** quantum number called *isospin* (*isobaric* or *isotopic*), denoted by $I = \frac{1}{2}$.

I_3	=	+1	0	-1
I	= 1	pp	np	nn
I	= 0		np	

32.7.4.4 **STRONG INTERACTION**

The Feynman graph of strongly interacting particles

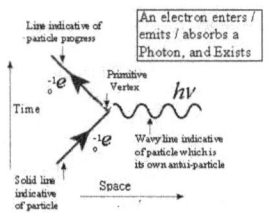

32.7.4.5 **EM INTERACTION**

The Feynman diagram

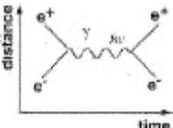

Feynman diagram

32.7.4.6 **WEAK INTERACTION**

Feynman graph below the neutrino – proton

32.7.4.7   Strangeness, S

The **Associated Production**

Particles produced by strong interactions but decay by weak interactions (*i.e.* they are observed to live 10^5 billion times longer than they should) are called strange particles

$$\pi^- + p^+ \xrightarrow{\ 10^{-22}s\ } \Lambda^0 + K^0$$

The total strangeness must remain constants in particle interactions governed by the strong and EM forces".

"On the other hand, the strangeness either remains the same or $\Delta S = \pm 1$, in processes where the weak force is involved".

32.7.4.8   Hypercharge, Y

$$\boxed{Y = B + S}$$

For all *strange* hadrons

$$\boxed{Q = e\left(I_3 + \frac{B+S}{2}\right)}$$

Strangeness (S) and *hypercharge* (Y) are the conserved in Strong interaction processes, but not always in Weak interaction decays

32.7.5   QUARKS

Murray Gell-Mann and George Zweig (1964) showed that the Eightfold Way patterns could be replicated if the mesons and baryons were composed of 'furthermore' elementary particles, which Gell-Mann called **quarks**. The most striking feature of the quarks is that they have *fractional electric charges.*

TABLE 18.7 . QUARKS - Participants in *Electro-weak, Strong* and *Gravitation*

	Particle	Nick name	Spin, s ($\hbar$)	Mass (MeV/c^2)	Charge (e)	Colour Charge
1	Up	u	$\frac{1}{2}$	~ 5	$+\frac{2}{3}$	r, g, b
2	Down	d	$\frac{1}{2}$	~ 10	$-\frac{1}{3}$	r, g, b
3	Strange	s	$\frac{1}{2}$	~ 200	$-\frac{1}{3}$	r, g, b
4	Charm	c	$\frac{1}{2}$	$\sim 1.5 \times 10^3$	$+\frac{2}{3}$	r, g, b
5	Bottom	b	$\frac{1}{2}$	$\sim 4.5 \times 10^3$	$-\frac{1}{3}$	r, g, b
6	Top	t	$\frac{1}{2}$	$\sim 180 \times 10^3$	$+\frac{2}{3}$	r, g, b

32.7.4.5   The Quark Structure of Hadrons

In the 1960s by the **Quark Model** which says that hadrons (Most of the more than 200 particles are mesons and baryons, or, collectively, hadrons) are made out of spin-$\frac{1}{2}$ particles called quarks.

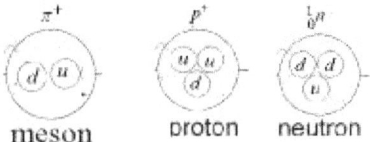

meson   proton   neutron

32.7.4.6   Interaction between a Pair of Quarks

The Feynman diagram

32.7.4.7 Quarkonium

Quark-anti-quarks $(q\,\bar{q})$ bound together is state of matter called *quarkonium*.
Similarly a strange and anti-strange quarks bound together $(s\,\bar{s})$ is called *strangeonium*.

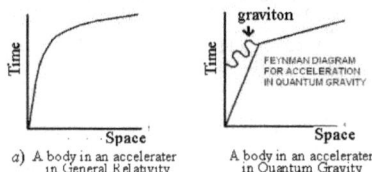

a) A body in an accelerater in General Relativity   A body in an accelerater in Quantum Gravity

32.7.4.8 **Proton Decay**

In a proton (*uud*), a *u* quark emits a virtual X-boson and then gets transformed into an anti-quark. A *d* quark absorbs the X-boson to become an $_{+1}^{0}e$, and the remaining quark-anti-quark pair convert to become a $\pi-$ meson. Such an event takes place once in $10^{30}\,yrs$, which is 10^{20} times the age of the Universe!!!

32.7.4.9 **The Higgs boson (God particle)**

The Higgs boson, nicknamed the **God particle**, is a hypothetical massive elementary particle (mass $\cong 125\,GeV$, spin 0, even parity) that is predicted to exist by the Standard Model (SM) of Particle physics. The Higgs boson is an integral part of the theoretical Higgs mechanism. It has been detected in the CERN's LHC, on Jul 2012 (Nobel Prize in Physics, 2013 for P. Higgs and F. Englert).

```
BOSONS (Force Carriers)(Exchange Particles)
------------------------------------------------------------
1) Photon γ    Massless, No charge, EM Force carrier, move at c, Long range
2) Gluon, b, g , r   High mass, Colour Charge, (B,G,R), Strong Force carrier, move at < c, Short range
2) W₊, W₋, Z₀   High Mass , Weak force carrier, move at < c, Short range
3) Graviton    Massless, No Charge, Gravity wave carrier, move at c, Long range
```

+&%&%&%&%&%&%&%+

Chapter 33

BASIC ELECTRONICS
Fundamentals, Discrete Devices, BJT Amplifier,
Diode Rectifiers, Oscillators, Analogue Circuits

"Whatever you see as duality is unreal" - Adi Shankara

33.1 INTRODUCTION

PASSIVE CIRCUIT ELEMENTS are:
1) Resistors, 2) Capacitors, 3) Inductors,
Capacitors are: paper, ceramic, electrolytic.

33.1 SOLDERING

Soldering is the process of using a filler material (solder) to join pieces of metal together. Soldering occurs at relatively low temperatures (around $200^{\circ}C$), using a soldering iron to solder. Most solder is made from a combination of tin and lead - it's about a 60% Sn, 40% Pb mix. Required tools consist of wire cutters, a wire stripper, needle nose pliers, and an automatic wire stripper. The solder should be a `rosin core` solder.

33.1.1 RESISTORS
Wire wound

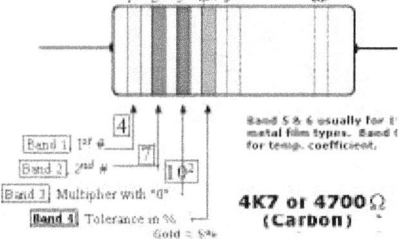

4K7 or 4700 Ω
(Carbon)

33.1.2. Carbon types
A new way of writing a resistor value [in practical electronics only]
1.5 Ω → 1R5
1.5 kΩ → 1K5
150 kΩ → 150 K or M15
1.5 MΩ → 1M5

33.1.3 **Resistor Colour**

Band	1,2,3
Colour	Value
Black	0
Brown	1
Red	2
Orange	3
Yellow	4
Green	5
Blue	6
Violet	7
Grey	8
White	9
Gold	0.1

33.1.4. What is a Solder-less Breadboard?

A breadboard is used to build and test quickly before finalizing any circuit design. It has many holes into which circuit components like PN Diodes, BJTs, ICs and resistors can be inserted. The holes are mostly spaced 0.1" apart to accommodate standard components. It is a 21x 30 grid sized to accommodate. It also provides a place to connect external power of $9V$ batteries. A typical bread board top and bottom power distribution rails is shown in Fig.

Solder-less breadboard

The breadboard has strips of metal sockets which run underneath the board, and connect the group of five holes on the board. The top and bottom rows of holes are connected horizontally, while the holes in the centre sections are connected vertically. A completed circuit will look like the following Fig.

Wired breadboard

33.2.1. RC CIRCUITS - Charging

An RC circuit contains a single resistor, R and a single capacitor C.

Time constant, τ

Electrical or Electronic circuits or systems suffer from some form of "time-delay" between its input and output, when a signal or voltage, either continuous, (DC) or alternating (AC) is firstly applied to it. This delay is generally known as the **time delay** or **Time Constant** of the circuit and it is the time response of the circuit when a step voltage or signal is firstly applied.

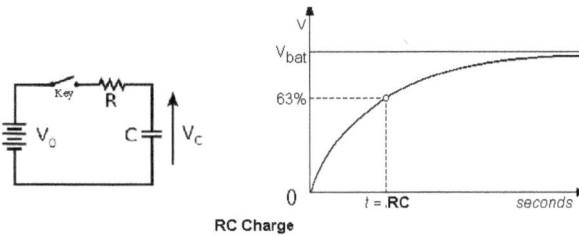

RC Charge

$$q(t) = EC \ [1 - e^{-t/\tau}]$$

$$q(t = 0) = Q_0 \ [1 - e^{-t/RC}] \ , \ \text{or}$$

$$V_C = V_0 \ [1 - e^{-t/RC}]$$

Time constant,

$$\tau = RC$$

Transient Period = 4τ = the time required to charge the capacitor to 0.99% of the Voltage input.

Steady state Period = 5τ.

33.2.2   Discharging RC Circuit:

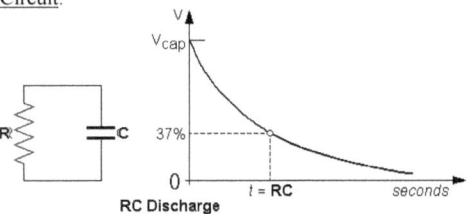

RC Discharge

$$q(t) = Q_0 \ e^{-t/RC}$$

$$V_C = V_0 e^{-t/RC}$$

33.2.3   RC Differentiator (HP Filter)

RC circuits work as filters (high-pass or low-pass filters), integrators and differentiators. Here we explain how, and give sound files examples of RC filters in action. For an introduction to AC circuits, resistors and capacitors,

When a high pass filter is used with a sine wave input, the output is also a sine wave. The output will be reduced in amplitude and phase shifted when the frequency is low, but it is still a sine wave. This is not the case for square or triangular wave inputs. For non-sinusoidal inputs the circuit is called a differentiator.

$$\boxed{\text{R and C = Small; } \boxed{\tau = RC = \text{small}}}$$

$$v_i = \tfrac{1}{C}\int_o^t i\, dt + i\, R$$

$$\boxed{v_o = i\, R = R\left(C\frac{dv_i}{dt}\right)}$$

$$\boxed{\tau_{RC} = RC > T}\ \text{Square wave out}$$

$$\boxed{\tau_{RC} = RC = T}\ \text{Pulse partially differentiated}$$

$$\boxed{Small\ \tau_{RC} = RC < T}\ \text{Pulse heavily differentiated}$$

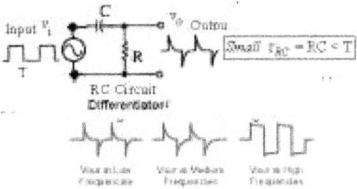

33.2.4   RC Integrator (LP Filter))

The RC circuit can act as a simple integrator or a first order low-pass filter

$$\boxed{\text{R and C = Large; } \boxed{\tau_{RC} = RC = \text{Large}}}$$

$$v_i = CR + \tfrac{1}{C}\int_o^t i\, dt$$

$$\boxed{v_o = \tfrac{1}{CR}\int_o^t i\, dt}$$

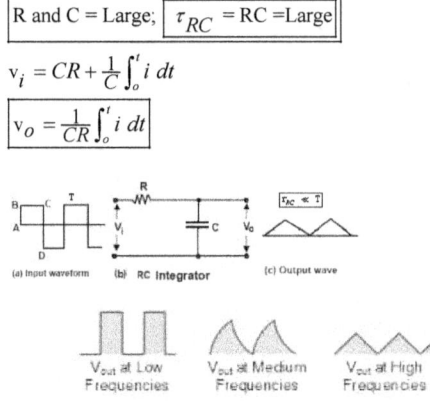

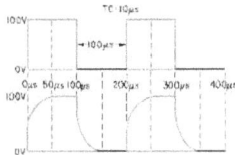

33.2.5 THERMIONIC EMISSION of electrons from metals

Metals at room temperature have a negative space charge cloud of electrons surrounding its surface which are cohesively attracted to the positive charged metal surface. 'Boiling off' of electrons is called 'Thermionic Emission' as the emission of electrons is produced by the heat.

Richard-Dushman Equation

No-space charge limited thermal electron current density of electrons, J

$$\boxed{J = AT^2 \, e^{-W_A / k_B T}}$$

where $A = \dfrac{4\pi m \, k_B T}{h^3}$

T = Absolute temperature and

k_B = Boltzmann constant.

"Schottky effect" causes reduction of work function of the metal by $\sqrt{\dfrac{e\vec{E}_{ext}}{4\pi\varepsilon_o}}$

33.3.1 Energy Band Diagram for Metals

In free electrons in a metal are free to move in the material and can carry electric current

RESISTIVITY of Materials

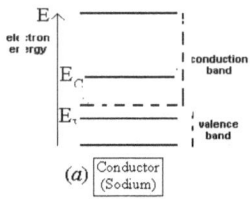

Material	Resistivity	Classification
Copper	$1.7 \times 10^{-8}\,\Omega\,m$	
Aluminium	$2.8 \times 10^{-8}\,\Omega\,m$	Conductors
▢	$\sim 10^{-5}\,\Omega\,m$	
Germanium	$0.65\,\Omega\,m$	Semi-conductor
Silicon	$2.0 \times 10^{3}\,\Omega\,m$	
▢	$\sim 10^{8}\,\Omega\,m$	
Glass	$1.7 \times 10^{11}\,\Omega\,m$	Insulators
Rubber	$1 \times 10^{16}\,\Omega\square\,m$	

The degree to which a material conducts electricity is represented by its **resistivity** ρ.

$$\sigma = 1/\rho$$

This is the resistance between opposite faces of a $1m$ cube of the material, measured in Ωm.

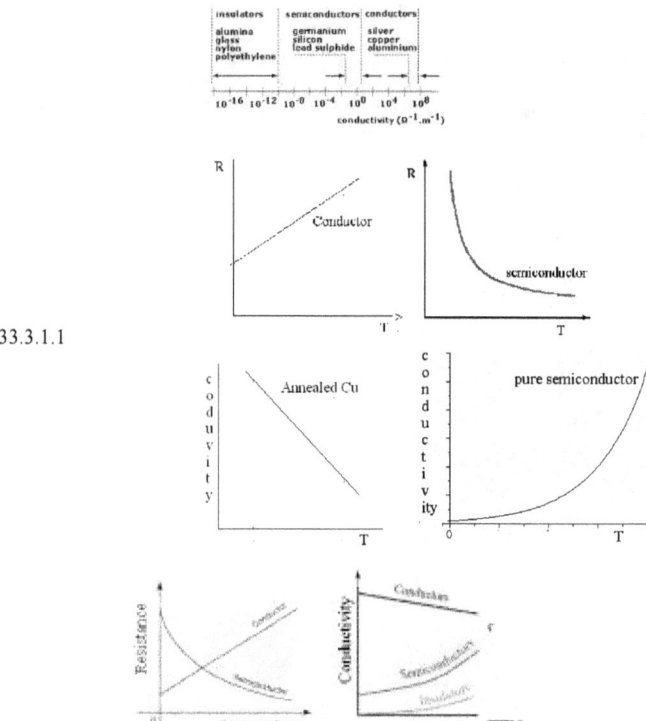

33.3.1.1

33.3.2 Insulators

The electrons in an insulator are tight bonded to atoms and cannot move. In insulators (glass, plastics, wood, *etc.*) all the electrons on the outer ring are held tightly by the strong forces of attraction of the nucleus.

Fermi Level, E_F.

"Fermi level" is the term used to describe the top of the collection of electron energy levels at absolute zero temperature It is the maximum energy that an electron can have. The average energy per electron that a free electron is 60% of E_F.

$$E_F = \frac{(hc)^2}{8mc^2}\left(\frac{3}{\pi}\right)^{2/3} n^{2/3}$$

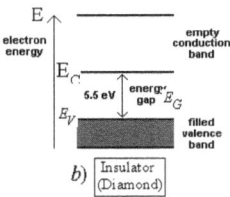

b) Insulator (Diamond)

n = # of conduction electrons per volume per unit energy

Material	Fermi energy (eV)
Cu	7.00
Ag	5.49
Fe	11.1
K	2.12

33.3.3 Semiconductors

They have electrical properties in between those of conductors and insulators. In free electrons in a metal are free to move in the material and can carry electric current/. The electrons in an insulator are tight bonded to atoms and cannot move. In a semiconductor the electrons are loosely bonded to atoms, and some movements are possible.

It is easy to understand through energy band diagrams. In a pure Si / Ge the electrons are enough to fill all the notches in the VB.

At room T a few electrons are transferred to the CB. The number of the electrons n_i depends on the band gap E_G and temperature T,

33.3.3.1   Boltzmann Criterion,

$$\boxed{\frac{N_{CB}}{N_{VB}} = e^{-(E_{CB}-E_{VB})/k_B T}}$$

$$E_G = E_{CB} - E_{VB}$$

Pure materials which are semiconductors are termed as intrinsic to distinguish between extrinsic other semi conducting materials which are formed by adding small quantities of impurities. Two elemental semiconductors, silicon (Si) and germanium (Ge), both exist in the same group (IV) of the Periodic Table. They are tetra-valent and covalent bonded. The Ge / Si atom has 8 electrons in its outermost ring and to covalent bond

Compound Semiconductor are formed, example binary compound like GaAs, resulting from III-V Group elements in the Periodic Table.

Elements III a	Elements V a
Al	N
Ga	P
In	As
	Sb

Ternary Semiconductors are formed from three elements,

eg., $Ga_{1-x}Al_x As$.

Quarternary compound

eg., $In_{1-x}Ga_x As_{1-y}P_y$

33.3.3.2   Energy Band Diagram for Semiconductors

Material	Energygap
Germanium	0.75 eV
Silicon	1.12 eV

$$E_G(eV) = \frac{1240}{\lambda_G(nm)}$$

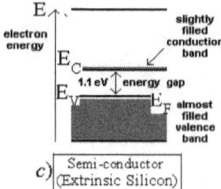

c) Semi-conductor (Extrinsic Silicon)

33.3.3.2 Fermi energy of semiconductors,

$$E_F = \frac{E_g}{2}$$

E_G = Energy band gap

33.3.4. Intrinsic Materials: (i-type)

Ge and Si, both have 4 valence electrons per atom. In the crystal they are corner-sharing covalent bonded giving tetrahedral lattice. In intrinsic semiconductors number of positive HOLES (positive charge carriers) and that of electrons (negative charge carriers) are equal.

Intrinsic concentration of charge carriers in a semiconductor varies as

$$T^{3/2}$$

The resistance R decreases with increasing temp. T

$$R = R_0 \, e^{-b/T}$$

The resistance of conductor increase uniformly (linearly) with T.

$$R = R_0[1 + \alpha \, T]$$

Intrinsic semiconductor:

$$n = p = n_i = 10^{13} \, cm^{-3}, at \, 400K, for \, Si$$

$$n = p = n_i = 10^{8} \, cm^{-3}, at \, 273K, for \, Si$$

the intrinsic carrier concentration. Electrons and the holes are created in pairs.

$$n^2 = AT^3 e^{-E_G/E_T}$$

E_G = band gap;

Electrostatic Voltage E_T

$$E_T = \frac{k_B T}{e}$$

33.3.4 Extrinsic Materials:
Doping

Doping means the introduction of impurities into a semiconductor crystal to the defined modification of conductivity.

The conductivity of a deliberately contaminated silicon crystal can be increased by a factor of 10^6.

By the process of doping, pentavalent (As, P, Sb) impurities (Donors) are added to intrinsic material Si or Ge to get n-type Si or n-type Ge. The negative electrons are the *majority charge carriers*. And holes are the *minority carriers*. $\boxed{n \gg p}$ /

Acceptor [Elements with 3 valence electrons (B, Al, In)] concentration is about 1 part in 10^8 ($10 - 100\ ppm$).

33.3.4.1 Illustration of Lattice impurities

$$\boxed{np = n^2}.$$

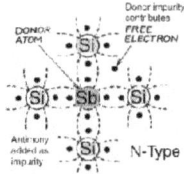 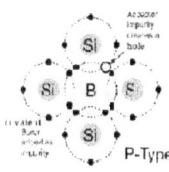

By the process of doping, trivalent (B, Al, Ga, In) impurities (Acceptors) are added to intrinsic material to get p-type Si or p-type Ge. The positive holes are *the majority charge carriers* and the electrons are the minority carriers. $\boxed{p \gg n}$

$$\text{Pure Si / Ge} + \frac{\text{Donar Impurity(P, As, Sb)}}{\text{Acceptor(B, Ga, In) [10-100 ppm]}} = \frac{\text{N-type Si / Ge}}{\text{P-type Si / Ge}}$$

33.3,5 Band gap Diagrams of Intrinsic

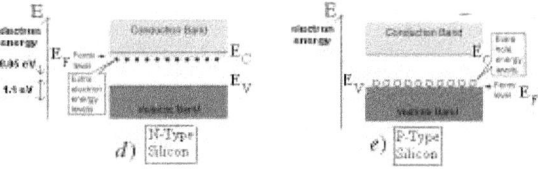

The barrier (dipolar, depletion or space charge) layer between the pn junctions has width around 5 *micron* for 0.01 *ppm* impurity concentration. (or, 10 *nm* for 0.001 impurity concentration) or proportional to (impurity concentration)$^{s-1/2}$.

Material	Voltage drop
Germanium	0.2 V
Silicon	0.6 V

33.4. **PN JUNCTION DIODE**:

Symbolized as

Thickness of a PN junction diode is about 1.3 *mils* (*i.e.*, 0.2 *mm* thick and 1 - 2 *mm*^{2} area).

33.4.1 Forward and reverse biasing:

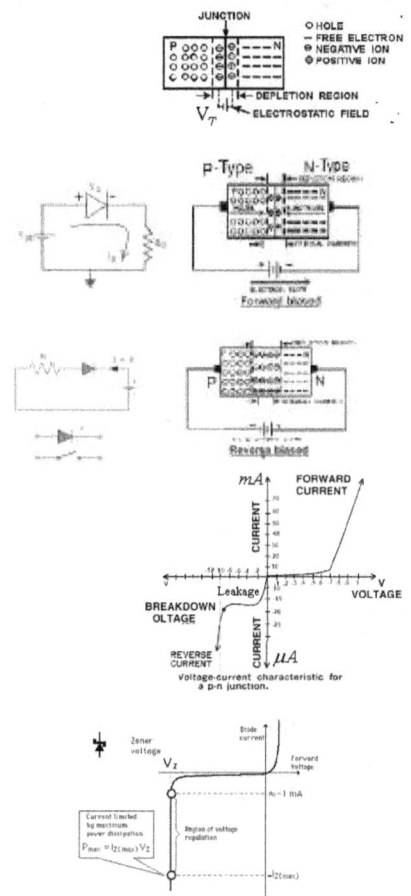

33.4.2 DIODE (Diffusion) EQUATION:
Current-Voltage Relationship

$$I = I_0 [e^{\frac{eV}{\eta k_B T}} - 1] \approx I_0 e^{V/V_Y}$$

$$I = I_0 [e^{eV_D / nk_B T} - 1]$$

$$I_D = I_S [e^{V_D / 0.026} - 1]$$

I_D = Diode current in mA.

I_S = Saturation current in A (typically, $1 \times 10^{-12} A$).

e = Euler's constant (~ 2.718281828).

V_0 = External Voltage applied in the circuit, in V.

$$V_T = \frac{k_B T}{e} = \text{Junction scale voltage}$$

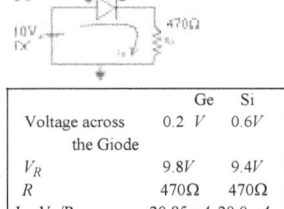

	Ge	Si
Voltage across the Giode	0.2 V	0.6V
V_R	9.8V	9.4V
R	470Ω	470Ω
I = V_R/R	20.85mA	20.0mA

33.4.5   ZENER DIODE

Zener diodes are a special type of semiconductor diode– devices that allow current to flow in one direction only.

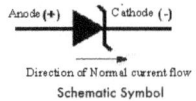

Direction of Normal current flow
Schematic Symbol

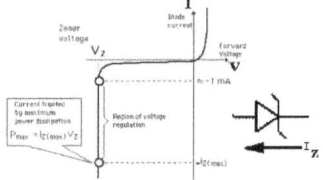

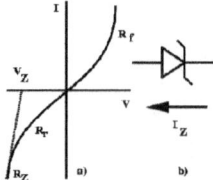

33.4.6   TUNNEL DIODE

A tunnel diode or Esaki diode is a type of semiconductor diode which is capable of very fast operation, well into the microwave region GHz, by utilizing quantum mechanical effects.

These diodes have a heavily doped p-n junction only some 10 nm (100 Å) wide. The heavy doping results in a broken band gap, where conduction band electron states on the n-side are more or less aligned with valence band hole states on the p-side.

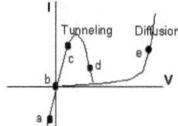

33.4.7   PHOTO DIODE

A junction photodiode is an intrinsic device that behaves similarly to an ordinary signal diode, but it generates a photocurrent when light is absorbed in the depleted region of the junction semiconductor. A photodiode is a fast, highly linear device that exhibits high quantum efficiency based upon the application and may be used in a variety of different applications.

eg., $GaAs$ doped with Si.

Material	Dopant	Peak λ or Range (nm)
GaP	N	550–590(*Green*)
Ga $As_{0.35}P_{0.65}$	N	589 (Yellow)
GaP	Zn, O	700 (red)
GaAs	Zn	900 (IR)
GAs	Si	910 - 1020 (IR)

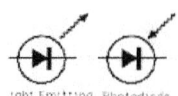

light Emitting Photodiode

$$I_{out} = I_{Dark} + I_{photodiode}$$

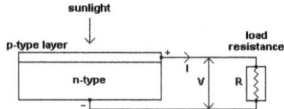

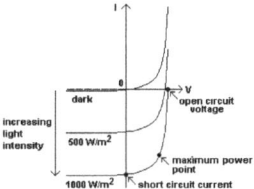

33.4.8. LED (Light emitting diode)

As seen above a P-N junction can relate the absorbed light energy into its proportional electric current. The same process is reversed here. That is, the P-N junction emits light when energy is applied on it. This phenomenon is generally called electro-luminance, which can be defined as the emission of light from a semi-conductor under the influence of an electric field. The charge carriers recombine in a forward P-N junction as the electrons cross from the N-region and recombine with the holes existing in the P-region. Free electrons are in the conduction band of energy levels, while holes are in the valence energy band. Thus the energy level of the holes will be lesser than the energy levels of the electrons. Some part of the energy must be dissipated in order to recombine the electrons and the holes. This energy is emitted in the form of heat and light.

The electrons dissipate energy in the form of heat for silicon and germanium diodes. But in Gallium-Arsenide-phosphorous (GaAsP) and Gallium-phosphorous (GaP) semiconductors, the electrons dissipate energy by emitting photons. If the semiconductor is translucent, the junction becomes the source of light as it is emitted, thus becoming a light emitting diode (LED). But when the junction is reverse biased no light will be produced by the LED, and, on the contrary the device may also get damaged.

The constructional diagram of a LED is shown below.

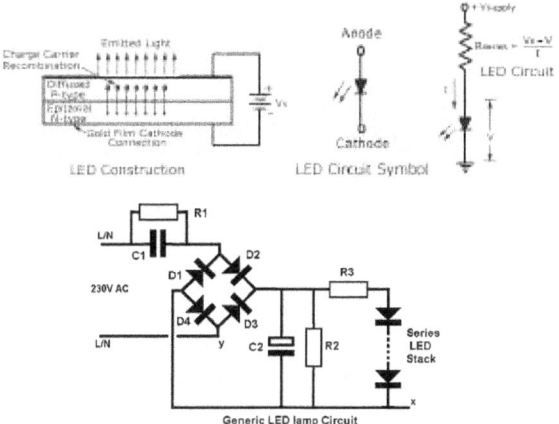

LED Characteristics

The forward bias Voltage-Current (V-I) curve and the output characteristics curve is shown in the figure above. The V-I curve is practically applicable in 'burglar alarms'.

Forward bias of approximately 1 V is needed to give significant forward current. The second figure is used to represent a radiant power-forward current curve. The output power produced is very small and thus the efficiency in electrical-to-radiant energy conversion is very less.

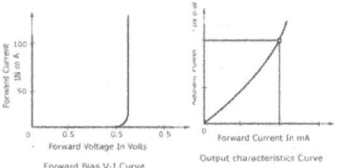

Forward Bias V-I Curve   Output characteristics Curve

33.5. BIPOLAR JUNCTION TRANSISTORS (BJTs)

Though J Bardeen and WH Brattain (1948), made the *first Point contact* transistor, it was W Shockley who formed the first *Junction transistor*.

33.5.1 Types of Transistors:
1. Point-contact transistor
2. Junction transistor: (i) Grown junction, (ii) Diffused junction, (iii) Alloy type and (iv) Epitaxial type.

Bipolar because transistors have both electrons and holes (whether majority or minority) are mobile at any time in the operation of the transistor.

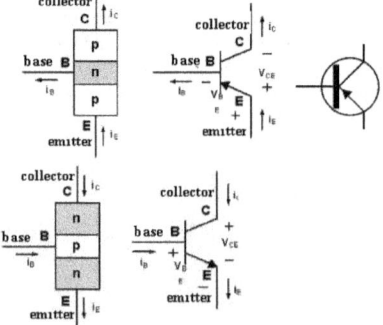

Transistor cases (housing) are assigned the so-called TO numbers, followed by number designating the physical dimensions; such as TO-3, TO-5, TO-92.

33.5.2.1 BASIC TRANSISTOR METHOD OF OPERATIONS:
a) Emitter-Base is Forward biased
b) Collector-Base is Reverse biased .

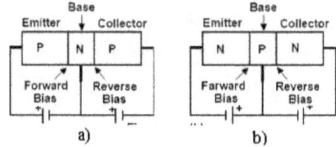

33.5.2.2 Basic Connection of a BJT

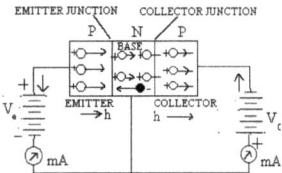

Basic connection of P-N-P Junction transistor

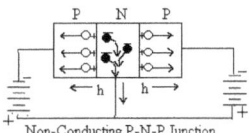

Non-Conducting P-N-P Junction

33.5.2.3 Currents in BJT

| Emitter-Collector current is Diffusion current | ,

| Emitter-Base current is Drift current |

| Collector-Base current is Drift current | .

| Base current is Recombination of both Electrons and Holes |

33.5.3   The three BJT configurations

1.   Common-Base (CB) or Ground Base

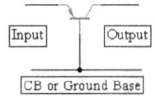

2.   Common-Emitter (CE)

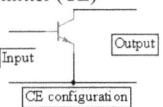

3.   Common-Collector (CC).

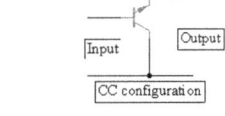

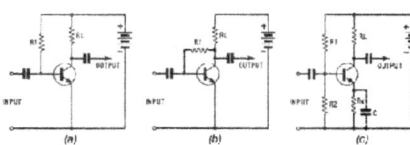

4.

Characteristic	Common Base (CB)	Common Emitter(CE)	Common Collector(CC)
1) Input Impedance	30 Ω (Low)	3.5kΩ (Medium)	580kΩ(High)
2) Output Impedance	3.1MΩ (Very High)	200kΩ(High)	35 Ω (Low)
3) Phase Angle	0°	180°	0°
4) Voltage Gain	High	Medium	Low
5) Current Gain	Low	Medium	High
6) Power Gain	Low	Very High	Medium

33.5.4 Current gains α and β of a transistor:

33.5.4.1 Fundamental Relation:

$$I_E = I_C + I_B$$

33.5.4.2 BJT Relations

$$I_C = I_C \text{(Major Carriers)} + I_{CO} \text{(Minor carriers)}$$

(I_{CO} = Leakage / Reverse saturation current, which is temp sensitive)

For Common Base

$$\alpha = h_{FB} = \left.\frac{\Delta I_c}{\Delta I_b}\right| < 1 \text{ always (0.95 to 0.999)}$$

For CE

$$\text{Current Gain, } A_I = \left.\frac{\Delta I_c}{\Delta I_b}\right|_{V_{EC}} \equiv \beta_{DC}$$

$$\beta_{DC} = h_{FE} = \frac{I_c}{I_b} > 1$$

$$\beta_{ac} = h_{fe} = \left.\frac{\Delta I_c}{\Delta I_b}\right|_{V_{EC}} > 1 \text{ always (20 to several 100)}$$

$$\beta = \frac{\alpha}{1-\alpha}$$

$$\alpha = \frac{\beta}{1+\beta}$$

$$I_{CEO} = (\beta+1)I_{CBO}$$

For CC

$$I_C = \beta I_B$$

for CE

$$I_C = \alpha I_E$$

I_C has magnitude of mA and I_B has μA.

33.5.5 V - I Output characteristics

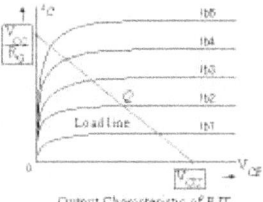

Output Characteristic of BJT

33.5.6   Three Regions of the Output V-I characteristic are:
(1) Cut-off region,
(2) Active region (i.e. region of safe operation), and
(3) Saturation region.

33.5.7 A practical BJT amplifier

A BJT is very versatile. It can be used in many ways, as an amplifier, a switch or an oscillator and many other uses.

Before an input signal is applied its operating conditions need to be set.

A bias circuit allows the operating conditions of a transistor to be defined. Bias design can take a mathematical approach or can be simplified using transistor characteristic curves. The characteristic curves predict the performance of a BJT. They are input characteristic curve, a transfer characteristic curve and an output characteristic curve

The output characteristic curves for a BJT are a graph displaying the output voltages and currents for different input currents. The linear (straight) part of the curve needs is utilized for an amplifier or oscillator

For use as a switch, a transistor is biased at the extremities of the graph, these conditions are known as "cut-off" and "saturation".

After the initial bend, the curves approximate a straight line. The slope or gradient of each line represents the output impedance, for a particular input base current. Take the middle curve. Draw the load line connecting point $\frac{V_{CC}}{R_L}$ in the I_C axis and point V_{CC} in the V_{CE}- axis. The operating point Q (Quiescent point) is the load line intersects the I_B curve. The V_{CE} is displayed up to 20 V For a single CE stage amplifier, and the $V_{CC} = 10$ V. Depending on whether the transistor used is a PNP or NPN, then one half-cycle will be amplified faithfully, the other cycle will approach the limits of the power supply and will "clip".

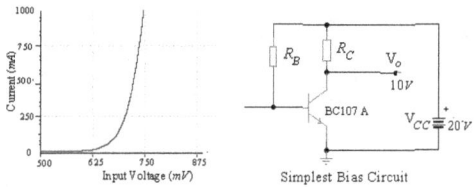

Simplest Bias Circuit

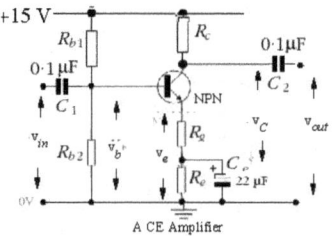

A CE Amplifier

Current gain

$$A_I = -\frac{h_f}{1+h_o R_L}$$

Voltage gain

$$A_V = -\frac{h_f}{h_i} R_L$$

$$R_o = -\frac{h_i + R_S}{h - h_o + R_S}$$

where $h = \begin{pmatrix} h_i & h_r \\ h_f & h_o \end{pmatrix}$

$$A_v = \frac{V_o}{V_i} = -\beta \frac{R_C \| r_o}{r_\pi + R_S}$$

The _class "A" amplifier_ amplifies the signal of the _same_ wave form as the input signal with usual amplifier. Operating point is the centre of the _load line._
The _class "B" amplifier_ is the way of being used for the push-pull amplifier. It combines a NPN-type transistor with one of PNP-type and the half is amplified. The output is bigger than the class "A" amplification.
The _class "C" amplification_ makes the bias point of the base electric current the side of the negative than the B point. It is used for amplification at high frequency distortion occurs to the output signal and can amplify the higher harmonic.

33.5.8   GAIN of an Amplifier:

$$dB = 10\log_{10}(P_{out} / P_{in})$$

$$dB = 20\log_{10}(I_{out} / I_{in})$$

$$dB = 20\log_{10}(V_{out} / V_{in})$$

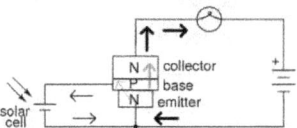

33.6.1 MOSFET

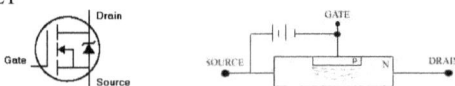

33.6.2 VARACTOR

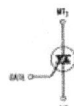

33.7 RECTIFIERS
33.7.1 Half wave rectifier

The single phase half-wave rectifier produces an output every half cycle and that it was not practical to produce a steady DC supply

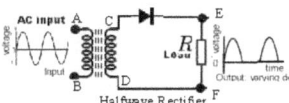

Halfwave Rectifier

V_{peak} = Peak value of the phase input voltage

V_{rms} = rms value of output voltage

$$V_{rms} = \tfrac{1}{2} V_{peak}$$

$$V_{dc} = \tfrac{1}{\pi} V_{rms}$$

$$V_O = V_{DC} + 1^{st} harm + 2^{nd} harm + ..$$

33.7.2 Full wave rectifier (Centre-tapped)

i) V_{dc} output voltage is higher than for half wave,

ii) the output is smoother and has much less ripple than that of the half wave rectifier.

$$V_{dc} = \tfrac{2}{\pi} V_{max} = 0.637 V_{max} = 0.9 V_{rms}$$

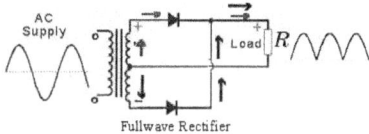

Fullwave Rectifier

$$V_O = V_{dc} - 1^{st} harm - 2^{nd} harm - ..$$

33.7.3 Bridge rectifier

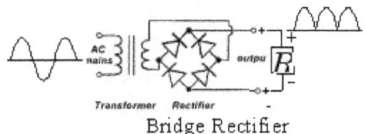

Transformer   Rectifier

Bridge Rectifier

$$V_{dc} = \tfrac{2}{\pi} V_{max} = 0.637 \, V_{max} = 0.9 \, V_{rms}$$

33.7.4 Block Diagram of a Power Supply

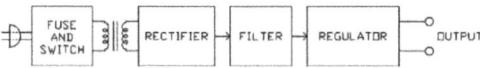

33.7.4.1 Filtering the Rectifier's Output (The Smoothing Capacitor)

The full-wave bridge rectifier however, gives less superimposed ripple than halfwave rectifier, while the output waveform is twice that of the frequency of the input supply frequency. To increase its average DC output level even higher by means of filters.

Filtering is another name for smoothing, such as LC (Inductor - Capacitor) and/or RC (Resistor - Capacitor) sections the critical part of most filters is the first capacitor.

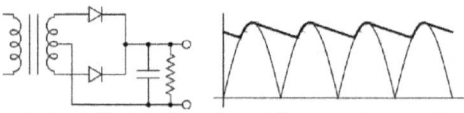

The amount of ripple (variation) in the voltage ΔV across the capacitor is given by

$$\Delta V = \tfrac{I}{\Delta t} C$$

Conventionally set $\Delta t = 8 \, ms$ approximately (1/120) for a full-wave rectifier or 16 ms approximately (1/60) for a half-wave rectifier

The DC is the average voltage,

$$V_{dc} \mid_{Av} = V_P - \left(\tfrac{1}{2} \Delta V\right) V_O$$

33.7.4.2 Ripple Factor (ρ):

$$\rho \approx \frac{V_{rms}(fundamental)}{V_{dc}}$$

For half wave rectifier, $\rho \approx \sqrt{(\pi^2 / 4) - 1} = 1.21 = 121\%$.

For a full wave rectifier, $\rho \approx \sqrt{(\pi^2 / 8) - 1} = 0.482 = 48\%$.

33.8. OSCILLATORS

An oscillator is a circuit that is capable of a sustained AC output signal obtained by converting input energy. Oscillators can be designed to generate a variety of signal waveforms, and they are convenient sources of sinusoidal AC signals for testing, control, and frequency conversion. Oscillators can also generate square waves, ramps, or pulses for switching, signaling, and control.

Many systems require an input in the form of a periodic, usually sinusoidal, waveform

(i)   to drive the heterodyne receivers.

(ii)   in domestic radios and TV.

(iii)   to apply for many other types of coherent signal source.

33.8.1 Heinrich Georg Barkhausen (1881–1956) criterion.

It is widely used in the design of electronic oscillators, and also in the design of general negative feedback circuits such as operational amplifiers, to prevent them from oscillating.

Regardless of its amplifier, an oscillator must meet the two *Barkhausen conditions* for oscillation:
1) The loop gain must be slightly greater than unity.
2) The loop phase shift must be 0° or 360°.

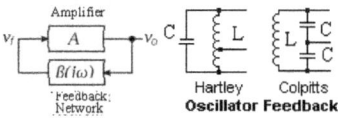

33.8.2   RC Phase shift Oscillator

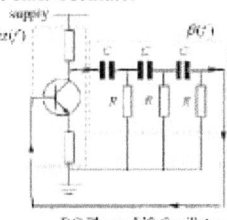

RC Phase shift Oscillator

$$f_o = \frac{1}{2\pi\sqrt{6}RC}$$

33.8.3   Hartley Oscillator

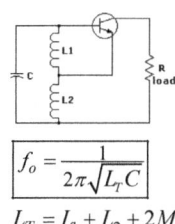

$$f_o = \frac{1}{2\pi\sqrt{L_T C}}$$

$$L_T = L_1 + L_2 + 2M$$

33.8.4.   Colpitts Oscillator

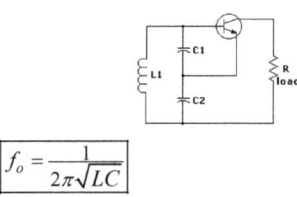

$$f_o = \frac{1}{2\pi\sqrt{LC}}$$

33.8.5.   Transformer coupled Oscillator

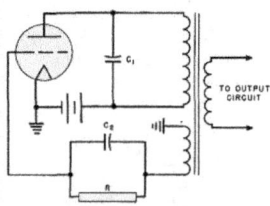

33.9. CATHODE RAY OSCILLOSCOPE (CRO)

The CRO is a common laboratory instrument that provides accurate time and amplitude measurements of voltage signals over a wide range of frequencies. Its reliability, stability, and ease of operation make it suitable as a general purpose laboratory instrument. The heart of the CRO is a cathode-ray tube

It is an evacuated glass tube containing an *electron gun* producing a beam of collimated electrons, *X- and Y-deflecting plates* and a *screen. A time base* is frequently applied to the X-plates so that the output is swept across the screen at a uniform speed with a very fast fly back return. The time base has a saw tooth wave form

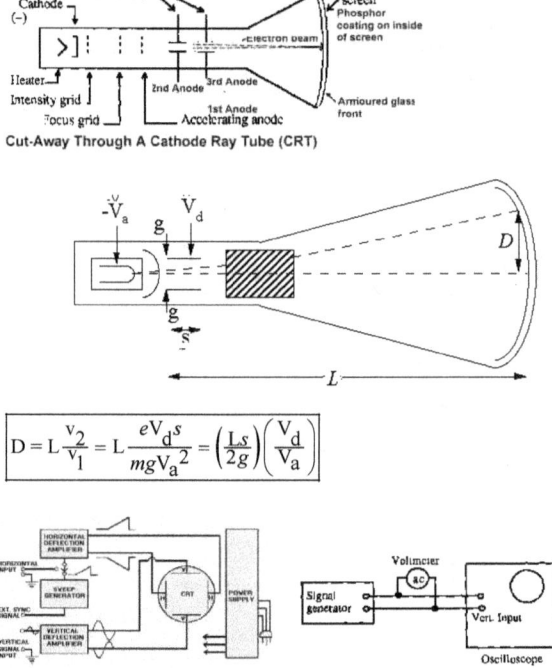

Cut-Away Through A Cathode Ray Tube (CRT)

$$D = L\frac{v_2}{v_1} = L\frac{eV_d s}{mgV_a^2} = \left(\frac{Ls}{2g}\right)\left(\frac{V_d}{V_a}\right)$$

Oscilloscope

33.9.2 APPLICATIONS OF CRO:
  (1) PD and EMF measurements, with the advantage that the CRO takes almost no current, since it has very high input impedance.
  (2) Wave form display of the Y-input signals
  (3) Frequency display and measurement
  (4) Time measurement

33.9.3 LISSAJOUS' FIGURES:
  With the tome base of the CRO switched off, a signal in the Y-input, and a signal generator to the X-input one can determine, from the pattern on the screen, the unknown frequency of the signal to the Y-plates.

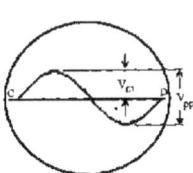

Trace seen on CRO

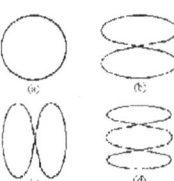

Lissajous Figures for ratios
a) 1:1, b) 2:1, c) 1:2, c) 3:1

33.10   ELECTRON (VACUUM) TUBE DEVICES

All modern vacuum tubes are based on the concept of the Audion--a heated "cathode" (thoriated filament: just a tungsten filament, coated with a mixture of barium and strontium oxides) boils off electrons into a vacuum; they pass through a grid (or many grids), which control the electron current; then strike the anode (plate) (of graphite) and are absorbed. By designing the cathode, grid(s) and plate properly, associating with a proper circuit, the tube will make a small AC signal voltage into a larger AC voltage, thus amplifying it.

The Richardson-Dushman theory of emission in the current-limited region of operation

is. $$J = AT^2 e^{-W_A/k_B T}$$

33.10.1.1   Child's Law (Child-Langmuir equation) (1911)

$$J = \left(\frac{4k}{9d^2} \sqrt{\frac{2e}{m}} \right) E_b^{3/2}$$

$$J = K E_b^{3/2}$$

K is *perveance,* depends on the geometry of the electrodes and their separation.
Child's law holds well in the space charge-limited region.

33.10.1.2   DIODE:

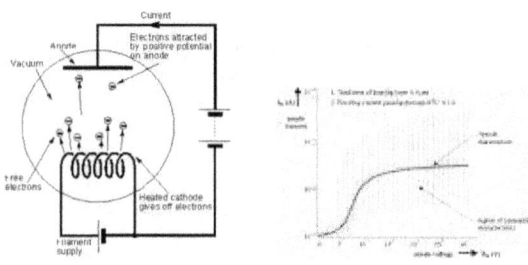

The first diode was by Fleming in 1904.

33.10.1.3 TRIODE

The triode (audion) was invented by de Forest in 1906.

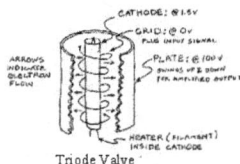

Triode Valve

Under space charge limited conditions with the plate voltage $E_P = 100$ V and permanence $K = 10^{-4}$ (in SI unit), the plate current in a vacuum diode will be

$$J = K[V_O + V_A / \mu]^{3/2} \, A \, .$$

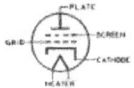

Circuit Symbols

33.10.1.4 TETRODE

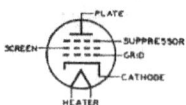

33.10.1.5 PENTODE

33.10.2 TRIODE

There are two major reasons to make grid bias generally negative:

Current flowing in the grid circuit can distort the shape of the output voltage with respect to the shape of the input voltage.

Positive grid voltage can cause excessive plate current and result in damage to the tube.

The characteristic curves are used to determine the performance of a tube under any operating condition. The characteristics can be obtained by using a suitable circuit. First adjust the grid voltage $V_g = 0V$. Now plate voltage increased in steps and the corresponding values of plate current are noted at each step. If we draw a graph between plate voltage and plate current, then a curve is obtain at $V_g = O$. Similarly obtain curve at $V_g = -2, -4V, -6V, -8V$ and so.

33.10.2.1 Plate (Dynamical or internal) Resistance (r_p)

It is the ratio small change in plate voltage V_p to the small change in plate current I_p when grid voltage V_g is kept constant, denoted by r_p.

$$r_p = \frac{\Delta V_p}{\Delta I_p}\bigg|_{V_g}$$

The value of r_p can be obtain from the plate characteristic and its value remains constant along the linear portion of the characteristic.

33.10.2.2 Amplification Factor (μ)

The ratio of small change in plate voltage V_p to the small change in the grid voltage V_g when plate current I_p is kept constant is denoted μ.

$$\mu = -\frac{\Delta V_p}{\Delta V_g}\bigg|_{I_p}$$

33.10.2.3 Mutual Conductance g_m (Transconductance)

The ratio of small change in plate current I_p to the small change in grid voltage V_g when plate voltages V_p is kept constant is called mutual conductance or transconductance and it is denoted by g_m.

$$g_m = \frac{\Delta I_p}{\Delta V_g}\bigg|_{V_p}$$

33.10.2.4. Relation between r_p, μ and g_m

$$\mu = (r_p)(g_m)$$

33.10.2.5. Cut off voltage, for $I_p \to 0$ is

$$V_c = -\frac{V_p}{\mu}$$

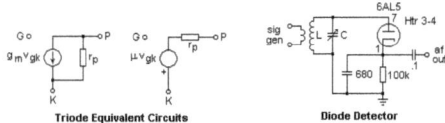

Triode Equivalent Circuits   Diode Detector

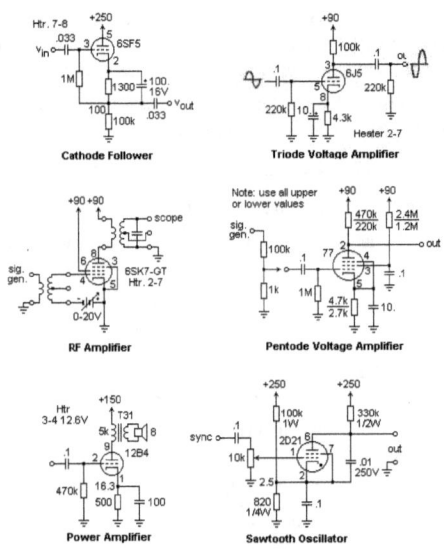

Cathode Follower | **Triode Voltage Amplifier**

RF Amplifier | **Pentode Voltage Amplifier**

Power Amplifier | **Sawtooth Oscillator**

33.10.2.6 Output voltage.

$$e_o = i_p R_L$$

$$e_o = i_p . R_L = \left\{ -\frac{\mu e_s}{R_L + r_P} \right\} R_L$$

33.10.2.7 Voltage amplification, A_V

$$A_V = \frac{\mu R_L}{R_L + r_P}$$

33.10.2.8 Power amplification, A_p

$$A_P = \frac{\mu^2 R_L R_g}{(R_L + r_P)^2}$$

33.10.2.9 TRIODE CHARACTERISTCS

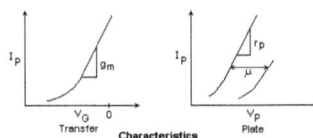

33.10.2.10 Load line

$$E_b = E_{bb} - I_b R_L$$

Slope $\boxed{\tan\alpha - \dfrac{1}{R_L}}$

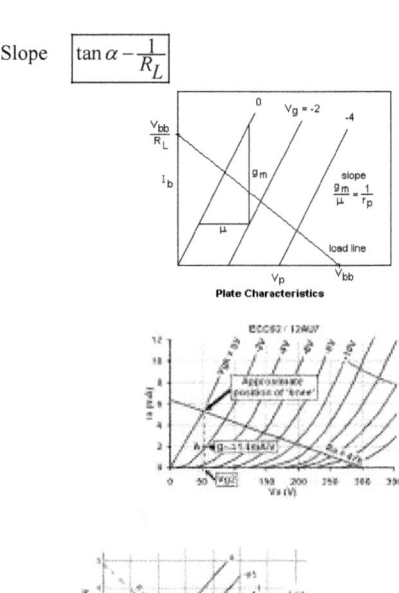

Plate Characteristics

Triode output Characteristic with Load line

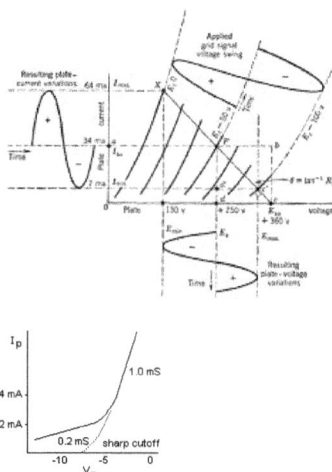

Transfer Characteristic

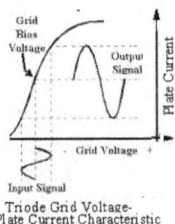

Triode Grid Voltage-
Plate Current Characteristic

33.10.3.   Negative Feedback in Amplifiers
Negative feedback is used in an AF amplifier to
1) Reduce amplitude distortion in gain resulting from variations in supply voltages and valve constants,
2) Reduce frequency distortion.

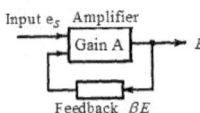

Effective gain, $\boxed{\dfrac{E}{e_s} = \dfrac{A}{1 - \beta A}}$

If $\beta A \gg 1$, $\dfrac{E}{e_s} = -\dfrac{1}{\beta}$.

i.e., the effective gain of the amplifier network is dependent only upon the fraction of the output voltage feedback, and substantially independent of the amplifier gain.

-o-0-o-0-o-0-o-0-o-

Chapter 34

DIGITAL ELECTRONICS
Pulse, Binary Algebra, Basic Logic Gates, Combinational Logic,
Half Adder, Full Adder, Sequential Logic Circuits,
Flip flops, Digital Computer Basics, Languages

*"When Nuclear Energy has been successfully applied for power production in,
say a couple of decades from now, India will not have to look abroad
for its experts but will find them ready at hand"*- in 1944, H. J. Bhabha

34.1.1   ELECTRICAL PULSE: (Pulse train)
Periodic signals are signals that repeat in time with a certain period. The most
fundamental periodic signal is the sinusoidal signal.

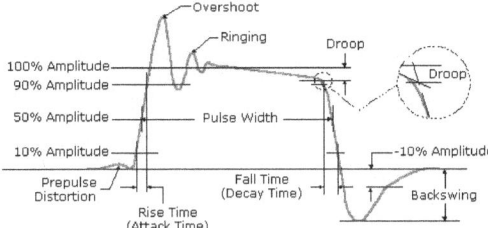

24.1.2.1   Positive going pulse

34.1.2.2   Negative going pulse.

Positive logic means that the more positive level, of the two voltage levels, called "1"and
the less positive level "0."

34.1.2.3 Period of cycle,

$$T_P = \frac{1}{f}$$

34.1.2.4 Period of a pulse

$$T_P = T_{ON} + T_{OFF}$$

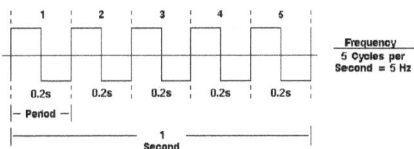

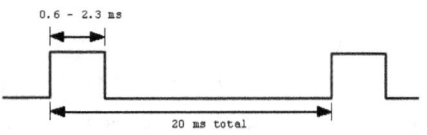

Duty cycle of a pulse:	(space ratio) 100%

34.2.1 RC differentiator

$$V_{in} \cong V_C \qquad v_{out} \cong RC \frac{d}{dt} v_{in}$$
$$(\omega \ll 1/RC)$$

Time constant
$$\boxed{\tau_{RC} = RC}$$

$$\boxed{T_{ON} \text{ or } T_{OFF} > 5\ \tau_{RC}}$$

34.2.2 RC Integrator

$$V_{in} \cong iR \qquad v_{out} \cong \frac{1}{RC} \int v_{in}\, dt$$
$$(\omega \gg 1/RC)$$

34.3.1.1 **Decimal Addition**:

$$
\begin{array}{r}
44 \\
-32 \\
\hline
12
\end{array}
$$

34.3.1.2 Digital number

32	16	8	4	2	1
0	0	1	1	1	1

34.3.1.3 Basic Rule for Calculations

34.3.1.4 BODMAS

Bodmas (Bracket Order of Divide Multiply Add Subtract) are the rules followed by calculators.

34.3.1.5 In general an n digit binary number can represent numbers from 0 to $(2^n - 1)$

For instance a byte is 8 bits and can represent numbers from 0 to 255 $(2^8 - 1)$.

34.3.2 Binary Addition

Base Place	2^2	2^1	2^0	
Carryover	1	1		
		1	1	3
	+1	1	1	+ 3
	1	1	0	6

34.3.3 Multiplication

Base 2 Place	2^3	2^2	2^1	2^0			
			1	1			3
	X		1	1		X	3
Carry	1	1	1	1			9
		1	1				
		1	0	0	1		

34.4.1. BOOLEAN OPERATIONS

Basic Boolean Algebraic Properties	
Additive	Multiplicative
$A + B = B + A$	$AB = BA$
$A + (C + C) = (A + B) + C$	$A(BC) = (AB)C$
$A(B+C) = AB + AC$	

34.4.2. BASIC LOGIC GATES symbols

Electronic circuits which combine digital signals according to the Boolean algebra are referred to as *logic gates*; gates because they control the flow of information. *Positive logic* is an electronic representation in which the true state is at a higher voltage, while *negative logic* has the true state at a lower voltage.

Name	Symbol
BUFFER	⊳
NOT	⊳∘
AND	⊐
OR	⊃
XOR	⊃
NAND	⊐∘
NOR	⊃∘
NOT-XOR	⊃∘

34.4.3. Logic circuits are grouped into families, each with their own set of detailed operating rules. Some common logic families are:

RTL: resistor-transistor logic,
DTL: diode-transistor logic,
TTL: transistor-transistor logic,
NMOS: N-channel metal-oxide silicon,
CMOS: complementary metal-oxide silicon and
ECL: emitter-coupled logic.

34.4.4 LOGIC CIRCUITS AND TRUTH TABLES:
34.4.4.1 NOT or INVERTER gate and Buffer

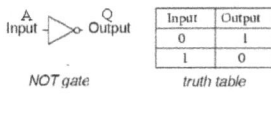

Input	Output
0	1
1	0

NOT gate   truth table

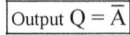

Output $Q = \overline{A}$

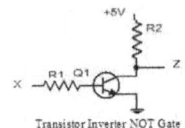

Transistor Inverter NOT Gate

34.4.4.2 OR gate

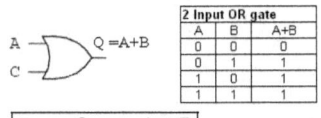

2 Input OR gate		
A	B	A+B
0	0	0
0	1	1
1	0	1
1	1	1

$$\boxed{\text{Output } Q_{OR} = A + B}$$

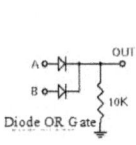

Diode OR Gate

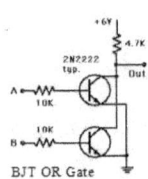

BJT OR Gate

24.4.3 AND gate

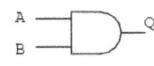

A	B	Output
0	0	0
0	1	0
1	0	0
1	1	1

$$\boxed{\text{Output } Q_{AND} = A \cdot B}$$

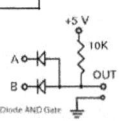

Diode AND Gate

34.4.4 NAND (Universal gate)

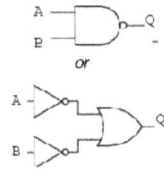

or

A	B	Output
0	0	1
0	1	1
1	0	1
1	1	0

Output

$$Q = \overline{A \cdot B}$$

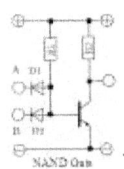

NAND Gate

34.4.5   NOR gate

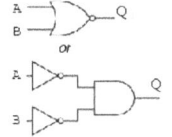

A	B	Output
0	0	1
0	1	0
1	0	0
1	1	0

Output

$$Q = \overline{A + B}$$

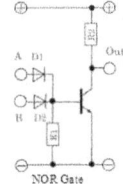

NOR Gate

34.4.6   XOR gate

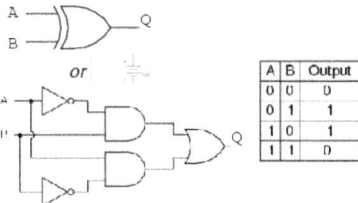

A	B	Output
0	0	0
0	1	1
1	0	1
1	1	0

Output

$$Q_{XOR} = A \oplus B = A\overline{B} + \overline{A}B$$

34.4.7   Karnaugh Graph
A graphical representation of a Truth Table that can be used to reduce a logic circuit to its simplest term.

34.5   **COMBINATIONAL LOGIC**
The outputs of **Combinational Logic Circuits** are only determined by the logical function of their current input state, logic "0" or logic "1", at any given instant in time
The output is dependant at all times on the combination of its inputs. So if one of its inputs condition changes state, from 0-1 or 1-0, so too will the resulting output as by default combinational logic circuits have "no memory", "timing" or "feedback loops" within their design.

34.5.1   HALF ADDER
With the help of half adder, we can design circuits that are capable of performing simple addition with the help of logic gates.

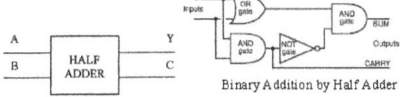

Binary Addition by Half Adder

Truth Table

Inputs A B	Outputs S C
0 0	0 0
0 1	1 0
1 0	1 0
1 1	0 1

34.5.2 Full Adder

The main difference between a half-adder and a full-adder is that the full-adder has three inputs and two outputs. The first two inputs are A and B and the third input is an input carry designated as CIN. When a full adder logic is designed we will be able to string eight of them together to create a byte-wide adder and cascade the carry bit from one adder to the next.

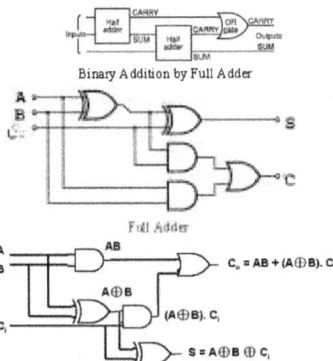

Binary Addition by Full Adder

Full Adder

$$C_0 = AB + (A \oplus B). C_i$$

$$(A \oplus B). C_i$$

$$S = A \oplus B \oplus C_i$$

34.5.3 Other important combinational logic circuits

These are Subtractors, Multipliers, Decoders, Encoders, Multiplexers.

34.5.4 Notion of Sets introduced first by Georg Cantor

$e \in S$	means element 'e' belongs to set S

$e \notin S$	means element 'e' does not belong to set S

$A \subset B$	means 'A' is a sub-set of set B or A is contained in set B

$A \cup B$	means set 'A' union set B

34.5.5. Venn diagram

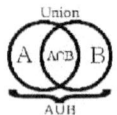

Union

A A∩B B

A∪H

| $A \cap B$ | means set 'A' Intersection set B |

Venn diagram

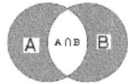

34.5.6 POINT SET THEORY IN DIGITAL LOGIC

1) Mathematically, AND gate is

| Output $Q_{AND} = A \cdot B$ |

Venn diagram for this is

| Output $Q_{AND} = A \cap B$ |

2) OR gate is

| Output $Q_{OR} = A + B$ |

Corresponding by Venn diagram

| $Q_{OR} = A \cup B$ |

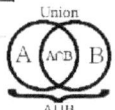

3) NOT gate is

| Output $Q_{NOT} = \overline{A}$ |

This corresponds to the Venn diagram,

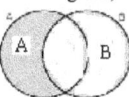

4) XOR gate

Output | $Q_{XOR} = A \oplus B = A\overline{B} + \overline{A}B$ |

The Venn diagram is

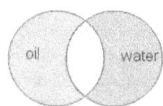

34.5.7. Logic in Basic Elevator (Lift) (L) System

Lift (L); Its Door (D); Floor Button (B);
a)   The conditions to be met are:
Lift Door is closed when D=1 and appropriate floor Button is B=1 when pressed.
b)   What is the logic expression?

$$L = D . B$$

c) Draw the Truth Table for it

D	B	L
0	0	0
0	1	0
1	0	0
1	1	1

d) What is the logic gate called?
It is an AND gate.

34.5.8. Application of Combinatorial logic: <u>Traffic Light System</u>

The traffic light sequence is
Red (R) is followed by
Red & Yellow (R & Y) followed by
Green (G) followed by
Yellow (Y) followed by Red, and so on, with the 4-line square repeating.
A and B are two-input of a binary system.

$$R = \overline{A} . \overline{B} + \overline{A} . B$$
$$Y = \overline{A} . B + A . B$$
$$G = A . \overline{B}$$

To implement these with suitable logic gates.

Input		Ouput			Basic
A	B	R	Y	G	expression
0	0	1	0	0	$R = \overline{A} . \overline{B}$
0	1	1	1	0	$R = \overline{A} . B$
1	0	0	0	1	$G = A . \overline{B}$
1	1	0	1	0	$Y = A . B$

$Y = \overline{A} . B$

34.5.9. Other Examples

1) A boiler shut down solenoid (S) operates when the temperature (T) reaches 50 °C and the circulating pump (P) is turned off, or the pilot lamp (L) goes out. The logic expression is $S = T . \overline{P} + \overline{L}$. The logic diagram is

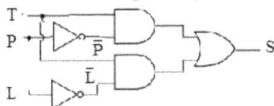

2) Control system using logic gates – Cooling Fan

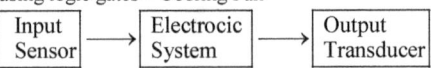

Input Sensor	Electrocic System	Output Transducer

34.6 **SEQUENTIAL LOGIC**:

The digital logic described thus far is called combinatorial logic because the output depends solely upon the presently existing combination of the inputs; *past* values of the inputs are not important. Sequential logic deals with the issue of time dependence and can get much more complicated than combinatorial logic -- much in the same way that differential equations are more difficult than algebraic equations. The fundamental building block of sequential circuits is the flip-flop.

34.6.1 FLIP FLOPS (FFs):
Flip flops (FFs) are used to make counters and registers.

34.6.2. RS FLIP FLOP:

The simplest Flip flop is this. It has two inputs Set (S) and Reset (R) and two outputs Q and $\bar{Q}$ gates:

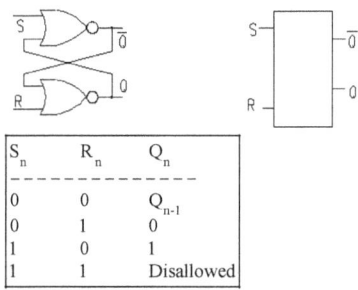

S_n	R_n	Q_n
0	0	Q_{n-1}
0	1	0
1	0	1
1	1	Disallowed

34.6.3. T- FLIP FLOP:
It divides the input frequency by two.

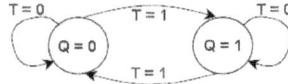

34.6.4. JK FLIP FLOP: (Universal FF)
It is a combination of a clocked RS Flip Flop and a T Flip Flop.

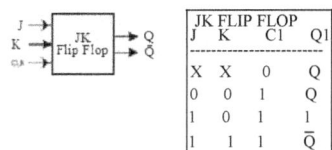

JK FLIP FLOP			
J	K	Cl	Ql
X	X	0	Q
0	0	1	Q
1	0	1	1
1	1	1	$\bar{Q}$

34.6.5 D FLIP FLOP:
It is a modification of the JK Flip Flop.

34.6.6 Asynchronous Counters

34.6.7 Synchronous Counters

34.7. COMPUTERS

34.7.1 Acronyms

BIT   The binary equivalent of $9_{10} = (1001)_2$ is a 4-bit.

BYTE A defined number of bits, say 8 bits

NYBBLE: is 4 bits.= ½ byte.

ASCII: is short for the American Standard Code for Information Interchange. One byte in ASCII is 7-bits.

$$(0110000)_{ASCII} = (0)_{10}$$

$$(0111001)_{ASCII} = (9)_{10}$$

Letters of alphabet in ASCII:

$$(1000001)_{ASCII} = A$$

$(1000010)_{ASCII} = B$

$(1000011)_{ASCII} = C$, etc.

1 K bytes: $1K = 1024 = 2^{10}$

$16Kb \rightarrow 16,384b$

34.7.2 INTEGRATED CIRCUITS (ICs)

Instead of making discrete transistors, resistors, capacitors and diodes the entire circuit can be made on one small piece of silicon by doping various parts and making appropriate connections. The circuit is then named a silicon chip, called an integrated circuit (IC). *Eg.*, a Microprocessor (μP). The first IC was developed in the 1950s by Jack Kilby of Texas Instruments and Robert Noyce of Fairchild Semiconductor

ICs are used for a variety of devices, including Microprocessors (μP), audio and video equipment, and automobiles. Mobile phone, Smart phones, iPods, *etc.* ICs are often classified by the number of transistors and other electronic components they contain:

34.7.2.1 Active Devices

(a) Discrete Transistor (1951)

(b) **SSI (small-scale integration):** Up to 100 electronic components per chip (1960)

Logic gates, TTL, Master slave flip-flop, JK flip-flop, *etc.*

(c) **MSI (medium-scale integration):** From 100 to 3,000 electronic components per chip (1966)

Adder, Multiplexer, Decoder, BCD.

(d) LSI **(large-scale integration):** From 3,000 to 100,000 electronic components per chip (1969).

I K bit RAM; a 4-bit Microprocessor (μP)(1971).

(e) VLSI **(very large-scale integration):** From 100,000 to 1,000,000 electronic components per chip (1975) Memories, Microprocessors (μP).

(f) ULSI **(ultra large-scale integration):** More than 1 million electronic components per chip.

Microcomputer: 256K bit RAM.

34.7.3 **MICROCOMPUTER (1970s)**

34.7.3.1 BASIC ANATOMY of a DIGITAL COMPUTER

1.   Input Unit.

2.   The CPU (Central Processing Unit) = ALU + Control Unit + Memory.

3.   Output Unit.

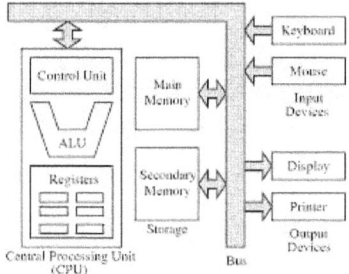

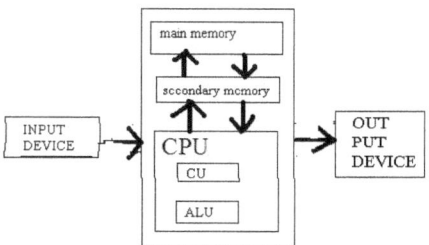

BLOCK DIAGRAM OF A DIGITAL COMPUTER

34.7.3.2   A <u>microcomputer</u> contains a microprocessor, a memory IC to store information, supporting circuits, input (say, a keyboard) and output devices (like a video screen.

CPU chipset constitutes a Microprocessor; Motorola 6800 Microprocessor had many similarities to the PDP-11 Microcomputer of DEL. IC PC/XT original PC bus and its derivatives such as PC/AT and compatibles are then developed.

CPU is the heart of the computer.

Computers do their computation in the CPU on chunks of data organized as computer words.

Word size can range from 4.bitsw to 32 bits or more; with a 16.bit word size being popular in Microcomputers (μP) of 1980s.

RAM: stands for Random Access Memory; is volatile, as when power is removed its information evaporates (Forgets).

ROM: stands for Read-Only-Memory; is non-volatile, to "bootstrap" the computer when the power is first turned ON. Additional ROM is often programmed with System routine, graphics routines, and other programs that one wants to be there for all the times. Computers which use, for example, μP 80386 or μP 68020 use a bus32 bits (4 *bytes*) wide.

Hard Disks and Floppy disks (*i.e.,* diskettes) or Compact Disks (*i.e.* CDs) or DVDs (Digital Video Disks) are the usual ones with storage capacities having different storage capacities (a few Mb to 7 *Gb*) are used to store data.

IBM has evolved improved buses in subsequent PC generations, first the PC/AT (uses μP 80286) with maximum BW 5.3 *Mb/s*, then the PS/2 series, then PC/AT (uses μP 80386), PC/ AT486; Pentium I. Pentium II, Pentium III, Pentium IV, *etc.*

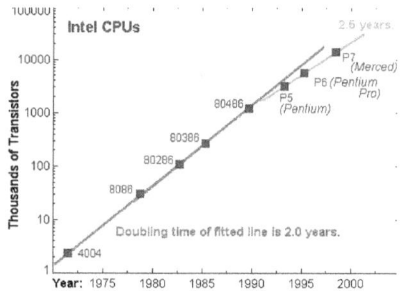

34.7.4. Language types

34.7.4.1 Machine and assembly languages

A <u>machine language</u> consists of the numeric codes for the operations that a particular computer can execute directly. The codes are strings of 0s and 1s, or binary digits ("bits"), which are frequently converted both from and to hexadecimal (base 16) for human viewing and modification. Machine language instructions typically use some bits to represent operations, such as addition, and some to represent operands, or perhaps the location of the next instruction. Machine language is difficult to read and write, since it does not resemble conventional mathematical notation or human language, and its codes vary from computer to computer.

34.7.4.2 Assembly language

It is one level above machine language. It uses short mnemonic codes for instructions and allows the programmer to introduce names for blocks of memory that hold data. One might thus write "add pay, total" instead of "0110101100101000" for an instruction that adds two numbers.

Assembly language is designed to be easily translated into machine language. Although blocks of data may be referred to by name instead of by their machine addresses, <u>assembly language</u> does not provide more sophisticated means of organizing complex information. Like machine language, assembly language requires detailed knowledge of internal computer architecture. It is useful when such details are important, as in programming a computer to interact with <u>input-output devices</u> (printers, scanners, storage devices, and so forth).

34.7.4.3 Algorithmic languages

Algorithmic languages are designed to express mathematical or symbolic computations. They can express algebraic operations in notation similar to mathematics and allow the use of subprograms that package commonly used operations for reuse. They were the first high-level languages.

34.7.4.4 FORTRAN

The first important algorithmic language was FORTRAN (*for*mula *tran*slation), designed in 1957 by an IBM team led by John Backus. It was intended for scientific computations with <u>real numbers</u> and collections of them organized as one- or multidimensional arrays. Its control structures included conditional IF statements, repetitive loops (so-called DO loops), and a GOTO statement that allowed nonsequential execution of program code. FORTRAN made it convenient to have subprograms for common mathematical operations, and built libraries of them.

FORTRAN was also designed to translate into efficient machine language. It was immediately successful and continues to evolve.

34.7.4.5 ALGOL

ALGOL (*al*gorithmic *l*anguage) was designed by a committee of American and European computer scientists during 1958–60 for publishing algorithms, as well as for doing computations. Like <u>LISP</u> (described in the next section), ALGOL had recursive subprograms—procedures that could invoke themselves to solve a problem by reducing it to a smaller problem of the same kind. ALGOL introduced <u>block structure</u>, in which a program is composed of blocks that might contain both data and instructions and have the same structure as an entire program. Block structure became a powerful tool for building large programs out of small components.

34.7.4.6 LISP

LISP (*list* *p*rocessing) was developed about 1960 by John McCarthy at the Massachusetts Institute of Technology (MIT) and was founded on the mathematical theory of recursive <u>functions</u> (in which a function appears in its own definition). A LISP program is a function applied to data, rather than being a <u>sequence</u> of procedural steps as in FORTRAN and ALGOL. LISP uses a very simple notation in which operations and their operands are given in a parenthesized list. For example, $(+ a (* b c))$ stands for $a + b*c$. Although this appears awkward, the notation works well for computers. LISP also uses the list structure to represent data, and, because programs and data use the same structure, it is easy for a LISP program to operate on other programs as data.

LISP became a common language for <u>artificial intelligence</u> (AI) programming, partly owing to the confluence of LISP and AI work at MIT and partly because AI programs capable of "learning" could be written in LISP as self-modifying programs. LISP has evolved through numerous dialects, such as Scheme and Common LISP.

34.7.4.7 C

The C programming language was developed in 1972 by Dennis Ritchie and Brian Kernighan at then AT&T Corpn for programming computer <u>Operating Systems</u>. Its capacity to structure data and programs through the composition of smaller units is comparable to that of ALGOL. It uses a compact notation and provides the programmer with the ability to operate with the addresses of data as well as with their values. This ability is important in <u>Systems Programming</u>, and C shares with assembly language the power to exploit all the features of a computer's internal architecture. C^{++}, along with its descendant C^{++++}, remains one of the most common languages.

34.7.5 Business-oriented languages

34.7.5.1 COBOL

COBOL (*c*ommon *b*usiness *o*riented *l*anguage) has been heavily used by businesses since its inception in 1959. A committee of computer manufacturers and users and U.S. government organizations established CODASYL (*C*ommittee on *D*ata *S*ystems and *L*anguages) to develop and oversee the language standard in order to ensure its portability across diverse systems.

COBOL uses an English-like notation—novel when introduced. Business computations organize and manipulate large quantities of data, and COBOL introduced the <u>record data structure</u> for such tasks. A record clusters heterogeneous data such as a name, ID number, age, and address into a single unit. This contrasts with scientific languages, in which homogeneous arrays of numbers are common. Records are an important example of "chunking" data into a single object.

+&^&^&^&^&^&^&^&^&^&^&^+

Chapter 35

COMMUNICATION ELECTRONICS
AM Modulation, FM Modulation, OpAmp

"The high destiny of the individual is to serve rather than to rule" -Albert Einstein

35.   INTRODUCTION

Two-wire lines (Telegaphy and Telephony)
Telegraphy (Samuel F.B. Morse, 1837, electric version)

Consists of a transmitter (telegraph key), Morse Relay (em amplifier), Local sounder (receiver) and Local Battery (1 to 1.5V).
Telephony (Alexander Graham Bell, 1875). The essential elements of a telephone are a **microphone** (*transmitter*) to speak into and an **earphone** (*receiver*) which reproduces the voice in a distant location. In addition, most telephones contain a ***ringer*** which produces a sound to announce an incoming telephone call, and a **dial or keypad** used to enter a **telephone number** when initiating a call to another telephone. Until approximately the 1970s most telephones used a rotator dial.

Telephone used for voice communication purposes (antenna feeds down-leads to receivers) have characteristic impedances Z_0 between $70 - 600$ Ω, and is useful for line frequencies and below 100 *MHz*. Propagation is essentially TEM wave. Specific resistance of the wire material, ρ, with radius a, separation d,

$$R_{dc} = \frac{2\,\rho}{\pi a^2}\;;\; L = \frac{\mu}{\pi}\ln d/a\, \text{H};\; C = \{\frac{\omega\varepsilon}{\ln d/a}\}\; F\;;$$

$$R_{ac} = \frac{1}{\pi\alpha}\{\frac{\omega\mu\,\rho}{2}\}^{1/2}\;;\; Z_0 \approx \{\frac{L}{C}\}^{1/2}\,.$$

Co-axial cables are used to propagate radiation waves (Radio- and TV) at higher frequencies, to avoid skin effect losses in the two-wire line conduction and radiation from the surface, and they use normally the TEM mode of propagation.

$$R_{ac} = \frac{1}{2\pi}\{\frac{\omega\mu\,\rho}{2}\}^{1/2}[\frac{1}{a} + \frac{1}{b}]\;\Omega/m\;;$$

$$L = \frac{\mu_0\mu_r}{2\pi}(\ln b/a)\;\text{H}/m\;;\; C = \{\frac{2\pi\varepsilon_0\varepsilon_r}{\ln b/a}\}\; F/m\;;$$

$$G = \omega C \tan\delta\; Siemen\; /\; m\,.$$

Wave-guide is used extensively for minimum losses and high transmissionat Microwave frequencies. And propagation is through either TE or TM waves.

35.1   WIRELESS COMMUNICATION

A signal of EM wave emitted by an antenna from a point can be received at another point by means of

(1) Ground wave which travels along the surface of the Earth FOR FRQUENCIES UP TO 1.5 *MHz* ($\lambda = 200$ *m*), since its attenuation increases with frequency.

(2) The sky wave only at frequencies > 1.5 *MHz* which is reflected by the ionosphere.

(3) For frequencies > 40 *MHz* ionosphere bends any incident EM wave and does not reflect back to Earth. Transmission is possible only by satellite communication.

35.1.1 MODULATION OF A SIGNAL:

The sine wave doesn't contain any information. There is need to **modulate** the wave in some way to encode information on it, for transmission There are three common ways to modulate a sine wave, *viz.*, Pulse, Amplitude and Frequency modulations.

35.1.2 **Pulse Modulation –**

In PM, you simply turn the sine wave on and off. This is an easy way to send Morse code. PM is not that common, but one good example of it is the radio system that sends signals to radio-controlled clocks in the USA. One PM transmitter is able to cover the entire United States

35.1.3 AMPLITUDE MODULATION (AM)

Amplitude Modulation (AM) is the process of imposing information contained in

a lower frequency (Audio) signal $\boxed{B_a \sin 2\pi f_a t}$

to a high frequency (Carrier) wave, $\boxed{A_C \sin 2\pi f_C t}$

by causing its amplitude to vary in accordance with the modulating signal

Modulation factor $\boxed{m = B_a / A_C}$

Percentage modulation, $\boxed{M = m(100)\% = \dfrac{B_a}{A_C}(100)\%}$

Frequency spectrum of the AM wave consists of: the carrier, the upper side band and the lower side band frequencies.

1. f_C,

2. $(f_C + f_a)$.

3. $(f_C - f_a)$

Power content of each side band = $m^2 P_c / 4$

Power transmitted (P):

$$P_r = P_C + P_{LSB} + P_{USB}$$

Band Width per station

$$n = [f_C - (f_C + f_a) + (f_C - f_a) - f_C]$$

Number of AM broadcasts accommodated by Broadcasting Body = (Total BW)/ (BW per station)

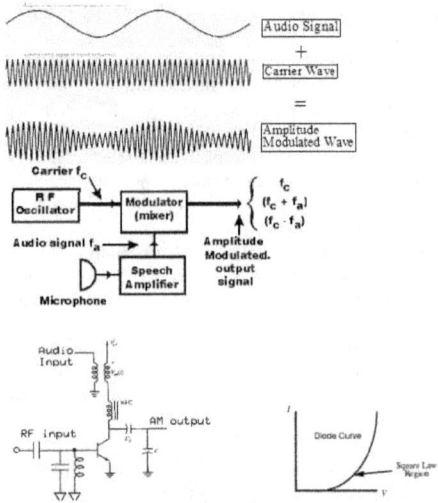

Determinations using the CRO:
(i) Saw tooth wave to the horizontal input, and the AM wave to the vertical input:

$$M = \frac{B_a}{A_C}(100)\% = \frac{\text{Max}_{\text{P-P}} - \text{Min}_{\text{P-P}}}{\text{Max}_{\text{P-P}} + \text{Min}_{\text{P-P}}} 100\%$$

(ii) With modulated signal to the Vertical input and the modulating one to the horizontal input

$$M = \frac{X-Y}{X+Y}100\%$$

The **maximum modulating frequency permitted by AM broadcast stations is** 5 *KHz* at carrier frequencies between 535 and 1,605 *KHz*.

35.1.3   FREQUENCY MODULATION (FM)
For sinusoidal modulating signal, FM has instantaneous amplitude

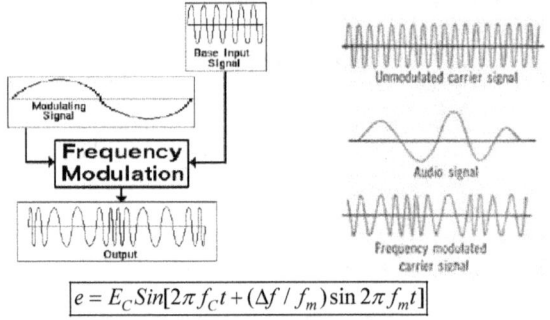

$$e = E_C Sin[2\pi f_C t + (\Delta f / f_m)\sin 2\pi f_m t]$$

Δf = frequency swing

Modulation index,

$$\boxed{\beta = \Delta f \,/\, f_m}$$

The most advantageous property of FM over AM is its improved S/N ratio.

The **maximum modulating frequency permitted by FM broadcast stations is 15 KHz** at carrier frequencies between 88 and 108 *MHz* .

Television transmitters use both AM and FM; the **video, or picture, signals are transmitted by AM** and the **sound by FM.**

35.1.4   ANTENNAS

An **AM broadcast antenna is vertically polarized**, requiring the receiving antenna to be located vertically also, like those found on automobiles. **Television and FM broadcast transmitters traditionally have used a horizontal polarization antenna**, although many FM and some TV stations are now circularly (horizontally and vertically) polarized.

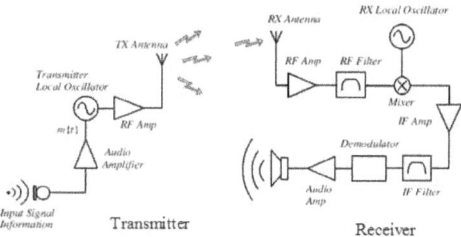

Transmitter   Receiver

35.15. Link behaviour of Ionosphere

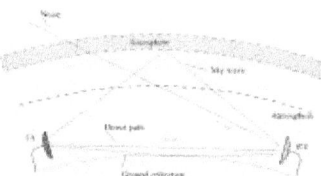

Realistic Link behaviour of Ionosphere

35.2.   **RADIO RECEIVERS**

35.2.1   RECEIVER REQUIREMENTS:

1.   Amplify the low power received signal from the aerial.
2.   Reject unwanted signals (noise & interference) which is outside the required BW.
3.   Detection of intelligence signal.
4.   Final amplification.

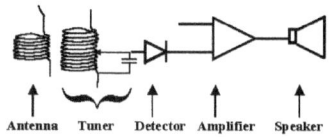

Antenna   Tuner   Detector   Amplifier   Speaker

To make an AM radio, we need to have <u>five basic parts</u>:

1) <u>An antenna</u>, to receive the electromagnetic waves and convert them back to electrical signals

2) <u>A tuner</u>, to select out the particular carrier frequency that we want, corresponding to a particular radio station that we are interested in listening to

3) A <u>detector</u> (diode), to get rid of the high-frequency signal but keep the low-frequency part.

4) An <u>amplifier</u>, to make the signal bigger

5) A <u>speaker</u>, to produce the sound that we can hear

Here are descriptions of the individual parts:

      1) <u>The antenna</u>: To make an antenna, all we need is a long piece of wire. Ideally, this should be   very long (like 50 *ft*). This coil is about 3 inches high and 1.5 *inches* in diameter.

      2) <u>The tuner</u>: The combination of inductor (about 200 turns of wire with a thin red insulation on it) and capacitor makes something called a "resonator"--The resonator resonates at a particular frequency that is determined by L and C.

      3) <u>The detector</u>: The detector is something called a "germanium diode".

      4) <u>The amplifier</u>: The signals that we pick up with the antenna and tuner maybe only a few thousandths of a volt. "Integrated circuits", or "chips", as "op-amps", or "operational amplifiers" can be used.

      5) <u>The speaker</u>. The speaker is the thing that actually makes sound.

      To summarize the entire process: The antenna picks up the signal and brings it to the antenna coil. The antenna coil is brought close to the tuner, and the electrical signal in the antenna coil transfers to the tuner coil. The inductor (coil) and capacitor that make up the tuner select out the particular carrier frequency of the radio station we want to listen to. The detector (diode) gets rid of the very high frequency, but keeps the low-frequency signal that corresponds to the "sound" that we want to hear. The amplifier makes the signal bigger, and the speaker converts the electrical signal back into sound!

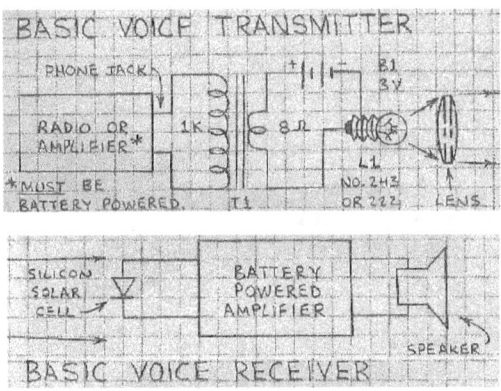

543

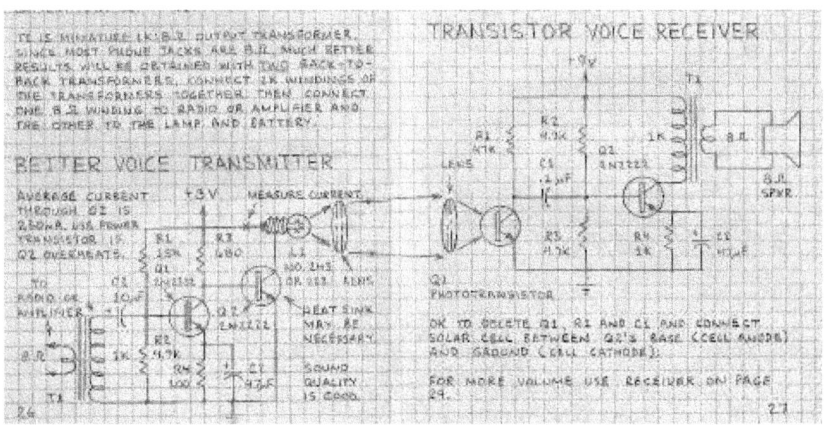

IT IS MINIATURE 1K:8Ω OUTPUT TRANSFORMER. SINCE MOST PHONE JACKS ARE 8Ω, MUCH BETTER RESULTS WILL BE OBTAINED WITH TWO BACK-TO-BACK TRANSFORMERS. CONNECT 1K WINDINGS OF THE TRANSFORMERS TOGETHER, THEN CONNECT ONE 8Ω WINDING TO RADIO OR AMPLIFIER AND THE OTHER TO THE LAMP AND BATTERY.

BETTER VOICE TRANSMITTER

AVERAGE CURRENT THROUGH Q1 IS 260MA. USE POWER TRANSISTOR IF Q1 OVERHEATS.

+3V MEASURES CURRENT

R1 15K R3 680

Q1 2N1302

TO RADIO OR AMPLIFIER

C1 10uF

R2 1K R4 100 C1 .33uF

Q2 2N1302

NO. 2H3 OR 222 LENS

HEAT SINK MAY BE NECESSARY.

SOUND QUALITY IS GOOD.

8Ω T1

Q1 PHOTOTRANSISTOR

R1 47K R2 47K Q1 2N1302 1K T1 8Ω 8Ω SPKR

LENS C1 .1uF

R3 4.7K R4 1K C2 47uF

OK TO DELETE Q1, R1 AND C1 AND CONNECT SOLAR CELL BETWEEN Q2'S BASE (CELL ANODE) AND GROUND (CELL CATHODE).

FOR MORE VOLUME USE RECEIVER ON PAGE 29.

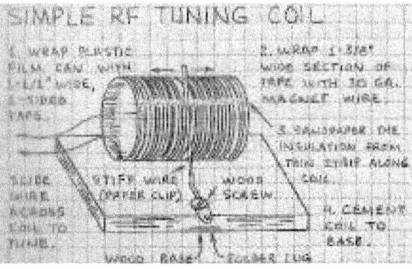

SIMPLE RF TUNING COIL

1. WRAP PLASTIC FILM CAN WITH 1-1/4" WIDE 2-SIDED TAPE.

2. WRAP 1-3/8" WIDE SECTION OF TAPE WITH 30 GA. MAGNET WIRE.

3. SANDPAPER THE INSULATION FROM THIN STRIP ALONG COIL.

4. CEMENT COIL TO BASE.

SLIDE WIRE ACROSS COIL TO TUNE.

STIFF WIRE (PAPER CLIP) WOOD SCREW

WOOD BASE SOLDER LUG

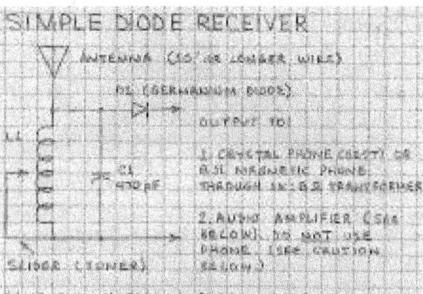

SIMPLE DIODE RECEIVER

ANTENNA (50' OR LONGER, WIRE)

D1 (GERMANIUM DIODE)

OUTPUT TO:

L1

C1 470 pF

1. CRYSTAL PHONE (2227) OR 8Ω MAGNETIC PHONE THROUGH 1K:8Ω TRANSFORMER

2. AUDIO AMPLIFIER (SEE BELOW). DO NOT USE PHONE (SEE CAUTION BELOW).

SLIDER (TUNER)

L1 IS COIL ON FACING PAGE. TUNING IS SENSITIVE. SOME STATIONS WILL COINCIDE WITH ONE WINDING.

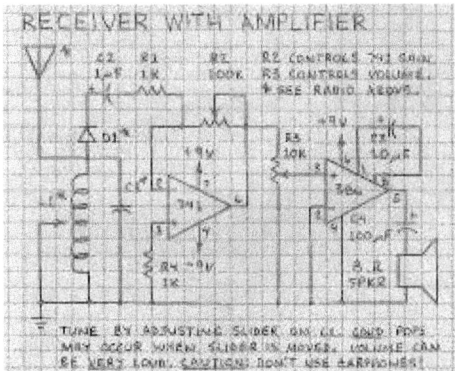

35.2.2 SIMPLE CRYSTAL RECEIVER

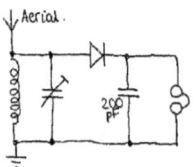

35.2.3 TUNED RADIO FREQUENCY SUPERHETERODYNE RECEIVER

A working definition of "heterodyne principle"

To generate new frequencies by mixing two or more signals in a nonlinear device such as a vacuum tube, transistor, or diode mixer. Reginald Aubrey Fessenden (1901) is credited with the invention of the heterodyne principle. The heterodyne method is to convert an incoming high frequency RF signal into one at a lower frequency, where high gain and selectivity could be obtained with relative ease. This signal, known as the intermediate frequency (IF), was then demodulated after much filtering and amplification at the IF had been achieved.

In the superhetrodyne receiver, basically the output from a variable "local" oscillator in the receiver is mixed or heterodyned with the signals from incoming radio transmissions. Superhet (known as supersonic heterodyne) implies that the oscillation is above sonic or audio frequencies, and is mixed or heterodyned with incoming signals.

While mixing an incoming RF signal with the local oscillator signal results in output signal containing 1) the original two signals plus 2) the sum and the difference signals of the two, plus 3) harmonics of these sum and difference signals.

For instance, in receiving an FM station (Station "A") on 99.7 *MHz*, the local oscillator could be tuned to 89 *MHz*. The output would consist of

The original radio station (A)	99.7 *MHz*.
The local oscillator (O)	89.0
The sum of the two	188.7 *MHz*
The difference of the two	10.7 *MHz*
Second harmonic of the sum	377.4 *MHz*
Second harmonic of the difference	21.4 *MHz*

The only one interested in is the difference frequency, 10.7 *MHz*, which is called the I.F. or intermediate frequency. 10.7 *MHz* is the normal FM receiver I.F., and is chosen because it is easy to amplify the wanted signal and because of the selectivity of the circuits, get rid of, or attenuate, all the other frequencies. All the stages of I.F. amplification take place at a fixed (and convenient) frequency, and it is this factor which gives it its superiority over the old TRF receivers.

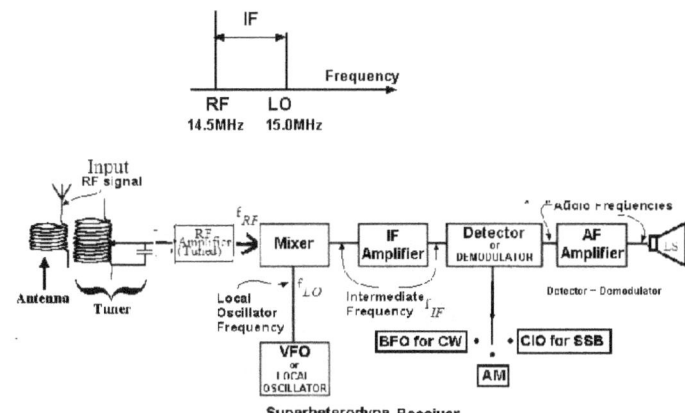

Superheterodyne Receiver

At the mixer stage, other radio signals will be present. For instance, the next FM channel or 99.9 *MHz* may well have a station present (Station B). Its difference frequency will be 10.9 *MHz*, or a difference of 2% (1.87%) in frequency, which puts it right outside the narrow pass band of the 10.7 *MHz* I.F. amplifier. The difference to a TRF operating at 99.7 *MHz*, would, however, have been only 0.2 %, which at those frequencies and with all stages needing to be individually tuned, poses very great selectivity problems.

35.2.4   Different Wave forms

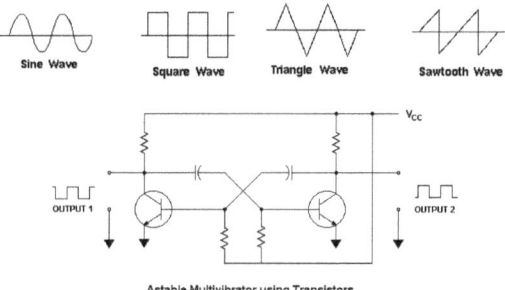

Sine Wave   Square Wave   Triangle Wave   Sawtooth Wave

Astable Multivibrator using Transistors

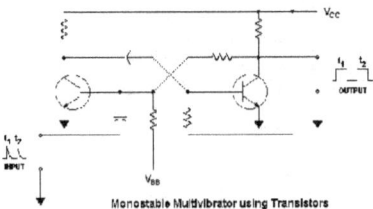

Monostable Multivibrator using Transistors

35.2.5 OPERATIONAL AMPLIFIERS (OpAmp):

One of the basic building blocks of Analogue Electronic Circuits, *Operational amplifiers* (OpAmp) are linear devices that have all the properties required for nearly ideal DC amplification and are therefore used extensively in signal conditioning, filtering or to perform mathematical operations such as add, subtract, integration and differentiation. An OpAmp has very high Z_{in} on both its inverting input and on its non-inverting input.

$$\boxed{A_V = \frac{V_0}{V_P}} = \text{very high}$$

= Open loop Gain depends on frequency.

Output $\boxed{V_0 = A_V(V_2 - V_i)}$ OPAMPs are IC amplifiers.

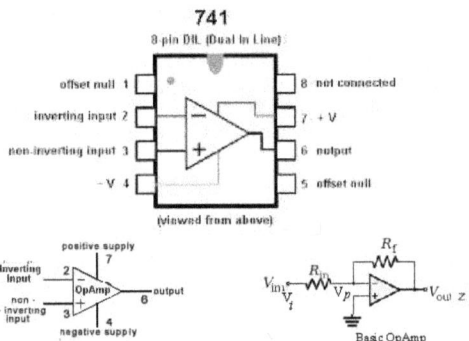

35.2.5.1 Inverting Amplifier

A negative feedback in an OpAmp is Gain can be made to depend on resistance value of the instance.

In the circuit since the inputs are virtual earth using the inverting input

$$\boxed{\text{Close loop Gain} = \frac{V_0}{V_i} = -\frac{R_f}{R_i}} \text{ for inverting input.}$$

Using the non-inverting input,

$$\boxed{\text{Close loop Gain} = \frac{V_0}{V_i} = 1 + \frac{R_f}{R_i}}, \text{ for non-inverting input.}$$

35.2.5.2 Summing Amplifier

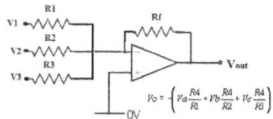

$$V_o = -\left(Va\frac{R4}{R1} + Vb\frac{R4}{R2} + Vc\frac{R4}{R3}\right)$$

35.2.5.3 Difference Amplifier

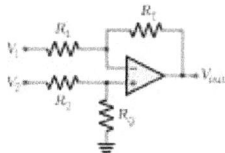

35.2.5.4 Differentiator

$$V_O = -RC\frac{dV_i}{dt}$$

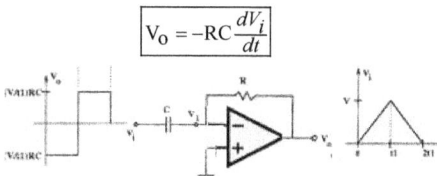

35.2.5.5 Integrator

$$V_O = -\int_0^t \frac{V_i}{RC}\,dt + K$$

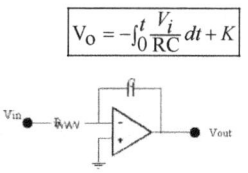

+*&*&*&*&*&*&*&+

Chapter 36

SPECTROCOPY –

INFRA RED, RAMAN, MOSSBAUER, NMR, ESR

*"The difference between stupidity and genius is that genius has its limits"*Albert Einstein.

*"All science is either physics or stamp collecting"*Ernest Rutherford

36.1 INTRODUCTION

(1) Joseph von Fraunhofer carried out the first spectroscopic experiments.(2) in 1859, the correct explanation for the Fraunhofer lines was suggested by G.R. Kirchhoff that sun light being absorbed at specific wavelengths by chemical species which was present in the solar atmosphere.(3) In 1864 James Clerk Maxwell published a paper entitled "Dynamical Theory of the Electromagnetic Field".(4) Heinrich Hertz in 1887 demonstrated that EM waves do exist.

36.1.1 An accelerating electric charge produces an EM Wave

36.1.2 CONVERSION FACTORS OF ENERGY UNITS

Conversion Factors for various Units in Spectroscopy

	cm^{-1}	GHz	eV	E/aJ
cm^{-1}	1	29.97925	1.2398×10^{-4}	1.9864×10^{-5}
GHz	3.3356×10^{-2}	1	4.1357×10^{-6}	6.6261×10^{-7}
eV	8065.54	2.41799×10^{6}	1	0.166022
E/aJ	50341.1	1.50919×10^{6}	6.24151	1

It is useful to note that

$$3 \times 10^8 \, MHz \leftrightarrow 10000 \, \overset{0}{A} \leftrightarrow 10000 \, cm^{-1}$$

$$10 \, k \, cm^{-1} \leftrightarrow 1.24 \, eV \leftrightarrow 28.6 \, kCal \, mole^{-1}$$

1 Kayser (K) $\equiv$ 1 cm^{-1}

36.1.3 Lambert-Beer Law:

$$I = I_0 \, 10^{-a \ell}$$

C = concentration of solute in moles per litre

$$-\log \frac{I}{I_o} = C \, a_m \ell$$

Absorbance, $\quad A = e \, b \, c$

where e = Beer's constant; b = path length

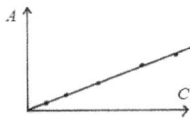

36.1.5 A Classification of Spectroscopic Methods: Types of Electromagnetic Radiation

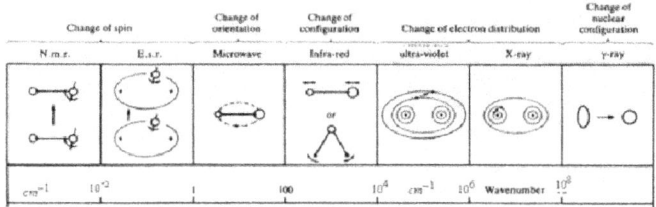

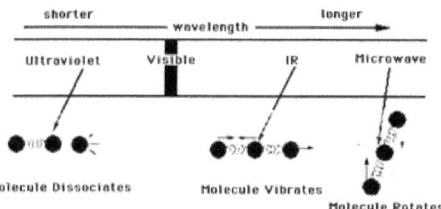

36.1.6 ENERGY LEVEL DIAGRAMS

Schematic energy level diagram for a typical molecule,

1) Firstly there are *discrete energy level bands* associated with electrons, shown as E0, E1, E2

2) Associated with each electronic level, there are associated *vibrational states* that the molecule can exist in and

3) Associated with each of these are the *rotational energy levels.*

The energies of the electronic, vibrational and rotational states are all governed by the rules of Quantum Mechanics.

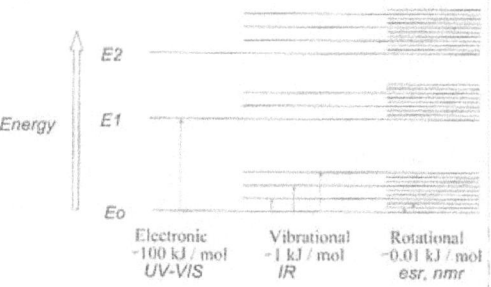

36.1.7.1 NATURAL LINE WIDTH (Γ) and Heisenberg's Uncertainty Principle

If a molecule is isolated for time, Δt seconds, in a particular state (n^{th}), the energy E_n of that state will be uncertain (blurred) to an extent ΔE_n (Section 4.6.1), where

$$\Delta E_n \cdot \Delta t \approx \hbar \approx 10^{-34} J\text{-}s$$

If one takes into account the Bohr frequency relationship, then

$$\Delta v \cdot \Delta t \approx \frac{1}{2\pi}$$

Since the energy of the ground state is known precisely, a transition from an excited state to the ground state will have an uncertainty called the natural line width, Γ

$$\Gamma = \Delta v \approx \frac{1}{2\pi \, \Delta t}$$

This is associated with <u>Lorentzian profile</u>.

36.1.7.2 DOPPLER BROADENING, $2\gamma_D$

The random motions of molecules in the gaseous and liquid states cause what is known as the Doppler broadening of spectral lines. The Full Width at Half Maximum (FWHM) is

$$2\gamma_D = 2\omega_D \sqrt{\frac{2k_B \, T \, \ln 2}{M \, c^2}}$$

M = mass of molecule. This is associated with <u>Gaussian profile</u>.

36.1.8. Franck-Condon Principle

According to Bohr model, an atom or molecule can exist only in a series of discrete states of electronic energy (Fig a). The energy levels are indicated usually in spectroscopy, by properly spaced horizontal lines. E_0 = ground state, E_1, E_2 denote higher electronic / excited levels. Fig a) shows absorption by vertical arrow lines, $E_0 \rightarrow E_1$ and $E_0 \rightarrow E_2$ between permitted levels. The wavy arrow in Fig b) $E_2 - E_1$ relates to another, radiation-less way in which a transition can occur by energy loss to surrounding molecules, or by its "internal conversion" into vibrational energy of the excited molecule. $E_1 - E_0$ is emission called **Fluorescence** by the downward arrow line (Fig b).

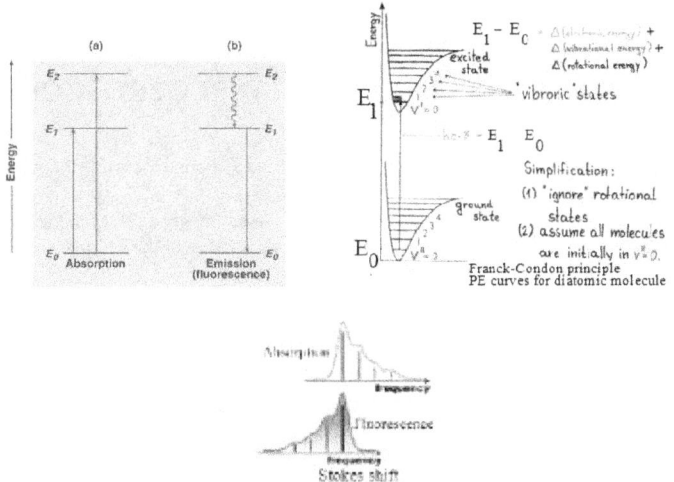

This experimentally observed <u>Stokes' shift</u> of the absorption energy $E_2 - E_0$ and emission $E_1 - E_0$ is due to the cycle absorption-emission thus contains two periods of energy dissipation (*i.e.*, lifetime of absorption, *viz.*, 10^{-9} *s* and that of radiationless transition, *viz.*, $< 10^{-12}$ *s*). This required vibrational levels to be included. The Franck-Condon Principle (James Franck, 1925) theoretically explained this as due to the nuclei do not move during an electronic transition, as nucleus has mass more than 2000 times that of electron. That is why two potential energy level diagrams, one for absorption and the other for emission is used (Fig).

-o-0-o-0-o-0-o-

INFRA RED (VIBRATIONAL) SPECTROSCOPY

36.2.1 Infra red (IR) spectroscopy

It deals with the interaction between a molecule and radiation from the IR region of the EM spectrum (IR region = 4000 - 400 cm^{-1}). Infrared energy is the EM energy of molecular vibration. The energy band is defined for convenience as the **near infrared** (NIR) (0.78 to 2.50 *microns*); the **infrared** (or mid-infrared) (IR) (2.50 to 40.0 *microns*); and the **far infrared** (FIR) (40.0 to 1000 *microns*).

The cm^{-1} unit is the wave number scale ; $\boxed{cm^{-1} = \dfrac{1}{\lambda \text{ in } cm}}$

IR radiation causes the excitation of the vibrations of covalent bonds within that molecule. These vibrations include the stretching and bending modes.

Early applications of IR spectroscopy were exclusively done in the near IR region because glass is transparent for the involved photon energies.

36.2.2 HOOKE'S LAW (THE SIMPLE HARMONIC POTENTIAL WELL)

36.2.2.1 To help understand IR,

Compare a vibrating bond to the physical model of a vibrating spring system, described by *Hooke's Law*,

$$\boxed{\vec{F} = -k \cdot \vec{q}}$$

$$\boxed{V(q) = -\frac{1}{2} k \, q^2}$$

where k = *force or spring constant*

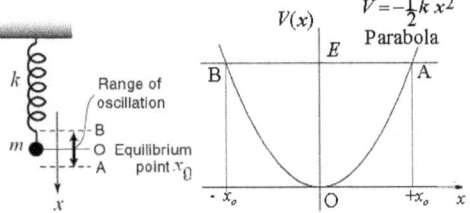

Typical values of k for different bonds

Force constants k for typical bonds	
Single bonds	4 to 6 x 10^6 *dynes / cm*
Double bonds	8 to 12 x 10^6 *dynes / cm*
Triple bonds	12 to 18 x 10^6 *dynes / cm*
50341.1	.50919 x10^6 6.24151

Vibrational motion is periodic concerted displacement of the atoms in a molecule, which leaves the centre of mass unaltered, in laboratory space, with vibrational coordinate, q.having r and r_o as instantaneous and equilibrium bond lengths

$$q = (x - x_0)$$

36.2.2.2.   The larger the force constant is the shorter the bond distance as shown in Table below.

Bond	Force constants, k milli $dyne$ / $\overset{0}{A}$	$\overset{0}{A}$
C - C	4.50	1.54
C = C	9.77	1.33
C ≡ C	17.2	1.20

36.2.2.4   The *vibrational energy* is

$$E_n = \left(n + \frac{1}{2}\right) \hbar\omega$$

Frequency of vibration, $v = \frac{1}{2\pi}\sqrt{\frac{k}{\mu}}$

$$\overline{v} = v / c$$

36.2.3   Criterion for IR Absorption

36.2.3.1 Dipole Moment:

Atoms are in general electrically neutral, *i.e.* in each of them the total negative charges are centred on the positive nucleus.

Electric dipole moment $\vec{\mu}$, a vector, is the product of the charge Q and their separation, d,

$$\vec{\mu} = Q.\,\vec{d}$$

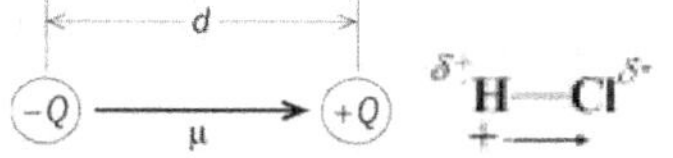

36.2.3.2   IR active

Thus for a *vibrational transition to be IR active there must be a change in the molecular dipole moment during a vibrational cycle.* This is illustrated schematically for a linear CO_2 molecule

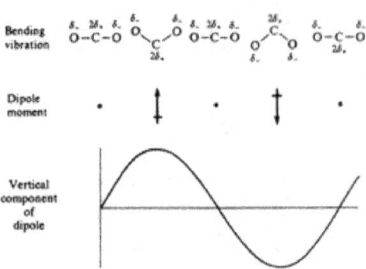

The bending motion of the CO_2
molecule and its associated $\bar{\mu}$ fluctuations.

36.2.4.1 FUNDAMENTAL FREQUENCY, $\bar{v}_{1\leftarrow0}$

Fundamental band occurs at a wave number $\bar{v}_{1\leftarrow0}$

$$\boxed{\bar{v}_{1\leftarrow0} = \omega_e - 2\,\omega_e x_e}$$

36.2.4.2 THE FIRST HOT BAND, $\bar{v}_{2\leftarrow1}$

The first hot band, $v = 2 \leftarrow 1$, occurs at

$$\boxed{\bar{v}_{2\leftarrow1} = \omega_e - 4\,\omega_e x_e}$$

36.2.4.3 THE FIRST OVERTONE, $\bar{v}_{2\leftarrow0}$

The first overtone band, $v = 2 \leftarrow 0$, occurs at

$$\boxed{\bar{v}_{2\leftarrow0} = 2\omega_e - 6\,\omega_e x_e}$$

This band <u>does not occur</u> at exactly twice the fundamental!

$$\boxed{\bar{v}_{2\leftarrow0} \neq 2\,\bar{v}_{1\leftarrow0}}$$

36.2.5 Characteristic IR Absorption Frequencies

36.2.6 THE *FUNCTIONAL* AND *FINGERPRINT* REGIONS OF AN IR SPECTRUM

As an example, the IR spectrum of Ethyl Alcohol (C_2H_6O) is shown in Fig.

In general terms it is convenient to split an IR spectrum into two approximate regions:

1) The <u>functional group region</u> [$4000\text{-}1000\,cm^{-1}$] and

2) The <u>fingerprint region</u> [$< 1000\,cm^{-1}$]

Most of the information that is used to interpret an IR spectrum is obtained from the functional group region.

In practice, it is the polar covalent bonds that are IR "active" and whose excitation can be observed in an IR spectrum.

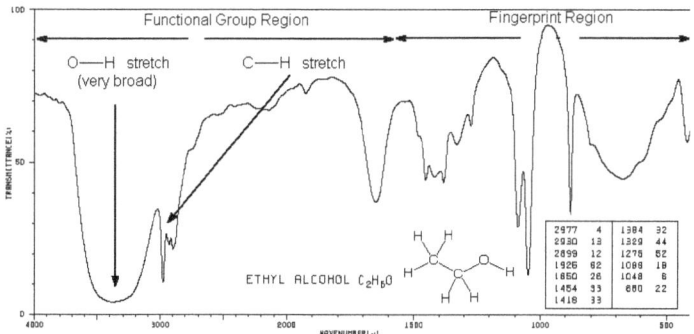

ETHYL ALCOHOL C_2H_6O

36.2.7 IR MODES OF TYPICAL MOLECULAR VIBRATIONS

Typical examples of H_2O and CO_2 molecules are depicted schematically.

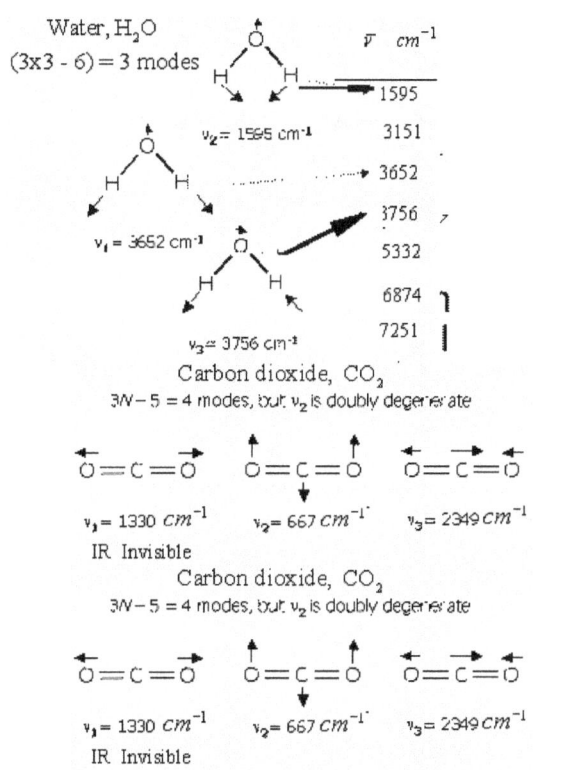

Water, H_2O

$(3 \times 3 - 6) = 3$ modes

$\bar{v}\ cm^{-1}$

1595

$v_2 = 1595\ cm^{-1}$ 3151

3652

$v_1 = 3652\ cm^{-1}$ 3756

5332

6874

7251

$v_3 = 3756\ cm^{-1}$

Carbon dioxide, CO_2

$3N - 5 = 4$ modes, but v_2 is doubly degenerate

$O = C = O$ $O = C = O$ $O = C = O$

$v_1 = 1330\ cm^{-1}$ $v_2 = 667\ cm^{-1}$ $v_3 = 2349\ cm^{-1}$

IR Invisible

Carbon dioxide, CO_2

$3N - 5 = 4$ modes, but v_2 is doubly degenerate

$O = C = O$ $O = C = O$ $O = C = O$

$v_1 = 1330\ cm^{-1}$ $v_2 = 667\ cm^{-1}$ $v_3 = 2349\ cm^{-1}$

IR Invisible

36.3.1 ENERGY OF PURE ROTATION

Reduced mass, $\mu = m_1 m_2 / (m_1 + m_2)$

$$\boxed{I = \mu\, a^2}$$

$$\boxed{E_\ell = \{\ell(\ell+1)\, \hbar^2 / 2I\}, \quad \ell = 0,1,2,3,....}$$

$\ell \geq m = an\ integer$

36.3.2 Energy Levels of a Rigid Rotator

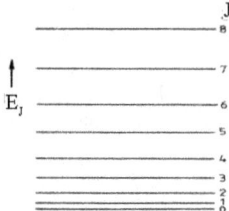

36.3.3 QUANTIZATION OF ANGULAR MOMENTUM, ℓ of the rotator

(Angular momentum) $\boxed{L^2 = E_J . \, 2I = \ell(\ell+1)\hbar^2}$

36.3.4 Change in Dipole Moment of a Rotating Molecule

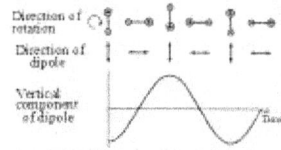

Fig. The rotation of a polar diatomic molecule, showing the fluctuation in the dipole moment measured in a particular direction

36.3.5 ROTATIONAL CONSTANTS OF SOME TYPICAL DIATOMIC MOLECULES UNDER PURE ROTATIONAL MOTION

Table 13.8
Examples of Rotational constants for some diatomic molecules

Molecule	$B_0 \ / \ cm^{-1}$	$B_0 \ / \ GHz$
H_2	59.3219	1778.43
HD	44.62	1338.93
HF	20.5567	616.274
HCl	10.4398	312.978
CO	1.92253	57.6360
N_2	1.98958	59.6461
CS	0.817085	24.4956

36.3.6 Interaction of Vibrational Transitions with Rotational or Electronic Transitions
The combination of a vibrational transition with rotatory transitions leads to a number of closely spaced bands that are only resolved in the **gas phase**. In solution the rotatory levels are unresolved, the net effect is a broadening of the IR bands.

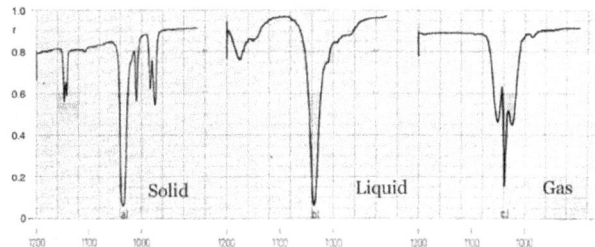

The IR spectra of a substance taken in the (a) solid, (b) liquid and (c) gas phases.

36.3.6 MOLECULAR SYSTEM AS A COMBINED NON-RIGID ROTOR AND MORSE OSCILLATOR
36.3.7 THE P-AND R- BRANCHES IN VIBRATION-ROTATION BAND OF IR SPECTRUM
The wave number of a vibration-rotation band is

$$\boxed{\overline{V}_{v'\leftarrow v''; J'\leftarrow J''} = G(v') - G(v'') + F(J') - F(J'')}$$

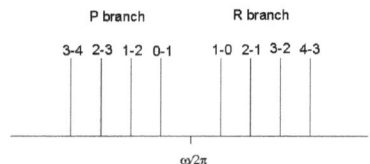

-o-0-o-0-o-0-o-

RAMAN SPECTROSCOPY

36.4.1 Raman Scattering

Also known as Raman Effect is not based on an absorption process but rather on a inelastic scattering of light, named after the Indian physicist Sir C.V. Raman, who was to first observe it in 1928, and received the Nobel Prize for Physics in 1930.

36.4.11 **Rayleigh scattering**.

A sample exposed to a beam of light with λ much higher than the size of the particle, the emitted photon has the same wavelength ($\bar{\nu}_0$) as the absorbing photon.

Stokes Formula:

Scattering of a large wave λ by smaller particles is in accordance with the *Stokes* formula,

Intensity, $\boxed{I \propto \dfrac{1}{\lambda^4}}$

36.4.1.2 Raman Scattering

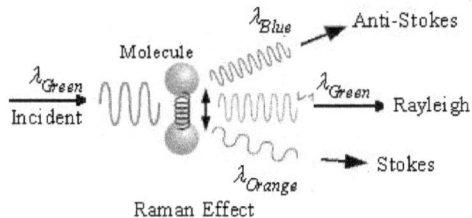

Raman Effect

36.4.1.3 Dipole moment, $\vec{P}$

The dipole moment of a molecule (a vector) and $\vec{E}_i$ is the electric field vector of the light, if α = polarizability of the molecule,

$$\boxed{\vec{P} = \alpha\, \vec{E}}$$

$$\boxed{p_i = p_{0i} \sum_n \left(\frac{\partial p_i}{\partial q_n}\right)_{q=0} q_n}$$

36.4.1.4 Criterion for Raman Activity

A change in **molecular polarizability** in the molecule is essential.

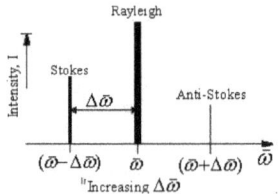

36.4.1.4 Why in Raman spectroscopy, only Stokes line is normally measured?
The ratio of the intensities of the Raman anti-Stokes and Stokes lines is predicted to be

$$\frac{I_{Anti-Stokes}}{I_{Stokes}} = \left(\frac{v_0 + v_{vib}}{v_0 - v_{bvi}}\right)^4 e^{\left(-h\,v_{vib}/k_B T\right)}$$

The Boltzmann exponential factor is the dominant term which explains why the anti-Stokes line is much *less intense* than the Stokes line. This occurs because only molecules that are vibrationally excited prior to irradiation can give rise to the anti-Stokes line.

36.4.1.5. RAMAN SHIFTS, $\bar{v}_{RSt}$

$$\bar{v}_{RSt} = \left(\frac{1}{\lambda_{incident}} - \frac{1}{\lambda_{scattered}}\right) cm^{-1}$$

gives the *Raman shifts*, which are

$$\bar{v}_{RSt} = (\bar{v}_{incident} \mp \bar{v}_{scattered})\ cm^{-1}$$

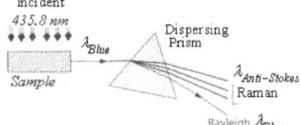

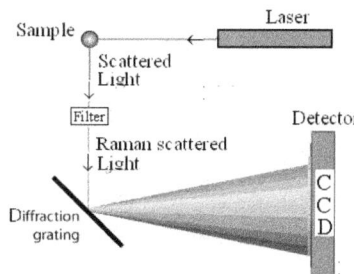

Simple Raman spectrometer - Schematic

For example a Stoke's Raman line for a sample with exciting line 253.6 *nm* was observed at 261.2 *nm*. The Raman shift of the corresponding anti-Stoke's line in wave number will be $\bar{v} = \left[\frac{1}{253.6\ nm} - \frac{1}{261.2\ nm}\right] = 39432.1\ cm^{-1} - 38284.8\ cm^{-1} = 1147.3\ cm^{-1}$.

Another example is in a sample a Raman spectral line was observes at 17652 cm^{-1} using 18303 cm^{-1}. a) The $\bar{v}_{vib} = 18303\ cm^{-1} - 17652\ cm^{-1}$. b) The corresponding IR absorption band, since $v_{vib} = c\ \bar{v}_{vib} = 1.95 \times 10^{13}\ s^{-1}$ and

$$\lambda = \frac{1}{\overline{v}_{vib}} = 15.4 \ \mu m .$$

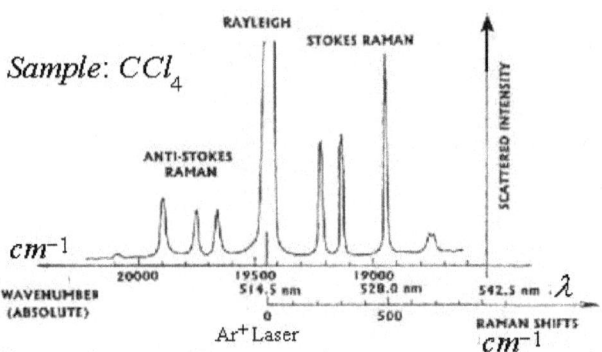

Sample: CCl_4

36.4.2. RAMAN AND IR SPECTRA DIFFER AND ARE COMPLEMENT TO EACH OTHER

36.4.2.1 **Rule of Mutual Exclusion**
If a molecule has a centre of Symmetry, it has been observed that those fundamental modes which are Raman Active are Infrared Inactive, and Infrared Active modes are Raman Inactive.

36.4.2.2 All homo-nuclear molecules have a Centre of Symmetry.

36.4.2.3 Infrared (IR) and Raman spectroscopy
Both measure the vibrational energies of molecules, but these methods rely on only different selection rules. Recall that for a vibrational motion to be IR Active, the dipole moment of the molecule must change. Therefore, the symmetric stretch in carbon dioxide CO_2 is not IR active because there is no change in the dipole moment. The asymmetric stretch is IR active due to a change in dipole moment.

$$C \Leftarrow O \Rightarrow C \qquad C \Leftarrow\Leftarrow O \Rightarrow\Rightarrow\Rightarrow C$$

No change in dipole moment   Change in dipole moment

36.4.2.3 How do the IR and Raman spectra differ?

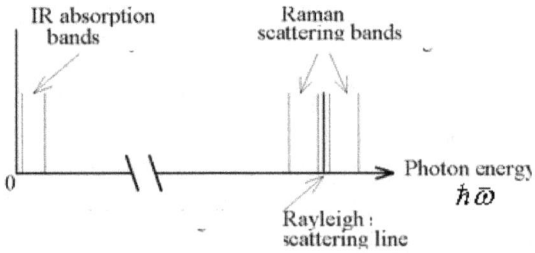

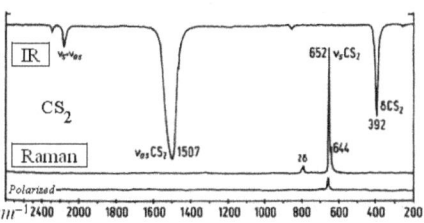

36.4.2 POLARIZATION OF RAMAN LINES

If the Exciting (Rayleigh) line is polarized as well as being monochromatic and an analyzer is inserted in between the sample and the spectrometer, the Raman lines will be observed to have different intensities for different directions of polarization of the incident beam. If $I_{\parallel}$ and $I_{\perp}$ are the intensities of a Raman line with the incident light parallel and perpendicular to the direction in which the analyzer passes the maximum amount of light, then the *depolarization ratio*, ρ is

$$\rho = \frac{I_{\perp}}{I_{\parallel}}$$

In the case of laser excitation, $\rho_{maximum} = 0.75$.

Fundamental vibrational modes of Acetylene

Normal Mode		Symmetry Species	Description	Activity Band Type	Frequency (cm⁻¹)	
					C_2H_2	C_2D_2
H−C≡C−H	$\bar{\nu}_1$	Σ_g^{+}	Sym.CH stretch	Rp, ‖	3372.8	2705.2
H−C≡C−H	$\bar{\nu}_2$	Σ_g^{+}	CC stretch	Rp, ‖	1974.3	1764.8
H−C≡C−H	$\bar{\nu}_3$	Σ_u^{+}	Asym CH stretch	IR, ‖	3294.8	2439.2
H−C≡C−H H−C≡C−H	$\bar{\nu}_5$	Π_g	Sym bend	Rdp, ⊥	612.9	511.5
H−C≡C−H H−C≡C−H	$\bar{\nu}_4$	Π_u	Asym. bend	IR, ⊥	730.3	538.6

36.4.3 Laser Raman spectrum of Diamond (DC Arc Jet Film)

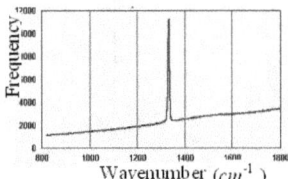

36.5. FOURIER TRANSFORM (FT) RAMAN SPECTROSCOPY

The foundations of the modern Fourier Transform (FT) spectrometers were laid by A.A. Michelson (1891), Lord Rayleigh (1892) and Peter Fellgett (1949), and a breakthrough came when two mathematicians named Cooley and Tukey who developed the famous computer algorithm bearing their names. Mathematically, an interferogram is defined as a sum of the cosine wav4es of all the frequencies present in the source

$$I(\delta) = \sum_{\overline{v}_i}^{\overline{v}_n} B(\overline{v}_i) \ Cos(2\pi\delta\overline{v}_i)$$

$$\delta = n\lambda = \text{retardation},$$

$$\overline{v} = \frac{1}{\lambda}$$

$$I(\delta) = \int_{0}^{\infty} B(\overline{v}) \ Cos(2\pi\delta\overline{v}) \ d\overline{v}$$

This represents a radiating source
The best resolution is

$$\Delta\overline{v} = \frac{1}{\delta_{max}} \ cm^{-1}$$

If the mathematical form of the interferogram $I(\delta)$, as a function of δ, is known then it is possible to calculate the corresponding spectrum by means of Fourier Transformation to give

$$B(\overline{v}) = \int_{0}^{\infty} I(\delta) \ Cos(2\pi\overline{v}\delta) \ d\delta$$

where $B(\overline{v})$ = intensity of the spectrum as a function of wave number $\overline{v}$.

Thus the Fourier Transform pair between the interferogram and its spectrum is employed in a FT spectrometer.

-o-0-o-0-o-0-o-

36.6 MOSSBAUER SPECTROSCOPY

36.6.1.1 OBSTACLES TO GET GAMMA RAY RESONANCE(THE MOSSBAUER EFFECT)
There are, however, two major obstacles in obtaining information:
a) The 'hyperfine' interactions between the nucleus and its environment are extremely small, and
b) The recoil of the nucleus as the gamma-ray is emitted or absorbed prevents resonance
36.6.1.2 RECOIL OF FREE ATOMS
A shot fired from a gun causes the gun to recoil with a speed. In the same way when an energy quantum (*i.e.*, a γ – radiation) gets emitted from a free (isolated) radioactive nuclide of mass M,
it acquires a recoil energy It is easily shown that the recoil energy, E_R is

$$E_R = \frac{E_\gamma^2}{2M\,c^2}.$$

Recoil of free nuclei in emission or absorption of a gamma-ray
This much energy is removed from the nuclear transition energy, E_t of the emitting nuclide, and results in the energy E_γ of the γ – radiation

$$E_\gamma^{emitter} = (E_t - E_R)$$

Similarly
$$E_\gamma^{abs} = (E_t + E_R)$$

36.6.1.3. THE DOPPLER EFFECT
The Doppler Effect was first analyzed by Christian Andreas Doppler in 1845. If the moving source is emitting waves with an actual frequency f_0, then an observer stationary relative to the medium detects waves with a frequency f given by:
a) <u>Source moving</u> toward (or away) observer

$$f' = \frac{V}{(V \mp v_S)} f_S \qquad \underline{\text{moving source}}$$

$$\Delta\lambda = (V - v_S)/f_S$$

where f_S = the frequency of the source,

V = the speed of sound.

v_S = the speed of the source

(positive if moving towards the observer, negative if moving away).

$$f = \frac{(V \mp v_0)}{V} f_0 \qquad \underline{\text{moving observer}}$$

$$\Delta f = \frac{V}{c} f_0$$

This is Doppler Effect.

$\Delta\lambda$ is $-ve$; (λ dectreases "Blueshifted")

Δf is $+ve$; (f increases)

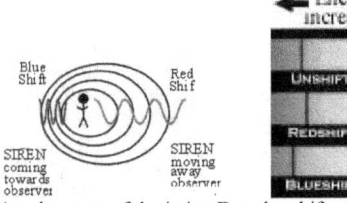

Another way of depicting Doppler shift.

36.6.1.4 Thermal Broadening (Maxwell Distribution Law)

The Doppler shift due to thermal motion of atoms .for Maxwell distribution of the speeds of atoms, the emission and absorption lines also have a shape of Maxwell distribution (Gaussian) with the Doppler width $2\delta_D$

$$\boxed{2\delta_D = 2\sqrt{E_R k_B T} \ eV}$$

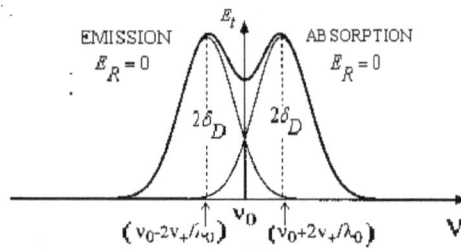

36.6.2.1 Energy profile

As atoms move due to random <u>thermal</u> motion, the gamma-ray energy E_γ has a spread of values E_D caused by the Doppler Effect. This produces a gamma-ray <u>energy profile</u> as shown overlap.

36.6.2.2 NATURAL LINE WIDTH Γ ($\equiv \Gamma_n$) OF GAMMA EMISSION AND ABSORPTION

A gamma ray is an extremely monochromatic energy quantum whose line broadening (width Γ of nuclear level) is determined only by its lifetime τ ($\equiv \tau_m$) with the Heisenberg's interval of uncertainty

$$\boxed{\Gamma \tau = \hbar}.$$

i.e., $\Delta v = \Gamma_n = \dfrac{1}{2\pi\,\Delta t} = \dfrac{1}{2\pi\,\tau_m}$

The mean life τ_m of the $I = \frac{3}{2}$ state in ^{57}Fe is $\tau = 1.4 \times 10^{-7} s$. The energy distribution is given by a Lorentzian (also known as Breit-Wigner) profile with a FWHM (Full Width at Half Maximum) of $\Gamma_{nat} = 4.7 \times 10^{-9} eV$.

36.6.2.3 Three different profiles of energy distributions

36.6.3. THE REQUIRED CONDITIONS TO OBSERVE THE MOSSBAUER EFFECT
a)   The atom (ion, molecule, *etc.*) must be in the **solid state**, to avoid recoil and thermal broadening.
b)   The gamma ray energy $E_\gamma \approx E_t$ must be **fairly low** (10 to 100 *keV*) to obtain an appreciable number of recoil-less events.

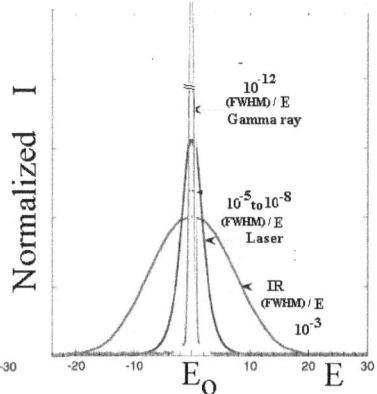

The life times (mean life) τ of nuclear excited states giving rise to such γ – rays must typically be in the range $\tau \cong 10^{-6} - 10^{-11} s$, since longer lived nuclear species emit lines which are extremely narrow for detection whereas the shorter lived species give lines which are broad and get lost in the counting statistics,
d)   The internal conversion electron coefficients α **should be as small** as possible ($\alpha = 0$ - 20), and

e)   The absorber sample may have the Debye temperature, Θ_D, **preferably high**.

Typical parameters for a low E_t value are listed in Table.

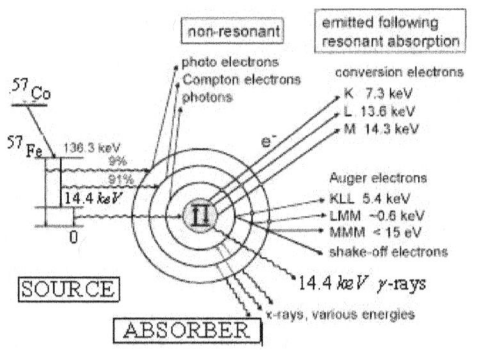

Typical parameters for Nuclei

Element	^{23}Na	^{191}Ir	^{198}Hg	^{57}Fe
Atomic weight, A in u	23	191	198	57
Transition energy, E_t	$2.11\,eV$	$129\,keV$	$412\,keV$	$14.41\,keV$
Recoil energy, $2\,E_R$ eV	$2.1\,x10^{-10}$	0.10	0.9	0.004
Doppler Width, $2\delta_D = 2\sqrt{E_R k_B T}$ eV	$6.6\,x10^{-6}$	0.2	0.4	0.02
Temperature, T	300 K	300 K	300 K	300 K
Natural width, Γ_n eV	$4.5\,x10^{-9}$	$6.5\,x10^{-8}$	$2.1\,x10^{-5}$	$4.5\,x10^{-9}$

$$E_R(Solid, \mu^3 volume) = E_R(Free\ atom)$$

$$E_R(bound\ atom) \cong 10^{-15} E_R(free\ atom)$$

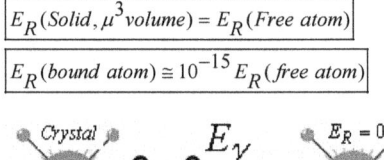

36.6.4.1 HOW TO OBSERVE MOSSBAUER EFFECT?
  1)   Mössbauer source must vary its energy over a significant range.
          By Doppler shifting the energy of the gamma beam.
  v and c are the frequency and velocity of γ-rays
  V is the relative velocity of the source of the observer,

Frequency shift, $\boxed{\Delta v = \dfrac{V\,v}{c}}$

$$\boxed{E_\gamma = (E_t \pm \tfrac{V}{c})}$$

Energy shift, ΔE_γ of the γ-ray transition energy E_t

$$\boxed{\Delta E_\gamma = (E_\gamma - E_t) = \dfrac{V E_t}{c}\ eV}$$

Moving the source at a velocity of $1\ mms^{-1}$ toward the sample will increase the energy of the photons by $\dfrac{V E_t (= 14.14\ keV)}{c} = 4.8\ x\ 10^{-8}\,eV$ or $10\,\Gamma$.

36.6.4.2 A convenient Mössbauer unit is $1\ mms^{-1}$

36.6.4.3 A Mössbauer spectrometer
consists of (1) source (2) Doppler drive system, and (3) a counter to monitor the intensity of the beam after it has passed through the sample.

36.6.4.4 EXPERIMENTAL SETUP

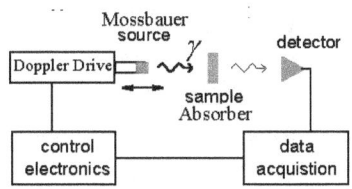

36.6.4.5 The Mössbauer spectrum is a plot of the counting rate against the source velocity, *i.e.*, the beam energy

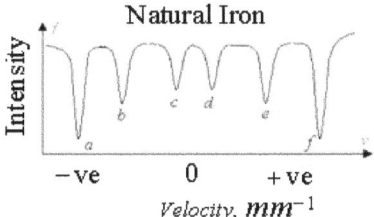

-o-0-o-0-o-0-o-

36.7 NUCLEAR MAGNETIC RESONANCE SPECTROSCOPY

36.7.1.1 NUCLEAR SPIN, I

Nuclei also have intrinsic spin angular momentum, and are characterized by a *nuclear spin quantum number I*. The electron spin quantum number is s $=\frac{1}{2}$, whereas nuclear spin quantum numbers range from $I = 0$, $\frac{1}{2}$, 1, $\frac{3}{2}$, 2, $\frac{5}{2}$, 3,....., 6. Nuclei are divided into three categories:

36.7.1.2 Nuclei having $I = \frac{1}{2}$ *integral value*, for those having mass number A = odd,

$$eg.., {}^{1}H, {}^{13}C, {}^{19}F, {}^{31}P, etc.$$

36.7.1.3. Nuclei having $I = 0$, for those with A = even and Z = even,

$$eg., {}^{16}O, {}^{12}C, etc.$$

36.7.1.3. Nuclei having $I = $ *integral value*, those with A = even and Z = odd,

$$eg., {}^{2}H, {}^{6}Li, {}^{14}N, etc.$$

Nuclei with an odd-A or odd-Z, *viz.*, ${}^{1}H, {}^{13}C, {}^{19}F, {}^{31}P$ have $I = \frac{1}{2}$.

36.7.1.4 NUCLEAR MAGNETIC MOMENTS, $\vec{\mu}_I$

Each nuclear spin (I) is a magnetic moment $\vec{\mu}_I$ which is associated with the angular momentum of the nucleus. It is common practice to express these magnetic moments in terms of the nuclear spin in a manner parallel to the treatment of the magnetic moments of electron spin (s) and electron orbital angular momentum.($\vec{\ell}$)

For the electron spin and orbital cases, the treatment of quantized angular momentum is required. *Bohr Magneton μ_B* is the unit defined as

$$\mu_B = {e\hbar}/{2m_e} = 9.2740154 \times 10^{-24} J\ T^{-1}$$
$$= 5.7883826 \times 10^{-5} eV\ T^{-1}$$

36.7.1.5 NUCLEAR g-FACTOR, g_I

The nuclear magnetic moment is expressed in terms of the nuclear spin in expression (16.2.3)

$$\begin{array}{|c|}\hline \textit{Nuclear magnetic moment} \quad \mu_I = g_I \dfrac{e}{2m_p} I \\ \hline \end{array}$$

$$\boxed{\mu_z = g_I \left(\dfrac{e\hbar}{2m_p}\right) m_I = g_I \mu_N m_I}$$

where the new unit called a Nuclear Magneton.($\mu_N \equiv nm$) defined as

$$\boxed{\mu_N = \left(\dfrac{e\hbar}{2m_p}\right) = 5.05084 \times 10^{-27} J\, T^{-1}}$$

$$= 3.15245 \times 10^{-8} eV\, T^{-1}$$

36.7.1.2 NUCLEAR MAGNETS

Since a nucleus is a charged particle in motion, it will develop a magnetic field. ^{1}H and ^{13}C have $I = \frac{1}{2}$ and so they behave in a similar fashion to a simple, tiny bar magnet. It is specified as in quantum state, $\left| I.\, m_I \right\rangle$, where the nuclear magnetic quantum number $m_I = \pm\frac{1}{2}$.

Proton

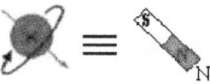

Proton magnet

36.7.2.1 BASIC PRINCIPLE OF NMR

In NMR, EM radiation is used to "flip" the alignment of nuclear spins from the low energy spin aligned state $\left| I.\, m_I \right\rangle \equiv \left| \frac{1}{2}.\, +\frac{1}{2} \right\rangle$ to the higher energy spin opposed state $\left| \frac{1}{2}.\, -\frac{1}{2} \right\rangle$. The energy required for this flipping transition corresponds to the radio frequency range of the EM spectrum. When the $\vec{\mu}_I$, associated with a I, is placed in an external magnetic field, the different spin states are given different *MAGNETIC POTENTIAL ENERGY E_m*. In the presence of the static magnetic field $\vec{H}_0$ which produces a small amount of *SPIN POLARIZATION*, a radio frequency signal of the proper frequency ν_L (Larmor frequency) can induce a transition between spin states. This "spin flip" places some of the spins in their higher energy state. If the radio frequency signal is then switched off, the relaxation of the spins back to the lower state produces a measurable amount of RF signal at the resonant frequency associated with the spin flip. This process is called *Nuclear Magnetic Resonance* (NMR).

36.7.2.2 Larmor Precession, ν_L

The allowed transitions of the dipole for interactions with EM radiation are as per the selection rule

$$\Delta m_I = \pm 1$$

the transition energy, ΔE_m,

$$\Delta E_m = g_I H_0 \mu_N$$

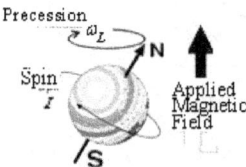

The transition frequency, known in NMR as the *Larmor precession frequency* (ν_L) is:

$$\nu_L = \frac{g_I H_0 \mu_N}{h}$$

Defining gyro-magnetic ratio, γ as

$$\gamma = g_I \mu_N / h$$

The **Larmor Equation**,

$$2\pi\nu = \gamma H_0$$

36.7.2.3 What frequency setting range that NMR signals occur?

It can be easily verified that all NMR signals fall within the range of frequencies 60 *MHz* to 750 *MHz*. Most of the chemical applications of NMR involve proton resonance spectroscopy (PMR).

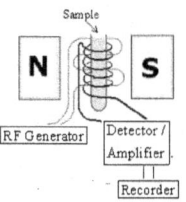

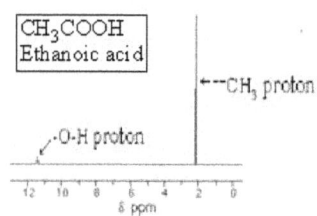

-o-0-o-0-o-0-o-

ELECTRON SPIN RESONANCE
(ESR or EMR or EPR) SPECTROSCOPY

36.8.1.1 Species that contain unpaired electrons (paramagnetic probes):
1. Free radicals
2. Odd electron molecules
3. Transition-metal complexes
4. Lanthanide ions
5. Triplet-state molecules
6. PRINCIPLE OF THE ESR

36.8.1.2 EPR
is based on the Zeeman Effect, which depends on energetic level splitting in paramagnetic molecules under the presence of a variable magnetic field. After induction a molecule with magnetic moment μ, in magnetic field, gains energy E...

36.8.1.3 BOLTZMANN CRITERION
In practice single paramagnetic probe never occurs but only a population of probes with many paramagnetic centres. If this configuration of probes is in thermal equilibrium at temperature T, statistical placing is described by Boltzmann distribution.

$$\Delta N = N_0 \frac{\Delta E}{2k_B T} = N_0 \frac{g_I H_0 \mu_N \Delta E}{2k_B T}$$

36.8.1.4 ELECTRON PARAMAGNET

The resonance condition for spin $-\frac{1}{2}$ electron is analogous to that of spin $-\frac{1}{2}$ nuclei (proton), treated in Section 16.2, except that both Bohr Magneton and g-factor for electron must be used. Thus for electron

For the electron spin and orbital cases, the treatment of quantized angular momentum is required. Bohr Magneton μ_B is the unit defined as

$$\mu_B = {eh}/{2m_e} = 9.2740154 \times 10^{-24} \, J \, T^{-1}$$

$$= 5.7883826 \times 10^{-5} \, eV \, T^{-1}$$

36.8.1.5   Energy levels (E) of a paramagnet in a magnetic field H

$$E_{m_s} = g_s \, \mu_B \, H_0 \, m_s$$

$$v_L = {g_s H_0 \mu_B}/{h}$$

$$\gamma = g_s \mu_B / \hbar$$

$$2\pi v = \gamma H_0$$

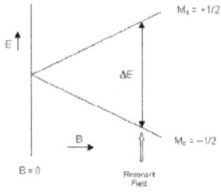

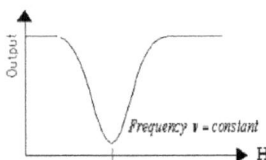

A Simple EPR Spectrum

36.8.2.1 ESR SPECTRUM

It s quite easy to understand that according to quantum mechanical treatment atomic hydrogen will have two lines in the ESR, centred at $g_s \, \mu_B \, H_0$. These are separated by $\hbar\omega$. This is confirmed quantitatively.

The degeneracy of the electron spin states, characterized by the quantum number, $m_s = \pm\frac{1}{2}$, is lifted by the application of a magnetic field.

##*#*#*#.

Chapter 37

EARTH'S ATMOSPHERE & ASTRONOMY

"Equipped with his five senses, man explores the universe around him and calls
the adventure Science. ~   win Powell Hubble, The Nature of Science, 1954

37.1.1   WHAT IS THE ATMOSPHERE?

The Earth's atmosphere is the part of the gaseous environment of Earth which is held close to
Earth by its gravity. The density (kg m $^{-3}$) of the atmosphere decreases with height above Earth's
surface, and the temperature and composition also vary with altitude. The atmosphere consists of
several regions, or altitude ranges, having different properties (temperature, pressure, and
composition) which vary with altitude in different ways.

37.2.1. Troposphere

The lowest layer of Earth's atmosphere, the weather and clouds occur in the troposphere, and is
the subject of the field of atmospheric science known as **meteorology.**

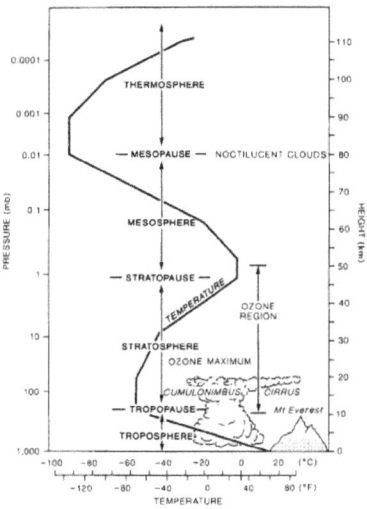

37.2.2. Stratosphere

This is the atmospheric layer between the troposphere and the mesosphere. The stratosphere is characterized by a slight temperature increase with altitude and by the absence of clouds. , is important because it contains the **ozone layer**, which shields life on Earth from harmful ultraviolet (UV) light from the Sun. The stratosphere is a zone of increasing temperature with altitude, due to absorption of solar UV radiation by ozone, and is the highest region in the atmosphere in which aircraft normally fly.

The narrow region between these two parts of the atmosphere is called the "Troposphere".

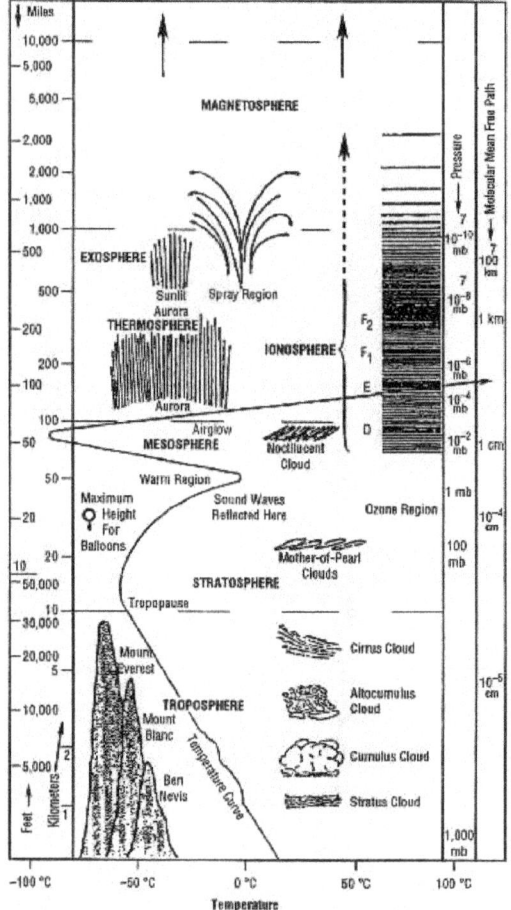

37.2.3 Mesosphere

The atmospheric layer between the stratosphere and the ionosphere js the mesosphere.

37.2.4 Ionosphere

The atmospheric layer between the mesosphere and the exosphere; it is part of the thermosphere. This region is electrically charged gas atoms and molecules.

37.2.5 **Thermosphere**
The layer of the atmosphere located above part of the ionosphere (starting at the coldest part of the atmosphere) and below outer space; it consists of the exosphere and part of the ionosphere. The ionosphere is of practical importance because it makes possible long-distance radio communications. The upper regions of the atmosphere are also of practical importance because, although the atmospheric density is very low compared to that in the lower atmosphere, it still acts to slow down artificial satellites and limit the length of time a satellite can stay in low-altitude orbits around Earth.

37.2.6 **Exosphere**
The outermost layer of the Earth's atmosphere is Exosphere, where atmospheric pressure and temperature are low.

37.2.7 Near the Poles
The Earth's surface gets very cold near the poles, where the solar energy strikes at a sharp angle.
    If the Earth did not rotate on its axis and if it had a uniform surface, a relatively simple flow would set up between the Polar highs and the Equatorial low.

37.3. **Frequently asked questions - Pressure and Vacuum**

37.3.1 Why does atmospheric pressure change with altitude?
Atmospheric pressure reduces with altitude for two reasons - both related to gravity.
(i) The gravitational attraction (Strictly it is the gravitational force minus the effect of the Earth's spin (an effect that is greatest at the equator) between the earth and air molecules is greater for those molecules nearer to earth than those further away.
(ii) Molecules further away from the earth have less weight (because gravitational attraction is less) but they are also 'standing' on the molecules below them, causing compression. Those lower down have to support more molecules above them and are further compressed (pressurized) in the process.

37.3.2 Variation of pressure with altitude

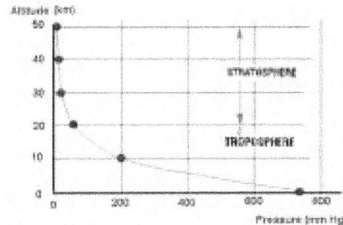

$$z = \frac{RT_m}{g} \, \ell n(\frac{P_0}{P})$$

$$P = P_0 e^{-g/RT_m}$$

    The properties of Earth's atmosphere are affected by the Sun, but in somewhat different ways at high altitudes than in the lower atmosphere. In the troposphere, the Sun heats the atmosphere, either directly or indirectly. At high altitudes, the atmosphere is affected by ultraviolet and X-ray radiations which do not penetrate to the lower atmosphere, and also by energetic **charged particles** (mainly, **electrons** and **protons**) produced, directly or indirectly, by the Sun. These radiations not only heat the atmosphere, but can directly affect the chemical composition (*e.g.,* by dissociating molecules and ionizing atoms and molecules).

37.3.3   WHY IS THE UPPER ATMOSPHERE IMPORTANT?

The high-energy electromagnetic and particle radiations produced by the Sun both heat and affect the composition of the upper atmosphere. These are much more variable with time, than is the visible light from the Sun. Since the ionosphere is of practical importance to radio and radar wave transmission, it is important to have ways of measuring both the composition and density of the ionosphere, and the variations in the high-energy solar radiation which affects it.

37.3.3.1   Magnetosphere -Van Allen Belts

Two giant donut-shaped swaths of radiation, known as the Van Allen Belts (after James A. van Allen), surrounding Earth were discovered in 1958. They are located at a distance extending from 650 - 6500 *kms* from the Earth's surface. These are called the Magnetosphere. The inner Belt trapped mainly protons (originate from **Cosmic Rays**) and some electrons, while the outer one consists of trapped mainly electrons (originate from **Solar wind**).

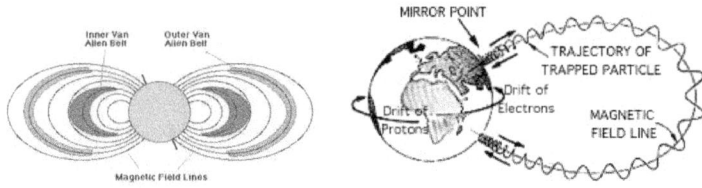

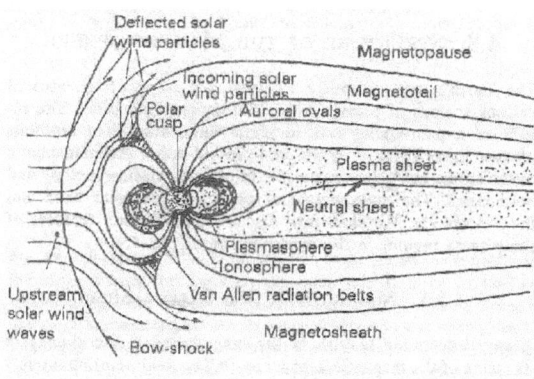

The Overall Structure of the Earth's Magnetosphere.

Particles from the solar wind become trapped by the Earth's magnetic field and are responsible for the **Aurora borealis** seen at Polar Regions. A part of a belt dips into the upper region of the atmosphere over the South Atlantic to form the Southern Atlantic Anomaly. This can present a dangerous hazard to satellites orbiting the Earth.

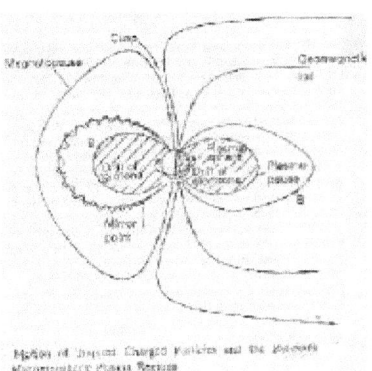

Motion of trapped Charged Particles and the differenti
Magnetospheric Plasma Regions

37.3.3.2.   Plasmasphere
This is the region of plasma in the magnetosphere with features of ionosphere.

37.3.3.3.   Alfen Wave
Basic magneto-hydrodynamic wave in a plasma containing a magnetic field. The plasma displacement is transverse to the magnetic field, with propagation along the axis of the field. Hannes Alfen (Nobel Prize in 1970) said this wave has very important role in the physics of various phenomena in space.

37.3.4 OZONE LAYER
What is Ozone and How is it formed?
Ozone (O_3: 3 oxygen atoms) occurs naturally in the atmosphere.
The ozone layer is a thin band in Earth's upper atmosphere. It blocks out the Sun's harmful ultraviolet (UV) rays. If it gets too thin, the harmful UV rays can damage crops, wild animals, and our skin.

37.3.5 **What causes the ozone hole?**
A large decrease in the total amount of ozone layer due to chemicals (called chlorofluorocarbons or CFCs) that are used in refrigerators and air conditioners are known this resulted a "hole".
The ozone layer is currently being destroyed by CFCs and other substances, its depletion progressing globally except in the tropical zone.

37.3.6 **What is CFC?**
CFC, chlorofluorocarbon, is the collective name for compounds made of carbon, fluorine, chlorine and hydrogen. Because of their stable, harmless and non-combustible properties, they are widely used in everyday applications such as cleansing agents for electronic components, coolants for air conditioners, foaming agents for the manufacture of insulating materials and so on. CFC variations include HCFC and HFC.

37.3.7 **What is El Niño?**
This means action of the wind and the sea and the abnormal warming of sea water in the East and Central Equatorial Pacific Ocean which leads to atmospheric changes. During a year when there is no El Niño, trade winds move surface water west across the Pacific Ocean and bring cold water from deep below to the top, such as plankton and algae. These are known to have adverse impact on the crucial summer Monsoon in India.

37.3.8 THE GREENHOUSE EFFECT

It results from "the dirty of the atmospheric infrared window" by atmospheric trace gases, permitting incoming solar radiation to reach the surface of the Earth unhindered but restricting the outward flow of infrared radiation. These greenhouse gases absorb and reradiate this outgoing radiation, producing a net warming of the surface.

i)   Greenhouse gases: the big three

These are Water vapour (H_2O), carbon dioxide (CO_2), methane (CH_4), and nitrous oxide

(N_2O). CFC and ozone are two less abundant gases.

ii)   Variability of Global Temperature -- Global Warming?

37.3.9   What is polar ice and how does it affect the Earth?

Polar ice is ice that covers the Earth's polar regions. It makes up 10% of the Earth's surface.

37.3.10   Are sea levels rising?

Oceans are rising on the average; rising by as much as 15 to 20 *cm* (about 6 to 8 *inches*) in the last 100 years.

This is due to melting of glaciers add water to the oceans, or by expansion of the existing ocean water due to slow warming. Rising ocean levels would make hurricanes and other storms more dangerous. More than half the U.S. population lives within 80 *kms* of a coastline. Some entire nations - like Bangladesh and the Netherlands - are at or near sea level.

37.3.11   What is the temperature in space?

Space contains atoms and ions, which have any temperature. Near Earth and the Moon, in direct sunlight, one heats up to $250\,°F$ ($121\,°C$). This is hotter than boiling water. In the shade, it can cool to around $25\,°F$ ($-156\,°C$). This is why astronauts must wear thermal space suits.

37.3.12.1   What happens to the Incoming Radiations from Space?

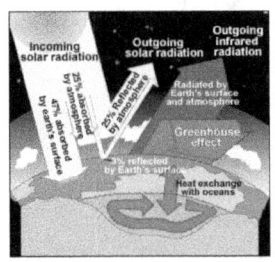

37.4.   THE SUN (The Closest Star) (Main Sequence Star)

37.4.1   The Solar Structure

It is composed of hydrogen (72%) and helium (27%), and core temperatures around 15 million$°C$, and hydrogen burning to form helium takes place at the rate of 600 billion hydrogen into helium per second, at 6000 $°C$ on the photosphere.

The Sun's Interior

a)   **Core**: 16 million K, 160 times density of water, and gaseous.

b)   **Radiative Zone**: includes around 85% of the Sun's radius. Energy transported outwards by photons.

c) **Convective Zone**: Energy in the outer 15% of the Sun's radius is transported by the bulk motions of gas in a process called convection.

d) **Photosphere**: ~ 500 *km* thick, at 5840 K, as per Wien's law.

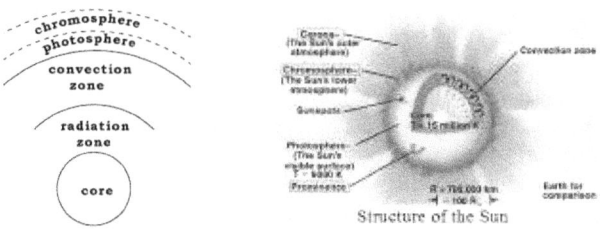

Structure of the Sun

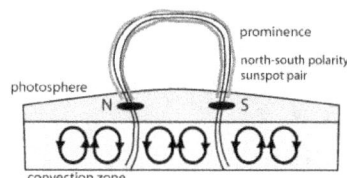

Tightly wound magnetic field lines loop into and out of the photosphere.

Sun spots are dark cool regions, of strong magnetic fields, on Sun's surface having $2000\,°C$. The magnetic field is somehow responsible for the 11-year sunspot cycle. **Solar flares** are huge bursts of energetic particles released in sunspot areas, which cause powerful magnetic fields. **Solar wind** consist of spray of charged particles, mostly of protons and electrons.

e) **Chromosphere**: 2000 to 3000 *kms* thick. Emission line of H 656.3 *nm* is seen.

f) **Corona**: The sparse outer atmosphere of the Sun is called **Solar corona**, and is a strong source of X-rays. Solar corona is visible during a total solar eclipse. Also expelled from it are solar **prominences**, luminous flares of hydrogen and helium. The corona extends outwards for more than a solar radius ($696,000\ kms$ = 109 earth radii).

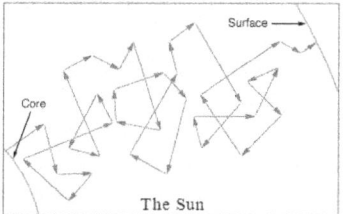

The Sun

It takes about 1×10^6 *yrs* for a photon to travel from the core to the solar surface!

37.4.2 Solar Luminosity, $L_\odot$

$$\boxed{L_\odot = 4\pi\sigma R_\odot^2 T^4}$$

$$\boxed{L_\odot = 4\pi^2 R_\odot \ell} = 3.90 \times 10^{26} W \ .$$

T= surface temperature of the Sun.

37.4.3. Source of Sun's Energy

Hans Bethe (1967 Nobel Prize) proposed the PP Chain reaction in 3 steps (Schematically shown below) in nuclear fusion for the source of energy.

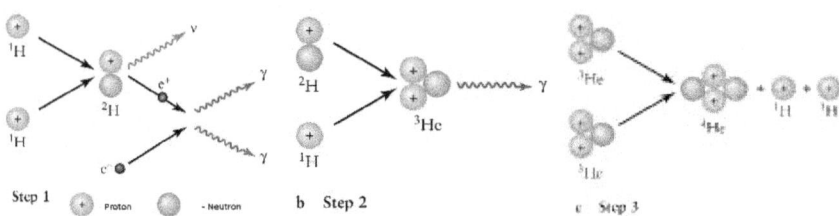

Step 1 $\quad$ $^1H_1 + {}^1H_1 \rightarrow {}^2H_1 + e^+ + \nu$

$e^+ + e^- \rightarrow \gamma + \gamma$, Step 2 $\quad {}^2H_1 + {}^1H_1 \rightarrow {}^3He_2 + \gamma$, Step 3 $\quad {}^3He_2 + {}^3He_2 \rightarrow {}^4He_2 + {}^1H_1 + {}^1H_1$,

37.5. METEOROLOGY

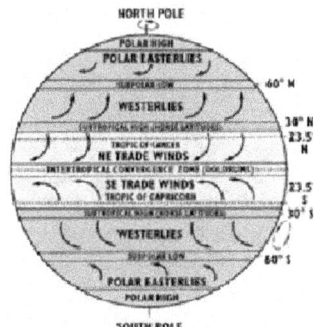

37.5.1 Coriolis Force

It is an artifact of Earth's rotation. Once air has been set in motion by the pressure-gradient force it undergoes an apparent deflection from its path, to an observer on Earth.

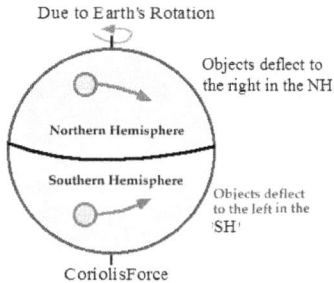

Hadley cell is a pattern of circulating winds in both the hemispheres. Air heated near the Equator expands, rises and travels towards the Poles. Subsequently it gets cooled and sinks towards the Equatorial region.

37.5.2.1 Air mass and Fronts

Air mass is a large mass of air which covers 1000s of *kms* with uniform characteristics throughout horizontally. The boundary between two air masses is called a **front**. A **cold front** is the boundary of a cold air mass moving over a warmer surface, whereas a **warm front** is the boundary of a warm air mass moving over a colder surface. On a weather map cold front are indicated as shown, so also warm fronts.

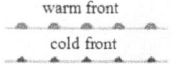

35.5.2.2   Polar-front Theory

The polar front is the semi permanent, semi continuous, boundary separating air masses of tropical and polar origin. Disturbances along the front in the form of cold-air outbreaks can generate low pressure vortices or storms, also known as mid-latitude cyclones. High pressure regions (anti-cyclones) are generally associated with fair weather. In modern meteorology weather forecasts could be prepared quickly by graphical methods. In the model of open wave cyclone, cold and warm fronts are produced along the boundaries of the air-masses.

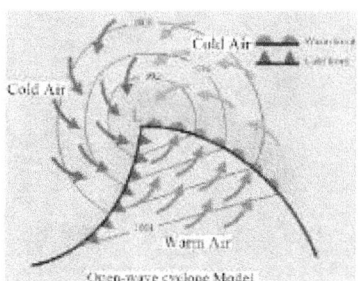

37.5.2.3.   Rotating Wave Tank

To demonstrate the applicability of polar-front theory to American weather Carl-Gustaf Rossby, in 1926, to dynamic meteorology, and to simulate atmospheric motion built this tank. 2 *m* in diameter, differentially heated, and filled with coloured paraffin waxes of different densities, the tank rotated around a vertical axis 3 – 4 rotations / minute. When the paraffin solidified it preserved dynamical features of the circulation.

37.5.2.4. Balloons

Helium filled balloons could fly to an altitude of 12 *km*, and launched at all weather conditions and radiosondes are adopted, for daily weather soundings. Large-scale flow patterns in the upper atmosphere that has direct influence on surface weather could be identified.

37.5.3. Rossby Waves (Planetary waves)

The atmosphere is filled with waves of various frequencies and amplitudes – short, shock, lee, mountain, Helmholtz, frontal and cyclone waves, along with thermal atmospheric tides – often made visible by cloud forms. Planetary waves (known as Rossby waves) determine surface weather in Earth's temperate zones. Rossby postulated a nearly friction-free atmospheric layer where the pressure-gradient balances with the Coriolis force. In that layer Rossby waves are the undulations in the upper-air wind flow and form as polar air moves toward the equator while the tropical air moves Poleward. He formulated the equation relating phase velocity v_φ to the mean zonal flow $\bar{u}$ (speed of air along a latitude line) as

$$v_\varphi = \bar{u} - \frac{\beta \lambda^2}{4\pi^2}$$

where Rossby parameter β accounts for the meridional (north-south) variation of the Coriolis force, and λ = wavelength. The equation implies that for short wavelengths the wave's phase moves eastward with zonal flow, whereas long wavelengths the phase moves retrograde (or westward). For stationary wave condition $v_\varphi = 0$. Typically λ is of the order of 5000 *km*. Only from a satellite planetary waves can be seen.

37.5.4. Clouds

Clouds are made of water droplets or ice crystals that are so small and light they are able to stay in the air. Water vapour gets into air mainly by evaporation – some of the liquid water from the ocean, lakes, and rivers turns into water vapour and travels in the air. When air rises in the atmosphere it gets cooler and is under less pressure. The vapour becomes small water droplets or ice crystals and a cloud is formed.

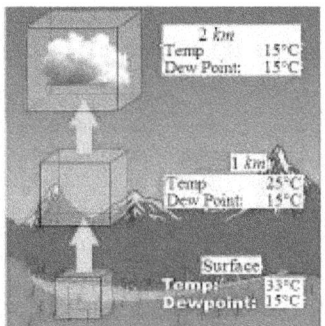

Process of Cloud Formation, credit: NCAR

37.5.4.1   Types of Clouds

Clouds are given different names based on their shape and their height in the sky. Some clouds are near the ground. Others are almost as high as jet planes fly. Some are puffy like cotton. Others are grey and uniform.

a) The highest clouds in the atmosphere are Cirro-cumulus & Cumulus, and Cirro-status & Cumulo-nimbus clouds can also grow to be very high.
b) Mid-level clouds include Alto-cumulus and Alto-stratus.
c) The lowest clouds in the atmosphere are Stratus, Cumulus, and Strato-cumulus.

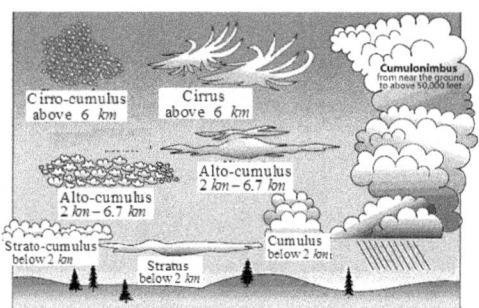

Common types of Clouds - Troposphere   credit: NCAR

37.5.5 Weather
Weather means rain and dull clouds, windy blue skies, cold snow, and sticky heat are very different conditions.
37.5.5.1 **El Nino** is a climate pattern. It means 'Christ child', and happens in around Christmas.
37.5.5.2. **La Nina** meaning 'little child' is another unusual weather pattern, and introduced in 1986..
It brings unusually cold temperatures to the eastern Pacific region.
Both these pattern affect the world weather.

37.5.6 **IONOSPHERE**
The Earth's atmosphere is transparent to visible light (while absorbing some of the light), astronomers who use telescopes can see things from far away using visible light to form images. However, it acts an opaque barrier to much of the EM spectrum. The atmosphere absorbs most of the wavelengths shorter than UV, most of the wavelengths between IR and MWs, and most of the longest radio waves. For radio astronomers use uses the 'radio window' to bring information about the universe to our Earth-bound instruments. The radio window consists of frequencies from about 5 MHz - 30 GHz. The low-frequency end of the window is limited by signals being reflected by the ionosphere back into space, while the upper limit is caused by absorption of the radio waves by water vapor and carbon dioxide in the atmosphere.

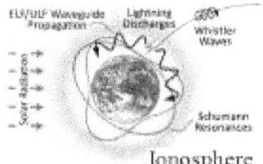

Ionosphere

The ionized part of the Earth's atmosphere is known as the ionosphere. UV light from the Sun collides with atoms in this region and ionize them. This results in a region Ionosphere. It is the free electrons that cause the reflection and absorption of radio waves.

Higher frequency waves are able to pass through the atmosphere entirely and reach the ground.

This process also works in reverse for radio waves produced on the earth. The high frequency waves pass through the ionosphere and escape into space while the low frequency waves reflect off the ionosphere and essentially "skip" around the earth (Radio broadcasting in Radio Stations use this principle).

Heinrich Hertz proved the existence of radio waves (1887).

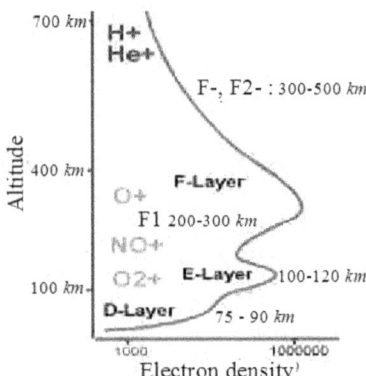

Electron density[1]

The ionosphere is composed of three main parts, named for obscure historical reasons: A typical electron density altitude profile, the most important ions, and the various ionospheric layers are indicated in the diagram above.

1) E-Layer (Electric layer)(Edward Appleton, 1927) named, after Guglielmo Marconi showed the bouncing of radio waves between America and Europe in 1901, to the layer 100-120 *km* altitude,

2) D- layer (75 – 90 *km*) and

3) F-layer (300 – 500 *km*)

The electron density is highest in the upper, or F region. The F region exists during both daytime and nighttime. During the day it is ionized by solar radiation, during the night by cosmic rays. The D region disappears during the night compared to the daytime, and the E region becomes weakened. The most abundant gas molecules are O_2 and N_2 below 200 *km*, atomic O above 200 *km*, and H and He above 600 *km*.

37.5.6.1.　　　Methods of measurement of Ionization os the Upper Atmosphere

(i)　　Langmuir Probe in which electrons and ions are counted

(ii)　　Change in the impedance of an antenna to determine the index of refraction of the medium and, hence, the electron density.

(iii)　　Use of satellite for an ionospheric sounder.

3.5.7.　Magnetosphere

It is located above the ionosphere, at the external limit of the Earth's magnetic field. It behaves like a giant magnet, retaining high energy particles and thus protecting the Earth. This layer has the lowest density of all, as air density gradually decreases as we go further up from the Earth.

3.5.8.　RADAR Probing Atmosphere

Microwave radars dominated in the tropospheric studies soundings.

MF/HF radars are one of the principal means to study mesospheric dynamics, through partial reflection drift (PRD) technique in 1984.

3.5.8.1 MST Radar for Atmospheric Research

High power VHF backscatter radar operating ideally around 50 *MHz* to explore the entire mesosphere-stratosphere-troposphere (MST) domain is known as the MST Radar. In India this is installed at the National MST Radar Facility, Tirupati (AP) since 1992. It comprises mainly of a) a high resolution 2-D phased array high power transmitter with appropriate feedback network, b) T/R switches, c) a phase coherent receiver with quadrature channels, d) a signal processor, and e) a super micro-computer (Ref: Rao *et al.*, *Radio Sci.*,1995).

3.5.8.2 Aurora

Aurora has long fascinated mankind. It is a natural phenomenon. After the spectral measurements in 1867 by Angstrom it is explained as the cause by the optical emissions of the atmospheric gases themselves.

37.6. The UNIVERSE

37.6.1 BIG BANG THEORY - The Premise

Discoveries in astronomy and physics have shown beyond a reasonable doubt that our universe did in fact have a beginning. Prior to that moment there was nothing; during and after that moment there was something: our universe. The big bang theory is an effort to explain what happened during and after that moment.

According to the standard theory, our universe sprang into existence as "singularity" around 13.7 billion years ago. What is a "singularity" and where does it come from? **Black holes** are areas of intense gravitational pressure. The pressure is thought to be so intense that finite matter is actually squished into infinite density (a mathematical concept which truly boggles the mind). These zones of infinite density are called "singularities." Our universe is thought to have begun as an infinitesimally small, infinitely hot, infinitely dense, something - a singularity.

37.6.2 The STEADY STATE THEORY

In 1948 by Hermann Bondi, Thomas Gold and Sir Fred Hoyle found the idea of a sudden beginning to the Universe philosophically unsatisfactory. Bondi and Gold suggested that in order to understand the universe the laws of physics were not different in the past. For Bondi and Gold not only would the laws of physics have to be the same in all parts of the universe, but at all times as well!! The Universe would also be the same, always static, always contracting or always expanding. The first two could be ruled out by the simple observation that the sky is dark at night.

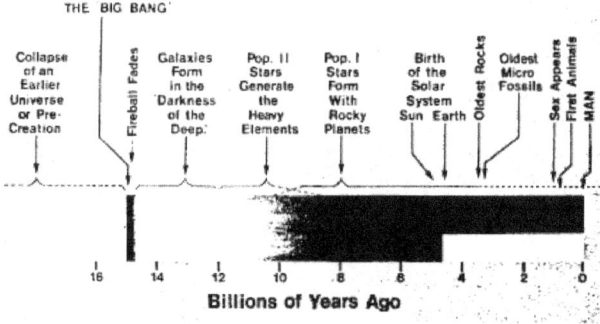

From "Xenology" by Robert A Freitos

The steady state theorists explained the hydrogen - helium abundance by the presence of supernovae. Originally the big bang theory suggested that all the heavy elements were produced at the start of the universe, but now it is accepted that only the helium and a little lithium was produced then and both theories now accept the role of supernovae in the creation of heavy elements.

The steady-state theory is now no longer accepted by most cosmologists, particularly after the discovery of "microwave background radiation" in 1965, for which steady state has no explanation.

37.6.3. CLASSIFYING STARS – The E Hertzsprung- HN Russell (H-R) Diagram (1914)

There are a few hundred billion stars in our galaxy, the Milky Way and billions of galaxies in the Universe. The basic H-R diagram is a temperature (T) *versus* luminosity (L) graph. T may be replaced or supplemented with spectral class. The main spectral class, in order from hottest to coolest, are O, B, A, F, G, K and M.

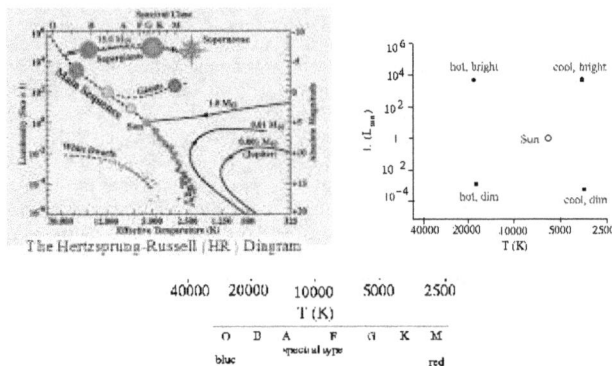

The Hertzsprung-Russell (HR) Diagram

For a Main Sequence star, its Mass M and luminosity L are correlated by the Mass-luminosity relation,

$$L \propto M^{3.5}$$

It is the Main Sequence stars spend their lives burning H to He in their cores.

37.6.4. BEYOND THE SOLAR SYSTEM
37.6.4.1 POLE STAR
is *Dhruwa.*
37.6.4.2 Ursa Major (Little Bear)
is *Saptharishi.*
37.6.4.3 Alpha Centauri
is the nearest star to the Earth.

Sun-to-Earth distance = 1 AU.($1.49598 \, x10^6 \, km$)

$1 \, Light \, year = 63290 \, AU$

$1 par \sec = 3.26 \, Light \, year$

$1 par \sec = 206,265 AU$

37.6.5 GLOBULAR CLUSTER

A cluster of stars contains $10^3 - 10^4$ stars. The size of a cluster is about 10^5 times the size of the solar system.

37.6.6 CLUSTER OF GALAXIES

It is a group of about 10^4 galaxies. A galaxy may contain about 250 billion stars and extends over 10^5 light years.

37.6.7 EINSTEIN'S GENERAL THEORY OF RELATIVITY VERIFIED
 a) Deflection of light near the Sun,
 b) Precession of the perihelion of the planet Mercury.

37.6.8. THE MILKY WAY (Akash Ganga) Galaxy
The celestial body containing about a few billion stars including our Sun. It has the shape of a disk (diameter 100,000 *light years*) with the centre bulged. The thickness of the central region = 5000 *light years*, and the galactic disk has thickness of 1000 *light years*. The Sun is located at about $2/3$ distance away from the galactic centre. The solar system is orbiting around this centre at v = 25 *km s*$^{-1}$; and its period is 250 *million years*. Mass of the Milky Way is about 150 billion $M_{\odot}$ (solar mass).

37.6.9. HUBBLE'S LAW:
$$\boxed{v = H\ r}$$
H is Hubble constant

37.6.10. Red spectral shift of a Galaxy
$$\boxed{z = (\lambda - \lambda_0)\,/\,\lambda_0 = v\,/\,c}$$

λ_0 = the observed wave length and λ = emitted wavelength of the body distance r from us.

37.6.11 **Size of a satellite**:
D = Distance of the planet from Earth
d = diameter of the planet and
α = angle subtended by d on Earth,
$$\alpha = d\,/\,D$$

37.7.1. **Mass of a Planet** (M):
A satellite orbiting a planet of mass M, with radius R and period T,
$$M = 4\pi^2 R^3\,/\,GT^2$$
$$G \approx 6.7\ x10^{-11}\ Nm^2 kg^{-2}.$$

37.7.2. **Brightness of a star**, ℓ
Magnitude, m
$$m = -2.5\log(\ell\,/\,\ell_0)$$
ℓ_0 = brightness of a standard star of zero magnitude.

37.7.3. **Stellar Spectra**
Photosphere emits continuous spectrum of radiation
Wien's Law:
$$\lambda_{max} T = 2.897\ x10^{-3}\ mK$$

37.7.4. **Chandrasekhar Limit**
Chandrasekhar Limit is the Maximum mass for a white dwarf star
$$\text{Chandrasekhar Limit} = 1.4 M_{Star}.$$

If the mass $< 1.4 M_{Star}$ the electron degeneracy pressure holds up the star the weight.

If the mass $> 1.4 M_{Star}$, the stars gravitational force causes it to collapse.

Low-mass stars ($< 3 \, M_\odot$) eject about half their mass in planetary nebula, so star ends up with $1.4 M_{Star}$.

Density in white dwarf is about $10^6 \, g \, cm^{-3}$ (1 teaspoon = wt. of truck).

37.7.5. <u>Densities of Interest</u> (given in two units, $g \, cm^{-3}$ and $kg \, m^{-3}$).

	$g \, cm^{-3}$.	$kg \, m^{-3}$.
Water	1×10^0	10^3
Lead	1.1×10^1	10^4
Core of Sun	1.50×10^2	1.5×10^5
White Dwarf	1×10^4	1×10^7
Neutron Star	1.0×10^{15}	10^{18}

A single teaspoon of the material from a White Dwarf would weigh tons.

37.7.6. <u>Deaths of Stars</u>

Low-mass (light weight) Stars ($< 3 \, M_\odot$).

Planetary Nebula,
White Dwarfs,
Type 1 Supernova.

High-mass (Heavy weight) Stars
Type II Supernova
Neutron Stars and Supernova remnant
Black hole and Accretion disk.

37.7.7. White Dwarf

After a low-mass star burns its core into carbon and oxygen, heat generation stops and the core collapses again.

The core will heat further as gravitational energy is released, but the collapse stops before the core is hot enough to burn carbon and oxygen because of the *Pauli Exclusion Principle* and *Electron Degeneracy*.

The density in the core will not collapse further and supports gravitational weight of star.

White dwarfs show up on the H-R diagram.

37.7.8. <u>Stars within about 5 % of the Sun</u>

Example of a White Dwarf: Sirius B.

Sirius A is the brightest star in the sky.

Sirius A is member of a binary system with a faint companion to Sirius B.

37.7.9. Dark Matter

Bullet Cluster is an example of a cosmic feature that indicates the presence of dark matter, a substance that doesn't interact with light (*i.e.*, invisible) or with itself. Gravity is universal and so dark matter can be influenced.

37.8.1. UNIVERSE – FACTS

1. How old is the 'Observable Universe'? 13.8 billion *yrs*.

2. What is the Life span? The Universe existed for just 1.377×10^{-89} % of its expected life-time.

3. Size of the expanding observable Universe has its edge 46 billion *lyrs* distant.

4. Universe has neither a centre nor an edge. Einstein's theory of Relativity says Gravity binds all the space time around into an endless curve.

5. Average colour of the Universe is 'cosmic latte'.

6. The Universe is a pretty empty space, only 4.2×10^{-21} % contains any matter.

7. There are 20 trillion Galaxies including our Milky Way Galaxy.

8. The future of the Universe depends on how much 'dark matter' it contains. If there is too much Gravity will eventually stop it expanding, and make it shrink again.

9. Physicists explain that our Universe may exist inside a 'Black hole', and every 'Black hole' in our Universe may contain a new Universe.

10. Life exists in only on Earth, but in 1986 NASA infers from a rock in Mars microscopic living things exist there.

11. There are 100 billion stars in our Milky Way Galaxy.

12. The Andromeda Galaxy is the nearest and 2.2 million *lys.* away from our Galaxy.

13. The age of our Moon is actually 4.527 billion *yrs.*

14. The age of our Sun is 5 billion *yrs.*, and will live for around 10 billion *yrs.*

15. The Earth formed approx. 4.54 billion *yrs.* ago.

+&+&+&+&+&+

Chapter 38

CLASSICAL MECHANICS
Newtonian, Lagrangian and Hamiltonian Formalisms,
Special Theory of Relativity, Statistical Mechanical Distributions

"Earth provides enough to satisfy everymans need,
but not every man's greed" - MK Gandhi

38.1 INTRODUCTION

The science of Mechanics, which dealt with the motions of bodies and with the forces that affected these motions, provided before 1900 a powerful example of the ability of a mathematical, scientific theory to predict, correlate and interpret observations on the nature of the physical world. Mechanics (or Classical Mechanics) was first based upon Newton's Laws of Motion. More general and powerful formulations were developed later by Lagrange and by Hamilton. Classical mechanics is an approximation of General Relativity in a weak gravitational field.

38.1.1 CLASSICAL EQUATIONS OF MOTION
Several formulations are in use
a) Newtonian
b) Lagrangian
c) Hamiltonian

38.1.2. Advantages of non-Newtonian formulations
i)   More general, no need for "fictitious" forces
ii) Better suited for multi-particle systems
iii) Better handling of constraints
iv) Can be formulated from more basic postulates
v)   Assume conservative forces
vi) Equations of Motion are prepared whose form is independent of a coordinate system
physical laws and their corresponding equations that describe/govern the motion and interaction of big bodies within the universe are Galilean invariant means they do not apply to non-inertial reference frames.

38.2. NEWTONIAN FORMULISM

Cartesian spatial coordinates $r_i = (x_i, y_i, z_i)$ are primary variables

38.2.1. What is the fundamental problem in Classical Mechanics?

The problem is to describe the motion of systems of particles under various kinds of forces and initial conditions. This means to solve the differential equations resulting from Newton's Second Law

$$\vec{F}_i = m\,\vec{a}_i$$

where $\vec{a}_i$ is the acceleration on the mass m when a force $\vec{F}_i$, acts on the body.

38.2.2 What are the two types of Systems?
(i)   Conservative systems and

(ii)   Non-conservative systems:

38.2.3 What is meant by a <u>Conservative system</u>?

i) A system in which the sum of the kinetic energy T and potential energy V

$$\boxed{T + V = \text{ constant, with time}}$$

ii) It is an isolated system and is not affected by an external force.

iii) A system one in which $\vec{F}_i$ can be represented by negative gradient $\vec{\nabla}_i$ of some potential function V

$$\boxed{\vec{F}_i = -\vec{\nabla}_i V}$$

i.e., $$\boxed{\vec{F} = m\frac{d^2x}{dt^2} = -\vec{\nabla}V = -\frac{dV(x)}{dx}}$$

$$-\int\left(\frac{dV(x)}{dx}\right)dx = -dV \; ; \; \int m\frac{d^2x}{dt^2}dt = m\int\frac{dv}{dt}dx == m\int v\,dv$$

i.e., $$\boxed{V(x) + \tfrac{1}{2}mv^2 = C, \text{ constant}}$$

38.2.4 What is a **Constant of Motion of a System**?

Any property of a system in motion, say E, which is independent of time t is called the constant of motion of the system.

$$\boxed{\frac{dE(= T + V)}{dt} = 0}$$

38.2.5 An Example of NEWTONIAN MECHANICS

A simple harmonic motion (SHM) is the most commonly considered motion of a particle., say the stretching of a ideal spring.(which obeys Hooke's Law) obeying

$$\vec{F}_x = -k\,x$$

Using Newton's Second Law,

$$\vec{F}_x = -k\,x(t) = m\frac{d^2x}{dt^2}$$

$$\boxed{E = \tfrac{1}{2}mv^2 + \tfrac{1}{2}kx^2 = C, \text{ constant}}$$

This proves that the sum of the kinetic and potential energies is conserved for conservative systems.

Example 2: A particle is moving in gravitational field $V(z) = m\,g\,z$.

38.3 THE LAGRANGIAN FORMULATION OF CLASSICAL MECHANICS

38.3.1 *Introduction*

The equations of motion in Newtonian form are most convenient to solve if the physical problem is such as to make Cartesian coordinates appropriate. They are dependent on coordinate system. On the other hand, coordinates independent general equations of motion were derived by Joseph Lagrange and William Hamilton, and are called the Lagrangian and Hamiltonian forms of the equations of motion.

38.3.2 Generalized Coordinates

The method of Lagrange multipliers enables one to find extrema of functions subject to constraints.

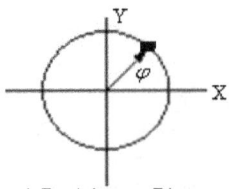

A Particle on a Ring

Example:

m = mass of a free particle of mass confined to move on the perimeter of a ring of radius R .
Constraints on a particle's motion arise from some set of unspecified forces. For the particle on a ring, imagine some force of infinite strength that limits the motion of the particle.
So a single coordinate φ is sufficient to locate the particle

$$x = R \cos\varphi$$

and $\qquad y = R \sin\varphi$

where angle φ is a generalized coordinate. The ith generalized coordinates symbolized as q_i , and $\dot{q}_i$ the time derivative of $\dot{q}_i$,

i.e., $\qquad \dot{q}_i = \dfrac{dq_i}{dt}$.

For a system containing N particles, it is required to specify 3N generalized coordinates q_i and 3N generalized velocities. $\dot{q}_i$. Such a system would have 6N degrees of freedom (in the case of a system without constraints).

38.3.4 Lagrangian function L The classical Lagrangian L is given by

$$\boxed{L(\dot{q}_i, q_i, t) = T(\dot{q}_i, q_i) - V(q_i, t)}$$

where T =KE is a function of $\dot{q}_i$ and q_i V = PE, a function of generalized coordinates q_i and time t .For a conservative system,

$$\boxed{L(\dot{q}_i, q_i) = T(\dot{q}_i, q_i) - V(q_i)}$$

38.3.5. Lagrange's Equations in Generalized Coordinates
For any set of generalized coordinates, with equation,

$$\vec{F}_i = -\vec{\nabla}_i V ,$$

Lagrange's equations take the form

$$\boxed{\dfrac{\partial}{\partial t}\left(\dfrac{\partial L}{\partial q_i}\right) - \dfrac{\partial L}{\partial q_i} = 0}$$

With the partial derivatives in this equation, the other $(6N-1)$ variables are held constant. The form of Lagrange's equations is invariant to the particular set of generalized coordinates chosen. Lagrange's equations look the same in any coordinate system.

38.3.6 Example: SHM $q_i = x_j$, $with\ j = 1$

$$\dot{q}_i = \frac{\partial L}{\partial \dot{x}} = m\dot{x}$$
$$L = T - V = \tfrac{1}{2}m\dot{x}^2 - \tfrac{1}{2}kx^2$$

and Lagrange's equation is

$$\boxed{\frac{d}{dt}\frac{\partial L}{\partial \dot{q}_i} - \frac{\partial L}{\partial q_i} = \frac{d}{dt}m\dot{x} + kx = 0}$$

or $\boxed{m\ddot{x} = -kx}$

which is just Newton's II law in generalized coordinates.

It should be noted that the Lagrange's equations of motion are a set of $3N$ second order differential equations.

38.3.7 One can use the following procedure that is general:

(i) Express the Lagrangian L in Cartesian coordinates;

(ii) Transform L to generalized coordinates;

(iii) Give Lagrange's equations in generalized coordinates.

38.3.8 Holonomic constraints.

The complete description of a system of N *free* particles requires 3N Coordinates. A constraint that can be described by an equation relating the coordinates (and perhaps also the time) is called a *holonomic*. In effect only($3N - k$)coordinates are needed to describe the system, given that the coordinates are connected by k holonomic equations *constraint* and the equation that describes the constraint is a *holonomic equation.*

38.4 HAMILTONIAN FORMULATION

38.4.1 What are conjugate momenta?

Newtonian and Lagrangian viewpoints take the q_i as the fundamental variables

*N-variable configuration space

*appears only as a convenient shorthand for $\frac{dq}{dt}$

*working formulas are 2nd-order differential equations

38.4.2 Hamiltonian formulation seeks to work with 1st-order differential equations

*2N variables

*treat the coordinate and its time derivative as independent variables

*appropriate quantum-mechanically.

Momentum p is an independent variable, p_i is the derivative of the Lagrangian with respect to $\dot{q}_i$, and replace with p_i .

38.4. Generalized momenta

The generalized momentum p_i conjugate to the coordinate q_i is defined by $p_i = \frac{\partial L}{\partial \dot{q}_i}$ But

momentum of a particle, p_i , is defined in terms of its velocity $\dot{r}_i$ by $p_i = m_i\dot{r}_i$.

The Lagrangian equation,

$$\frac{d}{dt}\frac{\partial L}{\partial \dot{q}_i} - \frac{\partial L}{\partial q_i} = 0$$

Becomes $\boxed{\frac{d}{dt}p_i - \frac{\partial L}{\partial q_i} = 0}$

i.e., $\boxed{\dot{p}_i = -\frac{\partial L}{\partial q_i}}$

38.4.4 The Hamiltonian function, H

For a system of particles each having masses m_i described by a set of generalized coordinates q_i, the classical Hamiltonian is defined by

$$H = \sum_i p_i \dot{q}_i - L\left(\{q_i\}, \{\dot{q}_i\}, t\right)$$

38.4.5 EQUATION OF MOTION: THE HAMILTONIAN FORM

For a conservative system,

$$\dot{q}_i = +\frac{\partial H}{\partial p_i}$$

$$\dot{p}_i = -\frac{\partial H}{\partial q_i}$$

$$\frac{\partial H}{\partial t} = -\frac{\partial L}{\partial t}$$

* The Hamiltonian H of a conservative system has the property that it is equivalent to the total energy E of the system.

* The solution of Hamilton's equations of motion will yield a trajectory in terms of positions and momenta as functions of time.

38.4.6 | Hamilton's equations can be easily shown to be equivalent to Newton's equations |

$$\frac{\partial H}{\partial p_x} = \frac{p_x}{m} = \dot{x} \Rightarrow \text{Vlocity}$$

$$\frac{\partial H}{\partial x} = \frac{dV}{dx} = -\dot{p}_x \Rightarrow \text{Newton's II Law}$$

38.4.7 | Hamilton's equations are just another formulation of Newton's Second Law. |

| The Hamiltonian can be directly obtained from the Lagrangian by a transformation known as a Legendre Transform. |

38.5 THE BASIC ASSUMPTIONS OF CLASSICAL MECHANICS

1) It is implied that an experimentalist can measure precisely and exactly the positions and velocities of the particles in a system at some initial time t in order to describe the state of the system.

2) Once the initial state is specified, the laws of Mechanics and a knowledge of the forces acting on the system enable the system to be characterized at any later time. In principle then an experimentalist can measure the position, velocity, momentum, energy, so on of any particle at any time and compare with the theoretical prediction. The inherent assumptions can be summarized as three statements given below:

(1) There is no limit to the accuracy with which one or more of the dynamic variables of as classical system can be simultaneously measured, except the limit imposed by the precision of instruments used for the measurement.

(2) *Precision and simultaneity* there is no restriction to the number of dynamic variables that can be accurately measured simultaneously.

(3) *Continuous spectra:* Since the expressions for velocity are continuously varying functions of time, the velocity, and hence the kinetic energy, can vary continuously.

When a particle is in the microscopic world, all these three assumptions are invalid. For these systems Classical Mechanics cannot describe their behaviour.

The new mechanics that was developed to describe these systems is known as Quantum Mechanics. Hamilton theory, or its extension the Hamilton-Jacobi equations, does have applications is Celestial Mechanics, and of course Hamiltonian operators play a major part in Quantum Mechanics.

38.6 SPECIAL THEORY OF RELATIVITY

Relativity is a widely used term. It is generally used to describe everything from the comical Galilean- version of $\boxed{E = mc^2}$ to concepts about time travel. The theory called the Special Theory of Relativity, first asserted by Albert Einstein.

20.6.1 Galilean Transformation

$$\boxed{\begin{array}{l} \text{Coordinates:} \\ x' = x - ut \\ y' = y \\ z' = z \\ t' = t \end{array}}$$

$$\boxed{\begin{array}{l} \text{Velocities:} \\ v'_{Ax} = v_{Ax} - u \\ v'_{Ay} = v_{Ay} \\ v'_{Az} = v_{Az} \end{array}}$$

The Michelson-Morley Experiment (1887) showed the invariance of the speed of light c in vacuum.

38.6.2. The Lorentz Transformations:

In his Special Theory of Relativity, Einstein laid down two postulates:

(i)   The laws of physics have the same mathematical form in all inertial reference frame,

(ii)   The speed of light through vacuum

$$\boxed{c = 2.99792458 \times 10^8\, ms^{-1}}\ \text{or } 186,000\ miles\ s^{-1},$$

is constant as observed by any observer, moving or stationary: This transformation is valid for all types of physical phenomena at all speeds u of inertial frame (primed) relative to an (unprimed) frame.

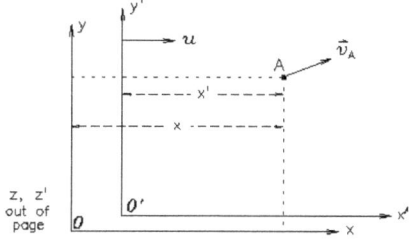

For two observers using two coordinate systems with origins separated by a fixed distance, a, the transformations are:

$$\boxed{\begin{array}{l} \text{Coordinates:} \\ x' = \gamma(x - ut) \\ y' = y \\ z' = z \\ t' = \gamma(t - \dfrac{x\beta}{c}) \end{array}}$$

$$\boxed{\begin{array}{l} \text{Velocities:} \\ v_A = \dfrac{v_A' + u}{1 + (v_A'\beta/c} \end{array}}$$

$$\boxed{\begin{aligned} \beta &= \frac{u}{c} \\ \gamma &= \frac{1}{\sqrt{1-\beta^2}} \end{aligned}}$$

$$\boxed{\begin{aligned} &\text{Coordinates:} \\ x' &= \frac{(x-ut)}{\sqrt{1-v^2/c^2}} \\ y' &= y \\ z' &= z \\ t' &= \frac{(t-xv/c^2)}{\sqrt{1-v^2/c^2}} \end{aligned}}$$

38.7 A NEW VIEW OF SPACE AND TIME

38.7.1   The Time Dilation equation for Relativity is
:

$$\boxed{t = \frac{t_0}{\sqrt{1-v^2/c^2}}} \text{ for a time interval } t_0 \text{ to } t.$$

Thus <u>moving clocks</u> run slow.

38.7.2   The FitzGerald length contraction L_0 to L :

$$\boxed{L = L_0 \sqrt{1-\frac{v^2}{c^2}}}$$

$$\boxed{\begin{aligned} \beta &= \frac{u}{c} \\ \gamma &= \frac{1}{\sqrt{1-\beta^2}} \end{aligned}}$$

i.e., the length of a moving object is shortened`

38.7.3   Equivalence of mass and energy relation.
Rest mass energy

$$\boxed{E_0 = m_0 c^2}$$

$$\boxed{E = mc^2}$$

38.7.4   Increase of rest mass m_o to m

$$\boxed{m = m_0 \sqrt{1-\frac{v^2}{c^2}}}$$

$$\boxed{E = \frac{E_0}{\sqrt{1-v^2/c^2}}}$$

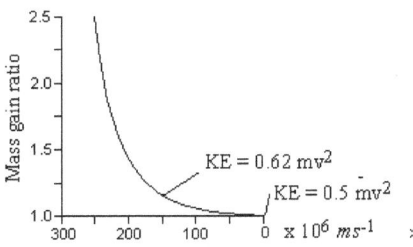

Relativistic mass vs velocity based on rest mass

38.7.5. Energy-momentum relationship:

$$E^2 = p^2 + m^2$$

$$E = c\sqrt{(m_0{}^2 c^2 + p^2)}$$

38.8. STATISTICAL MECHANICS

38.8.1. Statistical mechanics aims at understanding relating

The macroscopic properties of materials and the microscopic behaviour of their constituent particles

Examples include:

• specific heat capacity and its variation with temperature;

• the entropy of a sample of material, and its relationship with temperature and internal energy;

• the magnetic properties of materials.

A macrostate specifies a system in terms of quantities that "average" over the microscopic constituents of the system. Examples of such quantities include the pressure, volume and temperature of a gas.

A microstate specifies a system in terms of the properties of each of the constituent particles; for example, the position and momentum of each of the molecules in a sample of gas.

A key concept of statistical mechanics is that many different microstates can correspond to a single macrostate. Statistical mechanics explores the relationship between microstates and macrostates.

The principle of equal a priori probabilities can be used on its own to derive some interesting results in specific cases.

Statistical mechanics gives us the relationship between energy, the number of accessible microstates and the distribution parameter. The energy levels in two example cases:

• a collection of harmonic oscillators;

• a collection of magnetic dipoles in an external magnetic field.

Thermodynamics gives us the relationship between energy U, temperature T and entropy S:

$$\frac{\partial S}{\partial U} = \frac{1}{T}$$

Statistical mechanics gives us the relationship between energy, the number of accessible microstates and the distribution parameter

$$\frac{\partial \ln \Omega}{\partial U} = -\beta$$

$$S = k \ln \Omega$$

$$\beta = -\frac{1}{k_B T}$$

where $k_B = 1.3806 \times 10^{-23} J/K$.

The Stirling approximation

$$\boxed{\ln N! = N \ln N - N}.$$

20.8.2. The Maxwell-Boltzmann distribution

A molecule of mass m in a sufficiently dilute gas such that the intermolecular forces are negligible (*i.e.*, an ideal gas). The energy of the molecule in the Boltzmann distribution (which described how the number of <u>distinguishable</u> particles in different energy states varied with the energy of those states, at different temperatures) follows

Boltzmann distribution

$$\boxed{n_i = \frac{N}{Z} e^{-\varepsilon_i/k_{Bi}T}}$$

However, in systems consisting of collections of <u>identical fermions</u> or identical bosons, the wave function of the system has to be either *antisymmetric* (for fermions) or *symmetric* (for bosons) under interchange of any two particles. Instead, all the particles are "shared" between the occupied states. The particles are said to be indistinguishable.

38.8.3. Bose-Einstein Distribution Law

The most probable distribution can be written: the Bose-Einstein distribution for a collection of <u>indistinguishable</u> bosons

g_i = degeneracy of energy level ε_i

$$\boxed{n_i = g_i \frac{1}{e^{-\varepsilon_i/k_{Bi}T} - 1}}$$

obeyed by *Bosons*.

38.8.4. Fermi-Dirac Distribution

for a collection of indistinguishable fermions

$$\boxed{n_i = g_i \frac{1}{e^{-\varepsilon_i/k_{Bi}T} - 1}}$$

This is obeyed by *Fermions*.

38.9. GENERAL THEORY OF RELATIVITY

Einstein proposed his General Theory of Relativity in 1915, postulating bodies do not attract mutually, but the presence of massive bodies in space cause space to curve such that gravitational field is set up. Gravity is the property of space itself. The theory suggests that gravitation would travel in the form of waves at the speed of light, *viz.*, c. It also suggests that light should be bent by gravitation field and time should appear slower near a massive body like a planet. The underlying fundamental symmetry in the theory makes it difficult to formulate it as a dynamical theory. Further the theory is non-linear. Hence experimentally to create source of gravity waves and dynamically evolve it to infer theoretical predications are challenges of future. (Example, *Physics News*, Vol 46, Jan 2016).

38.9.1 Gravity Waves

Gravity waves are produced when matter gets accelerated, and they travel in space at speed, c, and carry energy as if it were emitted by a quadrupole.

38.9.2 Detection of Gravity Waves (LIGO Experiment)

The atomic force microscope (Nobel Prize 1986) enabled measurements of distances of about $10^{-10}m$, between separations of atoms. From high energy accelerators one can <u>infer</u> distances of about $10^{-15}m$, between high energy fundamental particles. Now LIGO detectors (in 14 Sep 2015) have proved measurements as small as about $10^{-21}m$.

Laser Interferometric Gravitational-Wave Observatory (LIGO) (designed at Caltech and MIT, and credited to NSF USA) has opened gravitational wave astrophysics. It is located at Livingstone in Lousiana and Hartford in Washington, USA. Gravitational waves arriving at the detector from a cataclysmic event in the distant Universe were detected on 14 Sep 2015 (just after $100\ yrs.$ after Einstein's suggestion!!). This event opens a new window onto the Cosmos.

+*+*+*+*+*+

Chapter 39

QUANTUM PHYSICS

"I become Death, shattered of worlds." --Shiva in "The Bhagavad Gita".

"Quantum Physics came from the Vedas????" by Bohr, Schrodinger, etc.
https://shar.es/1BwzIb

39.1. BOHR'S HYDROGEN atom

39.1.1 JELLIUM Model of atom

J.J. Thomson's $\frac{e}{m}$ experiment showed that the hydrogen atom to be 1836 times as heavy as the electron. In his atomic model electrons were embedded in a massive matrix of positive charge filling a volume of roughly one atomic diameter (~ 1 A).

39.1.2 NUCLEAR atom by Ernst Rutherford (1911)

The atom decays according to classical mechanics by its electron spiraling around it in a very short time by emitting electromagnetic radiation as per Larmor's expression,

$$\boxed{\frac{dU}{dt} = \frac{1}{4\pi\varepsilon_0} \frac{2}{3} \frac{e^2 a^2}{c^3}}$$

39.1.3 BOHR's theory of Hydrogen atom: (1913)

Niels Bohr extended Planck's quantum hypothesis, viz.,

$$\boxed{E = h\,\nu}.$$

39.1.4 Postulates:

(i)   Rutherford's nuclear model of the atom was adopted

Nuclear atom mode

(ii)   It assumed the Coulomb's law of force and Newton's laws of motion to be applicable in the atomic domain.

$$\vec{F}_{em} = \left(\frac{1}{4\pi\varepsilon_0}\right)\left(\frac{-Ze^2}{r^2}\right)$$

Centrifugal force,

$$F_C = \frac{mv^2}{r}$$

(iii) The path of the electron around the nucleus should be a conic section.

(iv) The conic section is a circle of radius r with the nucleus at the center of the circle

(v) **POSTULATE 1**. This relates to the mechanics of the atom (the idea of the stationary state). Only the electron orbits are allowed (or permissible) for which the angular momentum (L) of the electron is an integral multiple of $\hbar$,

$$L = n\frac{h}{2\pi} \equiv n\hbar,$$

where $n = 0, 1, 2, 3, 4, \dots(integer)$.

and that no energy is radiated while the accelerated electron remains in any of the permissible orbits,

(vi) An electron moving in one of the stable orbits does not radiate,

(vii) **P0STULATE 2:** relates to the electrodynamics of the atom (idea of quantum jump).

$$\boxed{E_i - E_f = h\nu}$$

$$\boxed{\Delta E = h\nu}$$

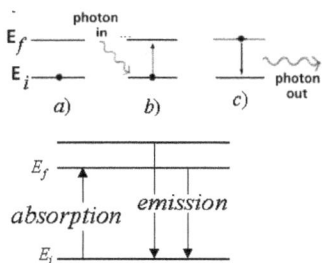

E_i and E_f denote the initial and final values of energy of the atom

ν = the frequency of the radiation emitted //absorbed.

39.1.4.2 ORBITS given by CLASSICAL PHYSICS:

$q_1 = +Z e = $ Nuclear charge,

$q_2 = e = 1.6021 \times 10^{-19} C$,

$M = $ nuclear mass,

$m = $ electronic mass $m_e = 9.11 \times 10^{-31} kg$

$\varepsilon_0 = \dfrac{1}{c^2 \mu_0}$, permittivity of free space $= \varepsilon_0 = 8.8542 \times 10^{-12} Fm^{-1}$

The COULOMB constant, $k = 8.9875 \times 10^9 Nm^2C^{-2}$, $k = 1/4\pi\varepsilon_0 = 8.9875 \times 10^9 F^{-1}m$.

Centrifugal force, $\vec{F} = \dfrac{m v^2}{r}$.

Coulomb force, $\vec{F} = kq_1q_2 \,/\, r^2$

$$\vec{F}_{em} = \left(\frac{1}{4\pi\varepsilon_0} \right)\left(\frac{-Ze^2}{r^2} \right)$$

39.1.5 QUANTUM number, n distinguish electrons
Velocity, v_n of electron in the nth orbit.

$$v_n = \frac{2\pi\, Z\, e^2}{n\, \hbar}$$

Radius, $\boxed{r_n = \frac{n^2\hbar^2}{m\, Z\, e^2}}$

39.2.1 Total energy, E = potential energy, V + kinetic energy, T .

$$E = V + T$$

$$\boxed{T = \tfrac{1}{2}mv^2 = \tfrac{1}{2}\frac{Ze^2}{r}}$$

The system being conservative, $\boxed{\vec{F} = -\dfrac{dV}{dr}}$, gives

$$\boxed{V = -\frac{Ze^2}{r}}\ .$$

39.2 **QUANTIZED (DISCRETE) energy levels**:

$$\boxed{E_n = -\frac{2\pi^2 m\, Z^2\, e^4}{n^2\, h^2}\,,\qquad n = 1,\ 2,\ 3,\ 4,\ \dots}$$

39.2.3 Energy level diagram:

39.2.4 The Fine Structure Constant, α :

$$\boxed{\alpha = \frac{ke^2}{\hbar c} = \frac{1}{137.0388}}\ .$$

39.2.5 RADIUS of the atom r_n

$$r_n = \frac{n^2}{Za_0}$$

39.2.6 *BOHR atomic radius constant*, a_0.

$$\frac{1}{r_i} = \frac{\alpha\, m\, c^2}{\hbar\, c}$$

$$r_1 \equiv a_0\, ;$$

$$a_0 = \frac{\hbar^2}{m\, e^2}$$

$$a_0 = 0.05292 nm$$

$$r_n = \frac{n^2}{Z} a_0\, .$$

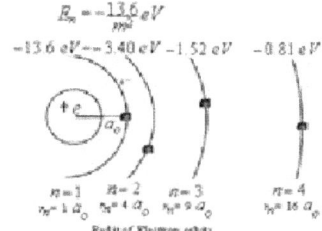

39.2.7 **The BINDING ENERGY,** E_n of the electron in a H-atom:

$$E_n = -\frac{e^2}{8\pi\, \varepsilon_0 a_0 n^2}$$

$$E_n = -13.6 \frac{Z^2}{n^2}\ eV\ .$$

$$E_1 = -13.6\ Z^2 eV$$

$$1 eV = 1.6021 \times 10^{-19}\, J$$

39.2.8 <u>Hydrogen SPECTRUM</u> Theoretically derived:

Wave number $\boxed{\bar{v} = 1/\lambda = v/c}$

$$\boxed{\bar{v} = 1/\lambda = v/c = R_H[1/2^2 - 1/n^2]}\quad ,n = 3, 4, 5,...$$

where R_H is the RYDBERG constant.

$$v = \frac{E_i - E_f}{h} = \frac{E_1}{h}\left(\frac{1}{n_i^2} - \frac{1}{n_f^2}\right)$$

$$\lambda(\text{in } nm) = \left(\frac{91.15}{Z}\right)\left(\frac{n_i^2 n_f^2}{n_f^2 - n_i^2}\right)$$

$$\bar{v} = (E_1/hc)[1/n_i^2 - 1/n_f^2] = R_H\left(1/n_i^2 - 1/n_f^2\right).$$

$$R_H = \frac{E_1}{hc} + \frac{1}{2}mc^2\alpha^2 Z^2 \; / \; hc = 1.0974 \; x10^2 \; nm$$

Wavelength, λ (nm)

$$\bar{v} = R \, [1 / n_i^2 - 1 / n_f^2]$$

Lyman   Balmer   Paschen

Energy level diagram and Spectral series for Hydrogen atom

39.2.9 SHELLS:

It is conventional that the electron is said to occupy a specific GROUP, ENERGY LEVEL or ATOMIC SHELL. These shells are given SYMBOLS (Roman capital letters) as follows:

$$
\begin{array}{llllll}
n = & 1 & 2 & 3 & 4 & 5 \dots \\
& K & L & M & N & O \dots
\end{array}
$$

39.3.1 **LINE SPECTRAL SERIES**:

$$\bar{v} = R_H \, [1 / n_i^2 - 1 / n_f^2] \qquad , \; n_i = 1$$

$n_1 = 1,\; n_2 = 2,\, 3,\, 4,..\, \&$	LYMAN series; Ultra Violet
$n_2 = 2,\; n_3\, 3 = 3,\, 4,\, 5,..\&$	BALMER series; Visible
$n_3 = 3,\; n_4 = 4,\, 5,\, 6,..\, \&$	PASCHEN series; Infra Red
$n_4 = 4,\; n_5 = 5,\, 6,\, 7,..\, \&$	BRACKETT series; Far Infra Red
$n_5 = 5,\; n_6 = 6,\, 7,\, 8,..\, \&$	PFUND series; far Infra Red
$n_6 = 6,\; n_7 = 7,\, 8,\, 9,..\, \&$	HUMPHREY'S series; far Infra red.

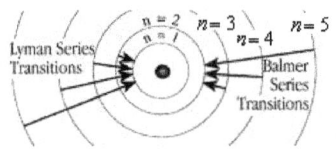

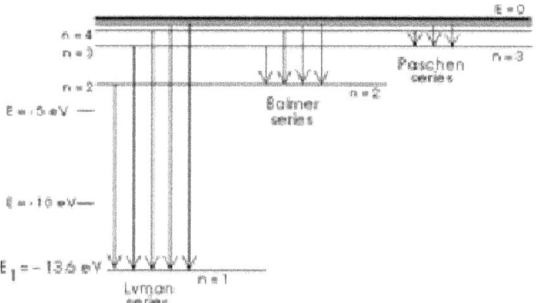

39.3.2.1 Series LIMIT, $\bar{v}_\infty$

$$\bar{v}_\infty = R_H (1/n_i^2)$$

39.3.2.2 IONIZATION energy: I

$$I = h\,c\,R_H = 2.179\ a\ J = 1312\ kJ\ mol^{-1} = 13.60\ eV$$

39.3.2.3. EXCITATION energy;

$$(E_2 - E_1) = 16.31\ x\ 10^{-19} J = 10.19\ eV$$

$$(E_\infty - E_1) = 21.76\ x\ 10^{-19} J = 13.58\ eV$$

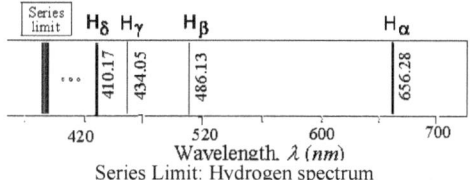

Series Limit: Hydrogen spectrum

59.3.3. Ritz COMBINATION Rule (1905):

When the frequency of the first line of the Lyman series is denoted by L_1 and other frequencies of the Lyman series and Balmer series L_2 and B_1 correspondingly, the empirical relation holds

$$\boxed{L_1 + B_1 = L_2}$$

39.3.4 Bohr CORRESPONDENCE Principle: (1923), in essence, states that results of classical physics should be contained as limiting cases (of quantum physics),

To illustrate in a Bohr atom:

$$v_{if} = c\,R\,2n\,\Delta n / n^4 = \left(\frac{1}{2\pi}\right)\sqrt{\left(\frac{e^2}{m\,r_n^3}\right)}$$

$$\Delta n = 1, n \to \infty, \qquad v_{QM} = \frac{v}{2\pi r} = \frac{\omega}{2\pi} = v_{CL}$$

39.3.5. BOHR–SOMMERFELD RELATIVISTIC MODEL OF H-ATOM

[Relative motion of the hydrogen nucleus or correction to the finite mass of the nucleus]

Bohr's theory *failed* to account for the spectrum of any atom having more than one electron (Z =1). A more general form of the atom was tried by N. Bohr & Arnold Semmerfeld, called the *relativistic elliptical atom model.*

The *reduced mass μ* of the atom is given by

$$\boxed{\mu = \frac{m\,M}{m+M}}$$

$$\boxed{J_i = \oint p_i \cdot dq_i = n_i h}$$

$$J_r = \oint p_r \cdot dr = p_\varphi \{\oint [\frac{1}{r^2}\left(\frac{dr}{d\varphi}\right)^2 d\varphi]\} = p_\varphi \mathfrak{I}$$

$$J_\varphi = \oint p_\varphi \cdot d\varphi = n_\varphi h$$

$$\boxed{n = (n_r + n_\varphi)}$$

$$E_n = E_{n\varphi} + E_{n_r} = -\frac{2\pi^2 \mu\, Z^2\, e^4}{n^2\, h^2}, \quad n = 1, 2, 3, 4, \ldots\ldots$$

$$E_n = \left[-\frac{2\pi^2 \mu\, Z^2\, e^4}{n^2\, h^2}\right]\left(1 + (\frac{Z^2\,\alpha^2}{n})\,(\frac{1}{n_\varphi} - \frac{3}{4n})\right), \quad n = 1, 2, 3, 4, \ldots\ldots$$

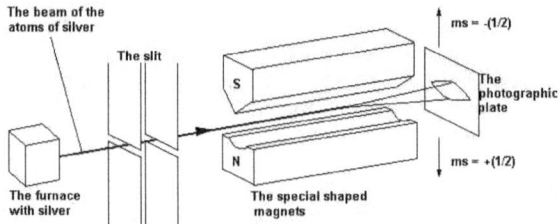

39.4 Electron INTRINSIC SPIN and FOUR Quantum numbers:

39.4.1 S.A. Goudsmit and G.E. Uhlenbeck (1925); electron possesses an intrinsic spin angular momentum, S independent of its orbital angular momentum, L. *INTRINSIC SPIN quantum number,*

$s = \frac{1}{2}$ for the electron.

The beam of the atoms of silver

The slit

$ms = -(1/2)$

The photographic plate

S

N

$ms = +(1/2)$

The furnace with silver

The special shaped magnets

The Stern-Gerlach experiment. On the photographic plate are two clear tracks.

The correct interpretation was given to the Stern-Gerlach observations (1922) (on neutral silver atoms in an inhomogeneous magnetic field)only after Goudsmit and Uhlenbeck (1925) were led by a wealth of spectroscopic evidence to hypothesize the existence of an electron spin and intrinsic magnetic moment

39.4.2 Space quantization of the electron spin momentum is given by the *Spin MAGNETIC quantum number*, m_s s.

$$m_s = \pm \frac{1}{2}$$

39.4.3 Pauli's EXCLUSION Principle (Wolfgang Pauli, 1925)

NO TWO ELECTRONS IN AN ATOM CAN POSSESS THE SAME FOUR QUANTUM NUMBERS, *viz.*, n, ℓ, m_ℓ, and m_s'

i.e.,: no two electrons in an atom can exist in the same quantum state.

39.4.4 **A QUANTUM STATE** of an electron in an atom is UNIQUE and is specified by the four quantum numbers, n, ℓ, m_ℓ, and m_s.

TOTAL Angular momentum, j:
$$j = \ell + s.$$

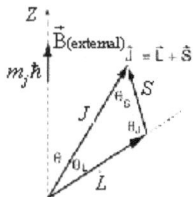

$$Cos(s,\ell) = \frac{[j(j+1) - \ell(\ell+1) - s(s+1)]}{2\sqrt{[\ell(\ell+1)s(s+1)]}}$$

39.4.5 PERIODIC TABLE of elements

The great success of the Pauli Principle is that it explains many aspects of the Periodic Table of Elements (Fig. 1.21) on the basis of quantum numbers, which originally were introduced for an entirely different purpose, *i.e.*, for the interpretation of spectra.

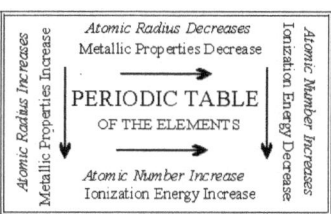

Abridged version of Periodic Table of Elements

1A				3A	4A	5A	6A	7A	NOBLE
H [1] $1s^1$	2A								He [2] $1s^2$
Li [3] $1s^2 2s^1$	Be [4] $1s^2 2s^2$		$1s^2$	B [5] $2s^2 2p^1$	C [6] $2s^2 2p^2$	N [7] $2s^2 2p^3$	O [8] $2s^2 2p^4$	F [9] $2s^2 2p^5$	Ne [10] $2s^2 2p^5$
Na [11] [Ne]$3s^1$	Mg [12] [Ne]$3s^2$	$1\overline{B}$	$2\overline{B}$ [Ne]	Al [13] $3s^2 3p^1$	Si [14] $3s^2 3p^2$	P [15] $3s^2 3p^3$	S [16] $3s^2 3p^4$	Cl [17] $3s^2 3p^5$	Ar [18] $3s^2 3p^6$
K [19] [Ar]$4s$	[Ar]$3d^9$	Cu [29] $4s$	Zn [30] $4s^2$	Ga [31] $4s^2 4p^1$	Ge [32] $4s^2 4p^2$	As [33] $4s^2 4p^3$	Se [34] $4s^2 4p^4$	Br [35] $4s^2 4p^5$	Kr [36] $4s^2 4p^6$
Rb [37] [Kr]$5s$	[Kr]$4d^n$	Ag [47] $5s$	Cd [48] $5s^2$	In [49] $5s^2 5p^1$	Sn [50] $5s^2 5p^2$	Sb [51] $5s^2 5p^3$	Te [52] $5s^2 5p^4$	I [53] $5s^2 5p^5$	Xe [54] $5s^2 5p^6$
Cs [55] [Xe]$6s^1$	$4f^{14}5d^n$	Au [79] $6s$	Hg [80] $6s^2$	Tl [81] $6s^2 6p^1$	Pb [82] $6s^2 6p^2$	Bi [83] $6s^2 6p^3$	Po [84] $6s^2 6p^4$	At [85] $6s^2 6p^5$	Rn [86] $6s^2 6p^6$

Periodic Table of the Elements

39.4.5.1 DISPACEMENT LAW

Developments during 1900 to 1927

According to the **Displacement Law**, any singly charged ion has the same type of spectrum as the neutral atom of the preceding element in the Periodic Table, but shifted to higher frequencies. One gets an *"isoelectronic sequence"* a row in the Periodic Table, which are reduced to the same number of external electrons.

39.5   ATOMIC STRUCTURE

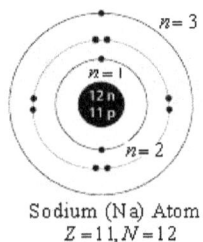

Sodium (Na) Atom
$Z = 11, N = 12$

39.5.1   ELECTRON CONFIGURATION of atoms

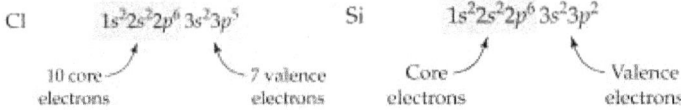

Cl $1s^2 2s^2 2p^6 3s^2 3p^5$ Si $1s^2 2s^2 2p^6 3s^2 3p^2$

10 core electrons   7 valence electrons   Core electrons   Valence electrons

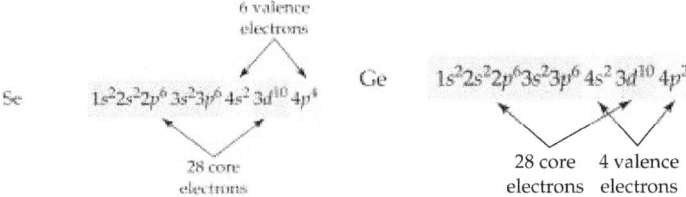

39.5.2 The outer electron configuration for the first 18 elements are shown

1A							8A
1 **H** $1s^1$	2A	3A	4A	5A	6A	7A	2 **He** $1s^2$
3 **Li** $2s^1$	4 **Be** $2s^2$	5 **B** $2s^22p^1$	6 **C** $2s^22p^2$	7 **N** $2s^22p^3$	8 **O** $2s^22p^4$	9 **F** $2s^22p^5$	10 **Ne** $2s^22p^6$
11 **Na** $3s^1$	12 **Mg** $3s^2$	13 **Al** $3s^23p^1$	14 **Si** $3s^23p^2$	15 **P** $3s^23p^3$	16 **S** $3s^23p^4$	17 **Cl** $3s^23p^5$	18 **Ar** $3s^23p^6$

Noble gases	Alkali metals	Alkaline earth metals	Halogens
2 **He** $1s^2$	3 **Li** $2s^1$	4 **Be** $2s^2$	9 **F** $2s^22p^5$

39.5.3 GROUP, ENERGY LEVEL, or ATOMIC SHELLS.
They are given symbols (Roman capital letters) as follows:

```
n = 1  2  3  4  5 . . . . .
    K  L  M  N  O . . . . .
```

39.5.3.1 **SUBGROUP, SUBLEVEL or SUBSHELL**.

a) Spectral NOTATION

```
n = 1  2  3  4  5 . . . . .
    K  L  M  N  O . . . . .
```

In this notation a state in which $n = 2, \ell = 0$, is a 2s state, one in which $n = 4, \ell = 2$ is a *4d state electron configuration* of sodium atom (Z = 11), in the Normal state,

$$1s^2 \ 2s^2 \ 2p^6 \ 3s^1$$

In order *to determine the angular momentum of an atom, only those electrons, which are external to the closed shells, need be considered.*

b) Capital letters are:

$$\boxed{\begin{array}{ll} L = & 0, \ 1, \ 2, \ 3, \ 4, \ 5, \ \& \\ & S, \ P, \ D, \ F, \ G, \ H, \ \& \end{array}}$$

$^2P_{1/2}$, $^2P_{3/2}$, read *"doublet P one half"*, etc.

or $\quad ^3P_2, \ ^3P_1, \ ^3P_0 \ $ read *"triplet P two"*, and so on.

The superscript is an indication of the **MULTIPLICITY** of the terms of the atomic configuration.

39.5.3.2 Electron Orbitals: <u>Chemistry depends on electron orbitals!</u>

3d orbitals

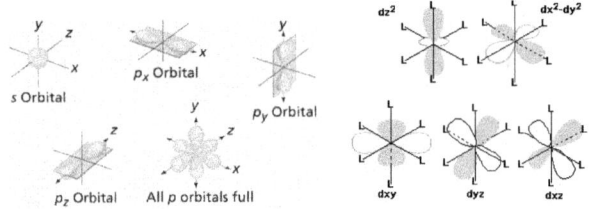

σ **- bonding orbitals**
π -bonding & anti bonding orbitals

30.5.4. FINE STRUCTURE AND HYPERFINE STRUCTURE of Spectral Lines

Normal spectral line, its Fine and Hyperfine structures are understood from the self-explanatory diagrams below.

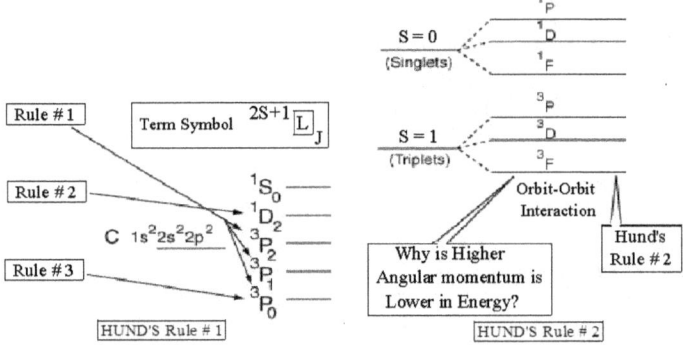

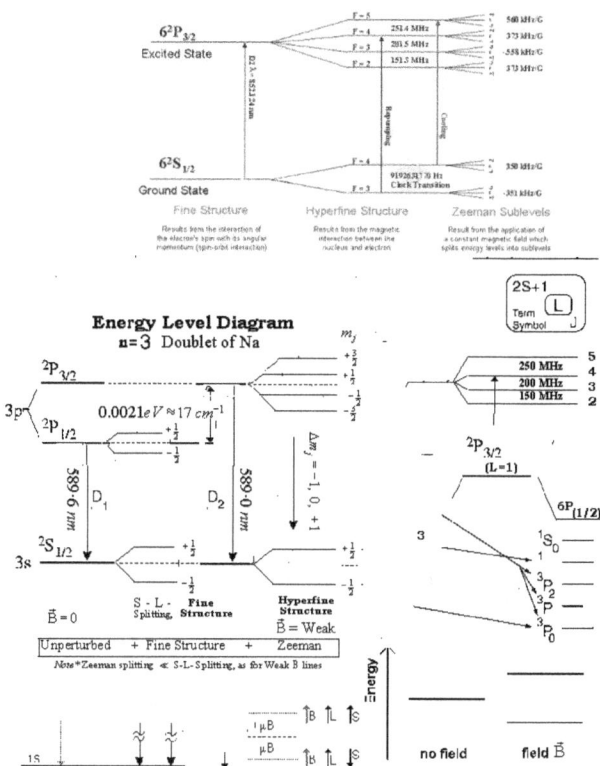

39.5.4.1 CHARACTERISTIC X-ray spectra (1913)

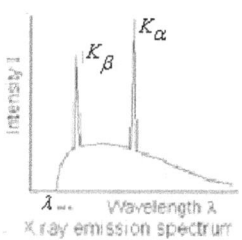

X ray emission spectrum

90.5.4.2 Moseley's Law.

$$\nu = C \ (Z - a)^2$$

where C and a For the K_α line, $C = (1/4)\ c\ R$, $a \approx 1$ and $R = 1.0967\ x\ 10^2\ nm$

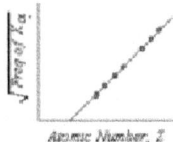

30.5.4.3 The FRANCK-HERTZ *Experiment*:

$$\frac{1}{2}m\,v^2 = eV_0 = h\,v = \frac{h\,c}{\lambda}$$

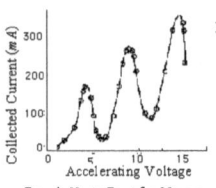

Franck-Hertz Expt for Mercury

39.5.5.2 **De BROGLIE HYPOTHESIS:** (1924, Louis Victor de Broglie):

$$p = \frac{h}{\lambda}$$

The two sets of quantities:
(1) particle E and p are characteristics of the particle nature and
(2) wave (v and λ) are connected through the Planck's constant h.

$$m = m_0\ /\ \sqrt{(1 - \frac{v^2}{c^2})}$$

$$\lambda = \sqrt{\left(\frac{150}{V(\text{in } kV)}\right)}$$

39.5.5.3. FUNDAMENTAL Wave-Particle relationships (Symmetric or useful forms of the Planck and de Broglie relation):

$$\omega = 2\pi v$$
$$E = \hbar\omega$$
$$p = \hbar k$$

39.5.5.4 ELECTRON WAVE EXISTS:
Electron diffraction (C. Davisson & L.H. Germer) experiment in 1927 confirmed.

$$\frac{1}{\lambda} = \frac{n}{2d\,Sin\theta} = \frac{p}{h} = \frac{\sqrt{2mE}}{h} = \frac{\sqrt{2meV}}{h}$$

Electron Wavelength	Bragg Law	deBroglie Wavelength	Accelerating through voltage V

The hypothesis by de Broglie.
Diffraction is a property that is only associated with wave motion, and the wavelength predicted for the electron was just given by equation

$$\lambda = \sqrt{\left(\frac{150}{V(\text{in } kV)}\right)}.$$

39.5.5.5. The Wave-Particle DUALITY is real and as listed below:

	Light	Electron
Particle aspect	Line spectra	(1) Cathode ray tube
	Photon emission	Electron deflection
	Photoelectric Effect	(2) Cloud Chamber
	Photon absorption	Electron Collision
	Compton Effect	
	Photon deflection	
Wave aspect	(1) Interference	(1) Electron diffraction
	(2) Diffraction of light	(2) Electron microscope
	(3) X-ray diffraction	

The two sets of quantities: (1) particle (E and p are characteristics of the particle nature) and (2) wave (ν and λ), are connected through the Planck's constant h.

$$m = \frac{m_o}{\sqrt{1 - v^2/c^2}}$$
$$\lambda = \sqrt{\frac{150}{V(mkV)}}$$
$$\omega = 2\pi\nu$$
$$p = \hbar k$$
$$\varepsilon = \hbar\omega$$

39.6.1   ELECTRON Wave:

39.6.1.1 WAVE PACKET Description of Material Particles:
WAVE FUNCTION, denoted by the Greek symbol, psi, Ψ:
Combination of two plane waves, in phase and interfere constructively,

$$U_1(z,t) = U_0 e^{i(k_1 z - \omega_1 t)},$$

$$U_2(z,t) = U_0 e^{i(k_2 z - \omega_2 t)}$$

with the angular frequencies ω_1 and ω_2; and

$$\omega_0 = <\omega> = (\omega_1 + \omega_2)/2;$$

and wave vectors / propagation constants $k_1 \approx k_2$, and

$$\omega_o = <\omega> = (\omega_1 + \omega_2)/2 \; ;$$

39.6.1.2 **WAVE and GROUP velocities**

The dashed- line curve is the *wave packet*

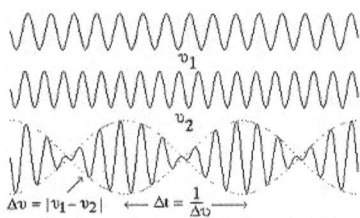

39.6.1.3 **PHASE (or wave-front) velocity,** *w*

$$w \equiv v_p \equiv v_\phi = \frac{dz}{dt} = \frac{\omega_o}{k_o}$$

39.6.1.4 **GROUP Velocity,** *u*

$$u \equiv v_p \equiv v_g = \frac{d\omega}{dk} ,$$

39.6.2. HEISENBERG'S UNCERTAINTY PRINCIPLE (1927)

39.6.2.1 The POSITION-MOMENTUM *Uncertainty relation*:

It is impossible to measure simultaneously and precisely both position and momentum of a particle".

$$\Delta z . \Delta p_z \geq \frac{\hbar}{2}$$

39.6.2.2 The TIME-ENERGY Uncertainty Relation:

$$\Delta E . \Delta t \geq \frac{\hbar}{2}$$

in the temporal part of a wave packet..

39.6.2.3 Gedanken experiments: (1) Gamma ray microscope and (ii) Electron Single-slit experiment

(1) Gamma Ray Microscope

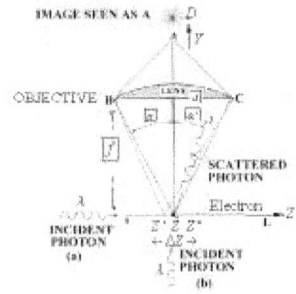

From Physical Optics, diffraction theory of light shows that the diameter, D, of the central disc of the diffraction patterns given approximately by

$D = \lambda / \mathrm{Sin}\alpha$

where λ = wavelength of the photon,

2α = angle subtended at the particle by the objective lens,

$\mathrm{Sin}\alpha = d / f$

d = diameter of the aperture of the objective.

$\Delta z = \lambda / \mathrm{Sin}\alpha$

$\Delta z \to$ small, when i) $\lambda \to$ small, and ii) $\mathrm{Sin}\alpha \to$ large,

$\Delta p_y = p_y \, \mathrm{Sin}\alpha = (\hbar / \lambda) \cdot (d / 2y)$

$\Delta z \cdot \Delta p_y = (\lambda / \mathrm{Sin}\alpha) \cdot (\hbar / \lambda) \cdot (\mathrm{Sin}\alpha) = \hbar \; \Delta z \cdot \Delta p_y = \hbar$.

2) Young's Single Slit Electron Wave Diffraction

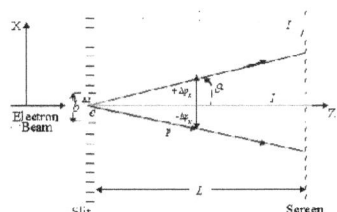

$\mathrm{Sin}\beta = d = \lambda / b$

$\Delta p_x = p \, \mathrm{Sin}\beta = p(\lambda / b) = (h / \lambda)(\lambda / b) = h / b$

$\Delta z \Delta p_x = b(h / b) = h$

39.7.1 **THE BASIC EQUATION** in Quantum Mechanics
The SCHRODINGER EQUATION (TISE, or Time Independent Schrodinger Equation)

$$\hat{H}\Psi(r,t) = \hat{E}\Psi(r,t)$$

$$\frac{d^2}{dt^2}\psi(z) + \frac{2m}{\hbar^2}[E - V(z)]\,\psi(z) = 0$$

where $H = (p^2 / 2m) + V(r) = $ *Hamiltonian of the system*

$$\hat{E} \to \{(-\hbar / i)\,(\partial / \partial t)\} = [i\hbar\,(\partial / \partial t)]$$

$$\Psi(z,t) = \frac{1}{\sqrt{2\pi}}\int \Phi(k_z)\, e^{\,i(k\,z\,-\,\omega\,t)}\,dk_z$$

$$\psi(z,t) = \psi(z) \cdot f(t)$$

$$E = (n + \tfrac{1}{2})\,\hbar\omega; \qquad n = 0,1,2,3,\ldots$$

Discrete / quantized levels

$$E = (\hbar / i)\frac{\partial}{\partial t} \text{ is the } \textit{energy}$$

39.7.2 **POSTULATES**
39.7.2.1 $\Psi(r,t)$ is the wave function of the system

(a) *FINITE* function everywhere ('Finite' means $\Psi(\bar{r},t)$ at the boundary points, to give energy values in conformity with the experimental results),

(b) *SINGLE-VALUED FUNCTION* [*i.e.*, $\Psi(\bar{r},t)$ has unique value], and

(c) *FUNCTION CONTINUITY*: Function $\Psi(\bar{r},t)$ must be *CONTINUOUS* everywhere of the 'configuration space' (*x, y, z*) of the system, in a region or bound space, under consideration. This is true only if the following so-called BOUNDARY CONDITIONS are satisfied:

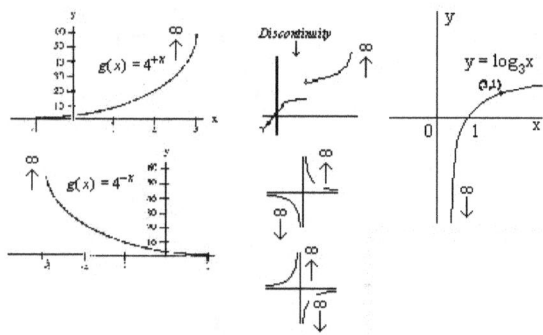

NOT Well-behaved Functions

(i)　　Amplitude continuity: Every $\Psi(\bar{r},t)$ must be continuous function of space,

(ii)　　Slope and curvature continuities: Derivatives of $\Psi(\bar{r},t)$, *viz.* $\Psi'(\bar{r},t)$, (*i.e.* the slope of $\Psi(\bar{r},t)$), and $\Psi''(\bar{r},t)$ (*i.e.* curvature of $\Psi(\bar{r},t)$), with respect to spatial co-ordinates, r, must also be continuous functions of r, for all r, *i.e.* everywhere, except where the potential $V(r)$ is finite. $\Psi(\bar{r},t)$ with $+\Psi''(\bar{r},t)$ looks like concave downward, and one with $-\Psi''(\bar{r},t)$ looks like convex upward. $\Psi''(\bar{r},t)$ is proportional to the amplitude of $\Psi(\bar{r},t)$. $\Psi(\bar{r},t)$ must be 'twice differentiable'.

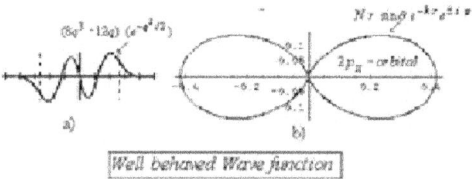

Well behaved Wave function

(iii)　　$\Psi(\bar{r},t)$ *must VANISH at infinity*, i.e. $\Psi(\bar{r},t) \to 0$, as $x, y, z \to \pm\infty$.

(iv)　　Born's Normalization condition, equation

$$\iiint\limits_{Entire\ space} |\hat{\Psi}(\bar{r},t)|^2\ d\tau = 1$$

$$|\Psi(\bar{r})|^2\ d\tau = \Psi(\bar{r})^* \cdot \Psi(\bar{r})\ d\tau$$

39.7.2.2　The motion of a particle, described by a wave function in Cartesian space

$\Psi(r, t)$, is also equally described by the wave function $\Phi(\vec{p}, t)$ in *momentum space* or $\Phi(\vec{k}, t)$ in *Fourier space*. If

$$\hat{\Psi}(\vec{r}, t) = \left(\frac{1}{\sqrt{2\pi}}\right)^3 \iiint \Phi(\vec{k}) \, e^{i \, (\vec{k} \cdot \vec{r} - \omega t)} dk$$

(4.2.4)

$$\Phi(\vec{k}, t) = \left(\frac{1}{\sqrt{2\pi}}\right)^3 \iiint \hat{\Psi}(\vec{r}) \, e^{-i \, (\vec{k} \cdot \vec{r} - \omega t)} d\tau$$

39.7.2.3 | TO EVERY OBSERVABLE, q THERE CORRESPONDS AN OPERATOR, $\hat{Q}$ |

$$\hat{p}_x \rightarrow [-i\hbar \, (\partial / \partial x)] = [-i\hbar \, \hat{\nabla}_x]$$

Momentum, $\hat{p}_z^2 \rightarrow -\hbar^2 \hat{\nabla}_z^2$ and

Energy, $\hat{E} \rightarrow (i\hbar) \, (\partial / \partial t)$

39.7.2.4 $$< \hat{Q} > \; = <q> = \left(\frac{\iiint \Psi^* \, \hat{Q} \, \Psi \, d\tau}{\iiint \Psi^* \, \Psi \, d\tau} \right)$$

39.7.2.5 $$\hat{Q} \, \psi_n = q_n \, \psi_n \; ,$$

$$q_n^* = q_n$$

$$\int \psi_m^* \psi_n \, dx = 0$$

$$\int (\hat{F} \psi_1)^* \psi_2 \, dx = \int \psi_1^* \, \hat{F} \psi_2) \, dx$$

$$\Psi(\vec{r}) = \sum c_i \, \hat{\psi}_i(\vec{r}) \neq 0$$

$$c_i = \int \hat{\psi}_i^* (\vec{r}) \, \Psi(\vec{r}) \, d\tau$$

$$\hat{Q} \left(\sum c_i \, \hat{\psi}_i(\vec{r}) \right) = q \left(\sum c_i \, \hat{\psi}_i(\vec{r}) \right)$$

39.7.2.6 <u>TDSE</u> (Time Dependent Schrodinger Equation)

$$(\frac{\partial^2}{\partial x^2} + \frac{\partial^2}{\partial y^2} + \frac{\partial^2}{\partial z^2})\Psi(\vec{r}, t) + \frac{2 \, m}{\hbar^2} (E - V)\Psi(\vec{r}, t) = 0$$

39.7.3 Ehrenfest Theorem

$$m \frac{d<x>}{dt} \; = \; < \hat{p}_x > \; = \; \int [\Psi^* \, (-i \, \hbar \nabla)\Psi] \, dx$$

39.7.4 Commutation Rules

$$[\hat{Q}, \hat{R}] = [\hat{Q} \hat{R} - \hat{R} \hat{Q}) \rightleftarrows \hbar \{ q, r \}$$

$$\Delta p_z = \sqrt{\left(< p_z^2 > - < p_z z >^2 \right)}$$

$$(\Delta q \cdot \Delta p_q) \geq \hbar / 2$$

39.8.1 **PARTICLE IN A BOX**:

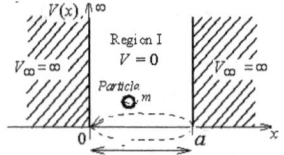

$V(x) = 0; \quad 0 < x < a),$
$V(x) = \infty; \; (x \le 0, x \ge a)$; *Square / rectangular potential*

Discrete / quantized levels

$$E_n = (\hbar^2 \pi^2 / 2m) \left(\frac{n^2}{a^2}\right); \quad n=1, 2, 3, ..$$ are <u>non-degenerate</u>.

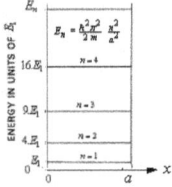

39.8.2 HARMONIC OSCILLATOR

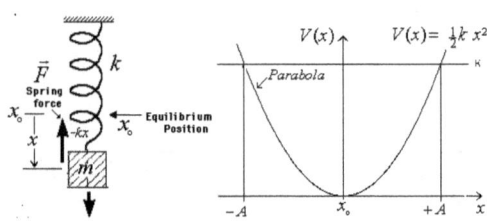

$$V(x) = \begin{cases} \frac{1}{2} k\, x^2, & |x| > 0, \\ 0. & x = 0. \end{cases}$$

$V(x) = \frac{1}{2} kx^2$; *Parabolic potential well*

$\omega = \sqrt{k/m}$

$$\hat{H} = \frac{p^2}{2m} + \frac{1}{2} m\omega^2 x^2$$

$$E_n = (n + \frac{1}{2}) \hbar\omega; \quad n = 0, 1, 2, 3,...$$

Zero-point energy,

$$E_0 = \frac{1}{2}\hbar\omega$$

$$\psi_n(q) = b^{1/4} \; [\pi^{\frac{1}{2}} \, 2^n n! \;]^{-\frac{1}{2}} \, H_n(q) \; (e^{-q^2/2})$$

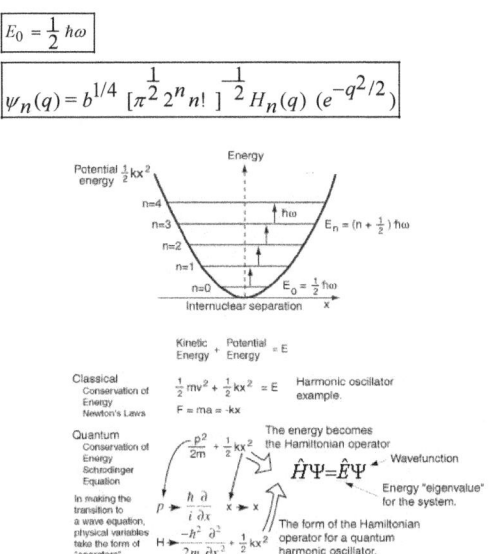

39.8.3. The Beginning of QED (Quantum Electro Dynamics)

By consistently applying quantum mechanics to Maxwell's EM Theory Dirac constructed the first known specimen of QFT (Quantum Field Theory).

39.8.4. SCHRODINGER'S CAT and Quantum Reality
39.8.4.1. ' IN RE ' Nature

The key problem in the theory of measurement is the reduction of the wave function and, in particular, the question of when it takes place. This problem is illustrated quite drastically by means of the "Schrödinger's cat".

Suppose that an ampoule of cyanide gas, a relay mechanism to break the ampoule and a Geiger Counter with radioactive atoms of $\tau_{1/2} = 1$ $hr.$ such that within one hour there is a $50 - 50$ chance of the gas being released from the ampoule. This device and the cat in a sealed box (let cat is alive in cyanide gas). Then after I hr there is equal chance of the cat being alive or dead.

Quantum mechanically, there is no way to achieve a $|\psi>$ for a living or dead cat. The cyanide gas, represented by $| \uparrow \; \& \downarrow >$, $i.e.$, $spin$-up and $spin$-$down$ particles, in a closed chamber with cat, is such that the $|\uparrow>$ gas particles would kill the cat, whereas the$|\downarrow>$ particles would not. Considering the effect of $|\uparrow>$ and $|\downarrow>$, which is analogous to the production of the $|\uparrow>$ and $|\downarrow>$ particles in the Stern-Gerlach experiment. Suppose the particle in the state $|\uparrow\&\downarrow>$ hits the cat, the state of spin and the cat makes a transition to $|\uparrow>|$ Dead cat $>$ + $|\downarrow>|$ Living cat $>$ - pure state. When is it decided whether the cat is '*dead*' or '*alive*'? The answer is just when the observer opens the box! – An objective statement, independent of the conscious mind of the observer would be impossible. – What is the state of the observer himself in the quantum mechanical description?

According to the point of view just presented, the cat (together with the mechanism for killing it, *i.e.* cyanide particles) is linked to other macroscopic objects. These are influenced differently in the final states so that their respective wave functions do not overlap. For everything that follows, these macroscopic consequences are not recorded; the trace (like $Tr\ \rho = Tr\ A(n)\ \rho(t)$ is taken over them. The final state of the cat is described by a mixture of states corresponding to dead cat and living cat: the cat is either dead or living and not in a pure state, | Dead cat $>$ + | Living cat $>$, which would include both possibilities. This is very counterintuitive but fundamental to quantum reality.

In many considerations people try to pass the intrinsic problems of uncertainty away on the basis that in the large real processes we witness individual quantum uncertainties cancel in the law of averages of large numbers of particles. Chaotic processes are potentially able to inflate arbitrarily small fluctuations, so molecular chaos may inflate the fluctuations associated with quantum uncertainty.

39.8.4.2. '*IN VOCE*' NATURE: "The Unexpected Hanging":

A prisoner is sentenced on one Sunday to be hanged at noon on one of the following 7 days and to be kept ignorant as to which day it will be until that day arrives. Prisoner says this will never happen – since he is alive till Saturday. Reasoning backwards this way he concludes that he cannot be hanged on any day (as per the sentence). Nevertheless he was hanged on Friday!

39.8.4.3. The EPR Paradox

Einstein, with his younger colleagues Boris Podolsky and Nathan Rosen in New Jersey, developed another challenge to Bohr that was not based on the Uncertainty principle, but on EPR Paradox. Accordingly, it is possible to obtain a pair of particles (say, electrons) A & B in a so-called singlet state where their spins cancel each other to give a total spin zero. If the A & B move widely apart, the spin of A along one direction is measured to be seen as "UP", it follows that the spin of B must have been "DOWN". This is not a problem to Classical Mechanics.

39.8.4.4. Bell's Inequality Theorem

Thirty years after EPR John S. Bell developed an ingenious 'Inequality Principle' to test the questions raised by EPR. Instead of electrons, his test was based on photons which has polarization instead of spin. But the principles are the same. The question remains – How do changes in A affect B? Einstein condition of locality was accepted as true. But experiments violated the Inequality, and gave the interpretation that nature is 'non-local'.

Experiments by John Clauser (1978) and Alain Aspect's group (1982) indicated experimental verification of the violation of Bell's inequality.

This means that in spite of the local appearances of phenomena, our world is actually supported by an INVICIBLE REALITY which is unmediated and allows communication FASTER THAN LIGHT, even **instantaneously**.

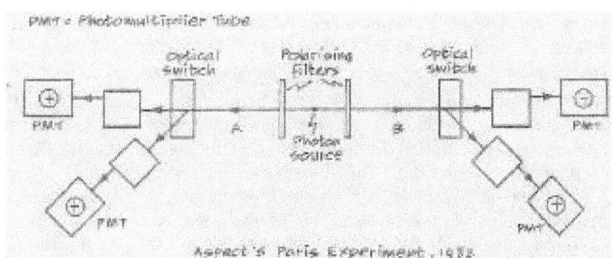

Aspect's Paris Experiment, 1982.

Interactions under Non-Local Reality
1. The interaction does not diminish with distance.
2. It can act instantaneously (faster than the speed of light).
3. It links up locations without crossing space.

The popular examples on non-locality are "Mind speed of humans" and extra-sensory perception.

39.8.5. QUANTUM INFORMATION: QUBITS (Quantum Bits):

By the 1990s advances in technology and experimental techniques led to a NEW PHYSICS (**Modern Physics** of the second half of the 20th Century) enable many of the thought experiments of the 1920s and 1930s to be carried out in the laboratory and opened the door to a variety of intriguing applications using the same counter intuitive ideas that challenged Einstein. Examples include *Quantum Cryptography*, **Quantum Computing**, *Quantum Teleportation* and Non-Interactive Measurements. As Steve Adams has put it, I quote; it is possible that we shall come to rely on quantum theory to secure our financial transactions and to allow rapid computations on **Super Quantum Computers**, or even to form images of objects we have never observed!

39.8.5.1. Quantum Computing

Quantum computing requires a 'quantum logic gate'. In conventional digital computing, data bits are zeros and ones, which must be stored. Any 'two-state device', such as on-off wall switch, could accomplish the task. Quantum Computing, however, uses **QBITS** in which the state is really a combination of two states. One of the ways being studied to produce a quantum logic gate uses a beryllium ion in an rf ion trap, the spin of the lone valence electron being the pair of states, either up or down *(Phys. Rev. Lett.*, Dec. 18, 1995, p 4714) *"Quantum Chaos"*.

The interface between chaos and Quantum Mechanics has become an area of challenge and interest because of the new ideas about the nature of chaos, introduced by quantum smoothing. It turns out that confined quantum systems from the nucleus to magnetically excited Atomic Orbitals (AOs) which should display chaos demonstrate a variety of subtle forms of repression of chaos which separates the energy levels, and converges in probability to the periodic repelling orbits called hidden in any chaotic system. Orbits with time thus tend to end on these periodic solutions. This phenomenon is called Quantum-'Scarring' of the wave function.

These constraints however begin to evaporate as soon as one leaves confined systems and begin to enter the domain of unbound systems, such as electrons traversing a free molecular medium. This raises the distinct possibility that quantum chaos expressed in biological systems is right at the transition between the classical and quantum worlds. Chaotic systems may thus be able to amplify quantum effects into global fluctuations.

A key idea here is the "**square root of not**". Suppose we know an atom is excited by a certain amount of energy, but only shine a laser on it for half the time needed to provide this energy. Then the atom is in a superposition of the ground state and the excited state. If one then collapses the wave function, squaring it to its probability, as in $P = \Psi^* \Psi$, it will be found to be in either the ground state or excited state with equal probability. One can use this uncertainty to perform a quantum calculation in the following way. Suppose one has a collection of such atoms which effectively form the 0s and 1s of a binary number 0 in the ground state and 1 in the excited state. If one then **partially** excites them all by giving them an energy only part way to the excited state, they each enter a superposition of excited and non-excited states and represent a superposition of all the binary numbers - *e.g.* for **two atom 'bits'** - 00, 01, 10 and 11.

One can now devise a problem to solve - decrypting by factorizing a large number. One has a number register in two parts. The left part L is excited to a superposition. The right half R is designed to give the results of a *quantum factorization remainder* of each of the possible numbers in L. These turn out to be periodic, so if we measure R we get one of the values.

To illustrate

a) Choose a random number x between 0 and n, then
b) Raise it to the power of the number in the L register.
c) Divide by n, and
d) Place the remainder in a second register.

It turns out that for increasing powers of x, the remainders form a repeating sequence. Because the number in the first register is different in each universe the result varies from universe to universe.

39.8.6. MATRIX MECHANICS (Mathematical Formulism)
LINEAR INDEPENDENCE

If λ, μ, v are 3 *non-zero* scalars such that given the three vectors of the set $V_3(F)$, *viz.*,$\vec{\alpha}$, $\vec{\beta}$, $\vec{\gamma}$ in the field F satisfy

$$\lambda \,\vec{\alpha} + \mu \,\vec{\beta} + v \,\vec{\gamma} = 0$$

then the set of vectors $\vec{\alpha}$, $\vec{\beta}$, $\vec{\gamma}$ are said to be <u>linearly independent set</u> in the field F, except for the trivial case,

$$\lambda = \mu = v = 0 .$$

If λ, μ, v are not of such a set, then

$$\lambda = \mu = v \neq 0$$

$$\boxed{\begin{array}{l} \lambda \,\vec{\alpha} + \mu \,\vec{\beta} + v \,\vec{\gamma} = 0 . \\ \lambda = \mu = v \neq 0 \end{array}} \text{ – linearly independent set}$$

This means $\vec{\alpha}$, $\vec{\beta}$, $\vec{\gamma}$ are said to be a set of <u>linearly independent vectors,</u> in F. In other words, none of these three vectors can be written as linear combinations of the other two. If a scalar a_{ii} is such that $a_{ii} = (\psi_i , \psi_i)$, then the vectors are linearly independent if the **gram determinant**, $|\Gamma| > 0$.

If $\qquad \lambda \,[\vec{\alpha} \,\vec{\beta} \,\vec{\gamma}] = \lambda \,\vec{\alpha} \bullet (\vec{\beta} \wedge \vec{\gamma})$

$$= \begin{vmatrix} \alpha_\lambda & \alpha_\mu & \alpha_v \\ \beta_\lambda & \beta_\mu & \beta_v \\ \gamma_\lambda & \gamma_\mu & \gamma_t \end{vmatrix}$$

then $\qquad |\Gamma| = [\vec{\alpha} \,\vec{\beta} \,\vec{\gamma}] > 0$

or, there is another way of dealing with linear independency of vectors.

Let $\vec{\delta} = (c_1\hat{u} + c_2\hat{v} + c_3\hat{w}) = 0$, means $\hat{u}, \hat{v}$ and $\hat{w}$ are linearly dependent; *i.e.* the *Wronskian determinant*, W

$$|W| = \begin{vmatrix} \bar{u} & \bar{v} & \bar{w} \\ \partial u/\partial x & \partial v/\partial y & \partial w/\partial z \\ \partial^2 u/\partial x^2 & \partial^2 v/\partial y^2 & \partial^2 w/\partial z^2 \end{vmatrix} = 0$$

39.8.6.2   Theorem # 1

The set of vectors $\vec{\alpha}$, $\vec{\beta}$, and $\vec{\gamma}$ are said to be <u>linearly dependent set</u>, then

$$\boxed{|\Gamma| = [\vec{\alpha}\ \vec{\beta}\ \vec{\gamma}] = 0}$$.

and $\vec{\alpha}$, $\vec{\beta}$, and $\vec{\gamma}$ are <u>coplanar</u> *with the origin.*

On the other hand, $\vec{\alpha}$, $\vec{\beta}$, and $\vec{\gamma}$ are **non-coplanar** *with the origin*, they are said to form a set of **linearly independent** vectors. $|\Gamma| > 0$, *i.e.*,

$$|W| = [\vec{\alpha}\ \vec{\beta}\ \vec{\gamma}] = \begin{vmatrix} \alpha_\lambda & \alpha_\mu & \alpha_\nu \\ \beta_\lambda & \beta_\mu & \beta_\nu \\ \gamma_\lambda & \gamma_\mu & \gamma_t \end{vmatrix} \neq 0$$

or $$\boxed{\begin{matrix} \bar{r} = \lambda\,\vec{\alpha} + \mu\,\vec{\beta} + \nu\,\vec{\gamma} \neq 0 \\ \lambda = \mu = \nu \neq 0 \end{matrix}}$$ – linearly independent set & Non-coplanar (11.2.27)

39.8.6.3. DIMENSIONS OF A VECTOR SPACE: Theorem # 2

A vector space is said to be three-dimensions (3-D) if it contains precisely three linearly independent vectors. Then if $\vec{\alpha}$, $\vec{\beta}$, and $\vec{\gamma}$ are linearly independent triad (set V_3) then any other (4th) vector **r** may be written as

$\bar{r} = \lambda\,\vec{\alpha} + \mu\,\vec{\beta} + \nu\,\vec{\gamma} \neq 0$,

such that

$$\boxed{\begin{matrix} \lambda = [\bar{r}\ \vec{\beta}\ \vec{\gamma}]/[\vec{\alpha}\ \vec{\beta}\ \vec{\gamma}] \\ \mu = [\bar{r}\ \vec{\gamma}\ \vec{\alpha}]/[\vec{\alpha}\ \vec{\beta}\ \vec{\gamma}] \\ \nu = [\bar{r}\ \vec{\alpha}\ \vec{\beta}]/[\vec{\alpha}\ \vec{\beta}\ \vec{\gamma}] \end{matrix}}$$

But the four vectors $\bar{r}$, $\vec{\alpha}$, $\vec{\beta}$, and $\vec{\gamma}$ are linearly dependent in 3- space. $\vec{\alpha}$, $\vec{\beta}$, and $\vec{\gamma}$ are known as <u>3-vectors</u> (*i.e.* 3-D vectors).

39.8.6.4.   Theorem # 3 : Reciprocal Vectors

If $\vec{\alpha}$, $\vec{\beta}$, and $\vec{\gamma}$ are linearly independent triad (set $V_3(F)$) then there exists a <u>reciprocal triad,</u> $\vec{\alpha}'$, $\vec{\beta}'$, and $\vec{\gamma}'$ defined by

$$\boxed{\begin{matrix} \vec{\alpha}' = [\vec{\beta} \wedge \vec{\gamma}] / [\vec{\alpha}\ \vec{\beta}\ \vec{\gamma}] \\ \vec{\beta}' = [\ \vec{\gamma} \wedge \vec{\alpha}] / [\vec{\alpha}\ \vec{\beta}\ \vec{\gamma}] \\ \vec{\gamma}' = [\ \vec{\alpha} \wedge \vec{\beta}] / [\vec{\alpha}\ \vec{\beta}\ \vec{\gamma}] \end{matrix}}$$

39.8.6.5.   BASIS (Coordinate System) TRIAD OF VECTORS

39.8.6.6.   PROPERTIES OF BASIS VECTORS: linear independency

A basis of a vector space is set of linearly independent vectors such that every vector in $V_3(F)$ is a linear combination of vectors in the basis.

i.e., $[\hat{i}\ \hat{j}\ \hat{k}] \neq 0$

In addition to the linear independency among basis vectors, there are three other properties. The introduction of the scalar product of vectors permits one to 'metrize' the vector space, i.e. to define length and distance. It is desirable to have the LENGTH (*i.e.* positive square root of norm or absolute magnitude) of a vector $\vec{X}$ is denoted as $\|\vec{X}\|$

$\|\vec{X}\| = +(\vec{X},\vec{X})^{1/2}$ = a real number with positive sign.

3N-Sace

$$\hat{e}_1 = \{1, 0, 0, \ldots\ldots 0, 0\ \}$$
$$\hat{e}_2 = \{0, 1, 0, \ldots\ldots 0, 0\ \}$$
$$\hat{e}_3 = \{0, 0, 1, \ldots\ldots 0, 0\ \}$$
$$\bullet$$
$$\bullet$$
$$\hat{e}_n = \{0, 0, 0, \ldots\ldots 0, 1\ \}$$

in $V_n(F)$. They are such that

$$\boxed{(\hat{e}_i,\ \hat{e}_j) = \delta_{ij}}\ .$$

Any arbitrary vector is expressed as

$$\boxed{\vec{X} = (x_1\hat{e}_1\ +\ x_2\ \hat{e}_2\ +\ \bullet+\bullet\qquad +\ x_n\ \hat{e}_n)}$$

Completeness Relation is in Hilbert Space is $\underset{n\to\infty}{Lt}\ \|\vec{\Psi} - \sum_1^n a_i\ \hat{e}_i\| = 0$.

PROPERTIES of a Hilbert space, $\mathfrak{A}$

(1) The space is linear; *i.e.* if a = constant and φ is any element of the space, then $a\varphi$ is also an element of the space, addition of any two elements of the space is also an element of the space.

(2) There is an inner product, (φ,Ψ), for any two elements φ and Ψ in the space.

(3) For functions, defined in any interval $a \leq x \leq b$.

$$(\Psi,\Psi) = \|\Psi\|^2 = \int_a^b \Psi(x)^* \Psi(x)\ dx = \text{finite, always.}$$

(4) An element of $\mathfrak{A}$ of has a norm, that is related to the inner product,

$$(\text{norm of }\Psi)^2 = \|\Psi\|^2 = (\Psi,\Psi) .$$

(5) $\mathfrak{A}$ is complete.

Unitary Matrix, U:

If the matrix A has complex elements such that A^{-1} and adj A are identical, then A is called

unitary, U matrix. $\therefore$ $\boxed{U^\dagger = U^{-1} = (U^*)'}$

$U^\dagger\ U^{-1} = \mathfrak{I} = U^{-1}\ U^\dagger$

If A is a (n x n) matrix, $\boxed{\det(-A) = (-1)^n\ \det A}$

39.8.6.7. DIRAC'S KET AND BRA NOTATION: (Dirac bracket notation)

$(\Psi_a, \Psi_b) \neq (\Psi_b, \Psi_a)$, but $\boxed{(\Psi_a, \Psi_b) = (\Psi_b, \Psi_a)^*}$

Although the absolute values of the two products are the same,

i.e., $|(\Psi_a, \Psi_b)| = |(\Psi_b, \Psi_a)|$

$\boxed{(\Psi_a, a\,\Psi_b) = a\,(\Psi_a, \Psi_b)}$

$\boxed{(a\,\Psi_a, \Psi_b) = a^*(\Psi_a, \Psi_b)}$

$\boxed{(\Psi_c, a\,\Psi_a + b\,\Psi_b) = a\,(\Psi_c, \Psi_a) + b\,(\Psi_c, \Psi_b)}$

Dirac KET $\quad \Psi_j(x) \equiv |\,j\,\rangle$

Dirac BRA $\quad (\Psi_j(x)) \equiv \langle\,j\,|$

Dual vector Space $\quad$ **Bra $-$ c $-$ ket** symbol $\langle a \| b \rangle$ = a scalar quantity

$\langle a \| b \rangle = (\Psi_a, \Psi_b) \equiv \langle a \mid b \rangle$

Hermitian Adjoint Operator (Observable),

$|\,b\,\rangle = \hat{A}\,|\,a\,\rangle \longleftrightarrow \langle\,b\,| = \langle\,a\,|\,\hat{A}^\dagger$

Eigen value Equation $\quad \boxed{\hat{\Omega}\,|\,j\,\rangle = \Omega_j\,|\,j\,\rangle}$

$|\,\Phi_j\,\rangle = \sum c_{ij}\,|\,\hat{e}_j\,\rangle = \sum c_{ij}\,|\,\hat{i}\,\rangle$,

where, $\boxed{c_{ij} = \langle\,\hat{e}_i \| \Phi_j\,\rangle = \langle\,\hat{i} \| \Phi_j\,\rangle}$.

Orho-normality Relation

$\boxed{\langle\,\psi_m \| \psi_n\,\rangle = 0}$

$$|\,\Phi_j\,\rangle \equiv |\,j\,\rangle = \begin{pmatrix} c_{11} \\ c_{21} \\ c_{31} \\ \cdot \\ \cdot \\ \cdot \\ c_{n1} \end{pmatrix} = \begin{pmatrix} \langle\,1 \| j\,\rangle \\ \langle\,2 \| j\,\rangle \\ \langle\,3 \| j\,\rangle \\ \cdot \\ \cdot \\ \langle\,\bar{n} \| j\,\rangle \end{pmatrix} \equiv \{\langle\,1 \| j\,\rangle, \langle\,2 \| j\,\rangle, \cdot, \cdot, \cdot, \cdot, \langle\,\hat{i} \| j\,\rangle\}$$

A ket vector is a <u>column vector.</u>

$\boxed{\langle\,j\,| \equiv \langle\,\Phi_j\,| = \text{A Row Vector.}}$

Closure Relation $\quad \boxed{\langle\,\hat{e}_i \| \hat{e}_i\,\rangle = \langle\,\hat{i} \| \hat{i}\,\rangle = \delta_{ii} = 1}$

Vector $\quad b_i = \sum \hat{\Omega}_{ij}\,a_j \quad$ im matrix form is

$$|\,b\,\rangle = \hat{\Omega}\,|\,a\,\rangle = \begin{pmatrix} b_1 \\ b_2 \\ \cdot \\ b_N \end{pmatrix} = \begin{pmatrix} \Omega_{11} & \cdots & \Omega_{1N} \\ & & \\ \vdots & & \vdots \\ \Omega_{N1} & \cdots & \Omega_{NN} \end{pmatrix} \begin{pmatrix} a_1 \\ a_2 \\ \cdot \\ a_N \end{pmatrix}$$

i.e., $|b\rangle = \hat{\Omega} |a\rangle$.

39.8.6.8.   Matrix Representation of operator

$$\hat{\Omega} = \left(\hat{\Omega}_{ij}\right) = \begin{pmatrix} \Omega_{11} & \Omega_{12} & \cdot & \cdot & \cdot & \Omega_{1N} \\ \hat{\Omega}_{21} & \Omega_{22} & \cdot & \cdot & \cdot & \hat{\Omega}_{2Nj} \\ \hat{\Omega}_{31} & \hat{\Omega}_{32} & \cdot & \cdot & \cdot & \hat{\Omega}_{3Nj} \\ \vdots & & \cdot & \Omega_{ij} & \cdot & \cdot & \vdots \\ \Omega_{N1} & \Omega_{N2} & \cdot & \cdot & \cdot & \Omega_{NN} \end{pmatrix}$$

Matrix element $\hat{\Omega}_{ij}$ is $\boxed{\hat{\Omega}_{ij} = \langle \hat{i} | \hat{\Omega} | \hat{j} \rangle = \langle \hat{j} | \hat{\Omega} | \hat{i} \rangle^{*}}$

Spectrum of an Observable operator $\hat{\Omega}$, such that $\hat{\Omega} | \hat{i} \rangle = \lambda_i | \hat{i} \rangle$.

The totality of the eigen values of $\hat{\Omega}$ is called the <u>Spectrum</u> of $\hat{\Omega}$.

If $|\alpha\rangle = \{x_1, x_2, x_2, \cdot \bullet, \bullet \bullet \bullet, \bullet \bullet, x_n\}$,

the set of equations can be written in short hand form, in the matrix algebra, as

$$\begin{bmatrix} x_1 \\ x_2 \\ x_3 \\ " \\ " \\ x_n \end{bmatrix} = \begin{bmatrix} a_{11} & a_{12} & a_{13} & \bullet \bullet \bullet \bullet \bullet & a_{1n} \\ a_{21} & a_{22} & a_{23} & \bullet \bullet \bullet \bullet \bullet & a_{2n} \\ a_{31} & a_{32} & a_{33} & \bullet \bullet \bullet \bullet \bullet & a_{3n} \\ " \\ " \\ a_{n1} & a_{n2} & a_{n3} & \bullet \bullet \bullet \bullet \bullet & a_{nn} \end{bmatrix} \begin{bmatrix} x^o_1 \\ x^o_2 \\ x^o_3 \\ " \\ " \\ x^o_n \end{bmatrix}$$

Or, $\boxed{|\alpha\rangle = \hat{A} |\alpha^o\rangle}$

39.8.6.9.   TRANSFORMATION OF BASIS (or Representation)

Let the coordinate equation $|\Psi_b\rangle = \hat{S} |\Psi_a\rangle$

describes any linear operator $\hat{S}$ transforming a vector $|\Psi_a\rangle$ into another vector, $|\Psi_b\rangle$.

UNITARY-SIMILARITY TRANSFORMATIONS and Operators transforming under a change of basis.

Let $\{|\hat{e}_i^o\rangle\}$ and $\{|\hat{e}_i\rangle\}$ represent the OLD and NEW sets of basis vectors in n-space.

$\langle \hat{e}_i^o \| \hat{e}_j^o \rangle = \delta_{ij}$, and $\langle \hat{e}_k \| \hat{e}_\ell \rangle = \delta_{k\ell}$

$|\hat{e}_k\rangle = \sum |\hat{e}_i^o\rangle S_{ki}$, $(S_{ki})^* = \langle \hat{e}_i^o \| \hat{e}_k \rangle^* = (S_{ik})$, $|\hat{e}_\ell\rangle = \sum |\hat{e}_j^o\rangle S_{\ell j}$

Unitary matrix , $\hat{S} = \hat{U}$; and so $\hat{S}^\dagger \hat{S} = \hat{U}^\dagger \hat{U} = \hat{\Im}$,

So $|\Psi_a^o\rangle = \hat{U} |\Psi_a\rangle$.

The two operators, $\hat{X}$ and $\hat{X}^o$, connected by $\hat{U}$ a Unitary Transformation are said to be

UNITARY EQUIVALENTS, *i.e.,* $\boxed{\hat{X} = \hat{U} \hat{X}^o \hat{U}^\dagger}$.

The SCHMIDT ORTHOGONALIZATION Procedure for Degenerate system

39.8.6.10.   TYPES OF PICTURES (EQUATIONS OF MOTION)

The EVOLUTION OPERATOR, $\hat{T}(t, t_0) \Rightarrow \hat{U}(t, t_0)$

$$\hat{U}(t-t_0) = e^{\left\{-i\,\hat{H}\,(t-t_0)/\hbar\right\}}$$

If the system were <u>non-conservative</u>, then $\boxed{\hat{H} = \hat{H}(t)}$.

$$\hat{U}(t,t_0) \neq e^{\left\{-i\,\hat{H}\,(t-t_0)/\hbar\right\}}$$

$$\hat{U}(t,t_0) = 1 - \left\{(-i/\hbar)\right\}\int\hat{H}.\,\hat{U}(t',\,t_0)\,dt'$$

which are FUNDAMENTAL LAW of Evolution in Quantum Mechanics. *Equivalently* an expression for this law is the <u>Schrödinger Equation</u>.

S-Picture:
$$\frac{d}{dt}\left\langle\hat{\mathbb{Q}}_S\right\rangle = \left\langle (\partial/\partial t)\hat{\mathbb{Q}}_S \right\rangle + (1/i\hbar)\left\langle \left[\hat{\mathbb{Q}}_S,\hat{H}_S\right]\right\rangle$$

H-Picture
$$(i/\hbar\,d/dt)\,\hat{\mathbb{Q}}_H(t) = \left[\hat{H}_H,\hat{\mathbb{Q}}_H\right] + (i\hbar)(\partial/\partial t)\,\hat{\mathbb{Q}}_H$$

I-Picture (Dirac Picture):
$$\left| a',\,t \right\rangle_S = e^{\left\{-i\,\hat{H}_o\,t/\hbar\right\}}\left| a',\,t \right\rangle_I$$

TABLE 12.1   Comparison of S-, H-, and I- Pictures

S-picture   Basis vectors = fixed.

State vectors $\left| a't_o,\,t \right\rangle_S$ = move,

$$\left| a't_o,\,t \right\rangle_S = \hat{U}(t,\,t_o)\left| a't_o \right\rangle_S \qquad (12.3.4)$$

$$\left| a't_o,\,t \right\rangle_S = (i\hbar\,d/dt)\left| a't_o,\,t \right\rangle_S \qquad (12.3.5)$$

(S- Equation).

Operators $\hat{\mathbb{Q}}_S$ are fixed, *i.e.* $(\partial/\partial t)\,\hat{\mathbb{Q}}_S = 0$,

$$\left[\hat{\mathbb{Q}}_S,\hat{H}_S\right] = 0,\ (\partial/\partial t)\left\langle\hat{\mathbb{Q}}_S\right\rangle = 0$$

$$(d/dt)\left\langle\hat{\mathbb{Q}}_S\right\rangle = (1/i\hbar)\left\langle\left[\hat{\mathbb{Q}}_S,\hat{H}_S\right]\right\rangle$$

H-picture Basis vectors = move.

State vectors $\left| a't_o, t \right\rangle_H$ = fixed, $(\partial / \partial t) \left| a't_o, t \right\rangle_H = 0$

$$\left| \psi_i(r, t) \right\rangle_H = \hat{U}^\dagger(t) \left| \psi_i(r, t_o) \right\rangle_S \qquad (12.4.13)$$

$$(i\hbar \, \partial / \partial t) \left| q', t \right\rangle_H = -\hat{H}_H \left| q', t \right\rangle_H \qquad (12.4.28)$$

(H-Equation).

$$\hat{E} \left| q', t \right\rangle_H = -\hat{H}_H \left| q', t \right\rangle_H \qquad (12.4.29)$$

Operators $\hat{Q}_H$ are time dependent

$$\hat{Q}_H(t) = \hat{U}^\dagger(t, t_o) \, \hat{Q}_S(0) \, \hat{U}(t, t_o) \qquad (12.4.14)$$

$$(d/dt) \, \hat{Q}_H(t) = \left[\hat{Q}_H, \hat{H}_H \right] + (\partial / \partial t) \, \hat{Q}_H \qquad (12.4.24)$$

If $\hat{Q}_H \neq \hat{Q}_H(t)$, explicitly, then $(\partial / \partial t) \, \hat{Q}_H = 0$

$$(i\hbar)(d/dt) \, \hat{Q}_H(t) = \left[\hat{Q}_H, \hat{H}_H \right] \qquad (12.4.25)$$

I - picture Basis vectors = move.

State vectors $\left| a't_o, t \right\rangle_I$ = move, $(\partial / \partial t) \left| a't_o, t \right\rangle_I \neq 0$

$$\hat{H}^I(t) = \hat{H}_o(0) + \hat{H}_1(t) \qquad (12.5.1)$$

$$\left| a', t \right\rangle_I = \hat{U}^\dagger(t) \left| a', t \right\rangle_S \qquad (12.5.3)$$

$$\hat{Q}^I(t) = \hat{U}^\dagger(t, t_o) \, \hat{Q}_I(0) \, \hat{U}(t, t_o) \qquad (12.5.8)$$

$$(i\hbar \, \partial / \partial t) \left| a', t \right\rangle_I = \hat{H}^I(t) \left| a', t \right\rangle_I \qquad (12.5.14)$$

$$(i\hbar)(d/dt) \, \hat{Q}_I(t) = \left[\hat{Q}^I(t), \hat{H}_o^I \right] + (i\hbar)(\partial / \partial t) \, \hat{Q}^I(t) \qquad (12.5.12)$$

(Tomonaga - Schwinger Equation)

Operators $\hat{Q}^I(t)$ change their form in time, as determined only by $\hat{H}_o^I$.

39.9. **Ladder Operators** (Creation & Destruction Operators) (Raising & Lowering Operators) (Non-Hermitian Operators) *i.e.* dimensionless complex dynamic variables, $\hat{a}$ and $\hat{a}^\dagger$,

$$\boxed{\hat{a} = \frac{1}{\sqrt{2}}(\hat{q} + i \, \hat{p})}$$

$$\boxed{\hat{a}^\dagger = \frac{1}{\sqrt{2}}(\hat{q} - i \, \hat{p})}$$

39.8.7.1. Boson Harmonic Oscillator (BHO)
Harmonic oscillators are bosons.

TISE is $-\nabla^2 \psi(q) + [2\varepsilon - q^2] \, \psi(q) = 0$

becomes $\boxed{(\hat{a}\hat{a}^\dagger + \hat{a}^\dagger \hat{a}) \, \psi(q) = 2\varepsilon \, \psi(q)}$

$[\hat{a}, \hat{a}^\dagger] = 1$; $[\hat{a}, \hat{a}] = 0$; $[\hat{a}^\dagger, \hat{a}^\dagger] = 0$; $[\hat{H}, \hat{a}] = -\hbar\omega_c\hat{a}$; $[\hat{H}, \hat{a}^\dagger] = +\hbar\omega_c\hat{a}^\dagger$; $\hat{H} = \hbar\omega_c[\hat{a}\,\hat{a}^\dagger - \frac{1}{2}]$;

$\hat{H} = \hbar\omega_c[\hat{a}^\dagger\,\hat{a} + \frac{1}{2}]$

$[\hat{a}^\dagger\psi_o(q) = 0]$; $\boxed{\{\hat{a}^\dagger\psi_n(q)\} = \psi_{n+1}(q)}$; $\boxed{\{\hat{a}\,\psi_o(q)\} = 0}$;

$\boxed{\psi_o(q) = \sqrt[4]{\pi}\, H_o(q)\, e^{-q^2/2}}$; $\boxed{\psi_n(q) = \sqrt{\frac{1}{n!}}\, \{\hat{a}^\dagger\}^n \psi_o(q)}$ or $\boxed{\{\hat{a}^\dagger\}^n \psi_o(q) = \sqrt{n!}\,\psi_n(q)}$

$\boxed{\hat{a}\,\psi_n(q) = \sqrt{n}\,\psi_{n-1}(q)}$

$$\hat{H} = \tfrac{1}{2}\,\hbar\omega_c \begin{pmatrix} 1 & 0 & 0 & 0 & . & . & . & 0 \\ 0 & 3 & 0 & 0 & . & . & . & 0 \\ 0 & 0 & 5 & 0 & . & . & . & 0 \\ ' & & 0 & 7 & 0 & . & & 0 \\ & & & & 9 & & & 0 \\ & & & & & & & \\ 0 & & & & & & & 0 \\ 0 & & & & & & (2n+1) \end{pmatrix}$$

$\hat{q}_{n,n} = \langle\, n \mid \hat{q} \mid n\,\rangle = 0$

$$\hat{q} = \begin{pmatrix} 0 & \sqrt{\tfrac{1}{2}} & 0 & 0 & . & . & . & 0 \\ \sqrt{\tfrac{1}{2}} & 0 & 1 & 0 & . & 0. & . & 0 \\ 0 & 1 & 0 & \sqrt{\tfrac{3}{2}} & 0. & . & . & 0 \\ 0 & 0 & \sqrt{\tfrac{3}{2}} & 0 & \sqrt{2} & 0 & & 0 \\ 0 & 0 & 0 & \sqrt{2} & 0 & & & 0 \\ 0 & & & & & & & 0 \\ 0 & & & & & & & \sqrt{n-\tfrac{1}{2}} \\ 0 & & & & & & \sqrt{n-\tfrac{1}{2}} & 0 \end{pmatrix}$$

$$\hat{x} = \left(\frac{\hbar}{m\,\omega_c}\right)^{1/2} \begin{pmatrix} 0 & \sqrt{\tfrac{1}{2}} & 0 & 0 & . & . & . & 0 \\ \sqrt{\tfrac{1}{2}} & 0 & 1 & 0 & . & 0. & . & 0 \\ 0 & 1 & 0 & \sqrt{\tfrac{3}{2}} & 0. & . & . & 0 \\ 0 & 0 & \sqrt{\tfrac{3}{2}} & 0 & \sqrt{2} & 0 & & 0 \\ 0 & 0 & 0 & \sqrt{2} & 0 & & & 0 \\ 0 & & & & & & & 0 \\ 0 & & & & & & & \sqrt{n-\tfrac{1}{2}} \\ 0 & & & & & & \sqrt{n-\tfrac{1}{2}} & 0 \end{pmatrix}$$

$$(\hat{p}_q)_{n,k} = \langle\, n \mid \hat{p}_q \mid k \,\rangle = i\sqrt{\frac{n}{2}}\,\delta_{n-1,k} - \sqrt{\frac{n+1}{2}}\,\delta_{n+1,k}$$

$$\hat{p}_q = \begin{pmatrix}
0 & -i\sqrt{1/2} & 0 & 0 & \cdots & & & 0 \\
+i\sqrt{1/2} & 0 & -i & 0 & \cdot\,0\cdot & & & 0 \\
0 & +i & 0 & -i\sqrt{3/2} & 0 & \cdots & & 0 \\
0 & 0 & +i\sqrt{3/2} & 0 & -i\sqrt{2} & 0 & & 0 \\
0 & 0 & 0 & +i\sqrt{2} & 0 & 0 & & 0 \\
0 & & & & & & & \\
0 & & & & & & & -i\sqrt{n-1/2} \\
0 & & & & & & +i\sqrt{n-1/2} & 0
\end{pmatrix}$$

39.8.7.2. The Ladder Operators of a harmonic oscillator $\hat{a}$ and $\hat{a}^{\dagger}$

$$\hat{a}^{\dagger}_{k,n} = \langle\, k \mid \hat{a}^{\dagger} \mid n \,\rangle = \sqrt{\frac{n+1}{2}}\,\delta_{k,n+1}, \quad \hat{a}_{k,n} = \langle\, k \mid \hat{a} \mid n \,\rangle = \sqrt{n}\,\delta_{k,n-1},$$

$$\hat{a}^{\dagger}_{n,n} = \langle\, n \mid \hat{a}^{\dagger} \mid n \,\rangle = 0, \quad \hat{a}_{n,n} = \langle\, n \mid \hat{a} \mid n \,\rangle = 0$$

$$\hat{a} = \begin{pmatrix}
0 & \sqrt{1} & 0 & 0 & \cdots & & & 0 \\
0 & 0 & \sqrt{2} & 0 & \cdot\,0\cdot & & & 0 \\
0 & 0 & 0 & \sqrt{3} & 0\cdot & \cdots & & 0 \\
0 & 0 & 0 & 0 & \sqrt{4} & & & 0 \\
0 & 0 & 0 & 0 & 0 & & & 0 \\
0 & & & & & & & 0 \\
0 & & & & & & & \sqrt{n-1} \\
0 & & & & & & 0 & 0
\end{pmatrix}$$

$$\hat{a}^{\dagger} = \begin{pmatrix}
0 & 0 & 0 & 0 & \cdot\,0\cdot & & & 0 \\
\sqrt{1} & 0 & 0 & 0 & \cdot\,0\cdot & & & 0 \\
0 & \sqrt{2} & 0 & 0 & 0\cdot & & & 0 \\
0 & 0 & \sqrt{3} & 0 & 0 & \cdots & & 0 \\
0 & 0 & 0 & \sqrt{4} & 0 & & & 0 \\
0 & & & & \sqrt{5} & & & 0 \\
0 & & & & & & 0 & 0 \\
0 & 0 & 0 & 0 & & 0 & \sqrt{n-1} & 0
\end{pmatrix}$$

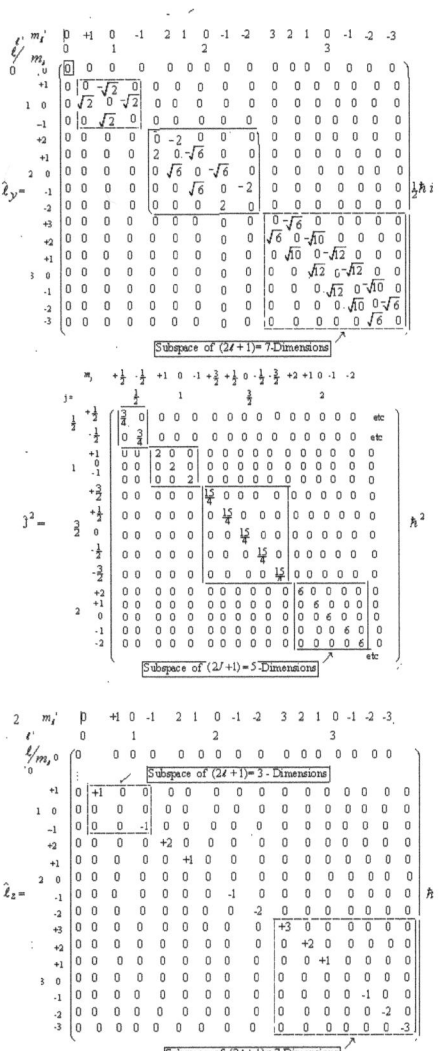

39.9.2. Second Quantization

First Quantization introduced the new constant, $\hbar$; second quantization does not introduce a new constant; hence it is best regarded as a mathematical manipulation. The latter is based originally on the *analogy* between the wave packets and the EM Field. In **Second Quantization**, this $\psi_n(q)$ is interpreted as an operator acting on an abstract space. $\psi_n(q)$ becomes <u>Annihilation Operator</u>, $\hat{a}$, and it on acting annihilates the state $\psi_n(q)$. Likewise $\hat{\psi}_n(q)^*$ becomes the <u>Creation Operator</u>, $\hat{a}^{\dagger}$.

For bosons a classical wave theory on quantization becomes a particle theory. In the same way, the quantum mechanical wave function for a fermion may be treated as a matter field variable and quantization the gives a particle theory for which creation and annihilation operators are appropriate.

$$[\hat{b}_r, b_s{}^{\dagger}]_+ = \delta_{rs} \; ;$$

$$[b_r, b_s]_+ = [b_r{}^{\dagger}, b_s{}^{\dagger}]_+ = 0 \, ,$$

where $[\hat{A}, \hat{B}]_+ = \hat{A}\hat{B} + \hat{B}\hat{A}$.

This procedure is known as <u>Second Quantization.</u>

+*+*+*+*+*+*+*+

Chapter 40

STATISTICAL MECHANICS

"Imagination is more important than knowledge"   Albert Einstein

40.1   Introduction

In a physical (Macro-) system composed of a large number of particles (atoms, molecules or other micro-particles),it is usually impossible to apply the basic physical laws directly to each particle. It is but proper to use a statistical approaching which one describes a distribution in various states in a statistical manner.

To predict macroscopic properties like Temperature T and pressure P:

i)   Apply the basic laws of mechanics to motion of individual particle,

ii)   Use the 'average' behavior to predict T & P.

40.1.1 Principal features of Molecular Kinetic Model:

i)   The gas consists of a very large number of molecules,

ii)   The molecules are identical entities,

iii)   The volume of the particles together forms negligibly small quantity, compared to the volume of motion

iv)   The molecules are constantly in motion

v)   The molecular collisions are perfectly elastic.

Pressure, P of a gas (of n moles) on the wall is the average force due to many collisions.

$$P = \frac{Nm <v_x^2>}{V}$$

$$P = \frac{1}{3}\frac{nN_A m <v^2>}{V}$$

40.1.2 Root Mean Square (RMS) velocity of molecules

$$v_{rms} = \sqrt{<v^2>} = \sqrt{\frac{1}{N} \sum_{i=1\text{to N}} v_i^2}$$

$$v_{rms} = \sqrt{\frac{3RT}{N_A\,m}} = \sqrt{\frac{3RT}{M}} = \sqrt{\frac{3k_B T}{m}}$$

Temperature, T of a gas is related to the average energy of the molecules.

40.1.3 Phase Space

The combined position and momentum of a particle is defined in a 6-D space called phase space.

40.1.4 Statistical Ensemble – Liouville's Theorem

It is the principle of conservation of density in phase space.

$w(X,t)$ = statistical ensemble's probability density in phase space.

40.1.5 Equation of Motion of a Statistical Ensemble:

$$\boxed{\frac{\partial w}{\partial t} = [H, w]}$$

The goal of statistical mechanics is to understand how the macroscopic properties of materials arise from the microscopic behaviour of their constituent particles.

Examples include:
• specific heat capacity and its variation with temperature;
• the entropy of a sample of material, and its relationship, with temperature and internal energy;
• the magnetic properties of materials.

40.1.6. The principle of equal a priori probabilities can be used on its own to derive some interesting results in specific cases

Statistical mechanics gives the relationship between the energy levels in two example cases:
• a collection of harmonic oscillators;
• a collection of magnetic dipoles in an external magnetic field.

Thermodynamics gives us the relationship between energy, temperature and entropy:

$$\boxed{\frac{\partial S}{\partial U} = \frac{1}{T}}$$

$$\frac{\partial \ln \Omega}{\partial U} = -\beta$$

$$\boxed{S = k_B \ln \Omega}$$

$$\beta = -\frac{1}{k_B T}$$

$$\boxed{k_B = 1.3806 \times 10^{-23} \, JK^{-1}}.$$

The Stirling approximation is

$$\boxed{\ln N! = N \ln N - N}$$

40.2   **The Maxwell distribution**

(Classical Statistical distribution of Distinguishable Particles)

Here | Identical particles of any spin are sufficiently widely separated to be distinguished. $(eg.,$ molecule of a gas$)$.

identical particles of any spin are sufficiently widely separated to be distinguished.$(eg.,$ molecule of a gas$)$.

40.2.1 Degeneracy, g_i

A system in which several physically distinguishable states having same energy is called DEGENERATE

in cell i.) Total number of microstates corresponding to a macrostate

$$\# = \boxed{n_i = N! \; \pi \frac{g^{n_i}}{n_i!}}$$

pq			pq	q	p	p	q		MB Statistics

where N = total number of molecules.

Total number of microstates corresponding a macrostate; n_i =A system in a collection of distinguishable particles follows the Boltzmann distribution for energies

$$n_i(\varepsilon)d\varepsilon = \frac{N}{Z}e^{-\varepsilon/k_BT}d\varepsilon = Ag_i\,e^{-\varepsilon/k_BT}d\varepsilon \,, \text{ or}$$

$$f_{MB}(E) = A\,e^{-\varepsilon/k_BT}$$

for having number of molecules with energy between ε and $(\varepsilon+d\varepsilon)$.
In the case of momentum,

$$n_i(p)dp = \frac{\sqrt{2\pi}N}{(\pi\,m\,k_B\,T)^{3/2}}\,e^{-p^2/2mk_BT}\,dp$$

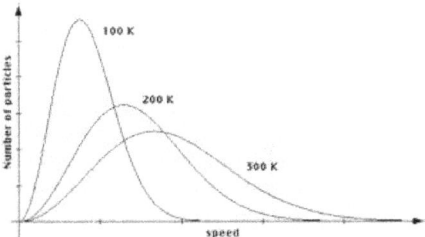

40.3. QUANTUM STATISTICS
(Identical Particles with spin or Indistinguishable Particles)
However, in systems consisting of collections of identical bosons and fermions or identical bosons, the wave function of the system has to be either antisymmetric (for fermions) or symmetric (for bosons) under interchange of any two particles. With the allowed wave functions, it is no longer possible to identify a particular particle with a particular energy state. Instead, all the particles are "shared" between the occupied states. The particles are said to be indistinguishable.

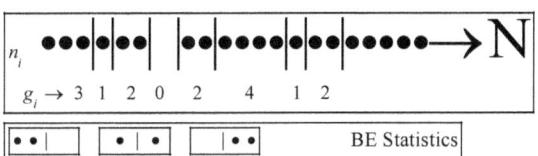

The n_i black balls and (g_i-1) partitions enough to provide g_i boxes.

Total number of particles and divisions possible in a row = $(n_i+g_i-1)!$

For all indistinguishable identical particles, total possible distinguishably different arrangements

$$\boxed{\frac{(n_i+g_i-1)!}{(n_i!)\,(g_i-1)!}} = \text{probability of N particle distributed}$$

$$= \boxed{P = \prod_{i=1}^{N} \frac{(n_i + g_i - 1)!}{(n_i!)\,(g_i - 1)!}}$$

40.4 BOSE-EINSTEIN DISTRIBUTION LAW

The most probable distribution can be written as the Bose-Einstein distribution for a collection of indistinguishable bosons

g_i = degeneracy of energy level ε_i

$$\boxed{n_i = g_i \frac{1}{A e^{\varepsilon_i / k_{Bi} T} - 1}} \text{ , or}$$

$$\boxed{f_{BE}(E) = \frac{1}{[A e^{\varepsilon / k_{Bi} T} - 1]}}$$

Obeyed by Bosons, which are integral spin (0 or 1) particles, like He nuclei, mesons, [4] He gas; H_2 gas.

40.5 FERMI-DIRAC DISTRIBUTION

This law applies to a collection of indistinguishable identical particles which are governed by the Pauli's Exclusion Principle, so that no two particles can be in the same dynamical state, *i.e.*, and the wavefunction of the whole system be anti-symmetric.

Such particle are called Fermions, and each has half-integral spin; electrons,

$$\boxed{n_i \le g_i}\; ;\; n_i \text{ occupied, and } g_i - n_i \text{ unoccupied.}$$

$$\boxed{{}^{g_i}C_{n_i} = P_i = \frac{g_i!}{(n_i!)\,(g_i - n_i)!}}$$

$$\boxed{P = \prod_{i=1}^{N} \frac{g_i!}{(n_i!)(g_i - n_i)!}}$$

$$\boxed{\begin{array}{|c|c|} \bullet & \bullet \end{array} \qquad \text{FD Statistics}}$$

i.e., the macrostates are NOT all equally likely.

$$\boxed{n_i = \frac{g_i}{[e^{(\varepsilon_i - \varepsilon_F)/k_B T} + 1]}} \text{ , or}$$

$$\boxed{f_{FD}(E) = \frac{1}{[e^{(\varepsilon_i - \varepsilon_F)/k_B T} + 1]}}$$

where ε_F = Fermi energy.

Chemical potential, μ

$$\mu =\varepsilon_F\left[1-\frac{\pi^2}{12}\left(\frac{kT}{\varepsilon_F}\right)^2-\frac{\pi^4}{80}\left(\frac{kT}{\varepsilon_F}\right)^4+\ldots\right]$$

FD statistics is successfully applied in the free electron theory of metals.

40.6   Comparison of MB, BE and FD Statistics

Comparision of MB, BE and FD Statistics			
	MB	BE	FD
1) Particles	Molecules in gas	Bosons	Fermions
2) Spin	any	Integral (0, 1)	$\frac{1}{2}$-spin
3) Name of particles	All mokecules	photons	Electrons
4) Law	$n_i=Ag_i\,e^{-\varepsilon/k_BT}$	$n_i=\dfrac{g_i}{[A\,e^{\varepsilon_i/k_{Bi}T}-1]}$	$n_i=\dfrac{g_i}{[e^{(\varepsilon_i-\varepsilon_F)/k_{Bi}T}+1]}$
5) Excusive Principle	Not Valid	Not valid	Valid
6) Applications	Specific heats of gases	a) Blackbody radiation b) Sp. heats of solids c) Gas degeneracy d) Bose condensation	a) Free electron s theory of metal b) Wtedeman-Franz law c) Photoelectric effect d) Paramagnetism e) Thermo-electric effect

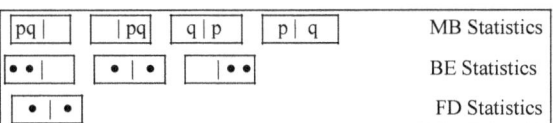

pq |	| pq	q | p	p | q	MB Statistics
• • |	• | •	| • •		BE Statistics
• | •				FD Statistics

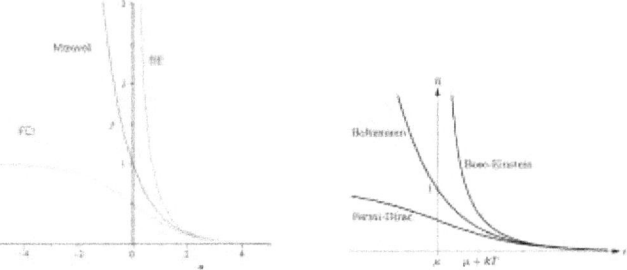

40.7.   Bose-Einstein Condensate (BEC) – New State of Matter

As early as 1924 the Indian physicist S. N. Bose carried out a statistical calculation for the kind of particles which have bear his name, bosons, as light particles termed photons. This led to an alternative derivation for the radiation law using Bose-Einstein statisics, predicting that when a given number of particles approach each other sufficiently closely and move sufficiently slowly they will together convert to the lowest energy state, now termed *Bose-Einstein condensation (BEC)* occurs.

This new fundamental state of matter, which was expected to have many interesting and useful properties, was achieved in 1995 by **Eric A. Cornell**, **Wolfgang Ketterle** and **Carl E. Wieman** (Nobel Prize, 2001). Manifestations of BEC have been observed in complicated systems: condensation of paired electrons in superconductors (loss of all electrical resistance) and suprafluidity (loss of internal friction in fluids. Spectacular demonstrations of BEC in for ^{87}Rb and ^{23}Na, small condensate of about 1000 atoms was obtained, exactly as predicted by the theory. BEC in dilute gases with the use of spin-polarized hydrogen and in meta-stable helium atoms are successful.

Revolutionary applications of BEC in lithography, nanotechnology and holography, *etc.*, appear to be just round the corner.

Chapter 41

SOLID STATE PHYSICS -2
ATOMIC BONDING, ENERGY BAND, BRILLOUIN ZONE, LATTICE DYNAMICS, DIELECTRICS, SEMICONDUCTORS, THERMAL EXPANSION, SPECIFIC HEAT, CRYSTAL DEFECTS, SUPERCONDUCTIVITY

|| Asesha-samklesa-samam vidhatte gunaanuvaada-sravanam Murare: |

Kuta: punastachharanaaravinda-paraaga-sevaaratiraatma-labdaa ||

(SirimadBhagavatam; Sk III, Ch 7, Sl 14)

41.1 ATOMIC BONDING

When the atoms are in equilibrium in a crystal lattice the minimum lattice potential energy

$$U = -\frac{A}{R^n} + \frac{B}{R^m}$$

R = Equilibrium separation between two neighboring atoms

Force,
$$F = -\frac{dU}{dR}$$

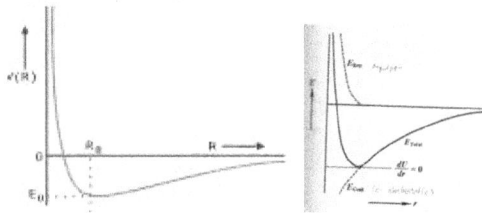

In equilibrium, there is minimum pE, when

$$R_{eq} = \left(\frac{mB}{nA}\right)^{1/(m-n)}$$

i.e.,
$$U_{eq} = U_{R=R_{eq}} = -\frac{A}{R_{eq}^n}\left(1 - \frac{n}{m}\right)$$

41.1.2 Ionic Bonds,

Ionic bonding is possible only between two unlike atoms, one electro-positive and the other electro-negative

$$U_{eq} = -\frac{e^2}{4\pi\varepsilon_o R_{eq}}$$

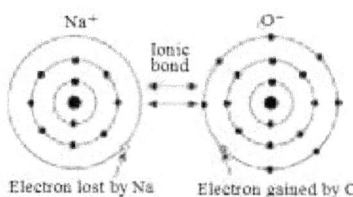

Electron lost by Na   Electron gained by O

or ionic alkali halide crystals, say NaCl, Born and Mayer obtained the lattice (cohesive) energy, *i.e.*, energy released at the time of forming the ionic bond is,

$$U_{eq} = -\frac{Ae^2 N_A}{4\pi\varepsilon_o R_{eq}}\left(1 - \frac{\rho}{R_{eq}}\right)$$

Bulk modulus of a solid,

$$\beta = -V\left(\frac{dp}{dV}\right),$$

and compressibility,

$$\left(\frac{1}{K}\right) = \beta = -V\left(\frac{dp}{dV}\right)$$

$$\frac{1}{K_o} = \beta = \frac{Ae^2}{18R_{eq}^4}\left(\frac{R_{eq}}{\rho} - 2\right)$$

For NaCl, R_{eq} = 0.2283 *nm*, ρ = 0.0345 *nm* (for all alkali halides),

$$U = -\frac{Ae^2 N_A}{4\pi\varepsilon_o R_{eq}}\left(1 - \frac{1}{n}\right)$$

Madulong constant., A

$$n = 1 + \frac{72\pi\,\varepsilon_o R_{eq}^4}{A\,e^2 K_o}$$

Dissociation energy, D can be obtains as

$$D = \frac{4A}{5R_{eq}^2}$$

41.1.3 **Covalent Bonds,**

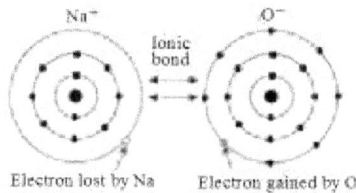

Electron lost by Na   Electron gained by O

41.1.4 **Metallic Bonds**

41.1.5 Van der Waals Bonds,

41.1.6 Hydrogen Bonds,

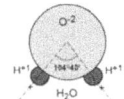

Types of Solieds				
Type	Units present	Characteristic Prioperties	Representative Crystal	Chesive energy (eV)
1) Ionic	Anion & Cation	Brittle, Insulator, High Melting Point	NaCl, LiF	184 244
2) Covalent	Atomic	Hard, High MP, Non-conducting	Diamond, SiC	170 244
4)Metallic	Anions & electron gas	High conductivity	Fe Na	94 26
5) H-bond	Molecules held by H-bonds	Low MP Insulators	H_2O (ice) HF	12 7.0
6)Vander Waals (Molecular)	Molecules Atoms	Soft, Low MP, Volatile Insulating	Ar	2.0

41.1.7. Properties of Bonding

1) Bond Length

2) Stored Energy

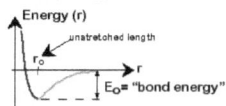

3) Melting temperature T_m

T_m increases as E_o increases

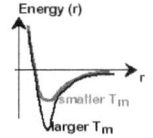

4) Elastic Modulus E

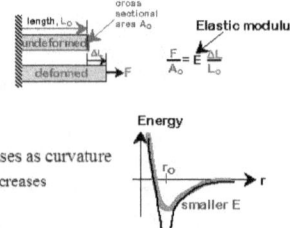

Elastic modulus

$$\frac{F}{A_o} = E\frac{\Delta L}{L_o}$$

E increases as curvature

$\left.\dfrac{\partial^2 E}{\partial r^2}\right|_{r_o}$ increases

smaller E

larger E

5) Coefficient of Thermal Expansion α:

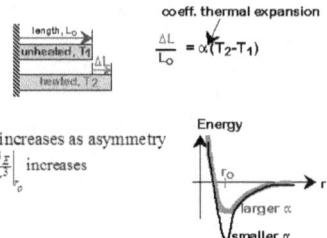

coeff. thermal expansion

$$\frac{\Delta L}{L_o} = \alpha(T_2 - T_1)$$

α increases as asymmetry

$\left.\dfrac{\partial^3 E}{\partial r^3}\right|_{r_o}$ increases

larger α

smaller α

Classification of Materials

Type of Solid	Crystal Units	Binding Force	Optical	Electrical	Thermal	Mechanical	Examples
Ionic	Simple and complex ions	Electrostatic attraction of oppositely charged ions (the ionic bond)	Transparent or coloured by characteristic absorption by ions	Insulators, forming conducting solutions in ionising solvents	Fairly high melting to form ions	Hardness increases with ionic charge; break by cleavage	Sodium chloride Calcite Magnesia
Covalent (giant molecular)	Group IV elements: III–V and II–VI compounds	Covalent, sometimes with some ionic character	Transparent, high refractive index; or opaque	Semiconductors except diamond; insoluble	Very high melting	Very hard; break by cleavage	Diamond Carborundum Rutile
Metallic	Positive ions and "free" electrons	Attraction between ions and electron gas (the metallic bond)	Opaque and reflecting	Electronic conductors; soluble in acids to form salts	Moderately high melting; good heat conductors	Tough and ductile except tungsten	Copper Iron Sodium
Molecular	Rare gas atoms; molecules	Dispersion and multi-pole forces (secondary bonds)	Transparent and like its molten form	Insulators; dissolve in non-ionising solvents	Fairly low melting points	Soft and plastically deformable	Argon Paraffins Calomel Ice Solid CO_2

41.2 ENERGY BANDS (Band theory of solids)

The atoms in the solid are very closely packed. The nucleus of an atom is so heavy that it electrons in an isolated atom have different and discrete amounts of energy according to their occupations in different shells and sub shells. These energy values are represented by sharp lines in an energy level diagram.

During the formation of solid, energy levels of outer shell electrons got split up. As a result, closely packed energy levels are produced. These are valence electrons. The band formed by a series of energy levels containing the valence electrons is known as Valence Band (VB). The next higher permitted band in a solid is the Conduction Band (CB). The electrons occupying this band are known as conduction electrons.

CB and VB are separated by Forbidden (FB)Energy Gap, E_g .

41.2.1. THE KRONIG-PENNEY MODEL of an Infinite Lattice

Infinite Lattice

L. Kronig and W.G. Penney (1931) made an important generalization of the square well potential, Here the number of interacting potential wells, N, is extremely large so that each of the single-well levels is split into N levels spaced so close together that they form nearly continuous energy levels.

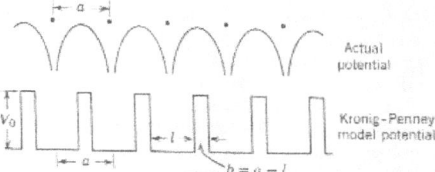

41.2.2 BLOCH THEOREM

Bloch has shown that the solution for such a periodic lattice, *viz.*,

$$\psi(x,t) = e^{i(k_n x - \omega t)} \psi(x)$$

is in the form of a plane wave, $e^{i(k_n x - \omega t)}$ and a function, $w(x)$ having the periodicity of the lattice, where

$$w(x+d) = w(x)$$

and $k_n = 2\pi n / N$

$$\nabla^2 \psi(x) + \frac{2m}{\hbar^2} [E - V(x)]\, \psi(x) = 0$$

is the Schrödinger Equation with periodic potential having the solution in the form of equation), substituting which one gets

$$(d^2/dx^2)\, w(x) + 2\, i\, k\, (dw/dx) + (2m/\hbar^2)[E - (\hbar^2 k^2/2m) - V(x)]\, w(x) = 0$$

This gives the **transcendental equation** for α , *viz.*,

$$Cos(kd) = Cos(\alpha d) + [P/\alpha d]\, Sin(\alpha d) ,$$

where $P = \lim_{c \to d} \left(\frac{1}{2}\beta_0^2 b\, c \right) = \left(m\, c V_0 b/\hbar^2 \right)$

$$= \frac{1}{2}\lambda \quad \alpha = \sqrt{2mE/\hbar^2} \quad = \text{constant.}$$

$$\beta_0 = \sqrt{2m\,(V_0 - E)/\hbar^2}$$

$$P = \frac{1}{2}\lambda$$

if $\quad V(x) = (\hbar^2/2m)(\lambda/d)\Sigma\,\delta(x - nd)$,

i.e., aseries of repulsive δ-function potentials. This means, one can obtain the allowed energy levels of an electron in a periodic potential by the following two interpretations of equation.

41.2.3. BAND STRUCTURE

Interpretation of the right hand side of equation

$$\boxed{Cos(kd) = Cos(\alpha d) + [P/\alpha d]\, Sin(\alpha d)}$$

will be satisfied only for those values of E for which the right hand side (RHS) such that -1 < RHS < +1, because

$Cos(kd) = \pm 1$

For real values of k, the physically meaningful solutions must be within these two limits by the heavy horizontal lines along the (αd)-axis. These energies are called ALLOWED BANDS (allowed regions or allowed ZONES) – which an electron in a periodic potential is allowed to take (shown shaded in Fig.), *i.e.*, in a crystal only certain energy bands are permitted for an electron. Between the allowed bands are the FORBIDDEN BANDS (forbidden regions or forbidden zones) of energies, which the electron cannot take while it is moving through a periodic potential.

Case (i) $V_0 b \rightarrow$ large, allowed zones become very narrow and the electron cannot move freely. This is what happens in the case of ELECTRONS that are TIGHTLY BOUND to the NUCLEUS.

Case(ii) On the other hand, as $V_0 b \rightarrow 0$, the allowed zones spread so much that the forbidden zones disappear. This is what happens in the case of VALENCE ELECTRONS IN AN ATOM.

Case (iii) For $V_0 b \rightarrow 0$, the situation reduces to the case of FREE ELECTRONS.

41.2.4. DISPERSION DIAGRAM; BAND STRUCTURE

Consider the term $Cos(kd)$. This function can assume only those values which correspond to allowed values of E.. The plot is the DISPERSION diagram of E *versus* $(k\ d)$ for electrons in the Kronig-Penney potential.

The graph of E *versus* k is known as the Band Structure AND STRUCTURE. The *dashed-line-parabola* corresponds to the case of a free electron for which

$$\boxed{E_n = \hbar^2 k_n^2 / 2m}$$

Note that the *heavy lines* depart slightly from the dashed line parabola only in the neighborhood of $\pm n\pi$; where $n =$ an integer. This means that the electron moving in a periodic potential behaves like a free electron for most values of k; except those near $\pm n\pi$. For large values of E, the two are almost similar.

41.2.5 BRILLOUIN ZONES

It is important to note that the DISCONTINUITIES in the allowed values of E occur at

$Cos(kd) = \pm 1$

i.e., $kd = \pm n\pi$, $n = 0, 1, 2, 3, \ldots\ldots$

If one substitutes $k = 2\pi/\lambda$ and $n\lambda = 2d$,

which corresponds to the Bragg condition in diffraction, provided that one substitutes $\theta = \pi/2$, in

$$\boxed{n\lambda = 2\ d\ Sin\theta}$$

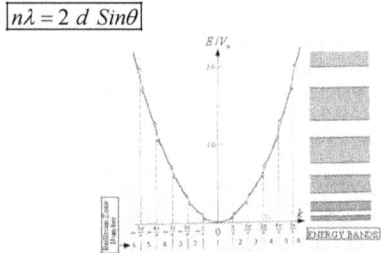

E versus k plot showing the Band structure of a Periodic Lattice

41.2.5.1   Brillouin Zones defined in Reciprocal Space around a lattice point

41.2.5.2   First Brillouin Zone (I B.Z.)

It is defined as the volume encompassed around a lattice point without crossing any Bragg planes, The region that contains electrons with momentum such that $0 < k < \pi / d$, *i.e.* $n = 1$, in equation (5.7.34), is called the FIRST BRILLOUIN ZONE (I B. Z.).

41.2.5.3   Second Brillouin zone (II B.Z.)is the volume obtained by crossing any one Bragg plane

Continue on to get higher order zones. II BZ contains electron with values of k, such that
$\pi / d < k < 2\pi / d$, so also the 3rd, 4th, *etc.*,

Periodicity of wave function mandates all unique information is centred within the first Brillouin Zone,

Wave functions in higher zones can be obtained by translating the 'pieces' back through the Bragg planes to the first Brillouin zone.

Within the B Z, the energy is a continuous function of α with a continuous derivative. A single continuous branch is called an ENERGY BAND. Discontinuities in energy occur only on the zone surfaces. Taking the allowed values of E from

$$\alpha = \sqrt{2mE / \hbar^2}$$

$$\boxed{E_n = \hbar^2 \frac{(\alpha d)^2}{2md^2}}$$

Outside the limits of ± 1, k must be complex with a non-zero imaginary part. The corresponding ranges of E are forbidden. Thus alternate regions of allowed and forbidden energy bands are formed. The grouping of the permitted energy values into these bands is one of the most important and characteristic features of the behavior of electrons in periodic lattices.

i)   As $K \to$ large (*i.e.* higher the potential separating zero potential regions) the energy bands become narrow.

ii)   ii) As $K \to \infty$, the crystal tends to become a series of independent square wells, and energy bands go over to the discrete eigenvalues as for a square well.

41.2.6   **Classical Free electron Theory of Metals** (Drude-Lorentz theory)

A metal consists of electrons which are free to move about in the crystal like molecules of a gas in a container. Mutual repulsion between electrons is ignored and hence potential energy is taken as zero.

41.2.6.1   Wiedemann-Franz Law

The ratio of thermal conductivity κ_T to electrical conductivity σ of a metal is directly proportional to absolute temperature T

$$\boxed{\frac{\Delta Q}{\Delta t A} = -\kappa_T \frac{\Delta T}{\Delta x}}$$

$$\boxed{<v> = \sqrt{\frac{8 k_B T}{\pi m}}}$$

$$\boxed{\kappa_T = \frac{n<v>\lambda C_v}{3 N_A}}$$

$$\boxed{\frac{\kappa_T}{\sigma} = LT}$$

$L =$, a constant called Lorentz number

$$\boxed{L = \frac{\kappa_T}{\sigma T} = \frac{\pi^2}{3}\frac{k_B^2}{e^2} = 2.45 \, x10^{-8} \, W\Omega K^{-2}}$$

41.2.6.2   Quantum Free electron Theory (A. Sommerfeld, 1928)

Classical free electron theory could not explain many physical properties.

The number of states per unit energy range, called the density of states

$$g(E) = \frac{dNs}{dE}$$

Fermi-Dirac statistics gives the probability E is occupied by an electron is given by,

$$f(E) = \frac{1}{1 + e^{(E - E_F / k_B T)}}$$

E_F =Fermi level

$dNs = f(E)g(E)dE$

All these results are depicted in the figure,

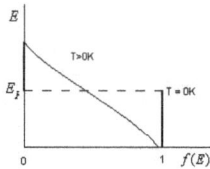

41.2.7 Hall Effect

The **Hall effect** refers to the potential difference (**Hall voltage**) on opposite sides of a thin sheet of conducting or semiconducting material in the form of a 'Hall bar' or a van der Pauw element through which an electric current is flowing, created by a magnetic field applied perpendicular to the Hall element. The ratio of the voltage created to the amount of current is known as the *Hall resistance*, and is a characteristic of the material in the element. Dr. Edwin Hall discovered this effect in 1879.

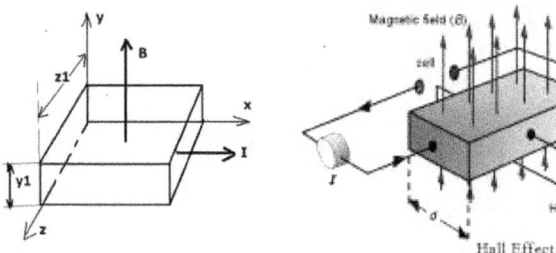

Hall constant $\boxed{R_H = -\frac{1}{ne}}$

n =density of charge carrier.

Hall field $\boxed{E_H = -\frac{1}{ne} B_z J_x}$

41.2.6 **Insulator**

Insulators are very poor conductors of electricity with resistivity ranges from $10^3 - 10^{17} \, \Omega m$.
$E_g = 6 \, eV$ .eg., Carbon.

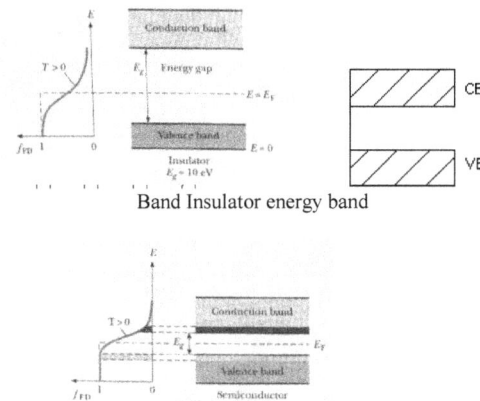

Band Insulator energy band

41.2.8 Conductor

Here valence band and conduction band overlap and there is no forbidden energy gap. Resistivity ranges between $10^{-9} - 10^{-4} \Omega m$ .The electrons are available for electrical conduction. The electrons from VB can freely enter the CB.

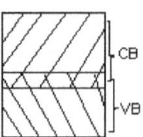

41.2.9 **Semiconductor**

A semiconductor material is one whose electrical properties lie between that of insulators and good conductors. The forbidden band is small and resistivity ranges between 10-4to 103Ωm. Ge and Si are examples with $E_g = 0.7eV$ and $1.1eV$, respectively. An appreciable number of electrons can be excited across the gap at room temperature. By adding impurities or by thermal excitation, we can increase the electrical conductivity in semiconductors

41.3 **Classification of Semiconductors on the basis of Fermi level and** E_F

41.3.1 Intrinsic semiconductors,

The E_F level lies exactly at the centre of the Forbidden Energy Gap. In n-type semiconductors E_F level lies near the Conduction Band. In p-type semiconductors E_F level lies near the Valence Band.

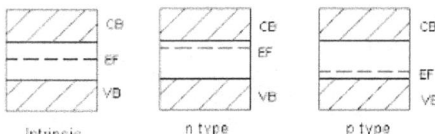

A pure semiconductor free from any impurity is called intrinsic semiconductor. Here charge carriers (electrons and holes) are created by thermal excitation. Si and Ge are examples. Both Si and Ge are tetravalent. $E_g = 0.7\ eV$. The number of free electrons is always equal to the number of holes.

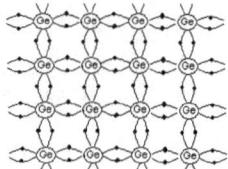

For intrinsic materials.

$$\boxed{n = p = n_i}$$

Electrical conductivity, $\boxed{\sigma_i = n_i e(\mu_e + \mu_h)}$

μ_e = electron mobilty, μ_h = hole mobility

Hall mobility, $\boxed{\mu_H = \mu_h - \mu_e}$

$$\boxed{\mu_H = \sigma R_H = \frac{p\mu_h^2 - n\mu_e^2}{n\mu_e + p\mu_h}}$$

$$\boxed{n = -\frac{1}{eR_H}}$$

$$\boxed{\mu_e = \frac{\sigma}{ne}}.$$

41.3.2 Extrinsic semiconductor

Doping, (adding suitable impurities to the intrinsic semiconductor),increases the electrical conductivity in semiconductors. The added impurity (about 1 ppm) may be pentavalent or trivalent. Depending on the type of impurity added, the extrinsic semiconductors can be divided into two classes: n-type and p-type. Addition of impurities introduces new allowed quantum energy states in the Forbidden energy Band. , $viz.$, E_d (donor in n-type) and E_a (acceptor in p-type).

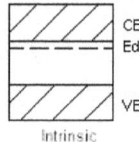 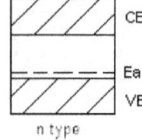 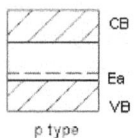

41.3.2.1 n-type semiconductor

Since current carriers are negatively charged particles, this type of semiconductor is called n-type semiconductor.

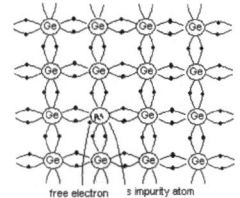

free electron   s impurity atom

41.3.2.2 p-type semiconductor

When trivalent impurity is added to pure semiconductor, it results in p-type semiconductor. Impurity atoms contribute <u>holes</u>.

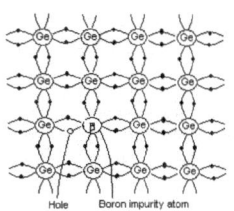

Hole   Boron impurity atom

Semiconductor Band Gaps		
Material	Energy gap (eV)	
	0K	300K
Si	1.17	1.11
Ge	0.74	0.66
InSb	0.23	0.17
InAs	0.43	0.36
InP	1.42	1.27
GaP	2.32	2.25
GaAs	1.52	1.43
GaSb	0.81	0.68
CdSe	1.84	1.74
CdTe	1.61	1.44
ZnO	3.44	3.2
ZnS	3.91	3.6

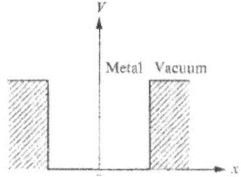

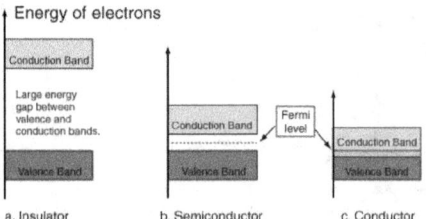

a. Insulator   b. Semiconductor   c. Conductor

41.4   PHONONS - Lattice Vibrations

Lattice vibrations can explain sound velocity, thermal properties, elastic properties and optical properties of materials. Lattice Vibration is the oscillations of atoms in a solid about the equilibrium position. For a crystal, the equilibrium positions of atoms form a regular lattice, The vibration of these neighboring atoms is not independent of each other. An ideal lattice has harmonic forces between atoms, and normal modes of vibrations are called lattice waves. Lattice waves range from low frequencies to high frequencies of the order of 10^{13} Hz or even higher. However, the wavelengths at extremely high frequencies are of the order of inter-atomic spacing. Due to the shortness of these wavelengths, the motion of the neighboring atoms is uncorrelated; with each atom moving about its average position in three dimensions with average vibrational energy, which is usually $3k_BT$. Lattice vibrations can also interact with free electrons in a conducting solid which gives rise to electrical resistance.

41.4.1   Monatomic 1-D Lattice.

Lattice dynamics offers two different ways of finding the dispersion relation within the lattice.

41.4.2   Quantum-mechanical approach

It can be used to obtain phonon's dispersion relation, the solution to the Schrodinger equation for the lattice vibrations must be solved.

41.4.3   Semi-classical treatment of lattice vibrations

This treatment gives classical mechanics the use of one additional postulate taken from quantum mechanics, mainly that the energy of lattice vibrations is quantized.

Newton's law of mechanics:

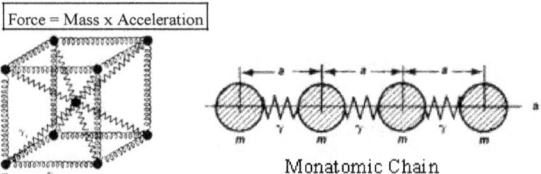

Monatomic Chain

Position $u(t)$ of an atom of mass m in a instantaneous potential $\varphi(u,t)$ gives

$$\boxed{\frac{d^2u(t)}{dt^2} = -\frac{1}{m}\nabla\varphi(u,t)}.$$

This potential energy is the interaction of the atom with the other atoms within the crystal. ,
a = interatomic distance.

41.4.4   Dispersion Relation

Solving, the velocity of the lattice wave $\boxed{v = \dfrac{\omega}{k}}$ plotted against $\boxed{k = \dfrac{2\pi}{\lambda}}$

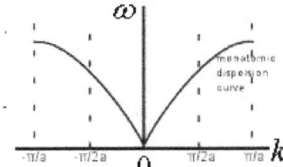

I Dispersion of Monatomic Chain
within First BZ

Within $\qquad \boxed{\omega = \pm\sqrt{\left(\dfrac{4\beta}{m}\right)}\ Sin\dfrac{ka}{2}}$

First Brillouin Zone, viz., $\boxed{-\dfrac{2\pi}{a} \le k \le +\dfrac{2\pi}{a}}$ is as shown.

41.4.5 Optical **Phonon**

Optical phonons are quantized modes of lattice vibrations when two or more charged particles in a primitive cell move in opposite directions with the center of mass at rest. This mode has highest energy for wavelength infinity or $k = 0$ when the two lattices move in opposing direction of each other.

41.4.6 Diatomic 1-D Lattice

Diatomic means the lattice with two kinds of atoms with masses m and M. The equations of motion are:

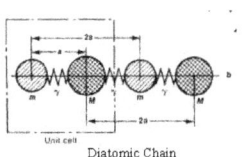

Diatomic Chain

$$\boxed{m\frac{\partial^2 u_{2n}}{\partial t^2} = \beta(u_{2n+1} - 2u_{2n} + u_{2n-1})}$$

$$\boxed{M\frac{\partial^2 u_{2n+1}}{\partial t^2} = \beta(u_{2n+2} - 2u_{2n+1} + u_{2n})}$$

Trial solutions are

$$\boxed{u_{2n} = A\, e^{i\,(2nka\,\pm\,\omega t)}}$$

$$\boxed{u_{2n+1} = B\, e^{i\{(2n+1)ka\,\pm\,\omega t)\}}}$$

β = spring constant
The solution of the diatomic lattice is

$$\boxed{\omega^2 = \beta\left(\frac{1}{m}+\frac{1}{M}\right) \pm \beta\sqrt{\left(\frac{1}{m}+\frac{1}{M}\right)^2 - \frac{4\,Sin^2 ka}{Mm}}}$$

41.4.7 Dispersion Relation

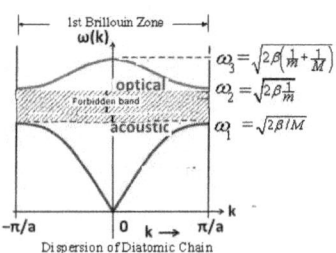

Dispersion of Diatomic Chain

41.4.8 Transverse Optic (TO) mode –

$$\boxed{\text{TO mode} \quad k \to 0, \quad \omega = \frac{2\beta}{\mu}},$$

$$\boxed{\text{Effective mass, } \mu = \frac{Mm}{M+m}}.$$

In the long-wavelength limit, optical modes interact strongly with electromagnetic radiation in polar crystals, hence the name.
Strong optical absorption is observed (Photons annihilated, phonons created).

$$\boxed{\omega \to \text{finite } as\ k \to 0}$$

Optical modes arise from folding back the dispersion curve as the lattice periodicity is doubled (halved in q-space).
Zone boundary
All modes are standing waves at the zone boundary,
$\frac{\partial \omega}{\partial k} = 0$: a necessary consequence of the lattice periodicity.
In a diatomic chain, the frequency-gap between the acoustic and optical branches depends on the mass difference. In the limit of identical masses the gap tends to zero.
Transverse Acoustic (TA) mode,

$$\boxed{\text{TA mode} \quad k \to 0, \quad \omega = \frac{2\beta a^2}{M+m}},$$

corresponds to sound waves in long wave limit, hence the name. $\omega \to 0\ as\ k \to 0$.

41.4.9 Origin of Optic and acoustic modes.

Effect of periodicity – of a diatomic chain is the result of that of monatomic
        The permitted waves are split into two branches called the optical and acoustical branches. The gap (forbidden band) between the optical and acoustic branch is the region where frequencies are not allowed to propagate. The width of this forbidden band depends on the difference of the masses of the two atoms. If the two masses are equal, the two branches join (become degenerate) at $\frac{\pi}{2a}$. The acoustical branch for the diatomic is similar to that of the monatomic lattice, but the optical branch is different. Pattern of Pattern of displacement of atoms
Zone boundary modes

$$\boxed{\text{Standing waves} \quad k = \frac{\pi}{2a},\ \lambda = \frac{2\pi}{k}}$$

Higher energy mode, only *light atoms move*,

Lower energy mode –only *heavier atoms move*.

The difference between the optical and acoustic branch is that the optical branch for the long wavelength limit both atoms in the unit cell move opposite to each other with an increase in the mass amplitude. The acoustical branch for the long wavelength limit, the

41.4.10   Phonons in 3-Dimension

In a 3-D crystal, the atoms vibrate in 3-Ds with three vibrational branches, one longitudinal and two transverse. For a 3-D Lattice with N atom per lattice point, there is $3(m-1)$ optical branches, of which $2(m-1)$ are TO phonons and the remaining LO phonons.

In a transverse wave, the atomic displacement direction is perpendicular to the direction of the propagated wave. (Fig.). The remaining two transverse waves will overlap if the two vibrational directions are symmetric.

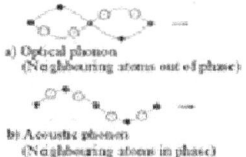

a) Optical phonon
(Neighbouring atoms out of phase)

b) Acoustic phonon
(Neighbouring atoms in phase)

In regards to electrons, the phonons are dispersed along different crystallographic directions.

Eg., NEON (FCC lattice)

Inelastic neutron scattering results in different crystallographic directions

Many features are explained by the 1-D model:

Dispersion is sinusoidal [Nearest Neighbour (NN) interactions]

All modes are acoustic (monatomic system).

41.4.11   NEON- An FCC Monatomic solid.

There are two distinct types of mode:

Longitudinal (L), with displacements parallel to the propagation direction. These generally have higher energy.

Transverse (T), with displacements perpendicular to the propagation direction. These generally have lower energy. They are often degenerate in high symmetry directions (not along $(\xi\xi 0)$).

Minor point (demonstrating that real systems are subtle and interesting, but also implicated):

1) mode along $(\xi\xi 0)$ has 2 Fourier components, suggesting next- NN interactions.

2) In fact there are only NN interaction.

The effect is due to the f.c.c structure. Nearest Neighbour interactions from atom, A (in plane I) join to atom C(in plane II) and to atom B(in plane III) thus linking nearest and next-nearest-planes.. Phonons in 3-D lattice, Diatomic solid eg., NaCl has sodium chloride structure! Two interpenetrating f.c.c. lattices.

Main Points The 1-D model gives several insights, as before. There are: Optical and acoustic modes (labels O and A); Longitudinal and transverse modes (L and T). Dispersion along $(\xi\xi 0)$ is simplest and most like our 1-D model.

$(\xi\xi 0)$ planes contain, alternately, Na atoms and Cl atoms (other directions have Na and Cl mixed).

NaCl Phonons

Note the energy scale. The highest energy optical modes are ~8 *THz* (*i.e.,* approximately30 *meV*). Higher phonon energies in NaCl than in Neon. The strong, polar bonds in the alkali halides are stronger and stiffer than the weak, van-der-Waals bonding in Neon.

<u>Minor point:</u>

Modes with same symmetry cannot cross, hence the avoided crossing between acoustic and optical modes in (00ξ) and $(\xi\xi0)$ directions.

Ignore the detail for present purposes

41.5 DIELECTRIC PROPERTIES

Dielectric materials are a special class of substances that, under almost all conditions are insulators.

41.5.1. POLARIZATION OF ATOMS AND MOLECULES

Electric dipoles are formed when a dielectric is inserted between the charged plates of a capacitor, and the electric dipole moment, p is given by

$$\boxed{p = q\,r}\,,$$

where q is the positive charge separated from a negative charge of the same size by the distance **r**,

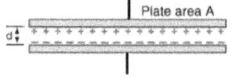

$$\vec{p} = q\vec{d}$$

41.5.2. Parallel Plate Capacitor

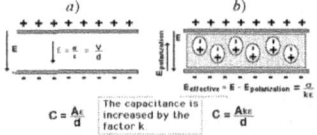

$$\boxed{C = \frac{\varepsilon A}{d} = \frac{k\varepsilon_o A}{d}}$$

41.5.3 LOCAL FIELD, E_{loc};

$$\boxed{p_{mol} = \alpha\, E_{loc}}$$

α = molecular polarizability.

Total dipole moment (Polarization),

$$\boxed{P = \sum e\, x_j = N\alpha\, E_{loc}}$$

In the case of a neutral atom in an electric field.

$$\vec{P}_{elec} = \alpha_{elec} \vec{E} = Ze\,\vec{x} = r^3\,\vec{E}.$$

i.e., displacement of the electron and nucleus, $\vec{x} = 10^{-15}\,\vec{E}$

In alternating electric fields the electronic polarizability is essentially constant up to UV frequencies . According to an empirical relationship by J.C. Slater and N.H. Frank, for each electron in an outer level,

$$\alpha_{elec} = r^3 = a\,(n^2 a_0)\,/\,(Z - S).\Big|,$$

a_0 = Bohr radius, n = quantum number.

41.5.4 The Clausius-Mossotti Relation

What a dielectric equation of state actually looks like?

The field at the molecule due to the surface charges on the sphere is $\vec{E} = \dfrac{P}{3\varepsilon_0}$.

The electric field at a distance r from a dipole $\vec{p}$ is

$$\vec{E} = -\frac{1}{4\pi\varepsilon_0}\left[\frac{\vec{p}}{r^3} - \frac{3(\vec{p}.\vec{r})\,\vec{r}}{r^5}\right]$$

$$\varepsilon_0 = 8.8542 x 10^{-12}\,Fm^{-1}$$

The net electric field seen by an individual molecule

$$\vec{E}_{loc} = \vec{E}_0 + \frac{P}{3\varepsilon_0}$$

$$\varepsilon = \frac{\varepsilon_0 E + P}{\varepsilon_0 E} = 1 + \chi$$

$$\chi = \frac{P}{\varepsilon_0 E} = \varepsilon - 1$$

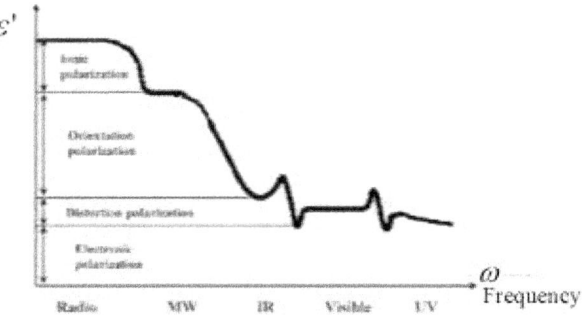

Frequency *versus* contributions to Polarizability

The Clausius-Mossotti relation between ε and α is

$$\frac{\varepsilon - 1}{\varepsilon + 2} = \frac{4\pi}{3}\sum N_j \alpha_j = \frac{1}{3\varepsilon_0}\sum N_j \alpha_j$$

Dielectric Constants at 20°C		Dielectric Constant ε (~300 K)			
Material	Dielectric Constant	Material			
		Air	1.005364		
Vacuum	1	Acetic acid	6.2	Alcohol, ethyl (grain)	24.55
Glass	5-10	Alcohol, methyl (wood)	32.70	Aamber	2.8
Mica	3-6	Bakelite	4.8	Calcite	8.0
Mylar	3.1	cellulose	3.7 - 7.5	Diamond, type I	5.87
Neoprene	6.70	Diamond, type IIa	5.66	Ebonite	2.7
Plexiglas	3.40	Epoxy	3.6	Germanium	316
Polyethylene	2.25	Glass	4 - 7	Glass, pyrex 7740	5.0
Polyvinyl chloride	3.18	Lucite	2.8	Mica, muscovite	5.4
Teflon	2.1	Mica, canadian	6.9	Nylon	3.5
Germanium	16	oil, linseed	23.4	oil, mineral 3	2.1
Strontiun titanate	310	oil, olive	3.1	oil, petroleum	32.0 - 2.2
Titanium dioxide	173 ⊥perp	oil, silicone	2.5	oil, transformer	2.2
(rutile)	86 ‖ para	paper	3.3, 3.5	plexiglas	3.1
Water	80.4	polyester	3.2 - 4.3	polyethylene	2.26
Glycerin	42.5	polypropylene	2.2 - 2.3	polystyrene	32.55
Liquid (-78°C)	25	polyvinyl chloride (pvc)	4.5	porcelain	6 - 8
ammonia(		water, liquid, 20 °C	80.2	Wax, beeswax	2.7 - 3.0
Benzene	2.284	wax, paraffin	2.1 - 2.5	Waxed paper	3.7
Air(1 atm)	1.00059	quartz, crystalline	4.60	Quartz, fused	3.8
Air(100 atm)	1.0548				

At optical frequencies, since $\varepsilon = n^2$,

$$\frac{n^2-1}{n^2+2} = \frac{4\pi}{3} \sum N_j \alpha_j (electronic)$$

The real part of the Clausius-Mossotti factor is a determining factor for the dielectrophoretic force on a particle, whereas the imaginary part is a determining factor for the electro-rotational torque on the particle.

41.6.1 SPECFIC HEATS OF SOLIDS

Performing a *normal mode analysis* of the oscillations, one gets $3N$ independent modes of oscillation of the solid. Each mode has its own particular oscillation frequency, and its own particular pattern of atomic displacements. Any general oscillation can be written as a linear combination of these *normal modes*. Let q_i be the (appropriately normalized) amplitude of the i the normal mode, and p_i the momentum conjugate to this coordinate. In *normal mode coordinates*, the total energy of the lattice vibrations takes the particularly simple form

$$E = \frac{1}{2} \sum_{i=1-3N} (p_i^2 + \omega_i^2 q_i^2)$$

ω_i =Frequency of normal mode, lattice modes are non-localized.

$\Delta E = \hbar \omega$ is the reason for lattice vibrations are more closely spaced than vibrational energy levels of vibrations of gaseous molecules. Lattice modes if obey classically, as per Equipartition of energy, mean value per mole, $\hat{E} = 3N k_B T$

Molar heat capacity at constant volume,

$$C_V = \frac{1}{V}\left(\frac{\partial \ddot{E}}{\partial T}\right)_V = 3R$$

41.6.2 Dulong and Petite's law
It is essentially a high temperature limit.

41.6.3 Einstein's approximation,
All vibrate at the same frequency.

$$C_V = -\frac{3N_A \hbar\omega}{k_B T^2}\left[\frac{\hbar\omega\, e^{\beta\hbar\omega}}{[e^{\beta\hbar\omega}-1]^2}\right]$$

Einstein Model

Einstein temperature $\boxed{\Theta_E = \dfrac{\hbar\omega}{k_B}}$

When $T \ll \Theta_E$

$$C_V = 3R\left(\frac{\Theta_E}{T}\right)^2\left[\frac{e^{\Theta_E/T}}{[e^{\Theta_E/T}-1]^2}\right]$$

$$C_V \sim 3R\left(\frac{\Theta_E}{T}\right)^2 e^{-\Theta_E/T}$$

In this model the specific heat approaches zero exponentially as $T \to 0$ Experimentally at low temperatures is more like

$$C_V \propto T^3 .$$

41.6.4 Debye approach.
In this model, choosing the total number of normal modes as, $3N$ define Debye frequency

$$\omega_D = c\left(6\pi^2 \frac{N}{V}\right)^{1/3}$$

leading to

$$C_V \sim 3R f_D(\beta\hbar\omega_D) = 3R f_D\left(\frac{\Theta_D}{T}\right)$$

$$f_D(y) = \frac{3}{y^3}\int_o^y \frac{e^x}{(e^x-1)^2}x^4 dx .$$

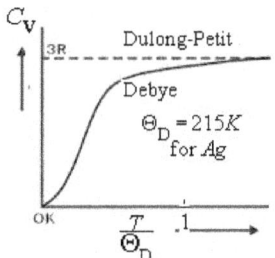

In asymptotic limit $T \geq \Theta_D$., for small y,

$$f_D(y) \to \frac{3}{y^3} \int_0^y x^2 dx = 1.$$

In the low temperature limit,

The Debye theory has seen valid with experiment for very low temperatures for non-metals.

For metals electron specific heat C_{vel} becomes significant at low temperatures, and has to be combined with the phonon value above in the Einstein-Debye heat capacity.

For metals: $\boxed{C_v = C_{vphonon} + C_{vel}}$

For non-metals: $\boxed{C_v = C_{vphonon}}$.

41.7   THERMAL EXPANSION OF SOLIDS

Ideal crystal has its each atom vibrate in a harmonic (parabolic) potential well,

$$\boxed{V(x) = \tfrac{1}{2} kx^2}$$

This means as temperature is increased the amplitude of vibration increases but the equilibrium position x_o does not change with T.

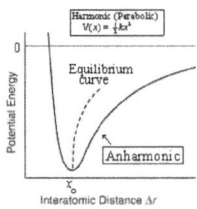

$\boxed{\text{For a harmonic solid thermal expansion is ZERO}}$.

Anharmonicity in the potential causes the solid to expand on increase on T.

41.7.1   Coefficient of linear thermal expansion, α

$\Delta \ell$ - change in length of sample of length L for temperature change from T_1 to T_2,

$$\boxed{\alpha = \frac{1}{L}\; \frac{\Delta \ell}{T_1 - T_2}}$$

For isotropic (Cubic) solids, there is only one expansion coefficient, α

For anisotropic solids, uniaxial crystals have two different values, whereas for biaxial crystals there are three different coefficients of expansion.

$$\boxed{\text{Volume expansion } \beta = \sum_{i=1,2,3} \alpha_i}$$

1) Thermal expansion of a crystal is a structure sensitive property of the crystal,
2) α is related intimately to the normal modes of vibration of crystalline lattice,
3) α is quite essential to convert the experimentally determined molar specific heat C_P value to C_V value required by theorists for theory of specific heats.

$$C_P - C_V = \frac{\beta^2 \, VT}{\chi_T \, J}$$

χ_T = compressibility of the solid,

V = molar volume of the solid,

J = Joules Mechanical equivalent of heat.

Thermal expansion is T (or $\theta^\circ C$) dependent.

$$\alpha_\theta = A + B\theta + C\theta^2$$

For non-metals: $\alpha = \alpha_{\text{lattice } phonon}$

For Metals: $\alpha = \alpha_{\text{lattice } phonon} + \alpha_{\text{electronic}}$

Coefficient of Thermal Expansion α:

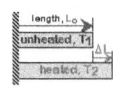
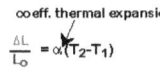

$\frac{\Delta L}{L_0} = \alpha(T_2 - T_1)$

α increases as asymmetry

$\left.\frac{\partial^3 E}{\partial r^3}\right|_{r_o}$ increases

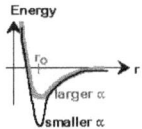

41.7.2 Gruneisen's Rule

γ = Gruneisen constant of a solid is

$$\frac{3\alpha}{\chi_T} = \gamma \frac{C_V}{V}$$

41.8.3. Thermal expansion is a second-rank tensor, $[\alpha_{ik}]$

$$[\alpha_{ik}] = \begin{bmatrix} \alpha_{11} & \alpha_{12} & \alpha_{13} \\ \alpha_{21} & \alpha_{22} & \alpha_{23} \\ \alpha_{31} & \alpha_{32} & \alpha_{33} \end{bmatrix}$$

41.8. CRYSTAL DEFECTS

A perfect crystal, with every atom of the same type in the correct position, does not exist

Crystal defects are results of thermal dynamic equilibrium contributed also by the increase in entropy $T S$ term of the Gibb's free energy.

41.8.1 Point Defects

Point defects are where an atom is missing or is in an irregular place in the lattice structure. Point defects include self interstitial atoms, interstitial impurity atoms, substitutional atoms and vacancies.

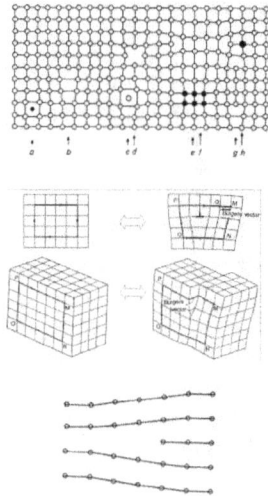

41.8.2 Interstitial impurity

These atoms are much smaller than the atoms in the bulk matrix. Interstitial impurity atoms fit into the open space between the bulk atoms of the lattice structure. An example of interstitial impurity atoms is the carbon atoms that are added to iron to make steel. Carbon atoms, with a radius of 0.071 nm, fit nicely in the open spaces between the larger (0.124 nm) iron atoms.

41.8.3 Vacancy or Schottky Defect

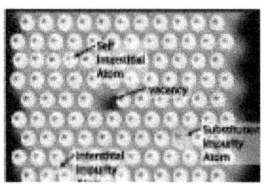

Vacancies are empty spaces where an atom should be, but is missing. They are common, especially at high temperatures when atoms are frequently and randomly change their positions leaving behind empty lattice sites. In most cases diffusion (mass transport by atomic motion) can only occur because of vacancies.

41.8.4 Lattice Defects

41.8.4.1 A dislocation

It is characterized by its Burgers vector: If you imagine going around the dislocation line, and exactly going back as many atoms in each direction as you have gone forward, you will not come back to the same atom where you have started. The Burgers vector points from start atom to the end atom of your journey (This "journey" is called Burgers circuit in dislocation theory).

41.8.4.2. Line defects

This weakens the structure along a one-dimensional space, and the defects type and density affects the mechanical properties of the solids.

41.8.4.3   Colour centres

These are imperfections in crystals that cause colour (defects that cause colour by absorption of light). Due to defects, metal oxides may also act as semiconductors, because there are many different types of electron traps. Electrons in defect region only absorb light at certain range of wavelength. The colours seen are due to lights not absorbed.

41.8.5.   **Nano crystals**

Crystallites with at least one dimension measuring $< 1000 \ nm$. Nano-crystals have a wide variety of proven and potential applications. Nanotechnology today is growing very rapidly and has infinite applications in almost everything we do, *viz.*, the medicine, food & chemicals we use, car we drive and very much more.

Synthesis: The traditional method involves molecular precursors, which can include typical metal salts and a source of the anion. Most semiconducting nano-materials feature chalcogenide (S^{S-}, Se^{S-}, Te^{S-}) and pnicnides (P^{3-}, As^{3-}, Sb^{3-}). Sources of these elements are the silylated derivatives such as bis (trimethylsilyl) ($S(SiMe_3)_2$ and tris (trimethylsilyl) phosphine

($P(SiMe_3)_3$.

Colloidal nanocrystals (NCs, *i.e.*, crystalline nanoparticles) have become an important class of materials with great potential for applications ranging from medicine to electronic and optoelectronic devices. Today's strong research focus on NCs has been prompted by the tremendous progress in their synthesis

41.8.5.1. **Quantum dots**

are microscopic nanocrystals that glow a specific wavelength (*i.e.*, colour) when given energy. The exact colour produced by the QD depends on its size: larger for longer wavelengths (red colours), smaller for shorter wavelengths (bluer).

Specific wavelengths of colour is what we need to great an image on a television. Using the three primary colours of red, green, and blue, we can mix a full rainbow of teals, oranges, yellows, and more.

Plasma and CRT televisions used phosphors to create red, green, and blue. All LCDs use colour filters to do the same.

There are multiple ways to use QDs in a display.

Nanotechnology will affect our lives tremendously over the next decade in very different fields, including medicine and pharmacy. Transfer of materials into the nanodimension changes their physical properties which were used in pharmaceutics to develop a new innovative formulation principle for poorly soluble drugs: the drug nanocrystals. The drug nanocrystals do not belong to the future; the first products are already on the market. The industrially relevant production technologies, pearl milling and high pressure homogenization, are reviewed. The physics behind the drug nanocrystals and changes of their physical properties are discussed. The marketed products are presented and the special physical effects of nanocrystals explained which are utilized in each market product. Examples of products in the development pipelines (clinical phases) are presented and the benefits for in vivo administration of drug nanocrystals are summarized in an overview.

Nanocrystals made with zeolite are reported to filter crude oil onto diesel fuel at Exxon Mobil Oil Refinery in Lousiana, USA, at cost less than convention methods.

41.9.   **GRAPHENE** (An amazing and strongest material in world)

Graphene is an allotrope of carbon, having 2-D hexagonal lattice, with one atom at each vertex.. Graphene is not graphite. It is the basic structural element of graphite, charcoal, carbon nano-tubes, fullerenes. As a material it is

i)   the thinnest ever

ii)   the strongest (200 times stronger than the strongest steel)
iii)   Best conductor of electricity on par with copper (a battery that charges in minutes)
iv)   Good heat conductor outperforming metals like silver and copper
v)   Almost completely transparent, but does not allow smallest gas to pass through it

The Nobel Prize in Physics 2010 was awarded jointly to Andre Geim and Konstantin Novoselov *"for groundbreaking experiments regarding the two-dimensional material graphene"*.

A great deal of interest has been spurred by graphene's conducting ability. Thus graphene transistors are predicted to be substantially faster than those made out of silicon today. Maybe we are on the verge of yet **another miniaturization of electronics** that will lead to computers becoming even more efficient in the future, like Sensors for 3-D cameras, supercars, super thin e-book, Night-vision for self driving cars, scanners for Smartphone, a battery for charging in minutes are in development in May 2017..

41.10 SUPERCONDUCTIVITY
41.10.1 Discovery: H. Kamerling Onnes (1911)

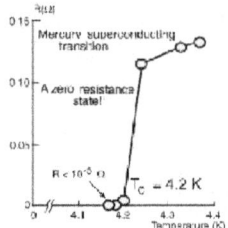

41.10.2. Typical features of superconductivity:

1) Transition to zero resistivity ; Sharp transition , $\Delta T < 10^{-4} K$
    A new state of matter.
2) Perfect diamagnetism (Exclusion of magnetic fields)

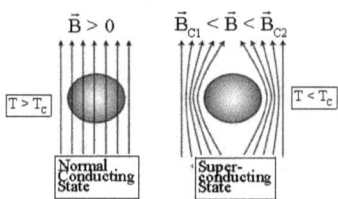

3) Flux quantization & Josephson junctions Phase coherence of electron wave functions in a superconductor leads to Flux quantization in B.
4) Critical temperature of some superconductors:
    1)   Al (1.2 K);
    2)   Hg (4,2 K);
    3)   Pb (7.2 K);
    4)   La-Ba-Cu Oxide (30 K);
    5)   Y-Ba-Cu Oxide (92 K)
    6)   Tl-Ba-Cu Oxide (125K)

5) Specific heat and suggested an energy gap

$$H_C(T) = H_C(0)\left[1 - \left(\frac{T}{T_c}\right)^2\right]$$

Empirically, Field

In the case of a superconducting wire of radius a, carrying current I,

$$H = 2\pi \frac{a}{I}$$

The effect of isotopic mass M on the T is

London penetration depth, $_c = M^{-\alpha}$

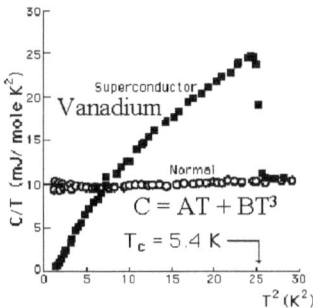

where α is the isotopic coefficient, $\alpha = \dfrac{\partial\, lnT_C}{\partial \ln M}$, and recent theories give,

$$\alpha = 0.5\,[1 - 0.01\{N(0)V\}^{-2}]$$

where M = isotopic mass.

41.10.3. London Equations:

1) $\dfrac{dJ_s}{dt} = \dfrac{n_s\, e^2}{m}\, E$

$n = n_n + n_s$, is the conduction electron density.

2) $\nabla \Lambda J_s = -\dfrac{n_s e^2}{m}\, B$

$J_s = -n_s e\, v_s$

London penetration depth, $\lambda = \sqrt{\dfrac{m}{\mu_0 n_s e^2}} = \lambda(0)$

$$\lambda(T) = \lambda(0)\left[1 - \left(\frac{T}{T_c}\right)^4\right]^{-1/2}$$

41.10.4.1. **Meissner-Ochsenfeld effect** (1934):

No magnetic field inside a superconductor.

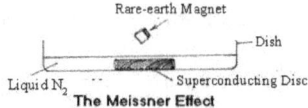

The Meissner Effect

41.10.4.2 The magnetic field can penetrate into the interior of a superconductor.

Perfect diamagnetism.

41.10.4.3 Type I and Type II superconductors:

Magnetic field B is excluded only up to a critical field.

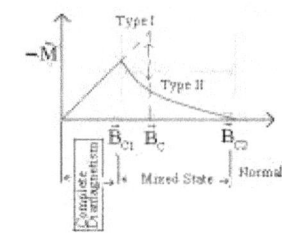

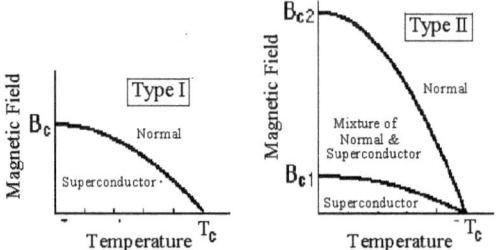

41.10.5 Theory

1) BCS Theory (Microscopic) of superconductivity, (1957).

2) High T_C superconductors were those used liquid nitrogen for the property.

3) K.A. Muller & J.G. Bednorz (Nobel Prize in 1988) experimented with a particular class of metal oxide ceramics called perovskites, In February of 1987, a perovskite ceramic material was found to superconducting at 90 K.

La-Ba-Cu Oxide (T_C =30 K);

Y-Ba-Cu Oxide $YBa_2Cu_3O_7$ (T_C = 92K),

Tl-Ba-Cu Oxide (T_C =125K).

-o-0-o-0-o-0-o-0-o-0-o-0-o-

Chapter 42

DATA COMMUNICATION

Internet, Cell Phone, Television and Antenna

" All science is either physics or stamp collecting"- Ernest Rutherford

42.1 **INTRODUCTION**

The distance over which data moves within a computer may vary from a few thousandths of an inch, as is the case within a single IC chip, to as much as several feet along the backplane of the main circuit board. Over such small distances, digital data may be transmitted as direct, two-level electrical signals over simple copper conductors. Except for the fastest computers, circuit designers are not very concerned about the shape of the conductor or the analog characteristics of signal transmission.

The basic idea behind a balanced circuit is that a digital signal is sent on two wires simultaneously, one wire expressing a positive voltage image of the signal and the other a negative voltage image. When both wires reach the destination, the signals are subtracted by a summing amplifier, producing a signal swing of twice the value found on either incoming line. If the cable is exposed to radiated electrical noise, a small voltage of the same polarity is added to both wires in the cable. When the signals are subtracted by the summing amplifier, the noise cancels and the signal emerges from the cable without noise:

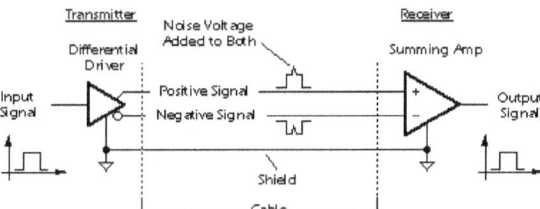

A great deal of technology has been developed for LAN systems to minimize the amount of cable required and maximize the throughput. The costs of a LAN have been concentrated in the electrical interface card that would be installed in PCs or peripherals to drive the cable, and in the communications software, not in the cable itself (whose cost has been minimized). Thus, the cost and complexity of a LAN are not particularly affected by the distance between stations.

A communications channel is a pathway over which information can be conveyed. It may be defined by a physical wire that connects communicating devices, or by a radio, laser, or other radiated energy source that has no obvious physical presence. Information sent through a communications channel has a source from which the information originates, and a destination to which the information is delivered. Although information originates from a single source, there may be more than one destination, depending upon how many receive stations are linked to the channel and how much energy the transmitted signal possesses.

In a digital communications channel, the information is represented by individual data bits, which may be encapsulated into multi-bit message units. A byte, which consists of eight bits, is an example of a message unit that may be conveyed through a digital communications channel. A collection of bytes may itself be grouped into a frame or other higher-level message unit. Such multiple levels of encapsulation facilitate the handling of messages in a complex data communications network.

42.2   **INTERNET**

A means of connecting a computer to any other computer anywhere in the world via dedicated routers and Servers. When two computers are connected over the Internet, they can send and receive all kinds of information such as text, graphics, voice, video and, computer programmes..

Web Browsing is possible by one of the following popular Browsers.

1) Microsoft Internet Explorer (MSIE), (1995), Internet Explorer (IE)
2) Mozilla Firefox
3) Apple's Safari
4) Opera, and
5) Google Chrome.

| No one actually owns Internet | ,

It is an infrastructure facility to connect networks to other networks.

IE supports JAVA, JavaScript, eyc.

The Internet Explorer10 is now integrated in MS Windows 8 and Windows Server 2012.

| *Internet* and *Web* , *i.e.*, *www* $\left(World\ wide\ Web\right)$ are NOT the same |

42.2.1   Internet Address (IP)

Refers to the IP of a web site (URL, *i.e.*, Uniform Resource Locator). It also represents e-Mail address. It has two protocols, *viz.*, FTP protocol and the http (hypertext transfer protocol) protocol.

Example: http://test.com/test.htm

42.2.2   Domain name

Domain names are used to identify one or more IP addresses. There are only a limited number of such domains. For example:

i)   **gov** - Government agencies
ii) **edu** - Educational institutions
iii) **org** - Organizations (nonprofit)
iv) **mil** - Military
v)   **com** - commercial business
vi) **net** - Network organizations
vii) **ca** - Canada
viii) **in** - India

Because the Internet is based on IP addresses, not domain names, every Web server requires a Domain Name System (DNS) server to translate domain names into IP addresses.

42.2.3   Internet Marketing

Internet marketing, or online marketing, refers to advertising and marketing efforts that use the Web and e-Mail to drive direct sales *via* e-commerce, in addition to sales leads from Web sites or emails. Internet marketing and online advertising efforts are typically used in conjunction with traditional types of advertising like radio, television, newspapers and magazines.

42.2.4   Modem and Router

Some High speed Internet Service providers (ISP) are listed

Verizon and AT&T (USA), BSNL (All over India, except Mumbai), MTNL (Mumbai and Delhi), Asianet SATCOM (Kerala in India0,Aircel, Tata DoCoMo, Vodafone, AirTel*etc.* are examples of service providers via cable, and a subscriber uses Modem to connect to a computer.

When subscribing to an ISP, the service provider would usually provide you with a box that connects to your phone line (or cable line) and to your computer. This box is usually both a router and a modem. A modem is a device that negotiates the connection with your ISP through your telephone line while a router is a device that is used to connect two networks together, in this case your network to your modem. It connects to the router via the standard RJ45 and with the telephone line via the smaller RJ11. Its job is simply to translate data from one protocol to another since telephone lines do not use the same signaling and transmission methods that are used in computer networks. Because of this, data isn't being screened by the modem and any potential threat would still go through to your network.

1. A router is used to connect two or more networks while a modem is used to connect to a phone line

2. A router only connects to RJ45 connectors while modems need an RJ45 and an RJ11 for the phone line

3. A router provides security measures to protect your network but a modem does not

4. A modem is essential to connect to the internet while a router is not.

Data communications through the telephone network can reach any point in the world. The volume of overseas fax transmissions is increasing constantly, and computer networks that link thousands of businesses, governments, and universities are pervasive. Transmissions over such distances are not generally accomplished with a direct-wire digital link, but rather with digitally-modulated analog carrier signals. This technique makes it possible to use existing analog telephone voice channels for digital data, although at considerably reduced data rates compared to a direct digital link.

Transmission of data from your PC to a timesharing service over phone lines requires that data signals be converted to audible tones by a modem. An audio sine wave carrier is used, and, depending on the baud rate and protocol, will encode data by varying the frequency, phase, or amplitude of the carrier. The receiver's modem accepts the modulated sine wave and extracts the digital data from it. Several modulation techniques typically used in encoding digital data for analog transmission are shown below:

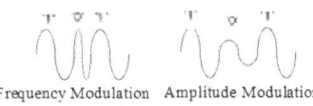

Frequency Modulation   Amplitude Modulation

Phase Modulation

Similar techniques may be used in digital storage devices such as hard disk drives to encode data for storage using an analog medium.

42.3 **IT Industry**

The policy of economic liberalization initiated by the government of India in 1991 and the rapid growth of the global software industry during the 1990s substantially contributed to its growth. In India Special economic Zones came up in different States in India. Many National and International IT Companies started with Professionals working form them. A few known centres are:

1) Hyderabad

2) Bangalore
3) Pune
4) Techno park, Thiruvananthapuram
5) M-City, Chennai,
6) New Delhi.
   Some of the very well-known IT Companies work in are:
   Infosys, Cognizant Technologies (CTS), TCS (Tata Consultancy Services), Oracle Corpn, Wipro, Accenture, US Global, CapGemini, IBS Software, *etc.*
E-Governance, e-Commerce, Net Banking, e-payment of Income Tax, are some of popular methods of living.

42.3.2 ATM Card

An ATM card issued by a Banking or financial institution to its customers which enables a customer to access an Automated Teller Machine (ATM) for transactions such as deposits, cash withdrawals, obtaining account information, and other types of banking transactions. The payment card may be any card which has that feature enabled, and may be a debit or credit.

Visa and MasterCard are brand names licensed to financial institutions to issue credit cards. American Express and Discover each issue their own credit cards. Some department stores also issue credit card

42.4 CELL PHONE – Overview

42.4.1 How does a cell phone work? What makes it different from a regular phone? What does the terms like GSM, *etc.*?

42.4.1.1 GSM, CDMA/TDMA

One of the most interesting facts about a cell phone is that it is actually a sophisticated radio.
The telephone was invented by Alexander Graham Bell (1876).
Wireless communication can trace its roots to the invention of the radio by Nikolai Tesla (1880s). formally presented by Italian named Guglielmo Marconi (1894).
Radio telephones were then started in cars. In this system, there was one central antenna tower, and around 25 channels available on that tower. This also meant that the phone in the car needed a powerful transmitter which should be big enough to transmit within the radius of about 50 - 70 *km.* without enough channels available.
The genius of the cellular system is the division of a city into small cells which allows extensive frequency reuse.
Cell phones have low-power transmitters in them. Many cell phones have two signal strengths: 0.6 *W* and 3 *W* (for comparison, most CB radios transmit at 4 *W*). The base station is also transmitting at low power.
All cell phones have a unique code associated with them (IMEI). These codes are used to identify the phone, the phone's owner and the service provider. Both walkie-talkies and CB radios are simplex devices.

42.4.2

A cell phone is a duplex device

A walkie-talkie typically has one channel , and

A CB radio has 40 channels .

A typical cell phone can
communicate on 1,664 channels or more!

A walkie-talkie can transmit about 1.6 *km* using a 0.25-W transmitter. A CB radio, because it has much higher power, can transmit about 8 *km* using a 5-W transmitter. Cell phones operate within cells, and they can switch cells as they move around. Cells give cell phones incredible range.

Someone using a cell phone can drive hundreds of miles and maintain a conversation the entire time because of the cellular approach.

42.4.3 Cell phone Technologies

There are three common technologies used by cell-phone networks for transmitting information:

1) Frequency division multiple access (FDMA)
2) Time division multiple access (TDMA)
3) Code division multiple access (CDMA)

   The first word tells you what the access method is. The second word, division, lets you know that it splits calls based on that access method.

4) FDMA puts each call on a separate frequency.
5) TDMA assigns each call a certain portion of time on a designated frequency.
6) CDMA gives a unique code to each call and spreads it over the available frequencies.

   FDMA separates the spectrum into distinct voice channels by splitting it into uniform chunks of bandwidth. To better understand FDMA, think of radio stations: Each station sends its signal at a different frequency within the available band. FDMA is used mainly for analog transmission. While it is certainly capable of carrying digital information, FDMA is not considered to be an efficient method for digital transmission.

   In FDMA, each phone uses a different frequency.

   TDMA is the access method used by the Electronics Industry Alliance and the Telecommunications

   Industry Association for InterimStandard 54 (IS-54) and Interim Standard 136 (IS-136). Using TDMA, a narrow band that is 30 kHz wide and 6.7 milliseconds long is split time-wise into three time slots.

Narrow band means "channels" in the traditional sense. Each conversation gets the radio for one-third of the time. This is possible because voice data that has been converted to digital information is compressed so that it takes up significantly less transmission space. Therefore, TDMA has three times the capacity of an analog system using the same number of channels. TDMA systems operate in either the 800MHz (IS-54) or 1900MHz (IS-136) frequency bands.

TDMA splits a frequency into time slots.

TDMA is also used as the access technology for Global System for Mobile communications (GSM).

42.4.4 GSM

GSM implements TDMA in a somewhat different and incompatible way from IS-136. Think of GSM and IS-136 as two different operating systems that work on the same processor, like Windows and Linux both working on an Intel Pentium III. GSM systems use encryption to make phone calls more secure. GSM operates in the 900 MHz and 1800 MHz bands in Europe and Asia, and in the 1900 MHz (1.9 GHz) band in the United States. It is used in digital cellular and PCS-based systems. GSM is also the basis for Integrated

42.4.5 Digital Enhanced Network (DEN),

A popular system introduced by Motorola and used by Nextel. TDMA enhances FDMA by further dividing the spectrum into channels by the time domain as well. A channel in the frequency domain is divided among multiple users. Each phone call is allocated a spot in the channel for a small amount of time, and "takes turns" being transmitted. In the figure above, each horizontal band represents the channel divided by the frequency domain. Within that is the vertical division in the time domain. Each user then takes turns occupying the channel. GSM, short for Global System for Mobile Communications is an open, digital cellular technology used for transmitting mobile voice and data services. It is one of the leading digital cellular systems. GSM uses narrowband TDMA, which allows eight simultaneous calls on the same radio frequency

42.4.6 FDMA and TDMA, CDMA transmission

It does not work by allocating channels for each phone call. Instead, CDMA utilizes the entire spectrum for transmission of each call. Each phone call is uniquely encoded and transmitted across the entire spectrum, in a manner known as spread spectrum transmission.

42.5 OPERATING SYSTEM (OS)

An operating system (OS) is a software program that manages the hardware and software resources of a computer – that is, a program that acts as an interface between the user and the computer hardware and controls the execution of all kinds of programs. The OS performs basic tasks, such as controlling and allocating memory, prioritizing the processing of instructions, controlling input and output devices, facilitating networking, and managing files.

Modern general-purpose computers, including PCs and mainframes, have an operating system to run other programs, such as application software. Examples of operating systems for PCs include

a) Microsoft (MS) Windows,
b) Mac OS (and Darwin),
c) Unix, and
d) Linux.

The lowest level of any operating system is its kernel. This is the first layer of software loaded into memory when a system boots or starts up. The kernel provides access to various common core services to all other system and application programs.

These services include, but are not limited to: disk access, memory management, task scheduling, and access to other hardware devices.

42.5.1 Types of Operating systems:

1) Batch Operating Systems:

The users of batch operating system do not interact with the computer directly. Each user prepares his job on an off-line device like punch cards and submits it to the computer operator. To speed up processing, jobs with similar needs are batched together and run as a group.

2) Time-Sharing operating systems:

Time sharing is a technique which enables many people, located at various terminals, to use a particular computer system at the same time. Timesharing or multitasking is a logical extension of multiprogramming. Processor's time which is shared among multiple userssimultaneouslyis termed as timesharing.

The main difference between Multi programmed Batch Systems and Timesharing Systems is that in case of multi programmed batch systems, objective is to maximize processor use, whereas in Timesharing Systems objective is to minimize response time.

3) Distributed Operating Systems

Distributed systems use multiple central processors to serve multiple real time application and multiple users. Data processing jobs are distributed among the processors according to which on 6e can perform each job most efficiently.

4) Network operating System

Network Operating System runs on a server and provides server the capability to manage data, users, groups, security, applications, and other networking functions. The primary purpose of the network operating system is to allow sharing of resources such as file/printer among multiple computers in a network, typically a local area network (LAN), a private network or to other networks. Examples of network operating systems are Microsoft Windows Server 2003, 2008, etc.

5) Real Time Operating System:

Real time system is defines as a data processing system in which the time interval required to process and respond to inputs is so small that it controls the environment. Real time processing is always on line whereas on line system need not be real time.

Real time systems are used when there are rigid time requirements on the operation of a processor or the flow of data and real time systems can be used as a control device in a dedicated application. For example Scientific experiments, medical imaging systems, industrial control systems, weapon systems, Air traffic control system, *etc.*

42.6   **MOBILE OPERATING SYSTEMS**:

Like a computer operating system, a mobile operating system is the software platform on top of which other programs run. When you purchase a mobile device, the manufacturer will have chosen the operating system for that specific device. The operating system is responsible for determining the functions and features available on your device, such as thumbwheel, keyboards.
Java ME Platform
Palm OS
Symbian OS
Linux OS
Windows Mobile OS

42.6.1 WAP
WAP synchronization with applications, e-mail, text messaging and more. The mobile operating system will also determine which third-party applications can be used on your device. Some of the more common and well-known Mobile operating systems include the following:

42.6.2 ANDROID (OS)
Android is a mobile operating system (OS) based on the Linux kernel that is currently developed by Google (2007), with a user interface based on direct manipulation, Android is designed primarily for touch screen mobile devices such as smartphones and tablet computers, with specialized user interfaces for televisions (Android TV), cars (Android Auto), and wrist watches (Android Wear). The OS uses touch inputs that loosely correspond to real-world actions, like swiping, tapping, pinching, and reverse pinching to manipulate on-screen objects, and a virtual keyboard. Despite being primarily designed for touchscreen input, it also has been used in game consoles, digital cameras, and other electronics.
Google Android Platform
Design and capabilities of a Mobile OS (Operating System) is very different than a general purpose OS running on desktop machines:
Mobile devices have constraints and restrictions on their physical characteristic such as screen size, memory, processing power, *etc.*
1) Scarce availability of battery power
2) Limited amount of computing and communication capabilities
3) A mobile OS is a software platform on top of which other programs called application programs can run on mobile devices such as PDA, cellular phones, smart-phone and etc. Below are the various layers in a mobile device:
4) Applications
5) OS Libraries
6). Devise OS Base, Kernel
7). Low-Level Hardware, Manufacturer Device Drivers

42.6.3 The following demonstrates the Symbian OS architecture:

Symbian OS
Libraries
KVM

Application Engines

Servers

Symbian OS Base- Kernel

Hardware

Despite being primarily designed for touch screen input, Android also has been used in game consoles, digital cameras, and other electronic. Fig. below displays the Android architecture.

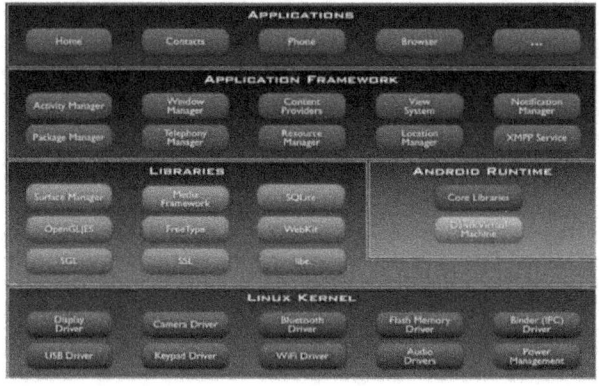

42.6.4 Macintosh Software, *eg.*, Apple,
Developed by Steve Jobs and used in iPod, iPhone, *etc.*
The logo of The Apple was introduced in 1978 by Rob Jaoff.

A free and open source Linux based operating system for touch screen mobile devices such as Smartphone's and tablet computers. Initially developed by Android, Inc., which Google backed financially and eventually bought.

42.6.5 IOS: Apple's proprietary operating system for iPhone, iPod Touch, iPad and Apple TV (2nd generation).

42.6.6 Blackberry OS/10:
A proprietary mobile operating system developed by BlackBerry Ltd. It is the predecessor of the BlackBerry 10 mobile operating system.

42.6.7 Windows Mobile:
This is a series of proprietary mobile Operating Systems developed by Microsoft and is a successor to the Windows Mobile platform.

42.6.8 Smartphone
In a nutshell, a Smartphone is a device that lets you make telephone calls, but also adds in features that, in the past, you would have found only on a personal digital assistant or a computer--such as the ability to send and receive e-mail and edit Office documents, for example.

42.6.9. **Wi-Fi**
Wi-Fi, also spelled **Wifi** or **WiFi**, is a local area wireless technology that allows an electronic device to exchange data or connect to the internet using 2.4 GHz UHF and 5 GHz SHF radio waves. The name is a trademark name, and is a play on the audiophile term Hi-Fi.
Mouse, key board or Internet facility can be possible with Wi-Fi, without having a physical cable / wire connection to the PC.

42.6.10. Summary – Comparison of Android OS *versus* iPhone OS
42.6.10.1 Android OS
i) Internal Memory
ii) Limited internal memory
iii) Big headache because apart from photos and media content, the default memory is already limited
iv) Lots of apps in the internal memory will eventually make your phones less smart
v) External Memory
*Yes. External SD card can be inserted to store photos, media, *etc.*
Notifications
Broadcasts: System-wide notifications and other application notifications (like a new mail/new tweet/new SMS) in cascade windows which can be pulled down to see details
42.6.10.2 iPhone OS
i) Internal Memory
ii) Good internal memory and you have choice of different internal memory sizes
 iii) External Memory
iv) No external expandable memory – since the internal memory itself is huge and good enough this is not an issue.
v) Notifications
a) Push notifications and individual notifications on updates
b) Saves power as server pushes data – there is no need to poll and pull data
42.6.10.3   Mobile Technology
Since Smartphones were introduced in the 1990s, the technology has progressed at an astounding pace and transformed daily life in the process. Today, millions of people walk around with an all-in-one computer, communication device and personal assistant in their pocket.
The computing capabilities of Smartphones and shareability of data made faster networks essential. 0G refers to pre-cell phone technology, for example radio-telephones. As we graduated from sending simple texts and pixelated photos over the 2G network to surfing the Web on 3G networks, it became possible for consumers to fully utilize the power of Smartphones.

Smartphone networks is in the process of have change. Current research in mobile wireless technology concentrates on advance implementation of 4G (fourth generation) to 5G. Rechargeable lithium-ion batteries are currently used in almost every Smartphone.

What is 2G?

Cell phones, began with 1G, received their first major upgrade when they went from 1G to 2G. This leap effectively took cell phones from analog to digital.

What is 2.5G?

Before making the major leap from 2G to 3G wireless networks, the lesser-known 2.5G was an interim standard that bridged the gap.

What is 3G?

Following 2.5G, 3G ushered in faster data-transmission speeds so you could use your cell phone in more data-demanding ways.

This has meant streaming video (i.e. movie trailers and television), audio and much more. Cell phone companies today are spending a lot of money to brand to you the importance of their 3G network. But what is 3G really and where did it come from?

What is 4G?

No technology would be complete without a looming upgrade for tomorrow. What's on the horizon for 4G, what improvements will it bring "beyond 3G" and when might we expect 4G to go live?

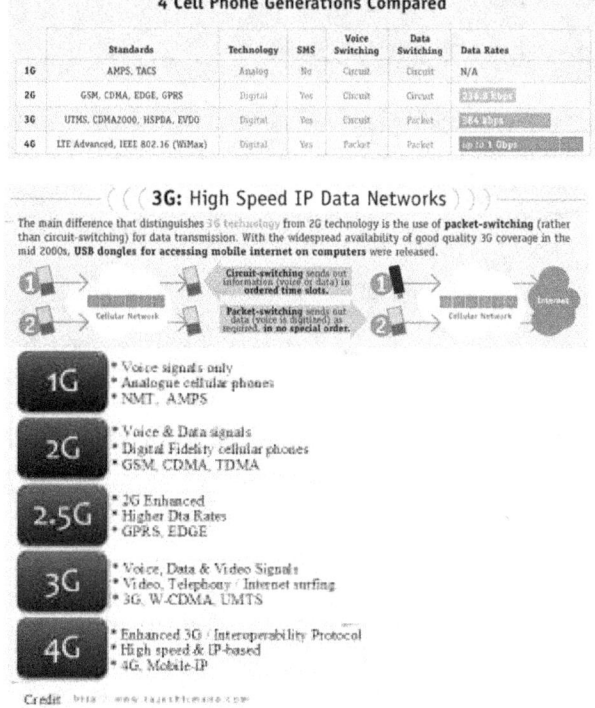

4 Cell Phone Generations Compared

	Standards	Technology	SMS	Voice Switching	Data Switching	Data Rates
1G	AMPS, TACS	Analog	No	Circuit	Circuit	N/A
2G	GSM, CDMA, EDGE, GPRS	Digital	Yes	Circuit	Circuit	236.8 Kbps
3G	UTMS, CDMA2000, HSPDA, EVDO	Digital	Yes	Circuit	Packet	364 Kbps
4G	LTE Advanced, IEEE 802.16 (WiMax)	Digital	Yes	Packet	Packet	up to 1 Gbps

(((**3G:** High Speed IP Data Networks)))

The main difference that distinguishes 3G technology from 2G technology is the use of **packet-switching** (rather than circuit-switching) for data transmission. With the widespread availability of good quality 3G coverage in the mid 2000s, **USB dongles for accessing mobile internet on computers** were released.

1G
- Voice signals only
- Analogue cellular phones
- NMT, AMPS

2G
- Voice & Data signals
- Digital Fidelity cellular phones
- GSM, CDMA, TDMA

2.5G
- 2G Enhanced
- Higher Dta Rates
- GPRS, EDGE

3G
- Voice, Data & Video Signals
- Video, Telephony / Internet surfing
- 3G, W-CDMA, UMTS

4G
- Enhanced 3G / Interoperability Protocol
- High speed & IP-based
- 4G, Mobile-IP

Credit http://www.rajeshkumare.com

42.7 TELEVISION

Greek 'Tele' means far plus Latin 'Visio' meaning sight.

Television, transmission and reception of still or moving images by means of electrical signals, especially by means of EM radiation using the techniques of radio and fibre-optic and coaxial cables.

42.7.1 Camera

The object image captured by the camera lens will be separated into three primary colours, viz., Red ®, Green (G) and Blue (B). Results will be emitted by the TV transmitterinto cromynance signal, luminence signal and synchornised signal. Example, a red filter transmits only red light.

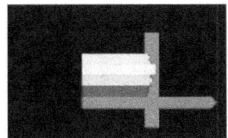

42.7.2 Colour Mixing

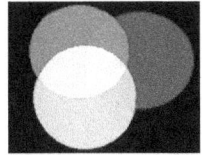

When the additive primary colours are mixed in proportion one gets white light.

42.7.3 Standard colour wheel

Standard colour wheel shows both primary (R,G,B) and secondary colours (Yellow, Magenta,and Cyan)

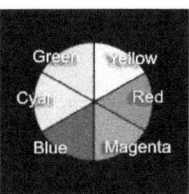

If any two colours exactly opposite each other on the colour wheel are mixed, the result is white. Note that instead of canceling each other as they did with subtractive colours, these complementary colours combine for an additive effect. it may be obvious at this point that by combining the proper mixture of Red, Green and Blue light, any colour of the rainbow can be produced.

> In colour TV only three colours $(R, G$ and $B)$ are needed
> to produce a full range of colours in a colour TV picture

42.7.4 One chip-colour CCD camera

One chip-colour CCD camera with an overlay of millions of tiny colored filters

The eye is much more sensitive to yellowish-green light than to either blue or red light

42.7.5 Modulating the signals
Besides pictures, television transmitter also carries voice signals. Voice signals are also transmitted along with video signlas.Images are transmitted by means of AM modulation, whereas the audio signals are transmitted with FM modulation. This is to eliminate noise and interference.

42.7.6 Channel
The group assined to the transmission frequency signal is called the Channel. Each has a 1 MHz channel in one field frequency bandare allocated to commercial TV broadcasters, viz.,
a) The field of low frequencr VHF channels, $2 - 6$ ($54 - 88 \ MHz.$)
b) The field of high frequency VHF channels $7 - 13$ ($174 - 216 \ MHz$)
c) UHF channels $14 - 83$ ($470 - 890 \ MHz$).

42.7.7 Transmission Systems
There are three TV transmitter systems,
i) National Television SystemCommittee (NTSC) in USA,
ii) Phases Alternating Line (PAL) in UK,
iii) Sequential Couleur a'Memorie (SECAM) used in France.
Theses Systems are distiguished by imageformat, the carrier frequency range, image and audio carriers.

42.8 **ANTENNA Fundamentals**
An antenna is a device for converting electromagnetic radiation in space into electrical currents in conductors or vice-versa, depending on whether it is being used for receiving or for transmitting, respectively. Passive radio telescopes are receiving antennas. It is usually easier to calculate the properties of transmitting antennas. Fortunately, most characteristics of a transmitting antenna (e.g., its radiation pattern) are unchanged when the antenna is used for receiving, so we often use the analysis of a transmitting antenna to understand a receiving antenna used in radio astronomy.

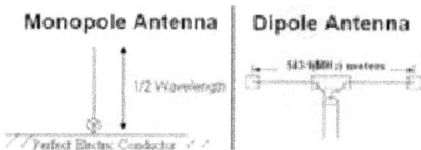

42.8.1.1 Dipole Antenna (Hertz Dipole)

Dipole antenna or aerial is a key element in RF types of antenna, used in ham radio and radio communications. The half wave ($\frac{\lambda}{2}$) dipole antenna consists of a conductive wire or rod that is half the length of the maximum wavelength λ the antenna is designed to operate at end to end, This wire or rod is split in the middle, and the two sections are separated by an insulator at the

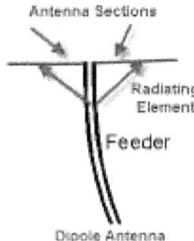

centre. Each wire or rod is connected electrically usually to a $50\ \Omega$ coaxial cable.

The approximate total dipole length L of a half wave($\frac{\lambda}{2}$) resonant dipole $= L = \frac{468}{f(MHz)}\ ft$.

Example: $L = \frac{468}{7(MHz)} = 66.85\ ft$

42.8.1.2 Radiation power distribution pattern

      The directivity of an antenna depends on the radiation pattern that it transmits. The power of the transmitting (radiating) signal hasdefinitive pattern, and the antenna in receiving end should be appropriately oriented (*i.e.*, its radiation pattern) has to be oriented parallel to the that

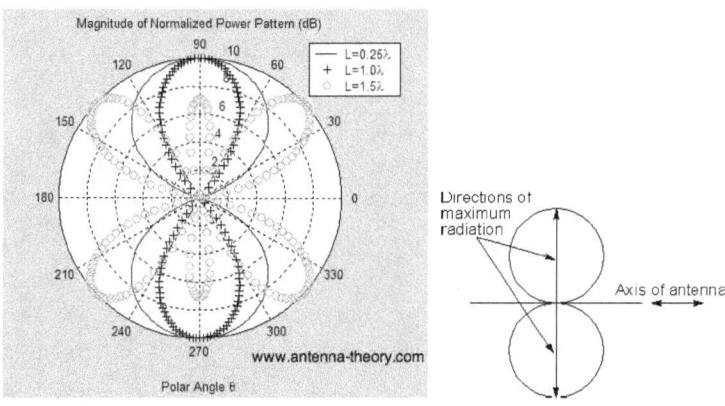

42.8.2 Other types of Antenna are

    1) Wire antennas

2) Travelling wave antennas
3) Reflectro antennas
4) Microstrip antennas
5) Log-periodic antennas
6) Aperture antennas

42.8.2.1 Long periodic Antenna - Yagi antenna

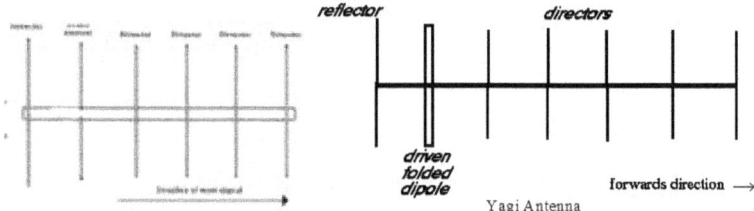

Yagi Antenna

42.8.2.2. Parabolic Reflector antenna

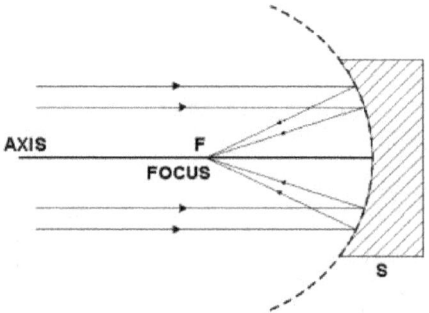

+% +%+^& +^+^%+

Appendix – 1, 2, 3

PRACTICE TESTS

Appendix – 1

PRACTICE TEST - 1

(JUNIORS) (Plus 2 candidates)
Entrance Test to B.Tech
[JEE (IIT), National Entrance tests NEET & CDS (UPSC)]

| Time 180 minutes |
| # Questions 100 |

DIRECTIONS:
Each of the items or incompleted statements or any other patterns of questions given below is
followed in general by 4 or 5 answer choices or completions. Select the best answer from the keys
for each of the 100 items.

1) Two measurements in an optical lever experiment gave values $c = (15.5 \pm .1)\, cm$ and
$d = (26.3 \pm 0.1)\, cm$. The error in the average value is
   a) 2.0%
   b) 1.9%
   c) 1.8%
   d) 0.95%
   e) 0.90%

2) In a Vernier caliper measurement, if 1 division in main scale $(msd) = 1mm$ and 10 vsd = 9 msd,
   and the plane surface A (or rod AB) coincide with the zero of the msd and B and zero of vsd
   coincide, and if the 6th vsd coincides with 1.6 cm of msd, length AB is
   a) 1.60 cm
   b) 1.06 cm
   c) 0.16 cm
   d) None of the above.

3) A spherometer has its fixed legs lie on the corners of an equilateral triangle. On a spherical
   surface (radius of the curvature 5.507 cm), the reading is 5.348 mm. Find the dimensions of the
   equilateral triangle (assume no zero error):
   a) 4.35 cm
   b) 4.10 cm
   c) 3.75 cm
   d) 3.173 cm
   e) 3.087 cm

4) A simple pendulum has a period of 10.0 s. Given, $g = 9.81\, ms^{-2}$ a calculation showed the length
   of the pendulum to be 24.8290289 m. The length is correctly expressed as:
   a) 25 m
   b) 24.8290 m
   c) 24.829 m

d) 24.83 *m*
e) 24.8 *m*

5) What is the dimension of centrifugal force per unit mass?
   a) LT^{-2}
   b) MLT^{-2}
   c) MLT^{-1}
   d) LT^{-1}
   e) None above.

6) With an initial velocity of $10\ ms^{-1}$, a dart player throws a dart at the target, kept $2.0\ m$ away and held at the same height as the hand of the player.. At what angle θ to the horizontal must the dart be thrown so that the dart strikes the target?

   a) $\sin^{-1}(0.1)$ above the horizontal.

   b) $\cos^{-1}(0.1)$ above the horizontal.

   c) $\frac{1}{2}\sin^{-1}(0.2)$ above the horizontal.

   d) $\frac{1}{2}\cos^{-1}(0.2)$ below the horizontal.

   e) $\tan^{-1}(0.1)$ below the horizontal.

7) The displacement of a particle from the origin is given by $x = \frac{1}{(1+t)}$, where t is the time. Calculate the acceleration of the particle.
   a) $2x^3$
   b) x^2
   c) $-x^2$
   d) $-x^3$
   e) $-2x^3$.

8) The plot given below represents a parabola, what should be the physical quantities representing Y and X so that the plot would indicate constant acceleration.

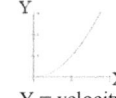

   a) X = time,   Y = velocity
   b) X = velocity,   Y= time.
   c) X = time,   Y = displacement
   d) X = time,   Y = acceleration
   e) X = velocity,   Y = displacement.

9) An electric field, $\vec{E}$ (in Vm^{-1}) is described by the vector drawn from point $(0,-2,3)$ to $(3,-2,-1)$. The unit vector along $\vec{E}$ is
   a) $(-0.6\hat{i} + 0.8\hat{k})$
   b) $(3\hat{i} - 0.8\hat{k})(0.6\hat{i} - 0.8\hat{k})$

c) $(0.6\hat{i} - 0.8\hat{k})$

d) $(3\hat{i} + 2\hat{k})$

e) None above.

10) A particle is displaced by $\vec{r} = (\hat{i}t^2 + \hat{j}t^{-2})$, find the time at which the velocity and displacement are mutually at right angles

a) $\frac{1}{4}s$

b) $\frac{1}{2}s$

c) $1s$

d) $2s$

e) $4s$.

11) What is the physical quantity which corresponds to the area under the acceleration *versus* time graph,

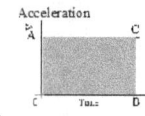

i.

a) Displacement.

b) Change in velocity

c) Force

d) Impulse

e) Work done.

12) An aircraft moves in a vertical circle of diameter $1km$. The speed of the aircraft at the bottom of this circle is $185\ ms^{-1}$. At this point of the aircraft, what would the apparent weight that the pilot would experience?

    a)   4.5 times heavier

    b)   3.5 times heavier

    c)   3.42 times heavier

    d)   (4.5)/2 times heavier

    e)   (3.5)/2 times heavier

13) A $0.5kg$ ball traveling at $40\ ms^{-1}$ stakes a target and comes to rest in a distance of $20\ cm$ If the time to stop is $4\ ms$, calculate the impulse..

    a)   2000 Ns

    b)   8 N

    c)   8 Ns

    d)   1.5 Ns

    e)   None above

14) An object of mass $50\ g$ traveling at $20\ ms^{-1}$ hits a vertical wall and rebounds back at $10\ ms^{-1}$. If the impact lasts for $0.02\ s$ calculate the average force on the wall

    a)   1.5 N

    b)   7.5 N

    c)   15 N

d)   75 N
e)   7.5 Ns

15) A 0.812 m long pendulum clock gains 2 minutes a day. By how much must its length be changed to make the clock run in time?

a)   2.25 mm
b)   22.5 mm
c)   225 mm
d)   None of above

16) It is required to remove a person of 50 kg completely from the earth's gravitational field. What quantity of energy is to be supplied to him?
a) 6000 MJ
b) 3000 MJ
c) 2200 MJ
d) 600 MJ
e) 300 MJ

17) A red ball (mass 30 g) moving at 4 ms⁻¹ collides with a stationary green ball (mass 50 g) which moves off with a velocity of 2 ms⁻¹ in the same direction as the red ball. What is the final velocity of the red ball? Is the collision elastic?

a) $0.67 \, ms^{-1}$, elastic

b) $0.67 \, ms^{-1}$, inelastic

c) $1.5 \, ms^{-1}$, elastic

d) $1.5 \, ms^{-1}$, inelastic

e) None above.

18) A solid ball of mass 0.2 kg is dropped from an elevated point of 2 m above the ground rebounds to a height of 1.8 m. What is the kinetic energy lost on impact?
a) 7.6 J
b) 4.0 J
c) 3.6 J
d). 0.9 J
e) 0.4 J .

19) A 10 m long steel mooring cable having and cross section $10^{-3} \, m^2$ is under a tension of $10^5 N$. If the Young's modulus of steel equals $2x10^{11} Pa$, find the strain energy stored in the cable.
a) 250 J
b) 125 J
c) 100 J
d) 25 J
e) None above

20) A stationery hunter shoots an arrow at a target moving directly away from him. When the arrow is shot, the target is 60 m away, and when the arrow strikes, the target is 80 m away. If

the arrow travels at $70\ ms^{-1}$, how fast was the target moving?. (Assume that the arrow travels horizontally, there is no air resistance, and there is no time arrow takes to accelerate from rest to $70\ ms^{-1}$)

   a) $70\ ms^{-1}$

   b) $52.6\ ms^{-1}$

   c) $17.5\ ms^{-1}$

   d) $7\ ms^{-1}$

   e) $5.25\ ms^{-1}$

21) An aircraft lands with a velocity of $55\ ms^{-1}$, reverse thrust from the engines is used to slow down it to a velocity of $25\ ms^{-1}$ in a distance of $240\ m$. If the mass of the aircraft is $3x10^4\ kg$, what is the measure of the reverse thrust supplied by the engines?
   a) $150\ kN$
   b) $110\ kN$
   c) $70\ kN$
   d) $35\ kN$
   e) $15\ kN$

22) A tennis player plays a ball which reaches him with a velocity of $20\ ms^{-1}$. If the maximum force he can exert on the wall is 200 N, how long must the ball be in contact with the racquet in order for the player to return it to his opponent with a velocity of $-30\ ms^{-1}$? (Mass of the ball $= 0.058\ kg$).
   a) $0.0029\ s$
   b) $0.0072\ s$
   c) $0.0145\ s$
   d) $0.145\ s$
   e) None above

23) Calculate the time required for a $0.6\ kW$ motor to raise a $50\ kg$ block vertically to a height of $8\ m$? (Assume 100% efficiency).
   a) $6.7\ s$
   b) $7.5\ s$
   c) $67\ s$
   d) $75\ s$
   e) $77\ s$

24) A mass of $2\ kg$ hangs by two equal cords making an angle of $60°$. What is the tension on each cord?
   a) $20\ N$
   b) $17.3\ N$
   c) $11.5\ N$
   d) $10\ N$
   d) $8.5\ N$

25) The distance between the Earth and the Sun is $1.5x10^{11}$ m. Find the mass of the Sun.

a). $20x10^{30}$ kg

b) $2.0x10^{30}$ kg

c) $73x10^{29}$ kg

d) $7.3x10^{29}$ kg

e) None of above.

26) One wishes to have a toy car (mass m) go in a loop-the-loop (Fig) around a circular track with radius R. What is the minimum speed v_{min} the car must have at the top of the loop?

a) $\sqrt{2gR}$

b) $\sqrt{gR/2}$

c) $gR/2$

d) $\sqrt{gR}$

27) A satellite (150 kg) is moving in a circular orbit of radius 7.3 Mm around the Earth. Calculate the escape velocity from this altitude.

a) 7.4 kms^{-1}

b) 10 kms^{-1}

c) 11.3 kms^{-1}

d) 11.1 kms^{-1}

e)   None above.

28) A motor cycle at rest is given a constant acceleration 'a' for some time, and then a constant retardation 'b' thereafter and comes to rest. If the total time elapsed is' t' seconds, the maximum velocity attained is

a) $\frac{ab}{a-b}t$.

b) $\frac{a+b}{ab}t$

c) $\frac{ab}{a+b}t$

d) $\frac{ab}{b-a}t$

e) None of above.

29) A spherical body executes simple harmonic motion with amplitude of 0.17 m and a period of 0.84 s. Its co-ordinate X is (Assume the body is initially at rest.)

a) $(0.17\ m)\ Cos[(7.5^e\ s^{-1})t]$

b) $(0.17\ m)\ Cos[(7.5^c\ s^{-1})t + \varphi]$.

c) $(0.17\ m)\ Cos(1.2t + \varphi)$.

d) $(0.17\ m)\ Cos(1.2Hz)t$.

30) Calculate the speed with which a person jumping from a height around $10\ m$ **without serious injury.**

    a) $0.14\ ms^{-1}$

    b) $1.4\ ms^{-1}$

    c) $14\ ms^{-1}$

    d) $7\ ms^{-1}$

    e) None of above.

31) Determine the terminal velocity of a hailstone of diameter $1.0\ cm$ falling from an altitude of $1.0\ km$. (Density of ice $0.92g - cm^{-3}$).

    a) $8.3\ ms^{-1}$

    b) $4.1\ ms^{-1}$

    c) $2.9\ ms^{-1}$

    d) $0.829\ ms^{-1}$

32) Calculate the angle of incline θ f a teak board on which a person of weight W can stand vertically without sliding down. The person wears leather-solid shoes.. $\mu_k = 0.6$ and $\mu_S = 0.5$

    a) $\theta = 27^\circ$

    b) $\theta = 31^\circ$

    c) $\theta = 42^\circ$

    d) $\theta = 44.3^\circ$

33) Find the distance between centres of copper atoms in the solid state, given that these atoms occupy cubic cells. ($\rho_{Cu} = 9x10^3\ kg.m^{-3}$, A = 64).

    a)   0.23 nm

    b)   0.31 nm

    c)   0.40 nm

    d)   0.47 nm

    e)   None above.

34) Find the angular momentum of the Earth relative to the Sun, given the earth's radius is $R_E = 1.5x10^{11}\ m$, and its mass is $M_E = 6.0x10^{24}\ kg$ ..

    a) $1.8x10^{40}\ kg.m^2 s^{-1}$

    b) $2.7x10^{40}\ kg.m^2 s^{-1}$.

    c) $10.8x10^{40}\ kg.m^2 s^{-1}$

d) $10.8x10^{39}\ kg.m^2s^{-1}$

e) $2.7x10^{39}\ kg.m^2s^{-1}$

35) A scientist sets up a simple pendulum of length $8.60\ mm$ in the Moon and measures its period for small oscillation to be $4.6s$.
What is the acceleration due to gravity at this location on the surface of the Moon?

   a) $10\ ms^{-2}$.

   b) $7.6\ ms^{-2}$.

   c) $1.6\ ms^{-2}$.

   d) $0.6\ ms^{-2}$.

   e) None above

36) Determine the pressure (in atm) at $3.00\ km$ deep from the surface of the ocean in a place. Assume ocean as a static incompressible fluid.

   a) $300\ atm$,

   b) $100\ atm$,

   c) $77\ atm$,

   d) $30\ atm$

37) What is the gauge pressure (in $torr$) at the bottom of a 6.2 m deep swimming pool?

   a) $230\ torr$,

   b) $290\ torr$,

   c) $394\ torr$,

   d) $460\ torr$

38) What volume of helium is required to float a balloon, if the empty balloon and its equipment have a mass of $390\ kg$?

   a) $3.51\ m^3$

   b) b)$35.1\ m^3$

   c) c) $75.2\ m^3$

   d) d) $175.1\ m^3$

   e) e)$351\ m^3$

39) Arteriosclerosis decreases the radius of the channel in an artery in the heart by a factor 2. By what factor must the heart increase the pressure gradient in the artery to keep the flow of blood constant? Assume the blood flow to be laminar.

   a) 32

   b) 16

   c) 8

   d) 4

   e) None above

40) When in thermal equilibrium at the Triple point of water, the pressure of helium in a constant volume gas thermometer is $1020\ Pa$. The pressure of helium is $288\ Pa$, when the thermometer is in thermal equilibrium with liquid nitrogen at its normal Boiling point. The normal Boiling point of liquid nitrogen is

a) $-269°C$

b) $-196°C$

c) $77.1K$

d) $-186°C$

e) $-183°C$

41) If one cylinder of a diesel engine, air initially at atmospheric pressure and is at 310 K, acquires a volume of $0.420\,l$. It is compressed quasi-statically and adiabatically using a compression ratio of 15. Determine the final temperature.
   a) 1020 K
   b) 920 K
   c) 870 K
   d) 790 K

42) Which of the accompanying PV diagrams best represents an isothermal (constant temperature) process?

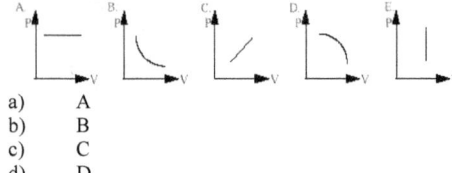

   a) A
   b) B
   c) C
   d) D
   e) E

43) At the top of Mt. Everest temperature is 250 K, with atmospheric pressure around $3.3x10^4\ Pa$. At sea level temperature is 300 K, atmospheric pressure is $1.0x10^5\ Pa$; density of air is $\rho_{air} = 1.2\ kg.m^{-3}$. Calculate the density of air at the top of Mt. Everest in $kg.m^{-3}$.
   a) 1.00
   b) 0.83
   c) 0.48
   d) 0.40
   e) 0.24

44) If $25\,J$ work is done by the person compressing the air in a bicycle pump, and $20\,J$ of thermal energy leaves the gas through the walls of the pump, what is the effect on the internal energy of the air in the pump?
   a) Increases by $5\,J$
   b) Decreases by $5\,J$
   c) Decreases by $45\,J$
   d) Increases by $45\,J$
   e) None above.

45) The limit to the efficiency of a modern power station is 53 %. It uses steam at a temperature T

as a heat source, with a heat sink ~ 373K (The temperature at which heat condenses at atmospheric pressure). T is

a) $527°C$
b) $485K$
c) $527K$
d) $213K$
e) $485°C$

46) A $1.0x10^{-3}$ m^3 of air at $20°C$ and $101.35 kPa$ is heated at constant pressure until its volume doubles. Obtain the final temperature of the air.

a) $586K$
b) $586°C$
c) $333°C$
d) $40°C$
e) None above

47) The output of a lamp is 10 W. What is the intensity of the light from the lamp at a distance of 2 km?

a) $2.5x10^{-6} W.m^{-2}$
b) $2x10^{-7} W.m^{-2}$
c) $8x10^{-7} W.m^{-2}$
d) $5x10^{-8} W.m^{-2}$
e) $2x10^{-7} W.m^{-1}$.

48) A representative transverse sinusoidal wave is expressed by :

$$y = 2x10^{-3} Sin(18x - 600t + 30°)m$$

One or more of the following is correct. Pick up the answer from the keys given'

1. The amplitude of the wave is 2 mm
2. The wavelength is 0.35 m.
3. The displacement at $t = 0s$
4. The speed of the wave is 33.3 ms^{-1}.

Answer key codes:

a) 1, 3, 4 are correct
b) 2, 3 are correct
c) 1 and 4 are correct
d) 4 only is correct.
e) 1,2,3,4 are correct

49) An organ pipe consists of open-ended pipes of different lengths ranging from 2.4 m to 30 mm. Determine the range of fundamental frequencies produced by the pipes.

a) $280 Hz - 2580 Hz$
b) $210 Hz - 16950 Hz$
c) $140 Hz - 11340 Hz$

d) 72 Hz – 5670 Hz

e) 35 Hz – 2835 Hz

50) An approaching car with 30 ms^{-1} sends its horn with a frequency of 300 Hz What note will you hear? (Given the velocity of sound is 340 ms^{-1}).

    a) 330 Hz to 2 SF

    b) 660 Hz

    c) 825 Hz

    d) 165 Hz

51) A sample of Silicon having cross section A = $3x10^{-6}$ m^2 carries a current of $100\mu A$. The number of free electrons for Silicon is $2.6x10^{18}$ m^{-3} Find the drift velocity.

    a) 80 ms^{-1}

    b) 71 ms^{-1}.

    c) 53 ms^{-1}.

    d) 20 ms^{-1}.

    e) None above

52) A capacitor with $10,000\mu F$, 25V is connected to a motor so that latter could lift a body of mass of 100 g . Calculate the maximum height to which the body could be raised.

    a) 32 mm

    b) 3.2 cm

    c) 32 cm

    d) 3.2 m

    e) 32 m

53) An aircraft with a wing span of 50 m flies horizontally with a velocity of 180 ms^{-1} at a location where the vertical component of the earth's magnetic field is $3.5x10^{-5}T$ .What is the magnitude of the emf induced across wingspan of the aircraft?

    a) 3.2 V

    b) 0.32 V (to 2 SF)

    c) 32 mV

    d) 3.2 mV

    e) None above.

54) The nuclide ^{226}R of mass 0.25 kg undergoes α – decay at a measured rate of $9.0x10^{-12} s^{-1}$. Find the half-life of ^{226}R ? (N_A = $6.1 x10^{23} mole^{-1}$).

    a) $5.1x10^{11} s$

    b) $5.1x10^{10} s$

    c) $5.1x10^{8} s$

    d) $5.1x10^{7} s$

    e) 162 yrs

55) Two deuterons (mass $2.014102\,u$) fuse to form a helium nucleus (mass $3.016030\,u$) and a neutron (mass $1.008665\,u$). Find the energy released in this process.

    a) $3.27\,MeV$.
    b) $0.327\,MeV$
    c) $0.033\,MeV$
    d) $0.35\,u$
    e) $0.035\,u$

56) The potential of $1\,MeV$ between two deuterons allow to fuse them to form deuterium at temperatures,

    a) $\sim 10^6\,K$
    b) $\sim 10^7\,K$
    c) $\sim 10^8\,K$
    d) $\sim 10^{10}\,K$
    e) $\sim 10^{12}\,K$

57) A wave is expressed by $y = 10\,Sin(\frac{2\pi}{5})(1800\,t - x)$. What is its period?

    a) $\frac{1}{360}$ s
    b) $\frac{1}{180}$ s
    c) $\frac{\pi}{1800}$ s
    d) $\frac{\pi}{900}$ s
    e) $\frac{2\pi}{900}$ s

58) Which material exhibits positive temperature coefficient of resistance?
    a)   Germanium
    b)   Insulators
    c)   Semiconductors
    d)   Silicon
    e)   Conductors

59) The depletion layer of a p-n junction
    a).   is of constant width
    b).   Acts like an insulating zone under reverse bias
    c).   has a width that increases with an increase in forward bias.
    d)   is depleted of ions

60) In a series resistance circuit
    a)   the impedance at resonance is high.
    b)   the applied voltage and current are in phase at resonance.
    c)   the voltage across the inductor and capacitor are 90° out of phase.
    d)   Minimum current flows at resonance.
    e)   the voltage across the inductor and capacitor are 45° out of phase.

61) An instrument has sensitivity of $50\,k\Omega.V^{-1}$. The internal resistance of the meter on the 10 V range will be

a) $500\ \Omega$
b) $5\ k\Omega$
c) $50\ k\Omega$
d) $500\ k\Omega$
e) $5\ M\Omega$

62) The graduated scale of the moving pointer–end measuring instrument on ac , is calibrated as
a) r.m.s. values
b) peak values
c) mean values
d) average values
e) peak-to-peak values

63) Elastic wave is propagated in a string made of brass $\rho_{Brass} = 8.5\ g.cm^{-3}$, and is expressed by $y = 10\ Sin(\frac{2\pi}{5})(1800\ t - x)$. The elastic modulus E of brass is
a) $(1800\ ms^{-1})^2 (8.5\ g.cm^{-3})$
b) $9\ GPa$
c) $0.9\ GPa$
d) $241\ GPa$
e) None of above

64) An immersion heater with electrical resistance $7\ \Omega$ is immersed in $0.1\ kg$ of water at $20^0\ C$ for 3 minutes. If the flow of current is $4A$, what is the final temperature of the water (Specific heat capacity of water $C = 4.2x10^3\ Jkg^{-1}K^{-1}$). Assume all the heat is absorbed by water with no heat loss.
a) $28^0\ C$
b) $48^0\ C$
c) $52^0\ C$
d) $68^0\ C$
e) $72^0\ C$

65) In a CRO, the input to the X-plate is $80\ Hz$ and a frequency $f = 160\ Hz$ is applied to the Y-plate. The CRO shows Lissajous pattern as

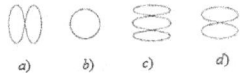

a)   b)   c)   d)

66) The refractive index of flint glass is 1.5 and that of alcohol is 1.36 (water is 1.33) with respect to air. Calculate the refractive index of glass with respect to (water) alcohol..
a) $(1.65 - 1.36)$
b) $\frac{1.36}{1.65}$
c) $\frac{1.65}{1.36}$
d) $\frac{1.65+1.36}{1.65}$
e) $\frac{1.65+1.36}{1.36}$

67) A wire potentiometer arrangement, used to measure the emf of a cell $\varepsilon_{l\,1}$, is shown in the circuit diagram below. AB is the potentiometer wire of length $1\,m$ and resistance $5\,\Omega$. The emf of a standard dc source (ε) is $\varepsilon\,V$ and rheostat R is adjusted to $5\,\Omega$. If the null point is obtained at C with AC = $75\,cm$, what is the emf of ε_l?

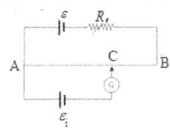

   a) $0.225\,V$
   b) $0.25\,V$
   c) $2.5\,V$
   d) $2.75\,V$
   e) $2.25\,V$

68) A thin lens is inside a tube AB, A sharp image of an illuminated object is formed on a screen when A is $90\,cm$ from the screen and also when A is $140\,cm$ from the screen. How far is the lens from A?
   a) $10\,cm$
   b) $25\,cm$
   c) $325\,cm$
   d) $50\,cm$
   e) None of above.

69) A beam of light with an unknown wavelength passes through two narrow slits $0.30\,mm$ apart and forms an interference pattern on a screen $2.0\,m$ away from the slits. If the distance between the fringes in the interference pattern is $3.0\,mm$, find the wavelength of the light.
   a) $590\,nm$
   b) $549\,nm$
   c). $450\,nm$
   d) $435\,nm$.
   e) $420\,nm$

70) The focal length f of a spherical mirror (concave / convex) is
   a) Half of its radius of curvature R
   b) R
   c) $(3/2)\,R$
   d) R
   e) R/4

71) Find the time required for a molecule with average speed $10^4\,cms^{-1}$ in a liquid a distance of $10^{-3}\,cm$. Given the mean free path is $10^{-8}\,cm$.
   a) $3x10^{-2}\,s$,
   b) $4.6x10^{-2}\,s$,
   c) $3.4x10^{-2}\,s$,

d) $1.7x10^{-2} s$,

e) $10^{-2} s$

72) The wheel on a grinder is a uniform disc of weight $0.90 kg$ and radius $8.0 cm$ In $30 s$ it attains from rest circular motion at $1400 rpm$ How large a friction torque brings it to rest?

    a) $1.2x10^{-2} N.m$

    b) $1.4x10^{-2} N.m$

    c) $-2.4x10^{-2} N.m$

    d) $-1.2x10^{-2} N.m$

    e) None above.

73) A wheel of mass $0.60 kg$ and radius of gyration $40 cm$ rotating at $300 rpm$. Find the rotational kinetic energy.

    a) $0.47 kJ$

    b) $0.58 kJ$

    c) $4.70 kJ$

    d) $5.80 kJ$

    e) $3.0 kJ$.

74) Four point charges are kept one each at the corners of a square of side $30 cm$ Two of the charges are $+2.0 \mu C$ and the other two are $-2.0 \mu C$. Find the potential at the centre of the square.

    a) $4.0x10^5$

    b) $0.40x10^3 V$

    c) $0.40x10^4 V$

    d) $3.4x10^5 V$

    e) $0 V$.

75) An α – particle falls from rest through a potential drop $3.6x10^6 V$. Find its kinetic energy.

    a) $3.0 MeV$

    b) $6.0 MeV$

    c) $9.0 MeV$

    d) $0.90 MeV$

    e)   None of above.

76) A dry cell has an emf of $1.52 V$. Its terminal potential falls to zero as a current of $25 A$ passes through it. Find the internal resistance.

    a) 0.61Ω

    b) $61 m\Omega$

    c) 0.061Ω

    d) $6.1 m\Omega$

77) Atom is the smallest indivisible particle of the matter. Atom is made of electron, proton and neutrons. The proton was discovered in the year by ------------.

    a) Goldstein

b) JJ Thomson
c) Rutherford
d) Millikan
e) None of above.

78) Match List I with List II and select the correct answer using the codes given below the Lists.

List I (Name)		List II (Examples)	
A.	Isobars	1.	$_1^1H$ & $_1^3H$
B.	Isotones	2.	$_{80}^{202}Hg$ & $_{82}^{202}Pb$ \
C.	Isotopes	3.	$_6^{13}C$ & $_7^{14}N$
D.	Isomer	4..	$_{17}^{37}Cl$ & $_{18}^{39}Ar$
		5.	$_{26}^{57}Fe$ & $_{26}^{57m}Fe$

Answer key Codes
a) 4   2   5   3
b) 2   3   1   5
c) 3   2   5   4
d) 3   2   1   4

79) Assertion (A): Classical Mechanics successfully describes the motion of macroscopic particles but fails in the case of microscopic particles.

Reason (R): Classical mechanics ignores the concept of dual behavior of matter, especially for sub-atomic particles and the Heisenberg's Uncertainty principle.

Codes

a) Both A and R are individually true, and R is the correct explanation of A'

b) Both A and R are individually tru, but R is not the explanation of A.

c) A is true but and R is false.

d) A is false but R is true

80) Match List I with List II and select the correct answer using the codes given below the Lists.

List I (Priciple)	List II (Explantion)
A. Aufbau Principle	1. Maximum multiplicity pairingof electrons in the orbital belonging to the same subshell. does not take place unless the ech orbital is singly occupied
B. Hund's Priciple	2. The completely filled or half-filled subshellshave symmtrical distribtion of electrons and are more stable.
C. (n+ℓ) Rule	3. In ground state of an atom the orbitals are filled inorder of increasing energies.
D. Exclusion principle	4. No two electrons in an atom can have the same set of 4 quantum numbers (n,ℓ,m,s),
	5. Orbitals with lower value of ℓ have lower energy, abnd lowlower n value

Answer key Codes
a) 4   2   5   3
b) 2   3   1   5
c) 3   2   5   4
d) 3   1   5   4

81) Neutrons can be found in all atomic nuclei except in one case. Which is this atomic nucleus ?

a) Deuterium
b) Tritium
c) Hydrogen atom

d) Helium ion

e) None of above.

82) When α – rays hit a thin foil of gold, very few α – particles is deflected back. What does it prove?

a) There is a very small heavy body present within the gold atom.

b) Within the gold atom a positively charged body with size of α – particle is present .

c) Within the gold atom a relatively very heavy body having positive charge with size of α – particle is present.

d) The target gold is of higher size than α – particle that it is a hindrance to the passage of incident α – particle beam.

83) How many electrons in an atom have the quantum numbers $n = 3, \ell = 0$?

a) 16

b) 12

c) 9

d) 4

e) 2.

84) An element with mass number 81 contains 31.7 % more neutrons as compared to protons. Assign the atomic symbol.

a) $^{81}_{35}Br$

b) $^{81}_{34}Se$

c) $^{81}_{38}Sr$

d) $^{81}_{33}As$

e) $^{81}_{36}Kr$.

85) Calculate the number of electrons which will together weigh one gram.

a) 6.01224×10^{26}

b) 6.01224×10^{27}

c) 1.098×10^{27}

d) 1.098×10^{26}

e) None of above.

86) What is the minimum product of uncertainty in position and momentum of an electron?

a) $\frac{h}{4}$

b) $\hbar$

c) $\frac{\hbar}{2\pi}$

d) $\frac{\hbar}{2}$

e) $2\hbar$.

87) What is the eigen value of operator $\left(\frac{d^2}{dx^2} - 2x\frac{d}{dx}\right)$ and function $(4x^4 - 12x^2 + 3)$?

a) -9
b) +9
c) 17
d) -17
e) None of above.

88) What is the eigen value of operator $\left(x\frac{d}{dx}\right)$ and function $5x^4$.

    a)   +4
    b)   +2
    c)   -3
    d)   -7
    e) None of above.

89) A molecule is known to give a vibrational IR absorption frequency. What do you interpret on the characteristic property of the molecule?
One or more of the following is correct. Pick up the right answer from the keys given'
    1. The molecule is centro-symmetric
    2. The molecule has a dipole moment.
    3. The molecule has no dipole moment
    4. The molecule has a change in dipole moment.
    Answer key codes:
        a) 1, 3, 4 are correct
        b) 2, 3 are correct
        c) 1 and 4 are correct
        d) 4 only is correct.

90) What is the explicit quantum mechanical operator corresponding to the Classical mechanical operator kinetic energy, in one-dimensions?

    a) $\left(-\frac{\hbar^2}{2m}\cdot\frac{\partial^2}{\partial x^2}\right)$

    b) $\left(\frac{\hbar^2}{2m}\cdot\frac{\partial^2}{\partial x^2}\right)$

    c) $\left(\frac{i\hbar}{2m}\cdot\frac{\partial}{\partial x}\right)$

    d) $\hat{H}$

91) In a series resistance circuit
    a)   the impedance of resonance is high.
    b)   the applied voltage and current are in phase at resonance.
    c)   the voltage across the inductor and capacitor are 90° out of phase.
    d)   minimum current flows at resonance.
    e)   the voltage across the inductor and capacitor are 45° out of phase.

92) The television tube normally employs
    a)   Magnetic deflection and electric focusing
    b)   Electric deflection and magnetic focusing.
    c)   Electric deflection and electric focusing.
    d)   magnetic deflection and magnetic focusing.
    e)   magnetic horizontal deflection, electrical vertical deflection and electric focusing.

93) Lenz' law of electro-magnetism is a consequence of the law of conservation of
   a) Energy,
   b) Momentum,
   c) Angular momentum,
   d) Charge
   e) Magnetic moment

94) The minimum force needed to slide a piano weighing $1000\ N$ on a horizontal plane surface is $350\ N$. Find the coefficient of friction involved.
   a) 0.25
   b) 0.30
   c) 0.35
   d) 0.40
   e) 0.42.

95) A certain galvanometer has a resistance of $10\ \Omega$ and maximum permitted current is $6\ mA$, which gives full scale deflection. How may currents up to $2\ A$ can be measured?
   a) A series resistor of $0.03\ \Omega$ is to be connected to the Galvanometer.
   b) A shunt of $0.03\ \Omega$ is to be connected in parallel to the galvanometer terminals
   c) A shunt of $0.30\ \Omega$ is to be placed across the galvanometer.
   d) A resistor of $300\ k\Omega$ is to be connected in series to the galvanometer.
   e) None of the above.

96) Determine the wavelength of a thermal neutron.
   a) 0.147 nm
   b) 1.47 nm
   c) 0.0147 nm
   d) 0.147 μm
   e) 0.29 μm

97) A sodium lamp of 20 W radiates light of 589 nm what is the number of photons the lamp radiates per second?
   a) $59x10^{17}$
   b) $5.9x10^{17}$
   c) $5.9x10^{18}$
   d) $5.9x10^{19}$
   e) None of above

98) Find the wavelength of the photon that produces at appropriate conditions an electron-positron pair each with kinetic energy 220 keV.
   a) $8.49x10^{-10}\ m$
   b) $8.49x10^{-13}\ m$,
   c) $2.49x10^{-13}\ m$,
   d) $2.49x10^{-10}\ m$,
   e) None of above.

99) A beam of electrons with $400 \, ms^{-1}$ is arranged to pass through a diffraction grating. Find the minimum separation between the slits in the grating so that the diffracted beam emerges at 25°
    a) 2.3 μm
    b) 4.6 μm
    c) 6.9 μm
    d) 9.2 μm
    e) 13.8 μm

100) Find the wavelength limit of the line for the hydrogen Paschen series.
    a) 821 nm
    b) 1.71 eV
    c) 921 nm
    d) 91 nm
    e) None of above

---------------END-----------------------

Appendix - 2

PRACTICE TEST - 2

(SENIORS) (Undergraduates)
[GATE, NET (UGC- CSIR), State SET, IAS (Preliminary)(UPSC),
GRE Physics (USA)]

Appendix - 2

PRACTICE TEST - 2

(SENIORS) (Undergraduates)
[GATE, NET (UGC- CSIR), State SET, IAS (Preliminary)(UPSC),
GRE Physics (USA)]

| Time 180 minutes |
| # Questions 100 |

> DIRECTIONS:
> Each of the items or incompleted statements or any other patterns of questions given below is followed in general by 4 or 5 answer choices or completions. Select the best answer from the keys for each of the 100 items.

1) A cylindrical rod and vernier caliper -1 $msd = 1$ mm, 10 $vsd = 9$ msd. When the jaws are pressed together the zero of the vernier scale is on the right of the zero of the main scale, and 2^{nd}vsd coincides with some msd. With the sample placed in between the jaws, it is found that 6th division of vernier scale coincides with 1.6 cm. What is the length measured?
   a) 1.66 cm
   b) 1.60 cm
   c) 1.08 cm
   d) 1.06 cm
   e) None of the above

2) A screw gauge has zero correction $= +5$ vsd. Pitch $= 1$ mm; Number of msd on head scale is 100.
   With a glass plate gently gripped, the reading of linear scale is 2 whereas 32 in head scale divisions on the reference line. Find the thickness of the glass plate.
   a) 2.32 mm
   b) 2.16 mm
   c) 2.15 mm
   d) 2.05 mm
   e) 2.08 mm

3) The fixed legs of a spherometer lie on the corners of an equilateral triangle. On a spherical surface the reading is 5.348 mm. The radius of the curvature of the surface is 5.507 cm. Compute the dimensions of the equilateral triangle is (assuming no zero error).
   a) 4.81 cm
   b) 4.58 cm
   c) 4.27 cm
   d) 4.21 cm
   e) 4.1 cm

4) What is dimension of Centrifugal force per unit mass?

a) LT^2
b) MLT^2
c) M^0LT^{-2}
d) $M^0L^{-1}T^2$
e) None of above

5) If a particle is displaced by $\vec{r} = \hat{i}\,t^2 + \hat{j}\,t^{-2}$. Compute the time at which the velocity and displacement are at right angles.
   a) 0.9 s
   b) 1 s
   c) 1.5 s
   d) 1.8 s
   e) 2.2 s

6) A particle of mass m obeys the Maxwell-Boltzmann distribution at temperature T. What is the most probable speed of the particle?.
   a) $\sqrt{\frac{2k_BT}{m}}$
   b) $\sqrt{\frac{3k_BT}{m}}$
   c) $\sqrt{\frac{k_BT}{m}}$
   d) $\sqrt{\frac{8k_BT}{\pi m}}$
   e) $\sqrt{\frac{3k_BT}{\pi m}}$

7) What is the maximum%- error in the measurement of area of a rectangle, of sides measured to $\pm 8\%$ and $\pm 5\%$?
   a) 3%.
   b) $\frac{8}{5}\%$
   c) $\frac{5}{8}\%$
   d) 13%
   e) $\frac{13}{8}\%$

8) Determine the maximum possible error in velocity that when $s = (25.0 \pm 0.5)\ ms^{-1}$ and $t = (18.0 \pm 0.1)\ s$.
   a) $0.04\ ms^{-1}$.
   b) $1.39\ ms^{-1}$
   c) $1.4\ ms^{-1}$
   d) $1.388\ ms^{-1}$
   e) $0.03\ ms^{-1}$

9) A body gets accelerated along x-axis according to $x = (2t^3 + 5t^2 + 5)\ m$, t is in sec.. Compute the average velocity between at t=2s and t=3s.

a) 44 ms^{-1}

b) 84 ms^{-1}

c) 63 ms^{-1}

d) 40 ms^{-1}

e) None of above

10) A body is accelerated at $a = (4x - 2)\ ms^{-2}$. Given the initial velocity $v_o = 10\ ms^{-1}$. Calculate the velocity at any other position x.

a) $v = \sqrt{(x^2 + x - 25)}\ ms^{-1}$

b) $v = \sqrt{(x^2 - x)}\ ms^{-1}$

c) $v = \sqrt{(x^2 - x + 5)}\ ms^{-1}$

d) $v = \sqrt{(x^2 - x + 25)}\ ms^{-1}$

e) $v = 2\sqrt{(x^2 - x + 25)}\ ms^{-1}$

11) A particle has radius vector $\vec{r} = (2\hat{i} + j)\ m$ and linear momentum $\vec{p} = (5\hat{i} + 2\hat{j})\ kg.ms^{-1}$. Compute its angular momentum in matrix form..

a) $\begin{pmatrix} 0 \\ 0 \\ -1 \end{pmatrix} kg.m^2 s^{-1}$

b) $\begin{pmatrix} 0 \\ 0 \\ 1 \end{pmatrix} kg.m^2 s^{-1}$;

c) $\begin{pmatrix} 0 \\ -1 \\ 1 \end{pmatrix} kg.m^2 s^{-1}$

d) $\begin{pmatrix} 0 \\ -1 \\ 0 \end{pmatrix} kg.m^2 s^{-1}$

e) $\begin{pmatrix} -1 \\ 0 \\ 0 \end{pmatrix} kg.m^2 s^{-1}$.

12) A balloon moves up with a velocity of $12\ ms^{-1}$ and drops an object at an altitude of $32.0\ m$. How long will it take to reach the ground?

a) 1.6 s .

b) 2.4 s

c) 2.0 s

d) 4.0 s

e) 0.8 s

13) A motor boat moving towards North at $15.0\ km.hr^{-1}$, when the water current is $5.0\ km.hr^{-1}$ and S $70°$ E . Obtain the resultant velocity of the boat.

a) 13.4 $km.hr^{-1}$, N $19.4°$ E .

b) 14.1 $km.hr^{-1}$, N $19.4°$ E

c) 14.1 $km.hr^{-1}$, S $19.4°$ E

d) $13.4 \, km.hr^{-1}$ S 19.4^o E

e) $13.4 \, km.hr^{-1}$, N

14) The displacement vector $\vec{r} = (2\hat{i} - 2j +\_5\hat{k}) \, m$ and linear momentum $\vec{p} = (-\hat{i} + 4j +\_2\hat{k}) \, kg.ms^{-1}$. Find the angular momentum, $\vec{r}$ in.

   a) $(26\,\hat{i} - 9j +\_5\hat{k}) \, kg.m^2s^{-1}$

   b) $(-6\,\hat{i} - 9j +\_5\hat{k}) \, kg.m^2s^{-1}$

   c) $(-6\,\hat{i} - 9j +\_5\hat{k}) \, kg.ms^{-1}$

   d) $(-26\,\hat{i} - 9j +\_5\hat{k}) \, kg.m^2s^{-1}$.

   e) $(-26\,\hat{i} - 9j +\_5\hat{k})$

15) What is the angle between the $\vec{F} = 2u_x + 3u_y - u_z$ units and $\vec{s} = -u_x + u_y + 2u_z$ units?

   a) 96.3^o

   b) 85.6^o

   c) 79.1^o

   d) 60°

   e) None of above

16) A body of $4 \, kg$, suspended from a spring, changes its length by $1.50 \, cm$. Now if the spring without the body is stretched by $2.0 \, cm$, find the work done.

   a) $5.22x10^{-2} J$.

   b) $7.52x10^{-2} J$

   c) $5.22x10^{-1} J$

   d) $19.32x10^{-2} J$

   e) $1.93x10^{-1} J$

17) A force $\vec{F} = 6t \, N$ acts on a particle of mass $2kg$ at rest. What is the work done during the first $2s$?

   a) $36.0 \, J$

   b) $18.0 \, J$

   c) $9.0 \, J$

   d) $4.5 \, J$

   e) $2.25 \, J$.

18) A rifle of mass $3.0 \, kg$ shoots a bullet of $0.025 \, kg$ at $100.0 \, ms^{-1}$ speed. Find the kinetic energies of the rifle and bullet.

   a) $0.52 \, J; \; 1.25x10^2 J$

   b) $2.08 \, J; \; 1.25x10^2$

   c) $1.04 \, J; \; 12.5 \, J$

   d) $1.04 \, J; \; 1.25x10^2 J$

   e) $1.04 \, J; \; 75 \, J$

19) From a horizontal hose pipe a jet of water is thrown out which hits horizontally a vertical wall at a speed of $20.0\ ms^{-1}$. The end of the pipe has a diameter of $2.0\ cm$. Find the force exerted on the wall.
   a) $1.3\ N$
   b) $9.6\ N$
   c) $75\ N$
   d) $125.7\ N$
   e) $12.6\ N$.

20) If an unpowered car of $1000\ kg$ takes a corner of radius $20.0\ m$. Given the coefficient friction between the wheel surface and road is 0.5. Find the maximum speed permitted.
   a) $10.0\ ms^{-1}$
   b) $13.3\ ms^{-1}$
   c) $39.9\ ms^{-1}$
   d) $3.9\ ms^{-1}$
   e) $5.0\ ms^{-1}$.

21) Calculate the tension in the string held by a person with a hammer of $7.0\ kg$ when swung around at $1\ rev.s^{-1}$ of the circle of radius $1.5\ m$.
   a) $26.3\ N$
   b) $52.5\ N$
   c) $105\ N$
   d) $414\ N$
   e) None of above

22) Calculate the depth below the surface of Earth where iron ($\rho = 7870\ kg\ m^{-3}$) will float, if in air at sea level has $\rho = 1.3\ kg\ m^{-3}$.
   a) $75\ km$
   b) $7.5\ km$
   c) $375\ km$
   d) $750\ km$
   e) $37.5\ km$

23) Water leaves the jet of a horizontal hose at $10.0\ ms^{-1}$. If the velocity of water leaving the hose is $0.4\ ms^{-1}$ calculate the pressure within the hose. Given the density of water as $\rho = 1000\ kg\ m^{-3}$, atmospheric pressure $1x10^5\ Pa$.
   a) $49,920\ Pa$
   b) $44,900\ Pa$
   c) $7.56x10^4\ Pa$
   d) $149,920\ Pa$
   e) $.1.0x10^5\ Pa$

24) Compute the maximum height over which water can be siphoned at atmospheric pressure of $75\ cm$ Hg.
   a) $102\ m$.

b) 70.2 m
c) 40.8 m
d) 20.4 m
e) 10.2 m

25) The barrel and receiver of a bicycle pump have volumes 0.2 ℓ and 0.8 ℓ, respectively. After how many strokes the pressure inside will increase to 6 times the original pressure?
   a) 4
   b) 7
   c) 20
   d) 26
   e) None of above.

26) What will be the energy stored in a 2 m long copper wire of cross section $A = 0.5$ mm^2 when a force $\vec{F} = 50$ N is applied between its ends?
   a) 0.04 J
   b) 0.01 J
   c) 0.11 J
   d) 0.41 J
   e) 0.20 J.

27) Calculate the gravitational attraction between two automobiles of mass 1000 kg and 1200 kg, at a separation of 5 m? Given $G = 6.67x10^{-11}$ $Nm^2kg^{-2} = 6.67x10^{-11}$ $m^3kg^{-1}s^{-2}$.
   a) $3.2x10^{-4}$ N
   b) $3.2x10^{-5}$ N
   c) $3.2x10^{-6}$ N
   d) $6.4x10^{-6}$ N
   e) $6.4x10^{-5}$ N.

28) Find the change in gravitational potential energy when a body of 2 kg is moved from the surface of the Earth to an altitude of $h = 100$ m. $m_E = 6.0x10^{24}$ kg., $R_E = 6400$ km, $G = 6.67x10^{-11}$ Nm^2kg^{-2}
   a) 1969 J
   b) 2114 J
   c) .196.9 J
   d) 370.3 J
   e) 211.4 J.

29) Calculate the change in pE of a satellite of 500 kg when it changes its altitude from 200 km to 199 km. Find also the retarding force on the satellite. Given $G = 6.67x10^{-11}$ Nm^2kg^{-2}, $m_E = 6.0x10^{24}$ kg,. $R_E = 6400$ km.
   a) $-8.0x10^{-6}$ J;';0.05 N
   b) $-4.0x10^{-6}$ J; 0.05 N
   c) $-4.0x10^{-6}$ J;0.1 N
   d) $-8.0x10^{-6}$ J; 0.1 N

e) $-2.0x10^{-6} J$; $0.1 N$

30) Obtain the terminal velocity of a rain drop of radius $0.2\ cm$, $\rho_{water} = 1000\ kgm^{-3}$, $\rho_{Air} = 1.3\ kgm^{-3}$, $\eta = 1.81x10^{-5}\ Pas$.
   a) $4.3\ ms^{-1}$
   b) $5.5\ ms^{-1}$
   c) $7.8\ ms^{-1}$
   d) $8.7\ ms^{-1}$
   e) $9.1\ ms^{-1}$.

31) Find the time taken for a Carbon particle of $r = 0.0001m$ and $\rho_C = 2300kgm^{-3}$ to fall $2m$ through air. Given $\eta = 0.001Pas$.
   a) 18 minutes
   b) 24 minutes
   c) . 48 minutes
   d) 72 minutes
   e) 77 minutes .

32) Find the excess pressure of air bubble of radius $0.1\ mm$ in water. What will be the total pressure if the bubble was $10\ cm$ below the surface. Given $T = 72.7x10^{-3}\ Nm^{-1}$, Atm. $P = 1.03x10^{5}\ Pa$.
   a) $1454\ Pa$; $1.039x10^{5}\ Pa$
   b) $114\ Pam^{-1}$; $1.039x10^{5}\ Pa$
   c) $1.14x10^{3}\ Pa$; $1.39x10^{5}\ Pa$
   d) $1454\ Pa$; $1.39x10^{5}\ Pa$
   e) $1.14x10^{3}\ Pa$; $1.39x10^{5}\ Pa$.

33) Total energy of SHM is $80\ J$. When the body is at $\frac{3}{4}$ of the amplitude of motion from equilibrium, what is the potential energy?
   a) $80\ J$
   b) $68\ J$
   c) $45\ J$
   d) $39\ J$
   e) None of above.

34) A sphere of $R = 1\ mm$ of water is sprayed into a million droplets of equal size. How much work was done? (Given surface tension $T = 72.7x10^{-3}\ Nm^{-1}$).
   a) $413.38\ ergs$
   b) $447.86\ ergs$
   c) $612.2\ ergs$
   d) $690.7\ ergs$
   e) $895.72\ ergs$

35) Calculate the radius of capillary in which water rises to $12.5\ cm$. Angle of contact $\theta = 0°$ for water and glass. Given $T = 72.7x10^{-3}\ Nm^{-1}$, $\rho_{water} = 1000\ kgm^{-3}$.

a) 0.15 *mm* .
b) 0.11 *mm*
c) 0.097 *mm*
d) 0.086 *mm*
e) None of above.

36) For two circular glass plates each of radius 5 *cm* separated with a 0.01 *mm* thick film of water between them, Find the force required to separate the plates.
a) 77.1 *N*
b) 102.3 *N*
c) 110 *N*
d) 118 *N*
e) 124 *N* .

37) Find the angular momentum of Earth around Sun; $M_\odot = 5.98x10^{24}$ *kg* , $R_\odot = 1.49x10^8$ *km* , and $T_E = 3.16x10^7$ *s* .

    a) $7.71x10^{41}$ *Js*
    b) $2.67x10^{41}$ *Js*
    c) $2.67x10^{39}$ *J*
    d) $2.67x10^{40}$ *Js*
    e) $7.71x10^{40}$ *Js*

38) A uniform ladder, 10 *m* in length and 400 *N* weight, rests in equilibrium against a vertical wall with frictionless, at 60° to horizontal. Find the reaction at both ends of the ladder.
a) 437.7 $N ; \theta = 78^o$ to vertical
b) 437.7 $N ; \theta = 78^o$ to vertical
c) 416.3 $N ; \theta = 78^o$ to vertical
d) 416.3 N , and at $\theta = 74^o$ to vertical
e) 485 $N ; \theta = 68^o$ to vertical.

39) Calculate the angular dispersion of a flint glass prism of refracting angle 20° Take $n_C = 1.6434$, $n_F = 1.6648$, $n_D = 1.6550$.

    a) 0.45^o
    b) 0.428^o
    c) 0.402^o
    d) 0.376^o
    e) 0.30^o .

40) A mirror forms an erect image 40 *cm* from the object and $\frac{1}{3}$ rd its height. Compute the position (*u* & v), the radius of curvature (*R*), and if mirror is convex or concave.
a) u = 30 *cm*, v = −10 *cm* ; R = −30 *cm* , convex .
b) u = 30 *cm*, v = −10 *cm* ; R = −30 *cm* concave
c) u = 30 *cm*, v = −15 *cm* R = −25 *cm* ; convex

d) $u = 25\ cm, v = -15\ cm$; $R = -25\ cm$; convex
e) $u = 25\ cm, v = -15\ cm$; $R = -25\ cm$; concave

41) A crown glass prism of refracting angle $6°$ is combined with a flint glass prism to form an achromat. Calculate the refracting angle of the flint glass prism. Given, $n_F = 1.6648$, $n_C = n_R = 1.6434$, $n_D = 1.6550$ (for Flint); $n_D = 1.5175, n_F = 1.5233, n_C = 1.5150$ (for Crown glass)

a) $1.93°$
b) $1.68°$
c) $1.54°$
d) $1.34°$
e) $1.18°$.

42) Light is refracted at the boundary between water ($n_1 = 1.333$) and an unknown medium. The angle of incidence is $25°$ and the angle of refraction is $20.6°$, Calculate the refractive index of the unknown medium.

a) $n_2 = 1.43$
b) $n_2 = 1.6$
c) $n_2 = 1.66$
d) $n_2 = 2.16$
e) $n_2 = 2.42$

43) An ideal gas with volume $0.3\ m^3$ expands at constant pressure of $2x10^5\ Pa$ to $0.45\ m^3$ What is the work done by the gas? Given $P = 1.5x10^5\ Pa$.

a) $3x10^4\ J$
b) $3.8x10^4\ J$
c) $5.7x10^4\ J$
d) $12.7x10^4\ J$
e) $23x10^4\ J$.

44) A light ray passes from water ($n_1 = 1.333$) to diamond ($n_2 = 2.42$) with an angle of incidence of $75°$. Calculate the angle of refraction. Discuss the meaning of your answer.

a) $\theta_2 = 23.7°$, light ray is bent towards the normal
b) $\theta_2 = 32.1°$, light ray is bent towards the normal
c) $\theta_2 = 29.4°$, light ray is bent away from the normal
d) $\theta_2 = 32.1°$, light ray is bent away from the normal
e) $\theta_2 = 32.1°$, light ray is bent towards the normal.

45) Given that the refractive indices of air and water are 1.00 and 1.33, respectively, find the critical angle.

    a) $Sin^{-1}1.00$

    b) $Sin^{-1}1.33$

    c) $Sin^{-1}\frac{1}{1.33}$.

    d) $Sin^{-1}\frac{1}{0.33}$

    e) $Sin^{-1}\frac{1}{2.33}$

46) A far-sighted person has near point at $100\ cm$. Reading glass of what power is required to see at $25\ cm$?

    a) $+1.8\ D$

    b) $+2.1\ D$

    c) $+2.6\ D$

    d) $+3.0\ D$

    e) $-2.5\ D$

47) A near-sighted person has near and fast points at $12\ cm$, respectively. What is the power of the conventional lens needed to see distant objects clear? What will be the near point? Assume that each lens is at $3\ cm$ from the eye.

    a) $P = -6.3\ D$, $d_i = -17cm$

    b) $P = -6.3\ D$, $30\ cm$ in front of the lens.

    c) $P = -6.7D$, $30\ cm$ in front of the lens

    d) $P = -5.9D$, $d_i = -17cm$.

48) A biconvex lens is made out of glass $n = 1.52$. One side has twice the radius of curvature of the other, and $f = 5\ cm$. Find the two radii.

    a) $R_1 = -3.9\ cm$; $R_2 = 7.8\ cm$

    b) $R_1 = -5.0\ cm$; $R_2 = 7.8\ cm$

    c) $R_1 = -4.8\ cm$; $R_2 = 7.0\ cm$

    d) $R_1 = 7.8\ cm$, $R_2 = -3.9\ cm$

    e) None of above.

49) Calculate the fringe-width in an Young's experiment with $\lambda = 550\ nm$, where the double slits are separated by $0.75\ mm$ and the screen positioned at $0.80\ m$ away from them.

    a) $0.30\ mm$

    b) $0.33\ mm$

    c) $0.27\ mm$

    d) $0.22\ mm$

    e) $0.39\ mm$.

50) In a Newton's ring experiment using $\lambda = 589\ nm$ was viewed from reflection. The diameter of the nth dark ring was found to be $0.28\ cm$, and that of the $(n+10)^{th}$ ring was $0.68\ cm$. Find the radius of curvature of the lens used.

    a) $0.5\ m$

b) 1.30 m
c) 1.66 m
d) 2.06 m
e) 2.18 m.

51) Calculate the ratio of the diameter of the 5th interference ring with and without water ($n = 1.33$) between the lens and the plate in a Newton's ring experiment. The radius of curvature of the lens is 0.5 m
   a) 0.5
   b) 0.68
   c) 0.87
   d) 1.02
   e) 1.12

52) A thin equi-convex lens of focal length 4 m and made of $n = 1.52$ rests on and in contact with an optical flat. Interference circular fringes were viewed normally using from reflection using $\lambda = 546$ nm. Find the diameter of the fifth bright ring.
   a) 0.33 cm
   b) 0.233 cm
   c) 3.19 mn
   d) 6.39 mn
   e) 0.123 cm

53) In a Fresnel's bi-prism experiment the refracting angles of the prism were 1.5^o and the refractive index of the glass was 1.5. With the single slit 5 cm from the bi-prism and using light of $\lambda = 580$ nm, fringes were formed on a screen 1.0 m from the single slit. Calculate the fringe width.
   a) 0.038 cm
   b) 0.044 cm
   c) 0.071 cm
   d) 0.096 cm
   e) 0.117 cm.

54) Calculate the drift velocity in a piece of cylindrical wire where the current is 1 A, the free electron density is $5x10^{28}$ m^{-3} and the diameter of the wire is 1 mm.
   a) $1.6x10^{-4}$ ms^{-1}
   b) $2.7x10^{-4}$ ms^{-1}
   c) $7.3x10^{-4}$ ms^{-1}
   d) $11.2x10^{-4}$ ms^{-1}
   e) $3.2x10^{-5}$ ms^{-1}.

55) A potential difference of 12V is causing electrons to flow through a wire so that $1.4x10^{20}$ electrons pass a point in the wire in 1 minute. Calculate the charge that passes a given point in 1 minute.
   a) 78.1 C
   b) 49.7 C
   c) 22.4 C

d) 14.9 C

e) 3.4 C

56) Electrons to flow through a wire so that $1.4x10^{20}$ electrons pass a point in the wire in 1 minute, when a potential difference of 12V is applied. Compute the resistance of the wire.

a) 4.3 Ω

b) 10.4 Ω

c) 18.4 Ω

d) 32.1 Ω

e) 123.1 Ω

57) In an RC charging circuit, $R = 47$ kΩ, $C = 1000$ μF and V = 5V DC, What value will be the voltage across the capacitor at 0.7 time constants?

a) 4.4 V

b) 2.5 V

c) 3.15 V

d) 5 V

e) 2.5 V

58) Which of the following elements would be expected to be paramagnetic and which exhibit diamagnetism? He, Be, Li, N? Given the electron configuration for each element. He $\rightarrow 1s^2$; ;
Li $\rightarrow 1s^2 2s^1$; Be $\rightarrow 1s^2 2s^2$; and N $\rightarrow 1s^2 2s^2 2p^3$.

a) Li & He are paramagnetic, while Be & N are diamagnetic.

b) Li & Be are paramagnetic, while N & He are diamagnetic

c) Li & N are paramagnetic, while Be & He are diamagnetic

d) Li, Be & He are paramagnetic, while N only is diamagnetic

e) Only N is paramagnetic, while Li, Be & He are diamagnetic

59) A series circuit consists of a resistance of 4 Ω, an inductance of 500 mH and a variable capacitance connected across a $100V, 50Hz$ supply. Calculate the capacitance required to give series resonance and the voltage generated across both the inductor and the capacitor.

a) 20.3 μF , 3927.5 V

b) 20.3 μF ; 39.3 V

c) 33.3 μF ; 39.3 V

d) 33.3 μF ; 392.7 V

e) None of above.

60) A series resonance network consisting of a resistor of 30 Ω, a capacitor of 2 μF and an inductor of 20 mH is connected across a sinusoidal supply voltage which has a constant output of 9 V at all frequencies.

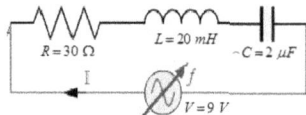

Calculate the resonant frequency, the quality factor Q and the bandwidth BW of the circuit.\

a) 796 Hz ; $Q = 4.01$; $BW = 717$ Hz

b) 796 Hz, $Q = 3.33$, $BW = 238$ Hz
c) 238 Hz, $Q = 3.33$; $BW = 238$ Hz
d) 1034 Hz; $Q = 4.01$; $BW = 717$ Hz
e) 1034 Hz; $Q = 3.33$; $BW = 717$ Hz

61) What is the emf induced in a coil of 200 turns placed in a magnetic field, and the rate of flux 0.01 Wbs^{-1}.?
    a) +4.1 V
    b) +2 V
    c) −2 V.
    d) −4.1 V
    e) +5.4 V

62) Calculate the emf generated between the wing-tips of an aircraft flying horizontally at 200 ms^{-1} in a region where the vertical component of Earth's magnetic field is $4.0x10^{-5}T$, if the aircraft has wing-span 25 m.
    a) 0.2 V
    b) 20 mV
    c) 50 mV
    d) 60 mV
    e) 90 mV.

63) Determine the inductance of a solenoid 0.5 m long, cross section area 20 cm^2 and having 500 turns.
    a) 1.25 mH
    b) 125 μH
    c) 25 μH
    d) 25 mH
    e) 0.25 H.

64) Consider a disc is rotating at 20 Hz inside a solenoid of 1000 turns and length 1 m carrying a current of $1A$. The radii of the disc and axle are 2 cm and 0.25 cm Find the emf generated.
    a) 1.36 mV
    b) 136 μV
    c) 13.6 μV
    d) 1.36 μV
    e) 0.14 mV

65) The coil of a galvanometer has area 4 m^2 and 200 turns Torque $\vec{\tau} = 2x10^{-7}$ Nm causes it to turn through $180°$ against the tension of the suspension. B = 0.2 T acts on the coil, Find the current that causes the spot on scale kept 1 m away to be deflected through 1 mm*
    a) 4 nA
    b) 0.40 nA
    c) 40 nA
    d) $0.05\mu A$
    e) $0.09\mu A$

66) An electron is travelling at 5% of the speed of light. Find the momentum and the deBoglie wavelength.

   a) 0.07×10^{-23} $kg.ms^{-1}$; 0.047 nm

   b) 2.8×10^{-23} $kg.ms^{-1}$; 0.47 nm

   c) 2.8×10^{-23} $kg.ms^{-1}$; 0.047 nm

   d) 1.4×10^{-23} $kg.ms^{-1}$; 0.47 nm

   e) 1.4×10^{-23} $kg.ms^{-1}$; 0.047 nm

67) By looking at the spectrum of the star Alpha Centauri, it is found that one of the calcium absorption lines has a wavelength $\lambda = 396.820$ nm. The same line in the spectrum of the Sun (which is close to the Earth) is $\lambda_\odot = 396.849$ nm. What is the speed of Alpha Centauri that is receding from us?

   a) -1.5×10^2 kms^{-1}

   b) -3.4×10^2 kms^{-1}

   c) -6.3×10^3 kms^{-1}

   d) -8.7×10^3 kms^{-1}

   e) -7.1×10^3 kms^{-1}.

68) A resistance R = 10 Ω and L = 10 μH and V = 30 V form a series circuit. At $t=0$. What is the energy stored in the inductor at $t = \infty$?

   a) 74×10^{-5} J

   b) 27×10^{-5} J

   c) 4.5×10^{-5} J

   d) 3.7×10^{-4} J

   e) 0.37 J.

69) Given the threshold wavelength of sodium for photo-electrons is $\lambda_{th} = 542$ nm, Calculate the work function of sodium for light of $\lambda = 450$ nm is incident on its surface.

   a) $0..91$ eV

   b) 2.29 eV

   c) 3.12 eV

   d) 5.20 eV

   e) None of above.

70) Before the advent of quantum mechanics atoms are considered to decay at very short time. How long the electrons in a hydrogen atom will spiral around the nucleus by emitting electro-magnetic radiation?

   a) 0.1 ns

   b) 7.0 ns

   c) 0.01 μs

   d) 0.71 μs

e) $4.6 \ \mu s$.

71) The two isotopes of potassium of atomic masses $38,975 \ u$ and , $40.974 \ u$ constitutes in nature in $93.4\% : 6.6\%$ ratio. Compute the atomic mass of potassium as found in nature.

    a) $39,97 \ u$
    b) $39,1 \ u$
    c) $40.1 \ u$
    d) $39,2 \ u$
    e) $40.0 \ u$

72) The electron and the μ-meson both have the same electric charge.. If the electron in a hydrogen atom is replaced by a μ-meson, what will be radius of the μ-mesonic atom in terms of a_o? Given the mass of μ-meson $= 207 \ m_e$

    a) $r_\mu = \dfrac{a_o}{\sqrt{207}}$

    b) $r_\mu = \dfrac{2a_o}{\sqrt{207}}$

    c) $r_\mu = \dfrac{a_o}{207}$.

    d) $r_\mu = \dfrac{2a_o}{207}$

    e) $r_\mu = a_o$

73) In one of the radioactive chains we come across the element $^{226}_{88}Ra$. The final product of the chain is $^{206}_{82}Pb$. Find the number of α and β emitted when $^{226}_{88}Ra$ decays into $^{206}_{82}Pb$

    a) 5α only
    b) 4α and 5β
    c) 3α and 6β
    d) 5α and 4β
    e) 6α and 6β

74) Compute the amount of $^{226}_{88}Ra$ in secular equilibrium with $1 \ kg$ of pure $^{238}_{92}U$ ($\tau_{1/2} = 1620 \ x10^9$ yrs and $4.5 \ x10^9$ yrs , respectively).

    a) $\left(\dfrac{238}{226} \dfrac{4.5 x10^9}{1620 x10^9} \right) gm$.

    b) $\left(\dfrac{226}{238} \dfrac{1620 x10^9}{4.5 x10^9} \right) gm$.

    c) $\left(\dfrac{238}{238+226} \dfrac{4.5 x10^9}{1620 x10^9} \right) gm$

    d) $\left(\dfrac{238-226}{238+226} \dfrac{4.5 x10^9}{1620 x10^9} \right) gm$.

    e) $\left(\dfrac{226}{238+226} \dfrac{1620 x10^9}{4.5 x10^9} \right) gm$

75) Write a nuclear equation for the decay where Z changes from $62 \to 60$ and N changes from $85 \to 83$.

  a) $^{147}_{82}X \to {}^{143}_{80}Y + 2{}^2_1He + {}^1_0n$ ..   ok

  b) $^{147}_{62}X \to {}^{143}_{60}Y + 4{}^1_0n$ ..

  c) $^{147}_{62}X \to {}^{143}_{60}Y + 2{}^2_1He$.

  d) $^{147}_{82}X \to {}^{143}_{80}Y + {}^4_2He$

  e) None of above

76) The nucleus of iron ^{56}Fe has matter density $2 \times 10^{17}\ kg\ m^{-3}$. What is its radius?

  a) $1.2\ fm$.

  b) $4.8\ fm$

  c) $6.0\ fm$

  d) $10.3\ fm$

  e) $13.3\ fm$

77) The binding energy per nucleon for elements near iron the Periodic Table is about $8.90\ MeV$. Calculate the atomic mass, including electrons, of $^{56}_{26}Fe$ '.

  a) $M = 55.5\ u$

  b) $55.1\ u$

  c) $56.3\ u$

  d) $M = 55.9\ u$

  e) $M = 56.7\ u$.

78) Find the binding energy of $^{107}_{47}Ag$, which has an atomic weight of $106.905\ u$.

  a) $915\ keV$

  b) $9315\ keV$

  c) $915\ eV$

  d) $931.5\ eV$

  e) $907\ eV$.

79) Gamma radiation from $^{191}_{77}Ir$ is used for a Mossbauer experiment. Find the velocity range required to scan the Mossbauer spectra. Given, absolute width $\Delta E = 4.7 \times 10^{-6}\ eV$ and $E = 129\ keV$.

  a) $0.022\ ms^{-1}$ to $+0.022\ ms^{-1}$

  b) $-2.2\ mms^{-1}$ to $+2.2\ mms^{-1}$

  c) $+2.2\ cms^{-1}$ to $-2.2\ cms^{-1}$

  d) $-22\ cms^{-1}$ to $+22\ cms^{-1}$

80) The half-life of ^{32}P is 14.3 days. $250\ \mu Ci$ ($= 9.25\ MBq$) is bought and used precisely after 43 days. Find the activity.

  a) $219.25\ \mu Ci$

  b) $118.5\ \mu Ci$

  c) $31.25\ \mu Ci$

d) $18.5\ \mu Ci$

e) $7.8\ \mu Ci$

81) To start with $1.0x10^{-2}\ gm$ of a pure radioactive substance, and $4\ hrs$ lapsed it was seen that only $0.25x10^{-2}\ gm$ remained. What is the half-life of the substance?

    a) $2.0\ hrs$

    b) $1.36\ hrs$

    c) $65.4\ mts$

    d) $48\ mts$

    e) $2.5x10^{3}\ s$

82) Find the Q-value of the reaction $\pi^{-} + p \rightarrow \Lambda^{o} + K^{o}$. Rest masses of $\pi^{-}, p, \Lambda^{o}\ and\ K^{o}$ are $140\ MeV$, $938\ MeV$, $1116\ MeV$ and $498\ MeV$, respectively.

    a) $-405\ MeV$

    b) $+405\ MeV$

    c) $+536\ MeV$

    d) $-536\ MeV$

    e) $-237\ MeV$

83) How much coal is required to run a $100\ W$ light bulb $24\ hrs$ a day for a year? Take the thermal energy content of coal is $6,150\ kWhr / ton$.

    a) $913\ kg$ of coal.

    b) $671\ kg$ of coal

    c) $325\ kg$ of coal

    d) $289\ kg$ of coal

    e) $51\ kg$

84) Protons and neutrons consist of two quark in different combinations. What are they?

    a) $p(uud)\ and\ n(udd)$

    b) $p(udd)\ and\ n(uud)$

    c) $p(ud\tilde{d})\ and\ n(u\bar{u}d)$

    d) $p(uu\breve{d})\ and\ n(u\tilde{d}d)$

    e) $p(ud\tilde{d})\ and\ n(uud)$

85) Calculate the total binding energy of $^{20}_{10}Ne$. Given, $^{20}_{10}Ne$ has a mass of $M = 19.992439\ u$, $M_{p} = 1.007825\ u\ M_{n} = 1.008665\ u$.

    a) $350.3\ MeV$

    b) $195.1\ MeV$

    c) $160.6\ MeV$

    d) $137.3\ MeV$

    e) $.124.3\ MeV$

86) Determine the amount of energy available if $1\ gm$ of $^{235}_{92}U$ is to completely fission.

    a) $13.1 x\ 10^{8}\ Ws$

b) 5.7×10^9 Ws

c) 35.7×10^9 Ws

d) 8.2×10^{10} Ws

e) 0.34 MWd

87) Nuclear fusion is actually a thermo-nuclear reaction. Find the temperature at which fusion of two protons will take place.

a) $T \approx 10^{13}$ K

b) $T \approx 10^{11}$ K

c) $T \approx 10^{10}$ K

d) $T \approx 10^9$ K

e) $T \approx 10^7$ K.

88) Find the wavelength of a neutron with $E \cong 0.08\ eV$.

a) $\approx 0.1\ nm$

b) $\approx 1\ nm$

c) $\approx 3\ nm$

d) $\approx 0.3\ \mu$

e) $\approx 2\ \mu$

f)

89) Consider the fusion reaction $2\ {}^2_1H\ \rightarrow\ {}^4_2He$ in a reactor to produce industrial power of $150\ MW$, with energy efficiency 30%. How much deuterium is required each day? The neutral atomic masses are $2.01402\ u$, and $4.002604\ u$, respectively.

a) $472\ g/d$

b) $148\ g/d$

c) $75\ g/d$

d) $48\ g/d$

e) $14\ g/d$.

90) A given metal contains a density of $10^{28}\ m^{-3}$ valence electrons and carries a current of $10^6\ Am^{-2}$. Find the drift velocity.

a) $6.25 \times 10^{-4}\ ms^{-1}$

b) $6.25 \times 10^{-3}\ ms^{-1}$

c) $6.25 \times 10^{-2}\ ms^{-1}$

d) $0.625\ ms^{-1}$

e) $0.125\ ms^{-1}$

91) At what fundamental frequency (f) the $C-C$ stretch vibration would be expected in the IR spectrum, if the force constant involved is $4.5 \times 10^5\ dynes/cm$?

a) $4.367 \times 10^{13}\ Hz$

b) $4.367 \times 10^{14}\ Hz$

c) 3.376×10^{14} Hz
d) 3.376×10^{13} Hz .
e) 8.245×10^{14} Hz

92) What is the frequency in wave numbers corresponding to $3.45\ \mu$?
   a) $3450\ cm^{-1}$
   b) $2898\ cm^{-1}$
   c) $289.8\ cm^{-1}$
   d) $345.0\ cm^{-1}$
   e) $3428.0\ cm^{-1}$.

93) What is the energy of 1 *mol* of photons with wavelength 400 *nm* ?
   a) $400.3\ kJ\ mol^{-1}$
   b) $1243.0\ kJ\ mol^{-1}$
   c) $783.2\ kJ\ mol^{-1}$
   d) $47.8\ kJ\ mol^{-1}$
   e) $299.3\ kJ\ mol^{-1}$

94) A free nucleus ^{57*}Fe emits a gamma ray of frequency $3.5 \times 10^{18}\ s^{-1}$. What is the recoil velocity of the nucleus?
   a) $0.82 \times 10^3\ m\ s^{-1}$
   b) $0.08 \times 10^3\ m\ s^{-1}$
   c) $8.2 \times 10^3\ m\ s^{-1}$
   d) $1.42 \times 10^3\ m\ s^{-1}$
   e) $0.82\ m\ s^{-1}$

95) The meta-stable state of ^{57}Fe has life time $1.5 \times 10^{-7} s$. Calculate the line width of the gamma ray emission and express in *Hz* .
   a) 7.12×10^8 Hz
   b) 5.92×10^8 Hz
   c) 5.92×10^7 Hz
   d) 1.52×10^7 Hz
   e) 2.07×10^6 Hz
   f)

96) Suppose E_n denotes the energy eigen values of a 1-D system, and $\psi(r,t)$, the corresponding energy eigen functions. If the normalized wave function of the system, at $t = 0$, is given by

$$\psi_n(x,t=0)=\frac{1}{\sqrt{2}},e^{i\alpha_1}\hat{\psi}_1(x)+\frac{1}{\sqrt{3}},e^{i\alpha_2}\hat{\psi}_2(x)+\frac{1}{\sqrt{6}},e^{i\alpha_2}\hat{\psi}_3(x),$$ where $\alpha_1,\ \alpha_2,\ \alpha_3$ are constants.

Find the probability that at time t a measurement of the energy of the system gives the value E_2 .

   a) $\frac{1}{6}$
   b) $\frac{1}{3}$

c) $\frac{1}{2}$

d) $\frac{1}{11}$

e) Unity

97) Calculate the energy of an electron confined in an atom, treating it as a particle in a box. Given the radius of the atom is the Bohr radius.

   a)   −3.40 eV

   b)   −13.6 eV

   c)   −37 eV

   d)   13.6 eV

   e)   37 eV

98) Consider a Radio Station operating on a frequency of 98 MHz and it radiates a power of 200 kW. Find how many quanta of energy are emitted per second.

   a) 3×10^{30}

   b) 13×10^{30}

   c) 13×10^{28}

   d) 3×10^{28}

   e) 6.2×10^{28}

99) A particle is trapped in an infinite square well of width '$2a$', such that the wave function is $\Psi_n(x) = C \left\{ Cos(\frac{\pi x}{2a}) + Sin(\frac{3\pi x}{a}) + \frac{1}{4} Cos(\frac{3\pi x}{2a}) \right\}$, inside the well. $\Psi(x) = 0$, outside the well. Find the energy of the particle in state represented by the second term.

   a) $-\left(\frac{\hbar^2 \pi^2}{8ma^2} \right)$

   b) $-\left(\frac{9\hbar^2 \pi^2}{2ma^2} \right)$

   c) $-\left(\frac{9\hbar^2 \pi^2}{8ma^2} \right)$.

   d) $-\left(\frac{\hbar^2 \pi^2}{2ma^2} \right)$

   e) $-\left(\frac{3\hbar^2 \pi^2}{2ma^2} \right)$

100)   Find the eigen values of the physical system represented by $\begin{bmatrix} -1 & 3 \\ 2 & 2 \end{bmatrix}$.

   a)   −3, 2

   b)   −2, 1

   c)   3, −2

   d)   −3, 1

   e)   3, -1

-o-0-o-0-o-0-o-END-o-0-o-0-o-0-o-

Appendix - 3

PRACTICE TEST - 3

(SENIORS) (Undergraduates)
[GATE, NET (UGC- CSIR), State SET, IAS (Preliminary)(UPSC),
GRE Physics (USA)]

Appendix - 3

PRACTICE TEST - 3

(SENIORS) (Undergraduates)
GATE, NET (UGC- CSIR), State SET, IAS (Preliminary)(UPSC),
GRE Physics (USA)]

| Time 180 minutes |
| # Questions 100 |

DIRECTIONS:
Each of the items or incompleted statements or any other patterns of questions given below is
followed in general by 4 or 5 answer choices or completions. Select the best answer from the keys
for each of the 100 items.

1) A piece of plane sheet is measured to be $16.2 \ cm$ by $9.8 \ cm$ by $1.1 \ mm$. What is the volume of the
sheet to the proper number of significant figure?

a) $17.4636 \ cm^3$

b) $17 \ cm^3$

c) $17.464 \ cm^3$

d) $17.46 \ cm^3$

2) The coordinates of a moving particle at time t is given by the two equations
$x = at^2$ and $y = b \ t^2$. What is the speed of the particle?

a) $2 \left(a + b \right) t$

b) $\left(a^2 + b^2 \right)^{1/2} t$

c) $2 \left(a^2 + b^2 \right)^{1/2} t$

d) $\left(a + b \right) t$

e) $2 \left(a + 2 b \right) t$

3) A mass of $4.0 \ kg$ is suspended at the end of a vertically held ideal light spring with its upper end is
found to stretch $2.0 \ cm$ from its equilibrium. What will be work done externally to stretch from
equilibrium to $4.0 \ cm$?

a) $3.14 \ J$

b) $2.91 \ J$

c) $1.57 \ J$

d) $0.39 \ J$

e) $0.20 \ J$

4) Take a look at the diagram v *versus* t of a body in motion. It indicates that

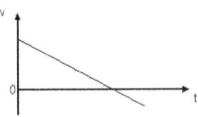

a) The amplitude of a simple pendulum has no constraint in its magnitude:
b) A body is falling down.
c) A body traverse at a uniform acceleration
d) A body is thrown up vertically
e) Illustrates Newton's First law of motion.

5) A car moving with acceleration $a_1 = [i\,(20) + \hat{j}(-36) + \hat{k}(73)]\ ms^{-2}$ and a motor cycle in motion specified by $a_2 = [i\,(20) + \hat{j}(-36) + \hat{k}(73)]\ ms^{-2}$, when one of them overtakes at a point.   Find the relative acceleration, a that will be felt by the person in the car.

a) $a = [i\,(-12) + \hat{j}(28) + \hat{k}(6)]\ ms^{-2}$

b) $a = [i\,(12) + \hat{j}(28) + \hat{k}(-6)]\ ms^{-2}$

c) $a = [i\,(-12) + \hat{j}(-28) + \hat{k}(6)]\ ms^{-2}$

d) $a = [i\,(12) + \hat{j}(-28) + \hat{k}(-6)]\ ms^{-2}$

e) $a = [i\,(12) + \hat{j}(-28)]\ ms^{-2}$

6) The resultant of two forces acting at an angle of 150° is $10N$ which is perpendicular to the force. What is the other force?
a) $20\sqrt{3}\ N$
b) $10\sqrt{3}\ N$
c) $20\ N$
d) $15\sqrt{3}\ N$
e) $15\ N$

7) An elastic collision conserves
a)   momentum but not kinetic energy.
b)   kinetic energy but not momentum.
c)   neither momentum nor kinetic energy.
d)   both momentum and kinetic energy.
e)   total energy.

8) Determine the torque applied to a body, where a force $\vec{F} = [i\,(5.196\ N) + \hat{j}(3\ N)]$ acts on the position vector $\vec{r} = [\hat{i}\,(0.289) + \hat{j}(0.345)]\ m$.

a) $+\hat{k}(0.925)\ Nm$

b) $-\hat{k}(2.659)\ Nm$

c) $-\hat{k}(0.925)\ Nm$

d) $+\hat{k}(2.659)\ Nm$

e) $+\hat{k}(1.653)\ Nm$

9) Which of the following holds good for a couple ($\vec{C}$)?

1. $\vec{C}$ produces translational motion only to the rigid body, due to the vector sum of the concurrent forces, all evaluated w.r.t. the same point.

2. $\vec{C}$ produces rotational motion only to the rigid body, due to the vector sum of the torques of the concurrent forces, all evaluated w.r.t. the same point. .

3. $\vec{C}$ produces both translational motion (due to the resultant of the forces) and rotational motion (due to the resultant of the torques of the forces).

4. $\vec{C} = -\vec{r} \wedge \vec{F}$

Codes	
a)	1, 2 & 3 are all correct.
b)	1 & 3 only are correct.
c)	2 & 3 only are correct.
d)	2 only is correct
e)	1, 2, 3 & 4 all are correct

10) A wheel of radius 1.0 m rolls forwards half a revolution on a horizontal ground. Find the magnitude of the displacement of the point of the wheel initially in contact with the ground.

   a) $\sqrt{2}\,\pi$
   b) π
   c) $\sqrt{3}\,\pi$
   d) $\sqrt{\pi^2+4}$
   e) 2π.

11) Which one of the following statements is NOT correct?
   a) The velocity of sound in air increases with rise in temperature.
   b) The velocity of sound in air is independent of pressure.
   c) The velocity of sound in air decreases as the humidity increases.
   d) The velocity of sound in air is not affected by the change in amplitude and frequency.

12) If the radius of the Earth were 5to shrink by 1%, if the mass remains the same, the value of "g" on the Earth's surface would
   a) increase by 0.5
   b) increase by 2%
   c) decrease by 0.5%
   d) decrease by 2%.

13) Match List I with List II, and select the correct answer using the codes given below the Lists

List I (Physical Law)	List II (Principle)
A. Kepler's II Law	1. otential energy
B. Beats	2. Conservation of angular momentum
C. Conservative force	3 . Conservation of linear momentum
D. NH_3 (non-linear)	4. Superposition principle
	5. Inertia

Codes :				
	A	B	C	D
a)	1	3	2	4
b).	2	4	3	5
c)	2	4	1	3
d)	3	2	5	1

14) Determine the divergence of the given vector, $\mathbf{A} = x^2\mathbf{i} - y^2\mathbf{j}$.

   a)   0

   b)   2x - 2y

   c)   x - y

   d)   1

15) A spherical body moves with uniform angular velocity ω around a circular path of radius r. Which one of the following statements is correct?

   a) The body has no acceleration.

   b) The body has a radial acceleration $\omega^2 r$ directed towards the centre of the path.

   c) The body has a radial acceleration $\frac{2}{5}\omega^2 r$ directed away from the centre of the path.

   d) The body has acceleration $\omega^2 r$ tangential to its path.

16) Consider two metal spheres of the same mass M, one solid and the other hollow, having radius R, are initially at rest. If the two start rolling down the same inclined plane, that reach the bottom of the inclined plane with speed ratio, v_{sol}/v_{hol} would be

   a) 1

   b) $\frac{\sqrt{12}}{7}$

   c) $\frac{\sqrt{10}}{7}$

   d) $\frac{\sqrt{25}}{21}$.

17) A weightless rubber balloon is filled with $200 \ cm^3$ of water. Its weight in water is

   a) $\frac{9.8}{5} N$

   b) $\frac{9.8}{10} N$

   c) $\frac{9.8}{4} N$

   d) Zero.

18)

   The following two statements are labelled 'Assertion' (A) and the other 'Reason'(R).
   Examine carefully the two, and decide if the Assertion and Reason are true individually.
   I so examine if the Reason is the is the correct explanation of the Assertion.
   Select the right answer from the codes given below;

a) Both 'A' and 'R' are individually true, and 'R' is the correct expalanation of 'A'

b) Both 'A' and 'R' are individually true, but 'R' is not the explanation of 'A'.

c) 'A' is true but and 'R' is false.

d) 'A' is false but 'R' is true.

Assertion (A) : Artificial satellites are always launched from the Earth in the eastward direction.
Reason (R): Earth rotates from West to East and so the satellite attains the escape velocity.

19) The following two statements are labelled 'Assertion '(A) and the other 'Reason' (R). Examine carefully the two, and decide if the 'A' and 'R' are true individually. If so examine if the 'R' is the correct explanation of the 'A'. Select the right answer from the codes given below

Assertion (A): If the gauge in pressure cooker is set to $0.2\ N.mm^{-2}$, water within it does not boil till $T < 120\ ^{\circ}C$, and takes less time to cook.

Reason (R): Saturated vapour pressure (svp) increases with rise in temperature T.

Codes

a) Both A and R are individually true, and R is the correct explanation of A'

b) Both A and R are individually tru, but R is not the explanation of A.

c) A is true but and R is false.

d) A is false but R is true

20) Match List I with List II, and select the correct answer using the codes given below the Lists

List I	List II	
(Type of Molecule)	(# of deg freedom of vibration)	
	(fundamental modes)	
A. CO_2	1.	7	$(3\times4-5)$
B. H_2O	2.	4	$(3\times3-5)$
C. C_2H_2 (linear)	3.	3	$(3\times3-6)$
D. NH_3 (non-linear)	4.	1	$(3\times2-5)$
	5.	6	$(3\times4-6)$

Codes::	A	B	C	D
a)	2	3	5	1
b).	3	2	1	5
c)	3	5	2	1
d)	2	3	1	5

21) Using the Dulong-Petit law calculate the specific heat of a small piece of copper.

a) $0.047\ Cal.g^{-1}K^{-1}$

b) $0.094\ Cal.g^{-1}K^{-1}$

c) $0.27\ Cal.g^{-1}K^{-1}$

d) $0.54\ Cal.g^{-1}K^{-1}$

e) $0.94\ Cal.g^{-1}K^{-1}$

22) Find the speed of sound in a diatomic ideal gas that has density $3.50\ kg.m^{-3}$ and at a pressure of $21.5\ kPa$.

    a) $293\ ms^{-1}$

    b) $333\ ms^{-1}$

    c) $303\ ms^{-1}$

    d) $240\ ms^{-1}$

    e)   None of above.

23) Light travels from a medium with $n=1.63$ into a medium of $n=1.42$. Which of the following statements are correct?

    1.   The speed of light increases as it enters the second medium.

    2.   The wavelength of the light remains the same. Wavelength is related to frequency and the frequency of light does not change as it moves from one medium to another

    3.   Away from the normal.

    a)   1, 2 & 3 are all correct.

    b)   1 & 3 only are correct.

    c)   2 & 3 only are correct.

    d)   1 & 2 only are correct

24) The following two statements are labelled 'Assertion '(A) and the other 'Reason' (R). Examine carefully the two, and decide if the 'A' and 'R' are true individually. If so examine if the 'R' is the correct explanation of the 'A'. Select the right answer from the codes given below;

    Assertion (A){ A bubble in a liquid at its B.P. expands as it rises to the surface.

    Reason (R): External pressure on the bubble inside a liquid decreases as it rises to the surface.

    Codes

    a) Both A and R are individually true, and R is the correct explanation of A'

    b) Both A and R are individually tru, but R is not the explanation of A.

    c) A is true but and R is false.

    d) A is false but R is true

25) In a Carnot engine with ideal gas, the quantity of heat flow from the hot thermal reservoir in one cycle is $40\ J$, while the efficiency of the engine is 0.25 per cycle. Calculate the mechanical work done during the cycle by the engine.

    a) $10\ J$

    b) $30\ J$

    c) $40\ J$

    d) $110\ J$

    e) $160\ J$.

26) An equi-convex lens of glass ($n=1.5$) and radius of curvature $1.0\ m$, will have a focal length

    a) $0.5\ m$

    b) $1.0\ m$

    c) $1.5\ m$

    d) $2.0\ m$

    e) None of above.

27) A combination of two lenses in contact may be achromatic. Choose the most correct answer from below:

 a) One of them is convex and the other concave and made from glasses of different refractive indices.

 b) A convex and the other concave and made from the same material.

 c) One of them is convex and the other concave and made from flint and crown glasses

 d) A convex (flint) with focal length f separated away by a distance f from the other concave (crown) with focal length f

28) Match List I with List II and select the correct answer using the codes given below the L the following two Lists

List I Instrument	List II Physical principle
A. Astronomical telescope	1. Magnified virtual image of close object
B. Simple Microscope	2. Magnified virtual image
C. Prism binocular	3. two convex lenses, one as Objective & other as eyepiece
D. Opera glass (Galilean telescope)	4. One convex lens & a divergent lens
	5 Internal reflection

Pick up the right answer from the codes given below:

Answer key Codes

a) 4 2 5 3
b) 1 2 5 3
c) 3 2 5 4
d) 3 2 1 3

29) The objective of focal length f and separated D from the eyepiece of a compound microscope is one of

 a) Convex lens of f = short

 b) Convex lens of f = long

 c) Convex lens of any focal length

 d) Lens of $(1+\frac{D}{f})$

 e) Lens of $(1-\frac{D}{f})$

30) Which of the following interfaces will have the largest critical angle?

 a) glass to water interface,

 b) diamond to water interface,

 c) diamond to glass interface.

 d) water to air

 e) glass to air

31) A beam of light of intensity unity and linearly polarized is incident on a polarizer at $45°$ to its optic axis. What will be the intensity of the emergent light?

 a) $\frac{1}{\sqrt{2}}$

b) $\frac{1}{4}$

c) $\frac{1}{2}$

d) $\sqrt{2}$

e) Unity.

32) A person has near point vision at 100 cm . How can he see clearly an object at normal distance of 25 cm ?

a) 20 cm

b) -33.3 cm

c) 33.3 cm

d) 100 cm

e) None of above

33) Find the other component of the light vector $\begin{bmatrix} 2 \\ -i \end{bmatrix}$ so they are mutually orthogonal.

a) $\begin{bmatrix} i \\ 2 \end{bmatrix}$

b) $\begin{bmatrix} -i \\ 1 \end{bmatrix}$

c) $\begin{bmatrix} 1 \\ i/2 \end{bmatrix}$

d) $\begin{bmatrix} 1 \\ i \end{bmatrix}$

e) $\begin{bmatrix} 1 \\ 2i \end{bmatrix}$

34) Compute the value of θ_C for a water-to-air interface, with $n = 1.33$.

a) $Sin^{-1}(0.752) = 48.8$ $^\circ$

b) $Sin^{-1}(0.715)$

c) $Sin^{-1}(0.615)$

d) $Sin^{-1}(0.534)$

e) None of above.

35) In an interference experiment of the Young type, the distance between the slits is ½ mm, $\lambda = 600.0$ nm. It is desired to have a fringe spacing of 1 mm at a screen. What is the screen distance?

a) 0.73 m

b) 0.83 m

c) 1.24 m

d) 1.5 m

e) 1.64 m

36) Calculate the length of coherence in the case of a laser source, emitting the IR line 1000 nm and spectral width 10^3 Hz.

a) 150 km

b) 250 *km*
c) 275 *km*
d) 300 *km*
e) 415 *km*

37) In a far-field diffraction set up, each aperture is 0.1 *mm* wide and the separation between them is 0.6 *mm*. Find the missing orders in the pattern.

a) 3,6,9,*etc*
b) 4,8,12,*etc*
c) 6,12,18,*etc*
d) 2,4,6,*etc*
e) None.

38) How many Bravais lattices (3D) are there ?
a) 36
b) 28
c) 14
d) 16
e) 7

39) How many lines should be ruled on a transmission grating so that it will just resolve the sodium doublet (D_1 = 589.592 *nm* and D_2 = 588.995 *nm*) in the 1st order spectrum?

a) 988
b) 588
c) 1358
d) 1541

40) What is a *primitive unit cell* in a lattice space?
a) A cell of volume unity,
b) Any parallelepiped with lattice points at its corners
c) A parallelepiped containing only one lattice point.
d) The cell volume with lattice vectors as edges.
e) A parallelepiped containing more than one lattice point.

41) Select the correct Maxwell's equations, which states that no magnetic monopole exists.

a) $\vec{\nabla}\cdot\vec{E} = \frac{\rho}{\varepsilon_0}$

b) $\vec{\nabla}\cdot\vec{B} = 0$

c) $\vec{\nabla}\wedge\vec{E} = \frac{\partial B}{\partial t}$

d) $\vec{\nabla}\wedge\vec{B} = \mu_0 J + \mu_0\varepsilon_0\frac{\partial E}{\partial t}$

e) None of above

42) In one of the radioactive chains one comes across the element $^{226}_{88}$Ra . The final product of the chain is $^{206}_{82}$Pb . The number of α and β emitted when $^{226}_{88}$Ra decays into $^{206}_{82}$Pb is

a)   3 α and 6 β .
b)   4 α and 5 β .

c)   5 α and 4 β .
d)   6 α and 6 β α .
e)   None of above.

43) The nuclear reaction

$$_{2}^{4}He + {}_{13}^{27}Al \rightarrow {}_{15}^{30}P + {}_{0}^{1}n$$
$$\downarrow$$
$$_{14}^{30}Si + e^{+}$$

is an example of
a)   nuclear chain reaction
b)   successive disintegration
c)   pair production
d)   associated production
e)   artificial radio-activity.

44) Consider the following particles: (1) Electron, (2) Proton, (3) Neutron.
Which of these are Baryons?
a)   1 and 2.
b)   2 & 3 .
c)   1 & 3 .
d)   1, 2, & 3

45) Regarding the atom of a chemical element, the magnetic quantum number refers to
a) Orientation
b) Shape
c) Size
d) Spin

46) The following two statements are labelled 'Assertion '(A) and the other 'Reason' (R). Examine carefully the two, and decide if the 'A' and 'R' are true individually. If so examine if the 'R' is the correct explanation of the 'A'. Select the right answer from the codes given below;
    Assertion (A): With increase of temperature the viscosity of glycerine increases.
    Reason (R): Rise of temperature increases the kinetic energy of molecules.
Codes
    a) Both 'A' and 'R' are individually true, and 'R' is the correct expalanation of 'A'
    b) Both 'A' and 'R' are individually tru, but 'R' is not the explanation of 'A'.
    c) 'A' is true but and 'R' is false.
    d) ' A' is false but 'R' is true.

47) Which one of the following statements is correct with reference to the Solar System?
    a) The Earth is the densest of all planets in the Solar System.
    b) The predominant element in the composition of Earth is Silicon.
    c) The Sun shares 75% of the mass of the Solar System.
    d) The diameter of the Sun is 190 times that of the Earth.

48) State which, if any, of the following operators are Hermitian, (i) zp_z, and $(z - ip_z)$.
    It is known that operators z and $p_z = -i\hbar(\partial / \partial z)$ are both Hermitian.
    a) Both are Hermitian

b) (zp_z) is Hermitian, but $(z - ip_z)$ is not Hermitian

c) (zp_z) is not Hermitian, but $(z - ip_z)$ is Hermitian

d) Both are anti-Hermitian

49) Which of the following statements are correct?

    1 Free electron theory in metals fails to explain distinction between metals, semi-conductors, semi-metals, insulators.

    2 Free electron model theory (free electron Fermi gas) must be extended to take account of the periodic lattice of the solid.

    3 In the case of semi-conductors, Hall coeff, $R_H = (-1/nec)$ is 'negative' for "free electrons".

    4 For metals like Na, K, Cu, Au, Mg, Al the R_H = +ve, whereas for Cd and Be, R_H = -ve, because they requires conduction of the effect of the periodic lattice potential as the conduction electron states.

    a)   1, 2 & 3 are all correct.

    b)   1 & 3 only are correct.

    c)   2 & 4 only are correct.

    d)   1, 2, 3, 4 all are correct

50) The SC lattice can be defined by the basis vectors $a(1,0,0)$, $a(0,1,0)$, and $a(0,0,1)$. Find the RL.

    a) The RL is defined by vectors: $(2\pi/a)(1,0,0)$, $(2\pi/a)(0,1,0)$, $(2\pi/a)(0,0,1)$;

    b) The RL vectors are: $(\pi/a)\,\mathbf{i}$, $(\pi/a)\,\mathbf{j}$, $(\pi/a)\,\mathbf{k}$

    c) It is an FCC lattice

    d) It is a BCC lattice

51) In Hydrogen spectrum, the wavelength of H_α line is $656.0\ nm$ and that for a distant galaxy $706.0\ nm$. Estimate the speed of the galaxy with respect to the Earth.

    a) $2x10^8\,ms^{-1}$

    b) $1.2x10^8\,ms^{-1}$

    c) $2x10^7\,ms^{-1}$

    d) $1.2x10^6\,ms^{-1}$

52) The internuclear distance for NO is 1.15Å. The reduced mass of NO is ($u = 1.66x10^{-27}\,kg$):

    a) $\frac{224}{16}u$

    b) $\frac{224}{30}u$

    c) $30\,u$

    d) $\frac{30}{224}u$

53) The RL to a SC lattice is

    a)   bcc

    b)   fcc

    c)   hcp

    d)   sc.

54) The interplanar spacing for (321) in a simple cubic lattice with interatomic spacing $a = 4.12$ Å is

    a) $(4.12/6)$ Å

    b) $(4.12/14)$ Å

c) $(4.12 / \sqrt{14})$ Å
d) $(4.12 / \sqrt{6})$ Å

55) An amplifier with an output resistance of 245 Ω is to be matched to a 5 Ω load. The turns ratio of the transformer used will be:

a) $1:49$
b) $7:1$
b) $49:1$
c) $1225:1$

56) At thermodynamic equilibrium and room temperature ($300K$), what is the ratio of populations at the upper and lower levels of a transition with photon energy of $0.1eV$? ($k_B = 8.6x10^{-5} eVK^{-1}$)

a) 0.001
b) 0.0207
c) 0.127
d) 1

57) In the first Bohr orbit of a hydrogen atom the total energy of the electron is $-21.76x10^{-19} J$, then the potential energy will be

a) $-43.52x10^{-19} J$
b) $-21.76x10^{-19} J$
c) $-13.60x10^{-19} J$
d) $-10.88x10^{-19} J$

58) In Galilean relativity, if the momentum of a body in motion is increased by 100%, the % increase in its k.E. is:

a) 400
b) 300
c) 200
d) 100.

59) The Duty Cycle of the waveform in the diagram below is

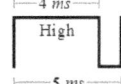

a) 80%
b) (5/9) 100 %
c) (4/9) 100%
d) none above

60) Under space charge limited conditions with the plate voltage E_p = 100 V and permanence K =10^{-4} (in SI unit), the plate current in a vacuum diode will be

a) $10^{-6} mA$
b) $10 mA$
c) $10^2 mA$
d) $10^6 mA$

61) Find the lower limit for the Paschen Series. Given the Rydberg constant
$R_{substance} = 109,677.6 \ cm^{-1}$. Select the right answer.
a) 2280 nm
b) 1876 nm
c) 1460 nm
d) 820.6 nm

62)   Find   the   upper   limit   for   the   Paschen   Series.   Given   the   Rydberg   constant
$R_{substance} = 109,677.6 \ cm^{-1}$. Select the right answer.
a) 2280 nm
b) 1876 nm
c) 820.6 nm
d) 91.2 nm

63) Match List I with List II, and select the correct answer using the codes given below the Lists

List I (Particle)	List II (Rest mass energy in MeV)
A. Proton	1. 150
B. Electron	2. 0
C. Neutrino	3. 0.5
D. π-meson	4. 940
	5. 273

Codes :

	A	B	C	D
a)	4	3	2	1
b).	4	3	1	5
c)	1	2	4	3
d)	1	3	2	5

64) The X-ray diffraction pattern of a crystalline powder sample data show that only reflection with Miller indices (h,k,l) such that only either all odd or all even are seen. This means the type of crystal lattice is
a) Simple Cubic,
b) Body-Centred-Cubic,
c) Diamond,
d) Face-Centred-Cubic.

65) In a proton NMR experiment what will be the frequency at which resonance takes place if the sweeping magnetic field is 0.6642 T ?
a) 3.51 MHz
b) 7.02 MHz
c) 14.0 MHz
d) 28.1 MHz
e) 32.8 MHz

66) Which of the following is NOT a correct quark assignment?

a) $p = uud$
b) $n = udd$
c) $\pi^+ = us$
d) $K^- = \bar{u}s$

67) Deep water waves have dispersion relationship $\omega^2 = gk + ak^3$, where g and a are constants, Find the phase velocity in terms of wavelength λ.

a) $\sqrt{\frac{g\lambda}{2\pi} + \frac{2\pi a}{\lambda}}$

b) $\sqrt{\frac{2\pi a}{\lambda}}$

c) $\sqrt{\frac{g\lambda}{2\pi}}$

d) $\sqrt{\frac{g\lambda}{2\pi} + \frac{\pi a}{2\lambda}}$

68) The following two statements are labelled "Assertion (A)" and the other "Reason ®R". Examine carefully the two, and decide if the Assertion and Reason are true individually. I so examine if the 'R'is the correct explanation of the 'A'. Select the right answer from the codes given below;

Codes

a) Both A and R are individually true, and R is the correct explanation of A'

b) Both A and R are individually tru, but R is not the explanation of A.

c) A is true but and R is false.

d) A is false but R is true

Assertion (A): The Universe is around 13.6 *billion yrs*, and the age of species of Sequoias trees of California exceeds 3,500 *yrs*...

Reason (R): Proton is the utmost stable particle

69) Consider $^7_3 Li$ accelerated to a kinetic energy of $50 MeV$ and a $^{208}_{82} Pb$ at rest elastically collide to undergo nuclear reaction. Under pure Coulombic interaction find the distance closest approach.

a) 2.24 *fm*

b) 3.54 *fm*

c) 7.08 *fm*

d) 8.2 *fm*

e) 10.7 *fm*

70) What do you understand by Bose-Einstein condensation?

a) Bosons are not physically acceptable particles.

b) For $T > T_c$ (Critical temperature) all particles dissolve gluons and quarks

c) For $T < T_c$ (Critical temperature, all particles remain lowest state.

d) As $T \rightarrow \infty$, all particles remain excited states.

71) Find the minimum energy necessary to destroy resonant absorption of γ – rays of 14.4 keV energy (life time $\tau = 9.8x10^{-8}s$) by a lattice of ^{57} Fe when the . γ – ray source is vibrated.

a) $0.10\ mms^{-1}$
b) $0.28\ mms^{-1}$
c) $0.14\ mms^{-1}$
d) $0.07\ mms^{-1}$

72) Obtain the normal recoil energy of a $^{191}_{77}$Ir γ – ray of energy $129\ keV$ ($\tau_{1/2} = 0.14ns$) in a Mossbauer experiment.
  a) $0.47eV$
  b) $4.68eV$
  c) $0.047eV$
  d) $4.68meV$

73) Find the eigen value of the angular momentum operator L_z, for Hydrogen-like atom of wavefunction
$\psi(r,\theta,\varphi) = Nr^2 e^{-Zr/3a_0} Sin^2\theta e^{2i\varphi}$.
  a) $\hbar$
  b) $2\hbar$
  c) $3\hbar$
  d) $\sqrt{6}\hbar$

74) Match List I with List II, and select the correct answer using the codes given below the Lists

List I (Types of Interactions)	List II (Field Quanta)
A. Electro-Magnetic	1. Graviton
B. Gravitational	2. Photon
C. Strong	3. Intermediate vector Boson
D. Weak	4. Pion
	5 Meson

Codes :

	A	B	C	D
a).	2	1	4	3
b).	4	2	3	1
c).	3	5	2	4
d).	5	3	2	1

75) The frequency response curve for a RC coupled amplifier is shown in the diagram below: The band width of the amplifier will be

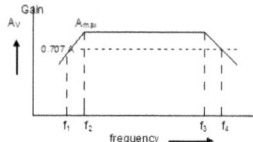

a) $f_3 - f_2$

b) $f_4 - f_1$

c) $(f_4 - f_2)/2$

d) $f_3 - f_1$

76) Match List I with List II, and select the correct answer using the codes given below the Lists

List I Particle	List II Propeerty
A. Proton	1. Spin $= \frac{1}{2}$
B. Electron	2. $\tau_{1/2} = 10^{32}$ yrs
C. Neutron	3. Spin $= 0$
D. Photon	4. $\tau_{1/2} = 10.2$ mts.
	5. Spin $= 1$

Codes :

	A	B	C	D
a)	1	4	2	3
b).	2	1	4	3
c)	2	4	3	2
d)	4	1	3	5

77) Match List I with List II, and select the correct answer using the codes given below the Lists

List I Particle	List II Event
A. Electron	1. Goldstein
B. Neutron	2. J.J.Thomson
C. Proton	3. Chadwick
D Photon	4. Einstin
	5. Planck

Codes :

	A	B	C	D
a)	2	1	3	3
b).	2	1	3	5
c)	1	2	3	4
d)	2	3	1	5

78) The following two statements are labelled 'Assertion '(A) and the other 'Reason' (R). Examine carefully the two, and decide if the 'A' and 'R' are true individually. If so examine if the 'R' is the correct explanation of the 'A'. Select the right answer from the codes given below;

Assertion (A): Earth's atmosphere does not contain Hydrogen.

Reason (R): Hydrogen molecules attain escape velocity by random collisions with other molecules in the atmosphere.

Codes

a) Both A and R are individually true, and R is the correct explanation of A'

b) Both A and R are individually tru, but R is not the explanation of A.

c) A is true but and R is false.

d) A is false but R is true

79) In a scattering experiment, a deuteron, accelerated by a $15MV$ device, is striking a lead target. Find the distance of closest approach of the particle and target nucleus.

    a) $15.74\,fm$

    b) $13.20\,fm$

    c) $7.87\,fm$

    d) $5.32\,fm$.

80) In an RC charging circuit, $R = 47\,k\Omega$, $C = 1000\,\mu F$ and $V = 3V$ DC, What value will be the voltage across the capacitor at 0.7 time constant?

    a) $2.0\ V$

    b) $2.5\ V$

    c) $3.15\ V$

    d) $5\ V$

    e) $0.02\ V$

81) Which one of the following diagrams $Y \rightarrow Log(J_s T^{-2})\ versus\ X \rightarrow T^{-1}$ illustrates the phenomenon of thermionic emission?

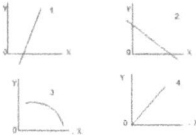

    a) 1

    b) 2

    c) 3

    d) 4

82) Intrinsic concentration of charge carriers in a semiconductor depends on temperature T as

    a) T^2

    b) $T^{3/2}$

    c) T

    d) T^{-1}

    e) $T^{2/3}$

83) When $20\ V$ is applied between the Y-plates of a CRO the spot on the screen is deflected by $4\ cm$. The deflection sensitivity of the plates is:

    a) $80\ Vcm^{-1}$

b) $5 \, Vcm^{-1}$

c) $2 \, Vcm^{-1}$

d) $0.2 \, Vcm^{-1}$

e) $0.02 \, Vcm^{-1}$

84) A triode amplifier has $R_L = 8 \, k\Omega$, and anode power supply voltage $V_{HT} = 200 \, V$, $r_p = 1 \, k\Omega$, $\mu = 30$. If a factor $\beta = -0.01$ of the output is fed back, what will be the gain of the amplifier?

a) $-\dfrac{80}{11}$

b) $\dfrac{300}{7}$

c) $-\dfrac{30}{1.3}$

d) $-\dfrac{80}{1.3}$

e) None of above.

85) An ideal OA has one or more of the following characteristics:
   1. Voltage gain (open loop gain) $A_{ol} = $ infinity.
   2. Input resistance $=$ finite.
   3. Output resistance $= 0$.

   Pick up you answer from the keys below:

   a) 1, 2 & 3 are all correct.

   b) 1 & 3 only are correct.

   c) 2 & 3 only are correct.

   d) 1 & 2 only are correct

86) In the inverting amplifier, what difference is the phase of the output voltage with the input voltage?

   a) 0°

   b) 90°.

   c) 180°.

   d) 270°.

   e) None of above.

87) The binary equivalent of the decimal 7.125 is:

   a) 111.111

   b) 111.101

   c) 111.010

   d) 111.001 .

   e) 110.111

88) The Boolean expression $F = \overline{A} \cdot \overline{B}$ represents

   a) NAND gate.

   b) OR gate.

   c) NOR gate.

   d) EX-OR gate.

   e) None of above

89) A common method of producing a triangular wave is to:
   a) integrate a square wave.
   b) differentiate a saw-tooth wave.
   c) differentiate a sine wave.
   d) integrate a saw-tooth wave.
   e) differentiate a square wave

90) Match List I with List II, and select the correct answer using the codes given below the Lists.

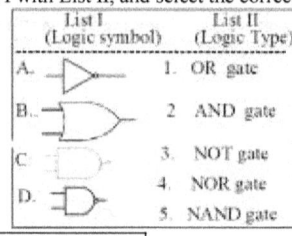

	List I (Logic symbol)		List II (Logic Type)
A.		1.	OR gate
B.		2.	AND gate
C.		3.	NOT gate
D.		4.	NOR gate
		5.	NAND gate

Codes:

	A	B	C	D
a).	1	2	3	4
b).	2	4	3	1
c).	3	1	2	5
d).	5	3	2	1

91) Plane y = 0 carries a uniform current of 30 a_z mAm^{-1}. At (1, 10, -2), the magnetic field intensity is

   a) $-15\ a_x\ mA-m^{-1}$

   b) $15\ a_x\ mA-m^{-1}$

   c) $477.5\ a_y\ mA-m^{-1}$

   d) $18.85\ a_y\ mA-m^{-1}$

   e) $-18.85\ a_y\ mA-m^{-1}$

92) In the scattering of light, the ratio of the intensities of light scattered at $300\ nm$ to that at $600\ nm$
   a)   2
   b)   4
   c)   8
   d)   9
   e)   16

93) A substance shows a Raman shift of $4000\ cm^{-1}$. If the mode is active in the Infra-red, then the corresponding mode in the Infra-red absorption band will be at
   a) 0.5μ
   b) 0.72μ.
   c) 1.0μ .
   d) 1.5μ.
   e) 2.5μ

94) If an electron has an orbital angular momentum quantum number $\ell = 7$, then it will have an orbital angular momentum equal to

a) $\sqrt{7}\hbar$
b) $7\hbar$
c) $3\hbar$
d) $\sqrt{56}\hbar$
e) $42\hbar$

95) A radio wave has a maximum electric field intensity of $10^{-4} Vm^{-1}$. on arrival at a receiving antenna. The maximum magnetic flux density of the magnetic field of such a wave is

a) $3.0x10^{-4} T$
b) Zero
c) $5.9x10^{-9} T$
d) $3.3x10^{-13} T$
e) $0.33x10^{-13} T$

96) What does the following circuit represent?

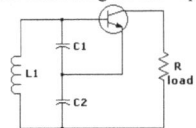

a)   Pierce Oscillator
b)   Hartley Oscillator
c)   ColpittsOscilator
d)   Phase shift Oscillator
e)   None of the above

97) A CE amplifier has a voltage gain of 50, an input impedance of $1\ k\Omega$ and an output impedance of $200\ \Omega$. The power gain of the amplifier will be

a) $24\ dB$
b) $41\ dB$
c) $250\ dB$
d) $125\ dB$
e)   None of above

98) For the following circuit, which among the following is the correct Boolean expression?

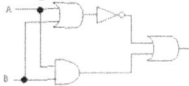

a) $Q = A\bar{B} + \bar{A}B$
b). $Q = A\bar{B} + AB + \bar{A}B$
c). $Q = AA + AB + \bar{A}B$
d). $Q = AA + AB + A\bar{B} + \bar{A}B$
e) $Q = AB + A\bar{B} + \bar{A}B$

99) In computers one comes across sizes of memories. What does 1 MB represent?
   a). 1024 x 8 BITS
   b) 2^{10} BYTES
   c) 2^{20} BITS
   d) 2^{20} BYTES

100) In a parallel resonant circuit:
   a) maximum current flows at resonance
   b) the impedance at resonance is high
   c) the impedance at resonance is low
   d) the applied voltage and current are 90^0 out of phase.

-o-0-o-0-o- END -o-0-o-0-o-

Appendix 4

Greek Alphabets

TRIGONOMETRIC FUNCTIONS

GREEK ALPHABETS

A	α	alpha	I	ι	iota	P	ρ	rho
B	β	beta	K	κ	kappa	Σ	σ	sigma
Γ	γ	gamma	Λ	λ	lambda	T	τ	tau
E	ε	epsilon	M	μ	mu	Y	υ	upsilon
Δ	δ	delta	N	ν	nu	Φ	φ	phi
Z	ζ	zeta	Ξ	ξ	xi	X	χ	chi
H	η	eta	O	ο	omicron	Ψ	ψ	psi
Θ	θ	theta	Π	π	pi	Ω	ω	omega

Trigonometric Functions

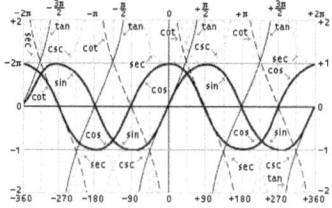

Trigonometric Values of Special Angles

ϑ	sinϑ	cosϑ	tanϑ
0	0	1	0
π/2	1	0	undefined
π	0	-1	0
π/4	$\sqrt{2}/2$	$\sqrt{2}/2$	1
π/6	½	$\sqrt{3}/2$	$1/\sqrt{3}$
π/3	$\sqrt{3}/2$	1/2	$\sqrt{3}$

Appendix 5

Table of fundamental constants

				SI unit	Standard error (parts in 10^6)

Principal constants

Speed of light in vacuum	c	2.997 924 580	$\times 10^8$	$m\,s^{-1}$	exact
Planck constant	h	6.626 0693	$\times 10^{-34}$	$J\,s$	0.17
Planck constant $(h/2\pi)$	$\hbar$	1.054 571 68	$\times 10^{-34}$	$J\,s$	0.17
Elementary charge	e	1.602 176 53	$\times 10^{-19}$	C	0.085
Mass of electron	m_e	9.109 3826	$\times 10^{-31}$	kg	0.17
Mass of electron in atomic mass units		5.485 799 0945	$\times 10^{-4}$	u	0.00044
Avogadro constant	N_A, L	6.022 14145	$\times 10^{23}$	mol^{-1}	0.17
Atomic mass unit, 10^{-3} kg mol^{-1} N_A^{-1}	u	1.660 540 2	$\times 10^{-27}$	kg	0.17
Faraday constant	$F\,(=N_A e)$	9.648 533 83	$\times 10^4$	$\frac{C}{mol^{-1}}$	0.086
Newtonian constant of gravitation	G	6.6742	$\times 10^{-11}$	$\frac{N\ m^2}{Kg^{-2}}$	150

Spectroscopy and atoms

Planck constant	h	4.135 667 43	$\times 10^{-15}$	$eV\,s$	0.085
Planck constant $(h/2\pi)$	$\hbar$	6.582 119 15	$\times 10^{-16}$	$eV\,s$	0.085
Charge/mass ratio of electron	$-e/m_e$	$-1.758\ 820\ 12$	$\times 10^{11}$	$C\,kg^{-1}$	0.086
Fine structure constant	α	7.297 352 568	$\times 10^{-3}$		0.0033
Fine structure constant, reciprocal	α^{-1}	137.035 999 11			0.0033
Rydberg constant (fixed nucleus)	R_∞	10 973 731.568 525		m^{-1}	0.0000066
Bohr radius $(4\pi/\mu_0 c^2)\hbar^2/m_e e^2$	a_0	5.291 772 108	$\times 10^{-11}$	m	0.00333
Compton wavelength of electron	λ_c	2.426 310 238	$\times 10^{-12}$	m	0.0067
Compton wavelength of electron $\div 2\pi$	$\hbar/m_e c$	3.861 592 678	$\times 10^{-13}$	m	0.0067
Classical 'radius' of electron $(\mu_0 c^2/4\pi)e^2/mc^2$	r_e	2.817 940 325	$\times 10^{-15}$	m	0.01
Thomson cross-section $8\pi r_e^2/3$	σ_e	6.652 458 73	$\times 10^{-29}$	m	0.02
Zeeman effect,	μ_B/hc	46.686 4507		$\frac{m^{-1}}{T^{-1}}$	0.086
Bohr magneton $e\hbar/2m_e$	μ_B	9.274 015 4	$\times 10^{-24}$	$J\,T^{-1}$	0.086
Nuclear magneton $e\hbar/2m_p$	μ_N	5.050 783 43	$\times 10^{-27}$	$J\,T^{-1}$	0.086
Ratio of masses	m_p/m_e	1 836.152 672 61			0.00046

proton/electron

Gyromagnetic ratio of proton	$\gamma_p = \mu_p / \frac{1}{2}\hbar$	2.675 222 05	$\times 10^8$	$s^{-1}\,T^{-1}$	0.086
in H_2O, sph., 25 °C	γ'_p	2.675 153 33	$\times 10^8$	$s^{-1}\,T^{-1}$	0.086
in H_2O (cycles), sph., 25 °C	$\gamma'_p/2\pi$	4.257 638 75	$\times 10^7$	$Hz\,T^{-1}$	0.086

age | 746 **Conversion factors for mass, energy and wavelength**

Energy and mass

Electron volt	1.602 176 53	$\times 10^{-19}$	J	0.17
Atomic mass unit	931.494 043		MeV	0.086
1 MeV	1.073 544 171	$\times 10^{-3}$	u	0.0086
Rest-mass of electron	0.510 998 9186		MeV	0.17
1 eV per molecule	9.648 533 83	$\times 10^7$	$\dfrac{J}{kmol^{-1}}$	0.086

Frequency, wavelength and energy

Quantum energy ÷ wave number	1.986 445 61	$\times 10^{-25}$	$J\,m$	0.18
Energy × wavelength	1.239 841 91	$\times 10^{-6}$	$eV\,m$	0.0085
Wave number ÷ energy	8.065 544 45	$\times 10^5$	$\dfrac{eV^{-1}}{m^{-1}}$	0.085
Quantum energy ÷ frequency	4.135 667 43	$\times 10^{-15}$	$\dfrac{eV}{Hz^{-1}}$	0.0085
Frequency ÷ energy	2.417 989 40	$\times 10^{14}$	$\dfrac{Hz}{eV^{-1}}$	0.085

Thermal constants

Molar gas constant	R	8.314 472	$\dfrac{J}{mol^{-1}\,K^{-1}}$	1.7
Loschmidt constant (number of molecules in 1 m^3 of ideal gas at STP)	n_0	2.686 7773	$\times 10^{25}\,m^{-3}$	1.8
Boltzmann constant R/N_A	k	1.380 6505	$\times 10^{-23}\,J\,K^{-1}$	1.8
Boltzmann constant		8.617 343	$\times 10^{-5}\,\dfrac{eV}{K^{-1}}$	1.8
Stefan-Boltzmann constant $(\sigma)^\dagger$	$2\pi^5 k^4/15c^2h^3$	5.670 400	$\times 10^{-8}\,\dfrac{Wm^{-2}}{K^{-4}}$	7
Constant in Planck formula $(c_1)^\dagger$	$2\pi hc^2$	3.741 771 38	$\times 10^{-16}\,W\,m^2$	0.17
Constant in Planck formula $(c_2)^\dagger$	hc/k	1.438 7752	$\times 10^{-2}\,m\,K$	1.7

Note: Magnetic moments are defined so that mechanical energy $= -\mu.\,\boldsymbol{B}$; the unit is $1\,J\,T^{-1} = 1\,J\,Wb^{-1}\,m^2 = 1\,A\,m^2$ (1 *tesla* = 1 *weber* $m^{-2} \equiv 10^4$ *gauss*).

Standard *values**

Von Klitzing constant R_{K-90} 25 812.807 $(1 \pm 0.2 \times 10^{-6})$ Ω

Josephson constant K_{J-90} 483 579.9 $(1 \pm 0.4 \times 10^{-6})$ GHz/V

* These standard values were recommended by the CCE in 1989 for adoption from 1.1.1990. The CCE considered later data than were available to Cohen and Taylor (1987) and they are not necessarily identical with h/e^2 and $2e/h$ respectively. They were defined by the CCE in order to help ensure global uniformity of standards of resistance and emf. (The suffix −90 is used here to indicate that they may be revised at a later date, although it is usually omitted.)

Reference

P. J Mohr and B. N. Taylor, *Rev. Mod Phys.*, **76**, no 4(Oct 2004).

Appendix 6

Table of **numerical** **constants**
Note: all values are approximate.

π	3.14159	26535	898
π^2	9.86960	44010	894
$\ln(\pi)$	1.14472	98858	494
$\log(\pi)$	0.49714	98726	941
e	2.71828	18284	590
e^2	7.38905	60989	307
$\log(e)$	0.43429	44819	033
e^π	23.14069	26327	793
$e^{-\pi}$	0.04321	39182	637(72)
$\sqrt{2}$	1.41421	35623	731
$\sqrt{3}$	1.73205	08075	689
$\sqrt{5}$	2.23606	79774	998
$\sqrt{10}$	3.16227	76601	684
$\sqrt[3]{2}$	1.25992	10498	949
$\ln(2)$	0.69314	71805	599
$\ln(3)$	1.09861	22886	681
$\ln(10)$	2.30258	50929	940
deg/rad	57.29577	95130	823
rad/deg	0.01745	32925	199(43)
Euler's constant γ	0.57721	56649	015

Appendix 7

Factors for Conversions to SI Units

Acceleration	$1\ ft/s^2 = 0.304\ 8\ m/s^2$		$1\ cal/s = 4.184\ W$
	$g = 9.807\ m/s^2$		$1\ ft \cdot lb/s = 1.356\ W$
Area	$1\ acre = 4047\ m^2$		$1\ horsepower\ (hp) = 746\ W$
	$1\ ft^2 = 9.290 \times 10^{-2}\ m^2$	**Pressure**	$1\ atmosphere\ (atm) =$
	$1\ in.^2 = 6.45 \times 10^{-4}\ m^2$		$1.013 \times 10^5\ Pa$
	$1\ mi^2 = 2.59 \times 10^6\ m^2$		$1\ bar = 10^5\ Pa$
Density	$1\ g/cm^3 = 10^3\ kg/m^3$		$1\ cmHg = 1333\ Pa$
Energy	$1\ Btu = 1054\ J$		$1\ lb/ft^2 = 47.88\ Pa$
	$1\ calorie\ (cal) = 4.184\ J$		$1\ lb/in.^2\ (psi) = 6895\ Pa$
	$1\ electron\ volt\ (eV) =$		$1\ N/m^2 = 1\ pascal\ (Pa)$
	$1.602 \times 10^{-19}\ J$		$1\ torr = 133.3\ Pa$
	$1\ foot\ pound\ (ft \cdot lb) = 1.356\ J$	**Speed**	$1\ ft/s\ (fps) = 0.304\ 8\ m/s$
	$1\ kilowatt\ hour\ (kW \cdot h) =$		$1\ km/h = 0.277\ 8\ m/s$
	$3.60 \times 10^6\ J$		$1\ mi/h\ (mph) = 0.447\ 04\ m/s$
Force	$1\ dyne = 10^{-5}\ N$	**Temperature**	$T_{Kelvin} = T_{Celsius} + 273.15$
	$1\ lb = 4.448\ N$		$T_{Kelvin} = \frac{5}{9}(T_{Fahrenheit} + 459.67)$
Length	$1\ angstrom\ (Å) = 10^{-10}\ m$		$T_{Celsius} = \frac{5}{9}(T_{Fahrenheit} - 32)$
	$1\ ft = 0.304\ 8\ m$		$T_{Kelvin} = \frac{5}{9}T_{Rankine}$
	$1\ m = 2.54 \times 10^{-2}\ m$	**Time**	$1\ day = 86\ 400\ s$
	$1\ light\ year = 9.461 \times 10^{15}\ m$		$1\ year = 3.16 \times 10^7\ s$
	$1\ mile = 1069\ m$	**Volume**	$1\ ft^3 = 2.832 \times 10^{-2}\ m^3$
Mass	$1\ atomic\ mass\ unit\ (u) =$		$1\ gallon = 3.785 \times 10^{-3}\ m^3$
	$1.660\ 6 \times 10^{-27}\ kg$		$1\ in.^3 = 1.639 \times 10^{-5}\ m^3$
	$1\ gram = 10^{-3}\ kg$		$1\ liter = 10^{-3}\ m^3$
Power	$1\ Btu/s = 1054\ W$		

Appendix 8

Periodic Table of the Elements (Long Form)

Reprentative Elements |← s-block →|

Representative Elements |←------- p-block -------→|

Transition Elements ←------- d-block -------→

Metals

| H 1 Hydrogen 1.00794 1s¹ | | | | | | | | | | | | | | | | | He 2 Helium 4.00260 1s² |

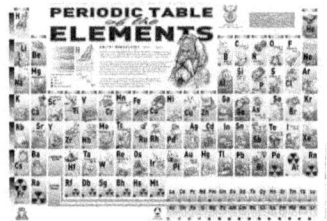

APPENDIX 9

List of Elements

LIST OF ELEMENTS

ATOMIC NUMBER	SYMBOL	NAME	ATOMIC WEIGHT	ATOMIC NUMBER	SYMBOL	NAME	ATOMIC WEIGHT
0	n	neutron	—	52	Te	tellurium	127.60
1	H	hydrogen	1.0079	53	I	iodine	126.9045
2	He	helium	4.00260	54	Xe	xenon	131.30
3	Li	lithium	6.941	55	Cs	cesium	132.9054
4	Be	beryllium	9.01218	56	Ba	barium	137.34
5	B	boron	10.81	57	La	lanthanum	138.9055
6	C	carbon	12.011	58	Ce	cerium	140.12
7	N	nitrogen	14.0067	59	Pr	praseodymium	140.9077
8	O	oxygen	15.9994	60	Nd	neodymium	144.24
9	F	fluorine	18.99840	61	Pm	promethium	—
10	Ne	neon	20.179	62	Sm	samarium	150.4
11	Na	sodium	22.98977	63	Eu	europium	151.96
12	Mg	magnesium	24.305	64	Gd	gadolinium	157.25
13	Al	aluminum	26.98154	65	Tb	terbium	158.9254
14	Si	silicon	28.086	66	Dy	dysprosium	162.50
15	P	phosphorus	30.97376	67	Ho	holmium	164.9304
16	S	sulfur	32.06	68	Er	erbium	167.26
17	Cl	chlorine	35.453	69	Tm	thulium	168.9342
18	Ar	argon	39.948	70	Yb	ytterbium	173.04
19	K	potassium	39.098	71	Lu	lutetium	174.97
20	Ca	calcium	40.08	72	Hf	hafnium	178.49
21	Sc	scandium	44.9559	73	Ta	tantalum	180.9479
22	Ti	titanium	47.90	74	W	tungsten	183.85
23	V	vanadium	50.9414	75	Re	rhenium	186.2
24	Cr	chromium	51.996	76	Os	osmium	190.2
25	Mn	manganese	54.9380	77	Ir	iridium	192.22
26	Fe	iron	55.847	78	Pt	platinum	195.09
27	Co	cobalt	59.0332	79	Au	gold	196.9665
28	Ni	nickel	58.71	80	Hg	mercury	200.59
29	Cu	copper	63.546	81	Tl	thallium	204.37
30	Zn	zinc	65.38	82	Pb	lead	207.2
31	Ga	gallium	69.72	83	Bi	bismuth	208.9804
32	Ge	germanium	72.59	84	Po	polonium	—
33	As	arsenic	74.9216	85	At	astatine	—
34	Se	selenium	78.96	86	Rn	radon	—
35	Br	bromine	79.904	87	Fr	francium	—
36	Kr	krypton	83.80	88	Ra	radium	—
37	Rb	rubidium	85.4078	89	Ac	actinium	—
38	Sr	strontium	87.62	90	Th	thorium	232.0381
39	Y	yttrium	88.9059	91	Pa	protactinium	—
40	Zr	zirconium	91.22	92	U	uranium	238.029
41	Nb	niobium	92.9064	93	Np	neptunium	—
42	Mo	molybdenum	95.94	94	Pu	plutonium	—
43	Tc	technetium	—	95	Am	americium	—
44	Ru	ruthenium	101.07	96	Cm	curium	—
45	Rh	rhodium	102.9055	97	Bk	berkelium	—
46	Pd	palladium	106.4	98	Cf	californium	—
47	Ag	silver	107.868	99	Es	einsteinium	—
48	Cd	cadmium	112.40	100	Fm	fermium	—
49	In	indium	114.82	101	Md	mendelevium	—
50	Sn	tin	118.69	102	No	nobelium	—
51	Sb	antimony	121.75	103	Lr	lawrencium	—

752

Answer Keys

Answer Keys

Appendix 1-A

Practice Test 1 (Juniors)

1. (c)	18. (e)	35. (c)	52. (d)	69. (c)	86. (a)
2. (b)	19. (a)	36. (a)	53. (b)	70. (a)	87. (a)
3. (b)	20. (c)	37. (d)	54. (b)	71. (e)	88. (a)
4. (e)	21. (a)	38. (e)	55. (a)	72. (d)	89. (d)
5. (a)	22. (c)	39. (b)	56. (d)	73. (a)	90. (a)
6. (c)	23. (a)	40. (c)	57.(a)	74. (e)	91. (b)
7. (a)	24. (c)	41. (b)	58. (e)	75. (b)	92. (a)
8. (c)	25. (b)	42. (c)	59. (b)	76. (c)	93. (a)
9. (c)	26. (d)	43. (c)	60. (b)	77. (a)	94. (c)
10. (c)	27. (b)	44. (a)	61. (d)	78. (b)	95. (b)
11. (b)	28. (c)	45. (a)	62. (a)	79. (a)	96. (a)
12. (a)	29. (a)	46. (a)	63. (a)	80. (d)	97. (d)
13. (c)	30. (c)	47. (b)	64. (d)	81. (c)	98. (b)
14. (d)	31. (a)	48. (e)	65. (a)	82. (c)	99. (b)
15. (a)	32. (b)	49. (d)	66. (c)	83. (e)	100. (a)
16. (b)	33. (a)	50. (a)	67. (e)	84. (a)	
17. (b)	34. (b)	51. (a)	68. (a)	85. (c)	

Appendix – 2 A

Answer Key

Practice Test 2 (Seniors)

1. (c)	18. (d)	35. (b)	52. (d)	69. (b)	86. (d)
2. (c)	19. (d)	36. (c)	53. (b)	70. (a)	87. (d)
3. (e)	20. (a)	37. (d)	54. (a)	71. (b)	88. (a)
4. (a)	21. (d)	38. (d)	55. (c)	72. (c)	89. (c)
5. (b)	22. (a)	39. (b)	56. (d)	73. (d)	90. (a)
6. (a)	23. (d)	40. (a)	57.(e)	74. (b)	91. (d)
7. (d)	24. (e)	41. (b)	58. (c)	75. (d)	92. (b)
8. (e)	25. (c)	42. (b)	59. (a)	76. (b)	93. (e)
9. (c)	26. (a)	43. (a)	60. (b)	77. (d)	94. (c)
10. (e)	27. (c)	44. (e)	61. (c)	78. (c)	95. (d)
11. (a)	28. (a)	45. (c)	62. (a)	79. (a)	96. (b)
12. (d)	29. (c)	46. (d)	63. (a)	80. (c)	97. (c)
13. (b)	30. (d)	47. (c)	64. (d)	81. (a)	98. (a)
14. (d)	31. (d)	48. (d)	65. (a)	82. (d)	99. (b)
15. (a)	32. (a)	49. (b)	66. (e)	83. (c)	100. (c)
16. (c)	33. (c)	50. (c)	67. (d)	84. (a)	
17. (a)	34. (e)	51. (c)	68. (c)	85. (c)	

Appendix – 3A

Answer Key

Practice Test 3 (Seniors)

1. (b)	18. (c)	35. (b)	52. (b)	69. (c)	86. (c)
2. (c)	19. (a)	36. (d)	53. (d)	70. (c)	87. (d)
3. (c)	20. (b)	37. (c)	54. (c)	71. (c)	88. (c)
4. (d)	21. (b)	38. (c)	55. (b)	72. (c)	89. (a)
5. (d)	22. (a)	39. (a)	56. (b)	73. (b)	90. (c)
6. (c)	23. (a)	40. (c)	57.(a)	74. (a)	91. (a)
7. (d)	24. (a)	41. (b)	58. (b)	75. (b)	92. (e)
8. (c)	25. (a)	42. (c)	59. (a)	76. (b)	93. (e)
9. (d)	26. (a)	43. (e)	60. (c)	77. (d)	94. (d)
10. (d)	27. (c)	44. (b)	61. (d)	78. (a)	95. (d)
11. (c)	28. (c)	45. (a)	62. (b)	79. (c)	96. (c)
12. (b)	29. (a)	46. (d)	63. (a)	80. (b)	97. (b)
13. (c)	30. (a)	47. (a)	64. (d)	81. (c)	98. (a)
14. (b)	31. (c)	48. (b)	65. (d)	82. (c)	99. (d)
15. (b)	32. (c)	49. (d)	66. (c)	83. (b)	100. (b)
16. (d)	33. (e)	50. (a)	67. (a)	84. (a)	
17. (d)	34. (a)	51. (b)	68. (a)	85. (b)	

BIBLIOGRAPHY

(For Further Reading)

1) Longhurst, RS., "Geometrical and Physical Optics" (Orient Longman, ND, 1973).
2) Zemansky, MW, & Dittman, RH., "Heat and Thermodynamics", MGH, 1981).
3) Kittel, C., "Introduction to Solid State Physics"(John Wiley, 1996).
4) Moffatt, WR., Pearsall, GW., & Wulff, J., "Structure and Properties of Materials" (Wiley Eastern, ND, 1968)
5) Gettys, GE., Keller, FJ, & Skove, MJ., "Classical and Modern Physics", (MGH, NY, 1989),
6) Halliday, D. & Resnik, R., "Physics", (Wiley Eastern, ND, 1991).
7) Hecht, J., "Understanding Lasers" (RadioShack Howard W Sams, 1988).
8) Wenyon, M. "Understanding Holography" (Aco, NY., 1978)
9) Rees, WG., "Physics by Example" (CUP, Mana Sakia for Foundation Books, ND, 1995.
10) Schurcliff, W. & Ballard, SS., "Polarized Light" (Affiliated East West Press, ND, 1964).
11) Gay, P., "An Introduction to Crystal Optics", (Longman, NY, 1967).
12) Katz, R., "An Introduction to Special Theory of Relativity"(D Van Nostrand, Affiliated East West, ND, 1964).
13) Bishop, DM., "Group Theory and Chemistry" (Dover, NY, 1993).
14) Bohm, D, "Quantum Theory" (Dover, NY, 1989).
15) Holzner, S., "Quantum Physics for Dummifies" (Wiley, Indianapolis, 2009).
16) Devanarayanan, S., "Quantum Chemistry" (SciTech, Chennai, 2013).
17) Krishnan, RS., Srinivasan, R. & Devanarayanan, S. "Thermal Expansion of Crystals" (Pergamon , Oxford, 1979).
18) Devanarayanan, S. "A Text Book on Nuclear Physics – For Graduate Students" (CreateSpace, & Kindle, Amazon, 2016).
19) Wood, EA., "Crystals and Light", (Affiliated East West, ND, 1964).
20) Fowles, GR., "Introduction to Modern Optics", (Dover, NY, 1989).
21) Ledermann, W., "Introduction to Group Theory"(Longman, UK, 1981).
22) Fermi, E., "Notes on Quantum Mechanics" (Univ. Chicago Press,1968).
23) Harris, R., "Non-Classical Physics Beyond Newton. View" (Addission Wesley CA, 1998).
24) Hughes, FW., "Basic Electronics-Theory and Experimentation"(Prentice Hall, NJ, 1984).
25) Alonso, M. & Finn, EJ., "Fundamental University Physics" Vol 1,2,3, (Addission Wesley. Reading, 1968).
26) Burcham, WE., "Elements of Nuclear Physics", (ELBS, NY, 1988).
27) Fermi, E., "Nuclear Physics" (*Reprint*, Univ. Chicago Press, 1974).
28) Burcham, WE. & Jobes, M., "Nuclear and Particle Physics" (Longman, 1994).
29) Hawking, S. & Mlodinow, L., "The Grand Design" (Bantam Books, NY, 2010).
30) Wood, EA., "Crystal Orientation Manual", (Colombia Univ. Press, NY, 1963).
31) Sagan, C., "Cosmos", (Ballantine Books, NY, 1985).
32) Barron, John D. "New Theories of Everything" (OUP, NY, 2007).
33) Bais, Sandor, "The Equations: Icons of Knowledge" (Harvard Univ. Press, Cambridge, MA, 2005).

About the Authors

Prof. (Dr.) S. DEVANARAYANAN, Ph.D. (IISc); M.Sc. (Kerala), D.Sc. (USA), Dip (Uppsala)

Dr. S. Devanarayanan was educated at the University College, Thiruvananthapuram (1961 – 63), Indian Institute of Science, Bangalore (1963 – 70), and Institute of Physics, Uppsala, Sweden (1970 – 71). He had a brilliant academic career throughout. He was in the Faculty of the University of Kerala (1971 – 2000) and was the Professor & HOD of the Department of Physics, University of Kerala (1993 – 2000); and has 37 years of teaching / research experience in Physics and Materials Science.. He was Professor in the University of Puerto Rico (1989 - 91), Some 21 students have completed Ph.D. and M.Phil. under his supervision. He has to his credit over 80 published research papers in standard scientific periodicals A Monograph entitled THERMAL EXPANSION OF CRYSTALS (Pergamon, Oxford, 1979) and books "QUANTUM MECHANICS" (SciTech, 2005), QUANTUM CHEMISTRY (SciTech, 2013), "A Text Book on Nuclear Physics" (CreateSpace, 2016; and Kindle, 2016) were authored by him. He has served as a Professor in Physics at the University of Puerto Rico, USA, (1989 – 91). He was awarded the SIDA Fellowship and worked at The Institute of Physics, Uppsala, Sweden (1970 – 71).

A Life Member of various academic bodies like the IAPA Indian Physics Association, American Physical Society, American Chemical Society, and Founder Fellow of the Indian Cryogenic Council, his biography has found place a number of times in the publications of Marquis' Who's Who (USA) (2013)(2014), IBC (UK)(2011), ABC(USA) (2005), etc. The Govt. of Kerala appointed him as a member of the Commission of Enquiry on the working of the University of Kerala, in Oct 2000 – Mar 2001.

He was in several Committees, Boards of Studies, Examinations and Selection Boards in different Universities, Faculty of Science, Academic Council of University of Kerala, Expert in State Level and National, UPSC competitive Tests.

Prof. Devanarayanan has found honour in finding a name in the Star Chart, - Cat #TYC-7882-99-1- Scorpio Constellation- NASA- Feb 2013 –png.

He had the special honour of being invited by the Royal Swedish Academy to submit proposals for the award of the Nobel Prize in Physics for 1995. Devanarayanan believes in Sir C.V. Raman's advice that one can become a good scientist only when one takes up research along with teaching at a University. He has made academic visits in Sweden, Finland, Leningrad (USSR), The Netherlands, Germany, France, Australia, Czech Republik, Hungary, Austria, England, and USA.

Mr. Ajith Shankar Devan, B-Tech. (E. E. & E, Kerala, India), MBA (U D, Newark, DE, USA)

After a receiving Distinction in B-Tech. in EE&E (1996 – 2000) from the College of Engineering, Thiruvanthapuram, he has been in US from 2002 onwards, working as a Software Developer. He got the coveted 'Honorable Mention' in the Third International Competition 'First Step to Nobel Prize in Physics (1995), from the Polish Academy of Sciences. He was also in the list of National Merit Scholarship based on the AISSCE (CBSE) Examination (1996).

-o-0-o-0-o-0-o-

www.ingramcontent.com/pod-product-compliance
Lightning Source LLC
Chambersburg PA
CBHW070319240526
45468CB00025B/1173